HYPERBOLIC PROBLEMS: THEORY, NUMERICS, APPLICATIONS

Proceedings of the Fifth International Conference on

HYPERBOLIC PROBLEMS: THEORY, NUMERICS, APPLICATIONS

University at Stony Brook, New York, USA

13 – 17 June 1994

Editors

J. Glimm **J. W. Grove**

M. J. Graham **B. J. Plohr**

Department of Applied Mathematics
University at Stony Brook, New York
USA

World Scientific

Singapore • New Jersey • London • Hong Kong

Published by

World Scientific Publishing Co Pte Ltd

P O Box 128, Farrer Road, Singapore 912805

USA office: Suite 1B, 1060 Main Street, River Edge, NJ 07661

UK office: 57 Shelton Street, Covent Garden, London WC2H 9HE

Library of Congress Cataloging-in-Publication Data

International Conference on Non-linear Hyperbolic Problems (5th : 1994 : Stony Brook, N.Y.)

Proceedings of the Fifth International Conference on Hyperbolic Problems : theory, numerics, applications : University at Stony Brook, New York, USA, June 1994 / J. Glimm . . . [et al.] (eds.)

xi, 494 p., 22.5 cm.

Includes index.

ISBN 9810224419 (alk. paper)

1. Differential equations, Hyperbolic--Congresses. 2. Differential equations, Nonlinear--Congresses. I. Glimm, James. II. Title.

QA377.I565 1994

532'.05'01515353--dc20 96-4840

CIP

British Library Cataloguing-in-Publication Data

A catalogue record for this book is available from the British Library.

The cover illustration shows a front-tracking computation by Dr. Richard L. Holmes of the Richtmyer-Meshkov instability. This instability is generated by accelerating the interface between two fluids with an outward-moving shock. Initial perturbations of the interface have evolved into the finger-like structures shown here.

Cover design/layout by Phyllis Graham.

This book is printed on acid-free paper.

Printed in Singapore by Uto-Print

5th ICHP – Committees and Supporting Organizations

Host Organization

Department of Applied Mathematics and Statistics
University at Stony Brook
Stony Brook, New York 11794-3600
USA

Scientific Committee

X. Ding	(China)	J. Glimm	(USA)	P. Lax	(USA)
A. Donato	(Italy)	R. Jeltsch	(Switzerland)	T.-P. Liu	(USA)
B. Engquist	(USA)	B. Keyfitz	(USA)	D. Serre	(France)

Local Organizing Committee

J. Glimm	J. W. Grove	E. Beltrami
M. J. Graham	B. Lindquist	Z. Xin

The Local Organizing Committee of Hyp94 thanks all of the dedicated graduate students, particularly Brian Boston, who generously contributed their time to help organize this conference.

Papers Committee

B. Boston	J. W. Grove	Z. Xin
G.-Q. Chen	R. L. Holmes	R. Young
X. Garaizar	B. Lindquist	Q. Zhang
J. Glimm	B. J. Plohr	
M. J. Graham	F. Tangerman	

Sponsoring Organizations

- Office of Naval Research
- United States Department of Energy
- United States Army Research Office
- National Science Foundation
- Department of Applied Mathematics and Statistics, University at Stony Brook

Proofreading and typesetting

B. Boston
C. Clark
M. J. Graham
J. Hurley
B. Lindquist
S. Mullhaupt
B. J. Plohr
R. Young

Contents

Contributed Lectures

Singularities and interfacial patterns in Hele-Shaw flows

Gregory Baker[1], Michael Siegel[1], and Saleh Tanveer[1]

Abstract

Interest in interfacial motion in Hele-Shaw cells stems from analogies with flows in porous media, dendritic growth, and bubble competition in Rayleigh-Taylor instabilities. Depending on the strength of surface tension effects, interfacial patterns may contain multiple fingers in competition, tip-splitting at the leading edge of a finger, branching along the sides of a finger, and even fractal behavior for very small surface tension coefficients. The usual fluid equations for the flow can be simplified by a conformal map of the physical domain into a unit disk. The presence of singularities in the conformal map outside the unit disk introduce specific structures in the physical domain. Depending on the location and nature of the singularities we find interfacial patterns such as fingers undergoing competition, tip-splitting and side-branching. Generally, singularities move towards the boundary of the unit disk, inducing the formation of these interfacial patterns. We present a new numerical method that can track the location of isolated singularities outside the unit disk in time efficiently. This allows us to watch the development of interfacial patterns as singularities approach the unit disk.

Key words: Hele-Shaw instability, ill-posed, singularities, side-branching, finger competition.

AMS(MOS) subject classification: 65M30, 76D99, 76M25.

1 Introduction

The displacement of a viscous fluid by a less viscous fluid in a Hele-Shaw cell has been the subject of intense investigation over the last decade, mainly due to newly discovered mathematical analogies with dendritic crystal growth, directional solidification, and electro-chemical growth. The original motivation behind the pioneering work of Saffman & Taylor [18] was the analogy to displacement in porous medium. Reviews by Bensimon, Kadanoff, Liang, Shraiman, & Tang [3] and Homsy [9] summarize the state of affairs as of the mid-eightees, while some of the the more recent developments are reviewed by Pelce [16] and Kessler, Koplik, & Levine [11].

In this paper, we shall limit our investigations to channel flow, although our numerical method is quite general, and can be applied to other geometries. A steadily advancing flat interface in a channel is unstable to perturbations when driven by the less viscous fluid. The perturbations grow into fingers, but the subsequent behavior depends on the relative strength of capillary effects, measured by the dimensionless number $B = \sigma b^2/(12\mu V a^2)$. Here σ is the surface tension coefficient, b is the gap width of the cell, a is the channel width, V is the speed of displacement well in front of the interface, and μ is the viscosity of the more viscous fluid (the viscosity of the less viscous fluid is assumed negligible). Numerical computations by Tryggvasan & Aref [23], DeGregoria and Schwartz [6], Bensimon [4], and Meiburg and Homsy [14] show competition between fingers resulting in the emergence of a single steady finger, provided B is greater than some critical number. This critical number lies in a range, typically somewhere in $0.0005 > B > 0.0002$, depending on the level of noise in an

[1]Mathematics Department, Ohio State University, Columbus, OH 43210, USA

experiment or numerical calculations. Below the critical value, a transition occurs between steadily moving fingers and continual unsteady motion. For still smaller B, DeGregoria and Schwartz [7] and Bensimon [4] find that fingers spontaneously split; this can be induced for higher values of B by introducing perturbations at the tip, even with small amplitude. Experiments also reveal the emergence of tip-splitting and side-branching instabilities for small enough values of B (Park & Homsy [15], Tabeling, Zocchi, & Libchaber [21]). The computations of Dai and Shelley [5] (in circular geometry) show great sensitivity to the level of numerical precision when the surface tension coefficient is small. For small enough B, even noise during experiments can be large enough to set off a pattern of continual tip-splitting and finger competition (Maxworthy [13], Arneodo, Couder, Grasseau, Hakim & Rabaud [1]). Indeed, when capillary effects are very small, it appears that the pattern is fractal (Maxworthy [13], Arneodo *et. al.* [1]).

Detailed understanding of this unsteady behavior is limited. In particular, the numerical work has not produced a clear understanding of the asymptotic trends as $B \to 0$. The difficulty is that the growth of round-off errors usually occurs on scales much too small for any accuracy in the numerical method. While the numerical results show similar large scale behavior, the details of their formation may be subject to the inaccuracies of each particular method. We have designed a numerical method for the case $B = 0$ that does not suffer from contamination by round-off errors, but we require additional information about the initial condition.

Our method is based on the use of a conformal map $z(\zeta, t)$ which maps the interior of a unit semi-circle in the ζ plane to the physical flow domain of a channel. The location of the free surface at a time t is given by $z(\zeta, t) = x(\theta, t) + iy(\theta, t)$ for $\zeta = e^{i\theta}$ on the arc of the semi-circle. The equation for the evolution of $(x(\theta, t), y(\theta, t))$ results from the usual application of the kinematic and dynamic conditions at the interface. The symmetry in the problem allows us to reflect the solution about the real axis. In essence, we include a mirror image of the channel so that the interface may be considered periodic. Thus, the conformal map must be analytic inside the unit disk, aside from a logarithmic singularity at $\zeta = 0$, but it may have singularities and zeros outside it. These can move towards the boundary of the unit disk, $|\zeta| = 1$. In particular, zeros may reach the boundary in finite time, causing a cusp to form on the interface. The origin of ill-posedness, then, is that small perturbations can introduce a zero or a singularity in z_ζ near the unit disk, which subsequently moves towards it and reaches it quickly. Following the work of Richardson [17] and Lacey [12], Tanveer [22] showed that all singularities, no matter what type, will move towards the unit disk, while preserving their type. Indeed, we surmise that round-off error in traditional numerical calculations introduces singularities at some distance from $|\zeta| = 1$; these subsequently approach the unit disk and lead to the random pattern of tip-splitting, side-branching and finger competition seen in computations for increasingly small B.

Recently, Tanveer [22] has demonstrated that analytically extending the initial value problem for $z(\zeta, t)$ into the region exterior to the unit disk leads to a well-posed evolution problem. Our method is based on this analytic extension. We decompose the map into isolated singularities of the form

$$z_\zeta \sim A(t)\,(\zeta - \zeta_s(t))^\alpha\,, \quad \alpha \neq 0, 1, 2, \ldots \tag{1}$$

and an analytic part. We track the motion of these singularities explicitly in the complex ζ-plane. Thus we are able to assess directly the impact of the close approach of pole and branch point singularities to $|\zeta| = 1$. We find that singularities induce tip-splitting and side-branching, and that the relative strength of the singularities control finger competition. Moreover, our studies draw connections between the parameters specifying the singularities and the resulting physical behavior. In other words, for given initial data we are able to *predict* the outcome from knowledge about the initial singularities in $|\zeta| > 1$. Within this framework, comparisons with experimentally observed features are possible by studying a random ensemble of initial conditions in $|\zeta| > 1$, subject to the constraint that they describe the same initial interface, up to some 'experimental' error.

Beside the explicit treatment of singularities, there is another crucial ingredient in our method. An analytic function, such as a conformal map, is determined completely by its values on a closed curve. Instead of advancing the conformal map on $|\zeta| = 1$ in time as in standard boundary integral methods, our method advances the conformal map on a much larger circle. The interpolation of the conformal map to $|\zeta| = 1$ is then a *well-posed* operation. In essence, knowledge of the conformal map on a much larger circle corresponds to knowledge of the conformal map on $|\zeta| = 1$ to a higher degree of precision. In contast, extrapolation of the conformal map to a larger circle is an *ill-posed* operation, an operation not to be attempted numerically. Since our method uses the discrete Fourier transform to evaluate Laurent series, our results are spectrally accurate and can be performed in $O(N \log N)$ operations. Furthermore, explicit treatment of the nearest singularities helps avoid point crowding typical of conformal maps.

There are several limitations to our method. First, there is the matter of the initial data. We require that all singularities of z_ζ in the extended domain be isolated branch points or poles. This rules out some types of initial data; for example, structures such as natural boundaries or essential singularities are not allowed in the extended domain. Nevertheless, the kinds of data we do allow produce a wide range of interfacial features which are similar to experimental observations. It is important to note that when the initial interface location is known only to a finite precision, initial data $z(\zeta, 0)$ in the extended region $|\zeta| > 1$ cannot be obtained in a well-posed manner. We require the data to be *specified* in $|\zeta| > 1$ (say, with $z(\zeta, 0)$ given in closed form).

Ideally, one would like to understand the Hele-Shaw dynamics for a small non-zero B perturbatively by exploiting the simpler $B = 0$ case. Unfortunately, the perturbative influence of small B is not always regular. Tanveer [22] has examined the effect of small B on isolated singularities in the initial data of the form (1). Capillary effects on singularities with $\alpha \leq -4/3$ cause only a regular perturbation. Their powers are unchanged but their speed and strengths $A(t)$ are modified slightly by terms proportional to B. Singularities with $-4/3 < \alpha < -1/2$ are immediately transformed into a localized cluster of $-4/3$ branch point singularities. Thus, for times much less than $1/B$ the interface behaves as though it is unaffected by surface tension, provided these singularities do not come within a distance of $O(B^{1/(2+\alpha)})$ from $|\zeta| = 1$ and zeros in z_ζ are far from the unit disk. For initial conditions with zeros present, recent evidence (Siegel, Tanveer, & Dai [20]) shows that small surface tension effects can significantly perturb the interface in $O(1)$ time. This can happen even when the zeros do not impinge on the unit disk. The method presented here provides a means to determine if and when the zero surface tension evolution of the interface deviates from the small surface tension evolution.

2 A Model Problem

We illustrate the method by considering a simple model problem that exhibits many of the salient features of the more complicated Hele-Shaw equations. The model problem is stated in terms of a function $u(\zeta, t)$ defined in the ζ plane, and takes the form

$$\frac{\partial u}{\partial t} - \zeta u_\zeta = 0 \qquad (2)$$

$$u(\zeta, 0) = u_0(\zeta). \qquad (3)$$

Here it is assumed that $u_0(\zeta)$ is analytic in a neighborhood of the unit circle $\zeta = e^{i\theta}$. In addition, we suppose that the restriction of $u(\zeta, t)$ to $|\zeta| = 1$ corresponds to some physical quantity of interest. For this reason, $|\zeta| = 1$ will be referred to as the physical domain.

The exact solution to the model problem can be easily obtained by integrating along characteristics $\zeta_0(t)$, defined by

$$\frac{\partial \zeta_0(t)}{dt} = -\zeta_0. \qquad (4)$$

Upon doing this, one obtains

$$u(\zeta, t) = u_0(\zeta e^t). \qquad (5)$$

We comment on several features of the exact solution (5). If

$$u_0(\zeta) = \sum_{k=-\infty}^{\infty} a_k \zeta^k$$

is a Laurent series expansion of the intial data, then the solution takes the form

$$u(\zeta, t) = \sum_{k=-\infty}^{\infty} a_k e^{kt} \zeta^k. \qquad (6)$$

In particular, we see that the kth mode (for $k > 0$) has a growth rate $\sigma(k) = k$, so that the solution on $|\zeta| = 1$ exhibits short-wavelength instability. In addition, it is clear that any singularities of $u_0(\zeta)$ that are present in $|\zeta| > 1$ will move along the characteristics given by (4) toward the unit disk. These reach $|\zeta| = 1$ in finite time, at which point the singularities actually appear in the physical solution.

As an example, consider the initial data

$$u_0(\zeta) = \delta \left(\frac{1}{(1 - \epsilon\zeta)^p} + \frac{1}{(1 - \zeta/\epsilon)^p} \right) \qquad (7)$$

where δ, $\epsilon > 0$ and $p > 0$ are constants. For this data, the exact solution to (2), (3) is given by

$$u(\zeta, t) = \delta \left(\frac{1}{(1 - \epsilon e^t \zeta)^p} + \frac{1}{(1 - (\zeta e^t)/\epsilon)^p} \right) \qquad (8)$$

This solution has two singularities: one inside the unit disk at $\zeta = \epsilon/e^t$ and one outside at $\zeta = 1/(\epsilon e^t)$. The inner singularity moves away from the unit disk, whereas the outer one reaches it at time $t = \ln(1/\epsilon)$. Note that by varying δ, the initial data (8) can be made arbitrarily small in the max norm on $|\zeta| = 1$, yet $u(\zeta, t)$ will diverge in this norm at a time $t = \ln(1/\epsilon)$ which can be made arbitrarily small. Thus, when considered on $|\zeta| = 1$, the initial value problem (2), (3) is ill-posed in the sense of Hadamard: small perturbations in $u_0(\zeta)$ for $|\zeta| = 1$ can introduce a singularity in $u(\zeta, t)$ near the unit disk, which moves towards it and quickly leads to large differences between the perturbed and unperturbed solutions. Unfortunately, such perturbations can be introduced by round-off error in traditional numerical calculations.

Garabedian [8] recognized that initial value problems such as (2), (3) can be made well-posed through analytic extension off of the physical domain. In particular, we consider the extension of (2), (3) to a time dependent annular region of the form $r(t) \le |\zeta| \le R(t)$. The values of $r(0)$ and $R(0)$ are chosen so that no singularities of u are present in $r(0) \le |\zeta| \le R(0)$. Of course, singularities in u may be present outside of this region, but as long as $r(t)$ increases and $R(t)$ decreases fast enough, these singularities will not intrude in $r(t) \le |\zeta| \le R(t)$.

Despite appearances, a numerical solution to the extended problem can be obtained without updating $u(\zeta, t)$ throughout the entire computational domain, an annulus $r(t) \le |\zeta| \le R(t)$. Knowledge of this function on the circles $|\zeta| = r(t)$ and $|\zeta| = R(t)$ is sufficient to determine it everywhere inside the annulus. Nevertheless, a special procedure must be used to update $u(\zeta, t)$ in a numerically stable way.

We introduce a decomposition for a function $g(\zeta)$ which is analytic in the annulus by writing it as the sum of a function $g_+(\zeta)$ which is analytic inside a circle $|\zeta| = R$ and a function $g_-(\zeta)$ which is analytic outside the circle $|\zeta| = 1$. The decomposition is made unique by requiring $g_+(0) = 0$. By considering the Laurent expansion for $g(\zeta)$, we note that

$$g_-(\zeta) = \sum_{k=-\infty}^{0} \hat{g}_k \zeta^k \tag{9}$$

$$g_+(\zeta) = \sum_{k=1}^{\infty} \hat{g}_k \zeta^k \tag{10}$$

where the Taylor series for $g_+(\zeta)$ will have a radius of convergence $R_+ > R$, and the Laurent series for $g_-(\zeta)$ will converge outside $|\zeta| = R_- < 1$. We call the function g_+ *inner analytic* and the function g_- *outer analytic.*

An important feature of an inner analytic function is that its values inside a closed curve can be determined from the values on the curve in a well-posed fashion. To see this, consider an inner analytic function $g_+(\zeta)$ evaluated on a circle of radius R. Let $\zeta = Re^{i\theta}$, then

$$g_+(\theta) = \sum_{k=1}^{\infty} \hat{g}_k R^k e^{ik\theta}$$

In practice, the coefficients $\hat{g}_k R^k$ are obtained by the discrete Fourier Transform. Since the sum converges, these coefficients must decay with increasing k. There will be a value, k_r say, where the coefficients reach the round-off levels of a computer. If sufficient resolution is used with the discrete Fourier Transform, then the coefficients with

$k > k_r$ will contain round-off levels and not their actual values. This has important consequences when we wish to evaluate g_+ on a different circle $\zeta = re^{i\theta}$. Then

$$g_+(\theta) = \sum_{k=1}^{\infty} \left(\frac{r}{R}\right)^k \hat{g}_k R^k e^{ik\theta}, \tag{11}$$

so each coefficient $\hat{g}_k R^k$ is multiplied by r^k/R^k prior to using the discrete Fourier Transform again to evaluate the sum. If $r > R$, then round-off errors in $\hat{g}_k R^k$ for $k > k_r$ will be significantly amplified, even to the point that all accuracy in $g_+(\zeta = re^{i\theta})$ will be lost. However, if $r \leq R$, round-off errors are decreased. For an outer analytic function the situation is reversed: evaluation on $r \geq R$ (here we will take $R = 1$) is numerically stable, while on $r < R$ it is not.

We obtain evolution equations for the inner and outer analytic components of u by applying the projection operators, $\mathcal{H}_+ f = f_+$ and $\mathcal{H}_- f = f_-$. In this case, the evolution equations for u_+ and u_- are identical to (2), so that the differences in the initial value problems for u, u_+ and u_- are solely a result of their differing initial data.

The function $u_-(\zeta, t)$ can be solved for, in a stable manner, on the circle $|\zeta| = r(0)$, i.e., there is no need to evolve $r(t)$. This fact is reflected by the decay of the outer analytic Laurent coefficients in (6). Although values of $u_{0-}(\zeta)$ for ζ inside the circle $|\zeta| = r(0)$ cannot be determined from data on the circle in a well posed manner, this "information" flows along the characteristics defined by (4) away from $|\zeta| = r(0)$. Thus, this information will not cross the boundary into the annulus $r(0) < \zeta < R(t)$. Note that to obtain the solution $u_-(\zeta, t)$ on the physical domain, it is sufficient to set $r(0) = 1$.

In contrast, the function $u_+(\zeta, t)$ is updated on the circle $|\zeta| = R(t)$. Values of $u_{0+}(\zeta)$ for $|\zeta| > R(0)$, which cannot be ascertained from data on $|\zeta| = R(0)$ in a well-posed manner, must be avoided in the determination of the solution (5) on $|\zeta| = R(t)$. This is accomplished by shrinking the radius $R(t)$ of the circle. If an advection term is included to account for the change in u_+ due to the change in $R(t)$, then the equation for u_+ on $\zeta = R(t)e^{i\theta}$ becomes

$$\frac{\partial u_+}{\partial t} - \zeta \left(1 + \frac{\dot{R}}{R}\right) u_{+\zeta} = 0. \tag{12}$$

It is easy to verify that (12) is well posed when $\dot{R} \leq -R$. In essence, for $R(t)$ satisfying this condition, information outside the circle $|\zeta| = R(0)$ is prevented from crossing the boundary into the annulus $r(0) < \zeta < R(t)$. Note that the values of $u_+(\zeta, t)$ for $|\zeta| = 1$ can be recovered from the solution $u_+(\zeta, t)$ on $\zeta = R(t)$ by a procedure like that shown in (11). Details on how to numerically implement the update of the upper and lower analytic components on their respective domains will be provided in the next section, in the context of Hele-Shaw flow.

Unfortunately, $R(t)$ must shrink faster than the approach of any singularity to $|\zeta| = 1$. Consequently, a numerical computation may terminate (when $R = 1$) before any interesting structures have developed in the physical solution. For initial conditions containing a finite number of isolated singularities, this obstacle may be removed by explicit treatment of the singularities. For example, when the only singularity present in $|\zeta| > 1$ is a branch point at $\zeta_s(t)$, we can write

$$u_+(\zeta, t) = E(\zeta, t)(\zeta - \zeta_s(t))^p + G(\zeta, t) \tag{13}$$

where $E(\zeta,t)$ and $G(\zeta,t)$ are analytic in $|\zeta|=1$, and $p \neq \ldots -1,0,1,\ldots$. Substitution of (13) into (2) yields the following set of equations for the quantites $\zeta_s(t)$, $E(\zeta,t)$, and $G(\zeta,t)$:

$$\begin{aligned} \dot{\zeta}_s &= -\zeta && (14)\\ E_t - \zeta E_\zeta &= 0 && (15)\\ G_t - \zeta G_\zeta &= 0 && (16) \end{aligned}$$

Equations (15) and (16) may be solved by decomposing into inner and outer analytic functions and treating as described earlier. However, since these functions are analytic in $|\zeta| > 1$, the initial radius $R(0)$ can be taken as large as is desired.

3 Application to Hele-Shaw flow

Consider a Hele-Shaw cell of infinite length and width $a=2$ in which air of negligible viscosity is pushing a viscous, incompressible liquid (see e.g. Figures 1–3). Introduce the conformal map $z(\zeta,t)$ which takes the interior of a unit semi-circle in the upper half ζ plane into the viscous fluid region of the channel, which lies in the z plane. The circular arc $|\zeta|=1$ is mapped to the interface, and the diameter is mapped to the channel walls.

The functional form of the conformal map is given by

$$z(\zeta,t) = -\frac{2}{\pi}\ln\,\zeta + i + f(\zeta,t) \tag{17}$$

where f is analytic in an open set which contains the unit semi-circle. The analyticity of f on the circular arc guarantees the smoothness of the interface. We require that

$$\mathrm{Im}\{f\} = 0 \tag{18}$$

on the real diameter of the semi-circle to satisfy the condition that z maps this diameter to the channel walls. In addition, we assume that $z_\zeta \neq 0$ in a region containing the unit semi-circle. The Schwartz reflection principle then implies that f is analytic and $z_\zeta \neq 0$ for $|\zeta| \leq 1$.

The equation which governs the evolution of the map in the extended domain $|\zeta| > 1$ was obtained by Tanveer [22] (see also Baker, Siegel & Tanveer [2]) and takes the form

$$z_t - q_1 z_\zeta = q_2 \tag{19}$$

where the functions $q_1(\zeta,t)$ and $q_2(\zeta,t)$ are given by

$$\begin{aligned} q_1 &= -\frac{\zeta}{\pi^2 i}\oint_{|\zeta'|=1} \frac{1}{z_\zeta(\zeta',t)\tilde{z}_\zeta(1/\zeta',t)}\frac{\zeta'+\zeta}{\zeta'-\zeta}\frac{d\zeta'}{\zeta'} && (20)\\ q_2 &= \frac{-2\zeta}{z_\zeta(1/\zeta)}. && (21) \end{aligned}$$

As in the model problem, we use the extended equation to develop a numerical procedure for calculating the interfacial motion in a well-posed fashion, for a restricted class of initial data.

It is easily seen that the functions q_1 and q_2 are analytic in $|\zeta| > 1$, except at infinity, where q_1 has a simple pole, i.e., grows as ζ. The analyticity of q_1 and q_2 has important consequences on the presence and motion of singularities of z. For the convenience of the reader, we summarize some of the known properties (Richardson [17], Lacey [12], Tanveer [22]):

1. There is no spontaneous generation of singularities or zeros of z_ζ in the finite complex plane. Furthermore, the form of a singularity which is present initially in the region $|\zeta| > 1$ is invariant with time.

2. We define a "characteristic" in the ζ plane by

$$\frac{d\zeta_c(t)}{dt} = -q_1(\zeta_c(t), t). \tag{22}$$

   Let $\zeta_c(t) = R_c(t)e^{i\nu_c(t)}$, then

$$\frac{\dot{R}_c}{R_c} = Re\left\{\frac{\dot{\zeta}_c(t)}{\zeta_c(t)}\right\} = -Re\left\{\frac{q_1(\zeta_c(t), t)}{\zeta_c(t)}\right\} \tag{23}$$

   Tanveer (1993) has shown that the right-hand side of (23) is less than zero when $|\zeta_c| > 1$. Consequently, "information" outside of the unit disc flows inward toward $|\zeta| = 1$, as in the model problem.

3. Each isolated singularity of the form (1) moves toward $|\zeta| = 1$. Singularities of this form with $\alpha < -1/2$ come indefinitely close to, but never reach the boundary $|\zeta| = 1$. Unlike singularities, the zeros of z_ζ do not generally move toward $|\zeta| = 1$, and it is hard to predict if a given zero will hit the unit circle.

Property 2 suggests that the design of a well-posed numerical method to solve (19) will mirror that discussed for the model problem. First, we assume that the initial data $z(\zeta, 0)$ contains isolated singularities of the form (1) in $|\zeta| > 1$. Since singularities of this type preserve their form in time, they can be treated explicitly, i.e., they are "subtracted" from the data and evolved separately. Proceeding as in Section 2, the solution $z(\zeta, t)$ is written in the form,

$$\begin{aligned} z(\zeta,t) &= \sum_{j=1}^{J_1} E^j(\zeta,t)\left(1-\frac{\zeta}{\zeta_j(t)}\right)^{\alpha_j+1} + \sum_{j=J_1+1}^{J}\left[E^j(\zeta,t)\left(1-\frac{\zeta}{\zeta_j(t)}\right)^{\alpha_j+1}\right. \\ &+ \left.\bar{E}^j(\zeta,t)\left(1-\frac{\zeta}{\bar{\zeta}_j(t)}\right)^{\alpha_j+1}\right] + G(\zeta,t) - \frac{2}{\pi}\ln\zeta \end{aligned} \tag{24}$$

where $|\zeta_j| > 1$; α_j are real constants (excluding $0, 1, 2, \ldots$); and $E^j(\zeta_j(0), 0) \neq 0$. Here $\zeta_j(t)$ for $j = 1, \ldots, J_1$ mark the location of singularities on the real line, and $\zeta_j(t)$ for $j = J_1 + 1, \ldots, J$ indicate singularities with non-zero imaginary part. In order to satisfy (18), the amplitudes $E^j(\zeta, t)$ for $j = 1, \ldots, J_1$ must be real. The amplitudes

$E^j(\zeta,t)$ for $j = J_1+1,\ldots,J$ are, in general, complex. Conjugate singularities $\bar{\zeta}_j(t)$ with amplitudes $\bar{E}^j(\zeta,t)$ are included to satisfy condition (18). The branch cut is chosen so that the argument of $(1-\zeta/\zeta_j)$ lies between π and $-\pi$: we have chosen this form to simplify the numerical evaluation of the interfacial location when $|\zeta| = 1$. Although it is not necessary, for the presentation here we assume that all singularities in the region $|\zeta| > 1$ are explicitly represented. Then, the functions $E^j(\zeta,t)$ and $G(\zeta,t)$ are analytic in the entire region exterior to the unit disk. Unfortunately, it is possible for E^j and G to have singularities in $|\zeta| < 1$ which cancel in (24) so that $z + (2/\pi)\ln\zeta$ is analytic in $|\zeta| \leq 1$. For initial data with logarithmic singularities (poles in z_ζ) the solution has a modified form; this is further discussed in [2].

As in the model problem, our numerical treatment involves decomposing both $E^j(\zeta,t)$ and $G(\zeta,t)$ into inner and outer analytic parts and using the method of lines to update E^j_+, G_+ on $|\zeta| = R$, and E^j_-, G_- on $|\zeta| = 1$. Evolution equations for these quantities are given in [2]; we restate them here for convenience:

$$E^j_{-_t} = H_-\left\{q_1 E^j_\zeta + (\alpha_j+1)\left(\frac{q_1(\zeta,t)-q_1(\zeta_j(t),t)}{\zeta-\zeta_j} - \frac{q_1(\zeta_j,t)}{\zeta_j(t)}\right)E^j\right\} \tag{25}$$

$$G_{-_t} = \mathcal{H}_-\left\{q_1 G_\zeta - \frac{2}{\pi\zeta}q_1(\zeta,t) + q_2(\zeta,t)\right\} \tag{26}$$

$$E^j_{+_t} = \mathcal{H}_+\left\{q_1 E^j_{+_\zeta} + (\alpha_j+1)\left(\frac{q_1(\zeta,t)-q_1(\zeta_j(t),t)}{\zeta-\zeta_j} - \frac{q_1(\zeta_j,t)}{\zeta_j(t)}\right)E^j_+ + \frac{\zeta}{R}\frac{dR}{dt}E^j_{+_\zeta}\right\} \tag{27}$$

$$G_{+_t} = \mathcal{H}_+\left\{q_1 G_{+_\zeta}\right\}. \tag{28}$$

To apply the method of lines to (25-26) and (27-28), we assume that we know E^j_+ and G_+ at N evenly spaced points on the circle $r = R(t)$, and E^j_- and G_- at N evenly spaced points on the unit circle. It is also assumed that q_1 and q_2 are known on the two circles; these are computed from (20), (21) using a procedure discussed in [2]. First, we describe how to evaluate the right-hand side of (25) and (26) on the unit circle.

We compute the coefficients $c_k R^k$ in the representation

$$E^j_+(\zeta = Re^{i\theta},t) = \sum_{k=1}^{N/2} c_k R^k e^{ik\theta}$$

by use of the discrete Fourier Transform. The coefficients c_k are obtained by a division by R^k and then used to evaluate E^j_+ and $E^j_{+_\zeta}$ on the unit circle by use of the discrete Fourier Transform. The function $E^j_{-_\zeta}$ is also determined on $|\zeta| = 1$ by using the discrete Fourier Transform. Then we have all the information needed to evaluate R_1, the right-hand side of (25), at N evenly spaced points on the unit circle. To execute the projection $\mathcal{H}_-$, we use the discrete Fourier Transform to calculate the coefficients in the representation

$$R_1(\zeta,t) = \sum_{k=-N/2}^{N/2} \hat{r}_k(t)\zeta^k.$$

Then we set to zero all the coefficients with $k \geq 1$. The result is an outer analytic function, which is the forcing term in (25). Balance of the Fourier coefficients of like modes on the left- and right- hand sides of (25) then yields a set of equations

$$\frac{d\hat{E}_k^j}{dt}(t) = \hat{r}_k(t)$$

for the Fourier coefficients $\hat{E}_k^j(t)$ $(k \leq 0)$ of E_-^j. In an equivalent procedure, we compute the forcing term in (26), and use this to obtain a set of equations for the Fourier coefficients $\hat{G}_k(t)$ $(k \leq 0)$.

A system of ODE's for the coefficients $\hat{E}_k^j(t)$ and $\hat{G}_k(t)$ $(k \geq 1)$ is obtained from equations (27), (28) in an analogous fashion. Details are given elsewhere (Baker, Siegel & Tanveer [2]). The value of dR/dt in (27) is chosen so that the circle collapses at a rate equal to the maximum characteristic speed on $|\zeta| = R$, i.e.,

$$\frac{1}{R}\frac{dR}{dt} = -\max_{|\zeta|=R(t)} \mathrm{Re}\left\{\frac{q_1(\zeta,t)}{\zeta}\right\} . \tag{29}$$

Then, information outside the circle $|\zeta| = R(t)$ will not cross the boundary into the annulus, $1 \leq |\zeta| \leq R(t)$.

Any suitable ODE Solver may be applied to the evolution equations for ζ_j, $\hat{E}_k^j$, and $\hat{G}_k$. For the results used in this paper, we use the standard fourth-order Runge-Kutta method with fixed step-size.

4 Numerical Results

A known exact solution due to Saffman [19] has been used to help validate the method; the results of this test are discussed in [2]. Most importantly, this test revealed no growth of round-off errors in the modes comprising $E^j(\zeta,t)$ and $G(\zeta,t)$.

There are no known exact solutions for the channel geometry when branch point singularities are located in the region $|\zeta| > 1$. We consider two branch point singularities of equal power and amplitude located symmetrically on the real axis of ζ. Our choice is motivated by the desire to understand the role of the power of a singularity in z_ζ on the shape of the interface. Thus we select initial data corresponding to (24) with $J_1 = 2$, $\alpha_1 = \alpha_2$, and $\zeta_1(0) = -\zeta_2(0)$. In Figure 1a, $\alpha_1 = -4/5$, and in Figure 1b, $\alpha_1 = -4/3$. Initial values for the singularity positions $\zeta_j(0)$ and amplitudes $E^j(\zeta,0)$ are determined through experimentation; those that lead to solutions in which the interface becomes well-deformed before zeros impinge on the unit disc are selected for presentation. In this manner, the initial singularity positions and amplitudes are chosen as $\zeta_1(0) = -\zeta_2(0) = 1.2$, $E^1(\zeta,0) = E^2(\zeta,0) = 1.94$ for Figure 1a, and $\zeta_1(0) = -\zeta_2(0) = 3.3$, $E^1(\zeta,0) = E^2(\zeta,0) = -2.2$ for Figure 1b. In these and all subsequent computations we set $G(\zeta,0) = i + c$, where c is a real constant selected to produce an initial profile with zero mean height. For constant initial data such as this, only negative Fourier modes in E^j and G are generated. Consequently, we only need to solve equations (25) and (26). The positive Fourier modes are set to zero after each time step in the calculation. We use 512 points on the unit circle in the ζ plane, and a time step of $\Delta t = 0.0005$. Using a larger number of points or a smaller time step produces no detectable differences in the solution within plotting accuracy.

Unlike the previous example, we cannot be sure that zeros do not approach the unit disk. So we monitor the presence of zeros by computing the integral

$$N_z = \frac{1}{2\pi i} \oint_{|\zeta|=1} \frac{z_{\zeta\zeta}}{z_\zeta} d\zeta \tag{30}$$

which, according to the argument principle, equals the number of zeroes minus the number of poles of z_ζ inside the unit disk. As long as there are no zeros of z_ζ in $|\zeta| \leq 1$, this integral will equal -1, due to the simple pole of z_ζ at $\zeta = 0$. For the results presented in Figures 1a and 1b, the value of N_z remains within 1% of -1 throughout the length of the computation.

The contrast between Figures 1a and 1b is quite striking. The higher value of $\alpha = -4/5$ in Figure 1a correspond to a singularity of weaker effect in that the bulge of fluid at the base of the finger is less rounded, causing a finger that is more pointed. The lower value of $\alpha = -4/3$ in Figure 1b produces a more spherically shaped bulge of fluid, causing a thinner neck at the base of the finger. In general, our experiments with various powers α show that for powers $\alpha > -1$, more pointed fingers are produced from the less pronounced bulges of fluid at their bases. For $\alpha < -1$, the fingers have parallel sides, but their bases have thinner necks as a consequence of more spherically rounded bulges of fluid there.

We turn now to a consideration of the influence of additional $-4/3$ singularities with weak amplitudes during finger formation. Figure 2a illustrates the interface evolution resulting from initial data of the form (24) with $J_1 = 2$, $J = 3$ and with singularity strengths $\alpha_j = -4/3$ for $j = 1, ..., 3$. The initial singularity positions were chosen as $\zeta_1(t) = -\zeta_2(t) = (1.65, 0.0)$, and $\zeta_3(t) = (-0.6, 1.83)$, and the initial amplitudes as $E^1(\zeta, 0) = E^2(\zeta, 0) = (-0.9, 0.0)$, and $E^3(\zeta, 0) = (-0.01, 0.03)$. The calculation uses N=512 points and a time step $\Delta t = 0.0005$. Two of the singularities (ζ_1 and ζ_2) reside on the real line, and these have large enough amplitude to produce the wide bulges of fluid centered at $y = 1$ and $y = -1$ which define the main finger. As the third singularity approaches the unit disk, it generates a small bulge of fluid that gives the appearance of the formation of a dimple on the evolving finger.

The dimple is clearly stationary in the laboratory frame, despite the overall growth of the main finger. Such behavior has been well documented in laboratory experiments (e.g. Park & Homsy [15], Tabeling, Zocchi, & Libchaber [21]) and in numerical calculations (e.g. DeGregoria and Schwartz [6, 7]) with $B \neq 0$. Our results show that it is the specific nature of the trajectory of the singularity $\zeta_3(t)$ that accounts for this behavior. We show the trajectories of all the singularities in Figure 2b. At first, a dimple starts to form as $\zeta_3(t)$ approaches the unit disk. As $\zeta_3(t)$ begins to move around the boundary of the unit disk towards $\zeta = -1$, the dimple continues to grows but remains stationary in the laboratory frame. With the assumption that the singularity is close to either $\zeta = \pm 1$, we can show that its speed is actually the correct speed for it to remain stationary in the laboratory frame. Details will be provided elsewhere. The point to be made here is only that certain physical properties can be understood in terms of the motion of singularities in the complex ζ plane.

Placing additional singularities in the complex plane produces additional dimples and leads to the appearance of side-branching. An example of side-branching due to multiple pole singularities is given in Figure 3. We use pole singularities, rather than $\alpha = -4/3$ singularities, since our ability to track them arbitrarily close to the unit

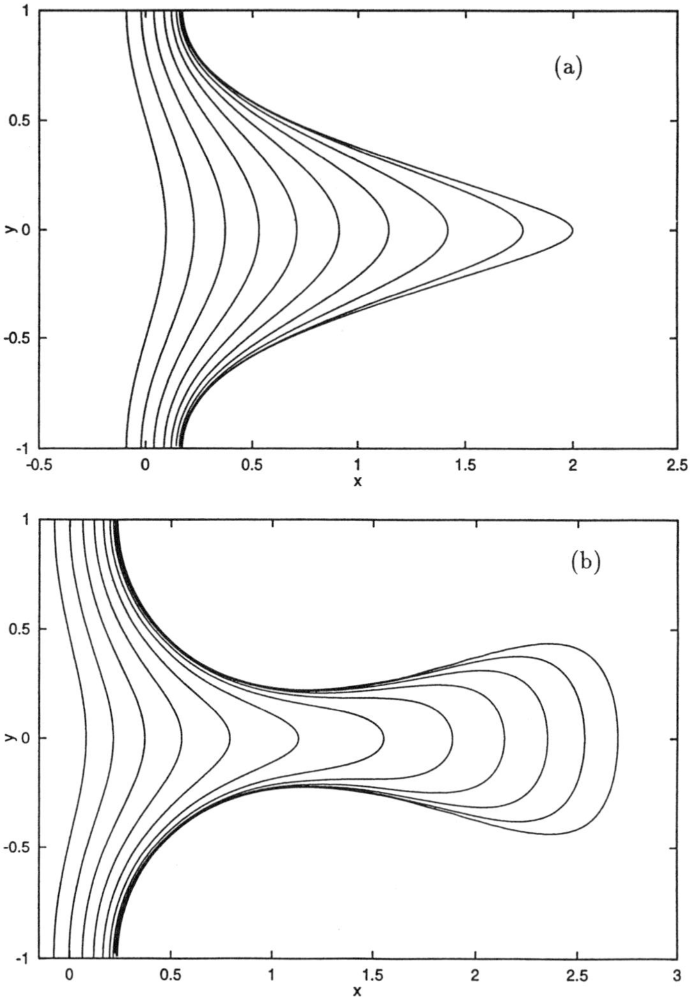

Figure 1: Fingers produced by two branch point singularities of strength α, located at symmetric positions on the real line. The fluid occupies the right part of the channel. (a) $\alpha = -4/5$. The profiles show the position of the interface from $t = 0$ to $t = 0.8$ in increments of .1, and at $t = 0.85$. In the last profile ($t = 0.85$) the singularities are located at $\zeta_1 = -\zeta_2 = 1.0081$. (b) $\alpha = -4/3$. The profiles show the position of the interface from $t = 0$ to $t = 1.1$ in increments of .1. In the last profile (t=1.1) the singularities are located at $\zeta_1 = -\zeta_2 = 1.0054$.

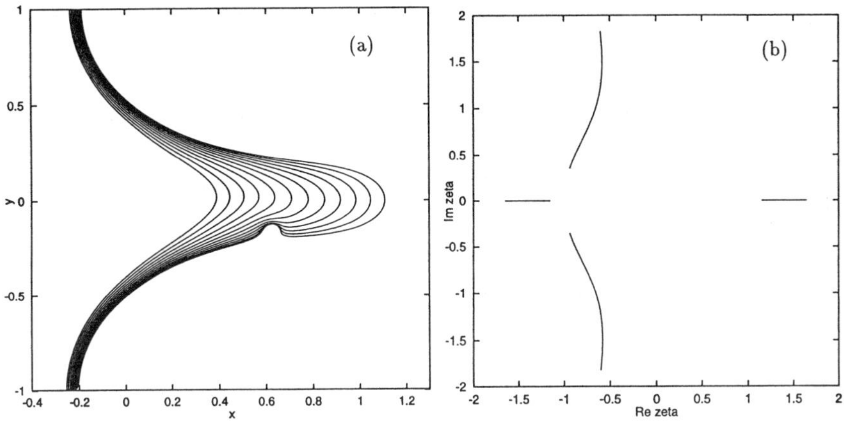

Figure 2: (a) An advancing finger with a side perturbation. The plots range from $t = 0$ to $t = 0.22$ in increments of .02. (b) The singularity trajectories corresponding to (a). After reaching the unit circle, the (upper) complex singularity ζ_3 moves in a counterclockwise direction.

disc leads to a more dramatic example of sidebranching, but in reality side-branching is more likely to occur from the patterns of $-4/3$ singularities created by the initial transformation of zeros in z_ζ. Initial data containing a total of 14 pole singularities in the upper half ζ-plane is used to generate the profiles. Additional singularities can be placed to allow side-branching to run the length of the finger. As expected, the dimples and the corresponding side-branches are stationary in the laboratory frame. The endless presence of singularities streaming in from infinity can generate fingers with side-branching patterns seen frequently in experiments.

We remark that in Figure 2a, the small amplitude of the singularity at ζ_3 causes the formation of a small dimple on the side of a well-developed finger. In contrast, when Re E^3 is the same order as Re E^1 and Re E^2, the close approach of ζ_3 to $|\ \zeta\ |= 1$ leads to a 'tip-splitting' event and the formation of two fingers which eventually compete. In this case, the direction the singularity moves around the circle determines which finger will dominate.

The authors would like to acknowledge support from NASA grant NAG 3-1415 (G.B.), Department of Energy contract DE-FG02-92ER142 70 (M. S. and S. T.), and an NSF Postdoctoral Fellowship (M. S.). S. T. was partially supported by NASA grant NAS1-18605 while in residence at The Institute for Computer Applications in Science and Engineering (ICASE). Some of our computing was done at the Ohio Supercomputer Center, and we gratefully acknowledge the support and computer resources we received.

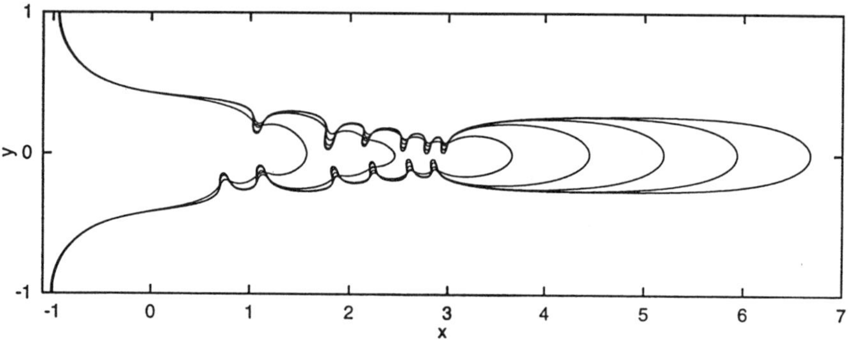

Figure 3: An example in which the approach of small amplitude poles causes the appearence of side-branching. Here N=512 and $\Delta t = 0.01$. The plots range from $t = 0$ to $t = 1.2$ in increments of .2.

References

[1] Arneodo, A., Couder, Y., Grasseau, G., Hakim, V. & Rabaud, M., Uncovering the analytical Saffman-Taylor finger in unstable viscous fingering and diffusion limited aggregation, *Phys. Rev. Lett.*, **63**, 984 (1989).

[2] Baker, G., Siegel, M. & Tanveer, S. (1995). A well-posed numerical method to track isolated conformal map singularites in Hele-Shaw flow, to appear in *J. Comput. Phys.*

[3] Bensimon, D., Kadanoff, L.P., Liang, S., Shraiman, B.I., & Tang, C., Viscous flows in two dimensions, *Rev. Mod. Phys.*, **58** , 977 (1986).

[4] Bensimon, D., Stability of viscous fingering, *Phys. Rev. A*, **33**, 1302 (1986).

[5] Dai, W. & Shelley, M., A numerical study of the effect of surface tension and noise on an expanding Hele-Shaw bubble, to appear in *Phys. Fluids A*.

[6] DeGregoria, A.J. & Schwartz, L.W., Finger breakup in Hele-Shaw cells, *Phys. Fluids*, **28**, 2313 (1985).

[7] DeGregoria, A.J. & Schwartz, L.W., A boundary-integral method for two-phase displacement in Hele-Shaw cells, *J. Fluid Mech.*, **164**, 383 (1986).

[8] Garabedian, P., *Partial Differential Equations*, (Chelsea Publishing Company, New York).

[9] Homsy, G.M., Viscous Fingering, *Ann. Rev. Fluid Mech.*, **19**, 271 (1987).

[10] Howison, S.D., Fingering in Hele-Shaw cells, *J. Fluid Mech.*, **167**, 439 (1986).

[11] Kessler, D., Koplik, J. & Levine, H., Pattern selection in fingered growth phenomena, *Advances in Physics*, **37**, 255 (1988).

[12] Lacey, A.A.., Moving boundary problems in the flow of liquid through porous media, *J. Australian Math. Soc.*, **B24**, 171 (1982).

[13] Maxworthy, T., The nonlinear growth of a gravitationally unstable interface in a Hele-Shaw cell, *J. Fluid Mech.*, **177**, 207 (1987).

[14] Meiburg, E., & Homsy, G.M., Nonlinear unstable viscous fingers in Hele-Shaw flows. II. Numerical simulation. *Phys. Fluids,* **31**, 429 (1988).

[15] Park, C.W. & Homsy, G.M., The instability of long fingers in Hele-Shaw flows, *Phys. Fluids,* **28**, 1583 (1985).

[16] Pelce, P. 1988, *Dynamics of Curved Fronts*, (Academic Press, San Diego).

[17] Richardson, S., Hele-Shaw flows with a free boundary produced by the injection of fluid into a narrow channel, *J. Fluid Mech.,* **56**, 609 (1972).

[18] Saffman, P.G. & Taylor, G.I., The penetration of a fluid into a porous medium of Hele-Shaw cell containing a more viscous fluid, *Proc. R. Soc. London Ser. A*, **245**, 312 (1958) .

[19] Saffman, P.G., Exact solution for the growth of fingers from a flat interface between two fluids, *Q. J. Mech. appl. Math.*, **12,** 146 (1959).

[20] Siegel, M., Tanveer, S. & Dai, W., Singular Perturbation of a smoothly evolving zero surface tension Hele-Shaw flow–a comparison between asymptotics and numerics, in preparation.

[21] Tabeling,P, Zocchi, G.& Libchaber, A., An experimental study of the Saffman-Taylor instability, *J. Fluid Mech.,* **177,** 67 (1987).

[22] Tanveer, S., Evolution of a Hele-Shaw interface for small surface tension, *Phil. Trans. R. Soc. London A.*, **343,** 1 (1993) (see also erratum 1994).

[23] Tryggvasson, G. & Aref, H., Numerical Experiments on Hele -Shaw flow with a sharp interface, *J. Fluid Mech.,* **136**, 1 (1983).

Phase Transitions in Self-Gravitating Relativistic Fluids

Demetrios Christodoulou[1]

General Relativity

General relativity presents us with a unified theory of space, time and gravitation, according to which spacetime is a 4-dimensional manifold M with a metric $g_{\mu\nu}$ of signature 2 whose connection represents the gravitational force. The fundamental law of the theory are the Einstein equations:

$$R_{\mu\nu} - \frac{1}{2}g_{\mu\nu}R = 8\pi T_{\mu\nu}$$

where $R_{\mu\nu}$ and R are respectively the Ricci curvature and scalar curvature of $g_{\mu\nu}$ and $T_{\mu\nu}$ is the energy-momentum tensor of matter. In view of the twice contracted Bianchi identities

$$\nabla^{\nu} R_{\mu\nu} - \frac{1}{2}\partial_{\mu} R = 0,$$

the Einstein equations imply the energy-momentum conservation laws:

$$\nabla^{\nu} T_{\mu\nu} = 0.$$

The energy-momentum tensor is required to satisfy the positivity condition – namely that, for each $x \in M$, the linear tranformation of $T_x M$:

$$v^{\mu} \to -T^{\mu}_{\ \nu} v^{\nu}; \quad T^{\mu}_{\ \nu} = g^{\mu\alpha} T_{\alpha\nu}$$

maps the closure of the open cone of future directed timelike vectors at x into itself.

Letting H_x^+ be the hyperboloid of unit future directed timelike vectors at x, the (*proper*) *energy density* ρ is defined by

$$\rho(x) = \inf_{u \in H_x^+} T(u, u).$$

It is a nonnegative function. The space H_x^+ being noncompact, the infimum may not be attained.

[1]Department of Mathematics, Princeton University, Princeton, New Jersey 08544, U.S.A.

Perfect Fluids

A *perfect fluid* is defined to be matter whose energy tensor satisfies the following two conditions:

1. The infimum above is attained at a unique $u \in H_x^+$. The vector field u^μ thus defined in the support of the energy tensor is called the *fluid velocity* and we have:

$$T_{\mu\nu}u^\nu = -\rho g_{\mu\nu}u^\nu.$$

The *stress* tensor $S_{\mu\nu}$ is then defined by:

$$S_{\mu\nu} = T_{\mu\nu} - \rho u_\mu u_\nu; \quad u_\mu = g_{\mu\alpha}u^\alpha.$$

The stress tensor at a given point $x \in M$ is a symmetric bilinear form in Σ_x, the spacelike hyperplane in T_xM orthogonal to u (local simultaneous space of fluid). Since the metric induced on Σ_x is positive definite, the eigenvalues p_1, p_2, p_3 of S are defined in the usual way and are called the *principal pressures*. The positivity requirement is equivalent to the condition that each of the principal pressures is dominated in absolute value by the proper energy density:

$$|p_1|, |p_2|, |p_3| \leq \rho.$$

2. The principal pressures coincide:

$$p_1 = p_2 = p_3 = p$$

to a nonnegative function p, called simply *pressure*. The stress tensor is then given by:

$$S_{\mu\nu} = p(g_{\mu\nu} + u_\mu u_\nu),$$

and the energy-momentum tensor takes the form:

$$T_{\mu\nu} = (\rho + p)u_\mu u_\nu + p g_{\mu\nu}.$$

Thermodynamics

According to the principles of thermodynamics, the energy density ρ and the pressure p are functions of the two thermodynamic variables, n, the *number density* of particles in the local simultaneous space of the fluid, and s, the *entropy per particle*, both taking values in the nonnegative real numbers. The differential of ρ is given by:

$$d\rho = \frac{(\rho + p)}{n} dn + n\theta ds,$$

that is,

$$p = n\frac{\partial \rho}{\partial n} - \rho, \qquad \theta = \frac{1}{n}\frac{\partial \rho}{\partial s}$$

Here θ is a nonnegative function, the *(absolute) temperature.* It is assumed to vanish when and only when s vanishes. The energy-momentum conservation laws are supplemented by the particle conservation law:

$$\nabla_\mu I^\mu = 0$$

where I^μ is the particle current:

$$I^\mu = nu^\mu.$$

Given ρ as a function of n and s (*equation of state*), the Einstein equations together with the particle conservation law form a complete system.

Barotropic Fluids

An important special case of a perfect fluid is one for which the pressure is a function of the energy density:

$$p = f(\rho).$$

Then ρ and p are functions of $\sigma = n\mu(s)$ and we have:

$$\sigma \frac{d\rho}{d\sigma} = \rho + p.$$

The energy-momentum conservation laws decouple from the particle conservation law; therefore, the Einstein equations form by themselves a complete system. We define:

$$V = ||V||u, \quad ||V|| = \frac{\rho + p}{\sigma}.$$

Assuming that $\lim_{\sigma \to 0}(d\rho/d\sigma) = 1$, the condition that

$$\frac{dp}{d\rho} = \eta^2 \geq 0; \quad \eta \, : \, \textit{sound speed}$$

implies: $||V|| \geq 1$. The components of the energy-momentum conservation laws along ($||$) and orthogonally ($\perp$) to u become:

$$|| : \; \nabla_\mu(G(||V||)V^\mu) = 0, \quad G = \frac{\sigma^2}{\rho + p},$$

$$\perp : \; \mathcal{L}_V \beta = d(||V||^2), \quad \beta_\mu = -g_{\mu\nu}V^\nu.$$

In the case of irrotational flow: $\beta = d\phi$, where ϕ is a *time function*, $||d\phi|| \geq 1$. The $\perp$ component is then trivially satisfied while the $||$ component becomes the nonlinear wave equation:

$$\nabla^\mu(G(||d\phi||)\partial_\mu\phi) = 0.$$

Discontinuities and Jump Conditions

Only non-spacelike hypersurfaces of discontinuity $\mathcal{B}$ can arise. Let N_μ be a covector at $x \in \mathcal{B}$, the null space of which is the tangent space of $\mathcal{B}$ at x. Then, denoting by $\triangle(\cdot)$ the jump across $\mathcal{B}$ at x, we have the jump conditions:

$$(\triangle T^{\mu\nu})N_\nu = 0, \quad (\triangle I^\mu)N_\mu = 0.$$

If $T_x\mathcal{B}$ is timelike, the normal vector $N^\mu = g^{\mu\nu}N_\nu$ is spacelike, and we can normalize it to be of unit length. Let N point from the side which we call *behind* and label 1 to the side which we call *ahead* and label 0. Defining $u_\perp = -g(u, N)$, the second jump condition reads:

$$n_1 u_{\perp 1} = n_0 u_{\perp 0} := f.$$

The quantity f is called the *particle flux*. If $f \neq 0$, the discontinuity is called a *shock*. The orientation of N is then chosen so that $f > 0$. Shocks are irreversible phenomena: the entropy per particle suffers a positive jump as we cross a shock from the side ahead to the side behind. The first jump condition implies the Hugoniot relation:

$$h_1^2 - h_0^2 = \left(\frac{h_1}{n_1} + \frac{h_0}{n_0}\right)(p_1 - p_0); \quad h = (\rho + p)/n \ : \ enthalpy$$

which gives the jump in entropy. In the barotropic case this reduces to:

$$\frac{\mu_1^2}{\mu_0^2} = \frac{\sigma_1^2}{\sigma_0^2}\frac{(\rho_0 + p_0)}{(\rho_1 + p_1)}\frac{(\rho_0 + p_1)}{(\rho_1 + p_0)}$$

If $f = 0$ the discontinuity is called a *contact discontinuity.* In this case the choice of orientation of N is merely conventional. Contact discontinuities are reversible. If $T_x\mathcal{B}$ is null, then the normal vector is a null vector hence belongs to $T_x\mathcal{B}$. We choose N to be future directed. Then $u_\perp > 0$ hence $f > 0$. A hypersurface of discontinuity which is everywhere null is called a *null shock.* There is no jump in entropy across a null shock.

The Two-Phase Model

A prime example in nature where relativistic fluid dynamics comes into play is the gravitational collapse of the degenerate core of a massive star. This leads to a spectacular explosion called a *supernova* and to the birth of a *neutron star* or a *black hole.* During the collapse, as long as the density of mass-energy remains below the nuclear saturation density, the equation of state remains soft, that is, the sound speed remains relatively low. However, as soon as nuclear density is exceeded, the strong nuclear repulsive forces come into play, and the equation of state turns hard. The sound speed at supranuclear densities is thought to approach the speed of light.

It is thus not unreasonable to consider the idealized model in which the sound speed is simply zero when the density of mass-energy ρ is less than a constant ρ_0 corresponding roughly to nuclear saturation density, and equal to unity, that is to the speed of light c, when ρ exceeds ρ_0:

$$\eta^2 = \begin{cases} 0 & \text{if } \rho < \rho_0 \\ 1 & \text{if } \rho > \rho_0 \end{cases}$$

We then have a barotropic fluid:

$$p = \begin{cases} 0 & \text{if } \rho \le \rho_0 \\ \rho - \rho_0 & \text{if } \rho > \rho_0 \end{cases}$$

with two phases, the *soft phase*, corresponding to $\rho < \rho_0$, and the *hard phase*, corresponding to $\rho > \rho_0$. We can choose the units so that $\rho_0 = 1$.

We study the dynamics of this model in the spherically symmetric case, in the hope of capturing some of the main qualitative features of the dynamics of stellar collapse. An understanding of the dynamics of our idealized model may serve as a guide to future investigations of more realistic models.

In terms of the variable σ,

$$\rho = \begin{cases} \sigma & \text{if } \sigma \le 1 \\ \frac{1}{2}(\sigma^2 + 1) & \text{if } \sigma > 1 \end{cases}$$

$$p = \begin{cases} 0 & \text{if } \sigma \le 1 \\ \frac{1}{2}(\sigma^2 - 1) & \text{if } \sigma > 1 \end{cases}$$

In the soft phase we have:

$$||V|| = 1, \quad G = \sigma,$$

while in the hard phase:

$$||V|| = \sigma, \quad G = 1.$$

In the irrotational case the equations of motion reduce in the soft phase to:

$$||d\phi|| = 1, \qquad \nabla^\mu(\sigma \partial_\mu \phi) = 0,$$

and in the hard phase to:

$$||d\phi|| > 1, \qquad \nabla^2 \phi = 0.$$

In the soft phase ϕ is the temporal distance function from a spacelike hypersurface, and

$$T_{\mu\nu} = \rho \partial_\mu \phi \partial_\nu \phi.$$

In the hard phase

$$T_{\mu\nu} = \partial_\mu \phi \partial_\nu \phi + \frac{1}{2}(||d\phi||^2 - 1) g_{\mu\nu}$$

differs from that of a massless scalar field by a "cosmological" term corresponding to a cosmological constant: $\Lambda = 4\pi\rho_0$.

Phase Transitions

Each of the two phases is by itself incomplete. Consider the irrotational case. Suppose that the initial conditions correspond to the soft phase. The function ϕ is in the soft phase the temporal distance function from a spacelike hypersurface Σ. For each point q of Σ at which Σ has negative mean extrinsic curvature, there is a focal point q_* of the timelike geodesic congruence orthogonal to Σ along the geodesic from q toward the future. The mass-energy density ρ blows up at q_*, unless ρ vanishes at q. However, before ρ can become infinite, it must exceed unity; therefore, there must be a phase transition to the hard phase.

Suppose next that the initial conditions correspond to the hard phase. The gradient of the function ϕ is timelike future directed and of norm exceeding unity along the initial Cauchy hypersurface Σ, and ϕ is a solution of the wave equation in the future of Σ. Now the gradient of ϕ may not remain timelike in the future of Σ; but before the gradient of ϕ becomes non-timelike, its norm must become less than unity; therefore, a phase transition to the soft phase must occur. Thus we may say that the soft phase turns in contraction into the hard phase, while the hard phase turns upon expansion into the soft phase. Only the two phases taken together constitute a complete model.

The requirement that a discontinuity arise naturally from smooth initial conditions implies that the hypersurface of discontinuity $\mathcal{B}$ is superpersonic relative to the state ahead and subsonic relative to the state behind. For our two-phase model this implies that the state behind cannot belong to the soft phase. Therefore, only contact discontinuities are possible within the soft phase. Also, the only type of discontinuity that may develop within the hard phase is a null shock. These null shocks have been studied in : D.C. "Bounded variation solutions of the spherically symmetric Einstein-scalar field equations", *Commun. Pure & Appl. Math.* **XLVI**, 1131 (1993). The remaining possibilities are a timelike shock with the state ahead in the soft phase and the state behind in the hard phase, or a contact discontinuity, as part of the boundary between the two phases.

To summarize, the phase boundary has both spacelike as well as timelike components. Across a spacelike component the thermodynamic variables and the fluid velocity are continuous, the final values of one phase providing the initial data for the next phase. However, across a timelike component the fluid velocity and the thermodynamic variables suffer discontinuities, determined by the integral form of the conservation laws, that is, by the jump conditions. These discontinuities are of an irreversible character, each point of a timelike component which is crossed by a flow line being a point of increase of the entropy.

Spherical Symmetry

In the case of spherical symmetry the group $SO(3)$ acts as an isometry group on the spacetime manifold M. The quotient of M by the group is a 2-dimensional manifold Q with boundary. The boundary of Q corresponds to the set of fixed points of the group action which is a timelike geodesic Γ (center of symmetry). The quotient spacetime Q is endowed with a metric g_{ab} of signature 0. The metric of M takes the form:

$$g_{ab}(x)dx^a dx^b + r^2(x) \overset{\circ}{\gamma}_{AB} (y)dy^A dy^B.$$

Here $\overset{\circ}{\gamma}_{AB} (y)dy^A dy^B$ is the metric of the standard unit 2-sphere expressed in arbitrary local coordinates, and $r(x) = \sqrt{\frac{area(x)}{4\pi}}$ is the *area radius* of the sphere corresponding to x. The energy tensor takes the form:

$$T_{ab}(x)dx^a dx^b + r^2(x)S(x) \overset{\circ}{\gamma}_{AB} (y)dy^A dy^B.$$

The Einstein equations become:

$$\nabla_a\nabla_b r = (1/2r)(1 - \partial^c r\partial_c r)g_{ab} - 4\pi r(T_{ab} - g_{ab}trT),$$

$$\nabla_b(r^2T^{ab}) = 2r\partial^a rS.$$

These imply the following equation for the Gauss curvature of Q:

$$K = (1/r^2)(1 - \partial^a r\partial_a r) + 4\pi(trT - 2S).$$

The *mass function* is defined by:

$$1 - \frac{2m}{r} = g^{ab}\partial_a r\partial_b r,$$

and satisfies the equations:

$$\partial_a m = 4\pi r^2(T_{ab} - g_{ab}trT)\partial^b r.$$

For a perfect fluid,

$$T_{ab} = (\rho + p)u_a u_b + p g_{ab}, \qquad S = p$$

$u = u^a \partial/\partial x^a$, the flow is irrotational: $\beta = d\phi$, and, in the barotropic case,

$$\nabla^a (r^2 G(||d\phi||)\partial_a \phi) = 0.$$

The Phase Transition from Soft to Hard

Consider initial data corresponding to the soft phase. In comoving coordinates (τ, χ) the metric of Q in the soft phase takes the form $-d\tau^2 + e^{2\omega} d\chi^2$, and we have $\phi = \tau$. Let $\mathcal{U}_*$ be the maximal developement of the soft phase solution. The soft phase is limited by the condition $\rho < 1$. Therefore, the set

$$\mathcal{K} = \{(\tau, \chi) \in \mathcal{U}_* \; : \; \rho(\tau, \chi) \geq 1\}$$

must be excised from $\mathcal{U}_*$. But this is not all, for causality demands that for each excluded point, the domain of its influence, i.e., its causal future, must also be excluded. Thus we must excise from $\mathcal{U}_*$ the set $\mathcal{J}^+(\mathcal{K})$, the causal future of $\mathcal{K}$ in $\mathcal{U}_*$. The boundary of $\mathcal{J}^+(\mathcal{K})$, consists of smooth spacelike segments Σ_i along which $\rho = 1$, and which form the spacelike part of the phase boundary, joined by pairs C_i^+, C_{i+1}^- of outgoing and incoming null segments. The end points N_i^+ and N_i^- of the spacelike segment Σ_i at which Σ_i turns null, and which are, at the same time, the past end points of the null segments C_i^+ and C_{i+1}^- respectively, we call *boundary null points*. These are analogous to the *branch points* of minimal surface theory.

The data induced by the soft phase along the spacelike segments Σ_i, provide the initial conditions for a subsequent hard phase, determined in the future domain of dependence of Σ_i. The extension of the solution to the causal future of the boundary null points requires the solution of a *free boundary problem*. Let us confine attention to a particular outgoing boundary null point N^+. Along C^+, the outgoing null curve issuing from N^+, we have data induced from the soft phase solution, while along C^{*+}, the incoming null curve issuing from N^+, we have characteristic initial data induced from the hard phase solution in $\mathcal{D}^+(\Sigma)$. Let C^+ be given by: $\tau = \tau_+(\chi)$ and let $\partial\mathcal{K}$ be given by: $\tau = \hat{\tau}(\chi)$.

The Problem of Formation of a Free Phase Boundary

Find a timelike curve $\mathcal{B}$ in the soft phase domain $\mathcal{U}_0 = \{(\tau,\chi) : \tau_+(\chi) \le \tau \le \hat{\tau}(\chi),\ \chi \ge \chi_0\}$ issuing from $N^+ = (\tau_0, \chi_0)$, and a hard phase solution in $\mathcal{U}_1 = \{(u,v) : s_1 \ge u \ge v \ge 0\}$, where u, v are canonical null coordinates, i.e., coordinates whose level curves are, respectively, outgoing and incoming null curves and $u = v = s$: arc length along $\mathcal{B}$, such that identifying $\mathcal{B}$ with $\{(u,v) \in \mathcal{U}_1 : u = v\}$ by identifying points corresponding to the same value of the arc length s. We have a development $\mathcal{U} = \mathcal{U}_1 \bigcup \mathcal{U}_0$ such that r is continuous across $\mathcal{B}$, $m_1(N^+) = m_0(N^+)$ and the jump conditions are fulfilled across $\mathcal{B}$. (It then folows that dr, m are continuous across $\mathcal{B}$, and g_{ab} is C^1 in the canonical null coordinate system).

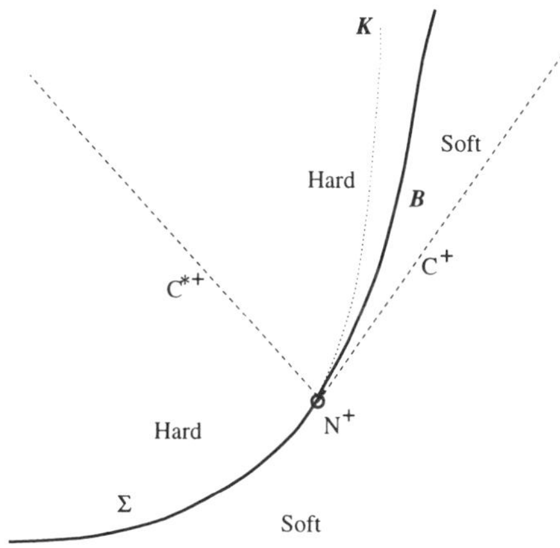

The ratio:

$$\gamma = \frac{[(\partial r/\partial u)/(\partial \phi/\partial u)]_1}{[(\partial r/\partial u)/(\partial \phi/\partial u)]_0} \qquad 0 < \gamma \le 1$$

plays a fundamental role in the problem. The velocity of $\mathcal{B}$ relative to the soft phase flow lines is given by:

$$\beta = \frac{1-\gamma^2}{1+\gamma^2-2\gamma^2\rho_*} \qquad 0 \le \beta \le 1,$$

where ρ_* is the soft phase mass-energy density along $\mathcal{B}$. Note that $\hat{\tau}(\chi_0) = \tau_+(\chi_0) = \chi_0, \gamma(\tau_0) = \rho_*(\tau_0) = 1$. Denoting by ϕ_* the hard phase wave function along $\mathcal{B}$, we have in general, $d\phi_*/d\tau \leq 1$ with equality at a point if and only if either $\rho_* = 1$ or $\beta = 0$ at that point.

Theorem: Let (r, ω, ρ) be a soft phase solution corresponding to smooth initial data and let $\mathcal{U}_0$ be the soft phase domain bounded by C^+, the outgoing null curve issuing from $N^+ = (0,0)$, and $\partial\mathcal{K}$, where, for the given soft phase initial data, $\rho(\tau, \chi)$ along each flow line first becomes equal to 1. Let also R be a function defined on an interval $[0, t_1]$ and representing r as a function of ϕ along C^{*+}, the incoming null curve issuing from N^+. Then there is a $\tau_1 > 0$ and a solution to the problem of formation of a free phase boundary such that $\mathcal{B}$ is a C^1 curve $[0, \tau_1] \mapsto \mathcal{U}_0$, $\tau \mapsto \chi_*(\tau)$, issuing from N^+, which has positive velocity β relative to the soft phase flow lines, is strictly timelike (i.e., $\beta < 1$) and contained in the interior of $\mathcal{U}_0$ except for its origin N^+, becoming null outgoing at N^+ ($\beta(0) = 1$). Also, on $\mathcal{U}_1$, $s_1 = s(\tau_1)$, r, m, ϕ are C^1 functions defining a genuine hard phase solution, i.e., $\rho > 1$ in $\mathcal{U}_1$ except at N^+ where $\rho = 1$.

The Form of the Free Phase Boundary in a Neighborhood of the Null Point

Theorem: Consider any solution of the problem of formation of a free phase boundary in the phase transition from soft to hard. Let ρ as a function of τ along C^+ be given, in a neighborhood of N^+, by:

$$\rho_+(\tau) = 1 - k\tau^2 + o(\tau^2)$$

where k is a positive constant. Then, in a neighborhood of N^+, ρ as a function of τ along $\mathcal{B}$ satisfies:

$$\rho_*(\tau) = 1 - \frac{3}{4}k\tau^2 + o(\tau^2),$$

and γ is given by:

$$\gamma(\tau) = 1 - \frac{3}{2}b_0\tau + o(\tau)$$

where b_0 is the positive constant:

$$b_0 = \left(\frac{\partial \rho}{\partial \tau}\right)(N^+).$$

Moreover,

$$1 - \beta(\tau) = \frac{k}{2b_0}\tau + o(\tau),$$

and:

$$\phi_*(\tau) = \tau - \frac{9}{128}b_0 k\tau^4 + o(\tau^4).$$

Theorem: Uniqueness of the solution of the formation problem and continuous dependence on the data of the problem.

The Phase Transition from Hard to Soft

In this case the phase boundary has corners at the null points.

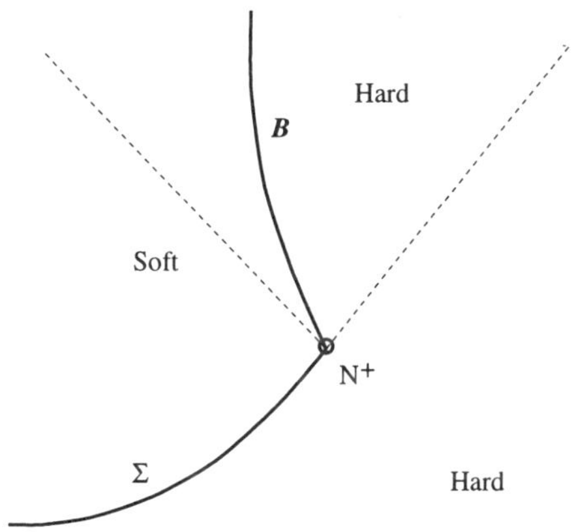

The Termination of a Free Phase Boundary

Define $\overline{\tau}$ to be the supremum of the set of real numbers τ_1 such that on $(0, \tau_1]$. $\mathcal{B}$ is a timelike curve of positive velocity β relative to the soft phase flow lines contained in the interior of $\mathcal{U}_0$, $\gamma < 1$ on $(0, \tau_1]$ and in $\mathcal{U}_1 \setminus (0,0)$ we have genuine hard phase, i.e., $\rho > 1$ in $\mathcal{U}_1 \setminus (0,0)$. Then the solution extends to $\overline{\tau}$ and we have the following 3 cases:

Case 1: $\overline{\rho} < 1, \overline{\gamma} < 1$; a spacelike transition from hard to soft with a null end point lying on the boundary of the causal past of the point of termination.

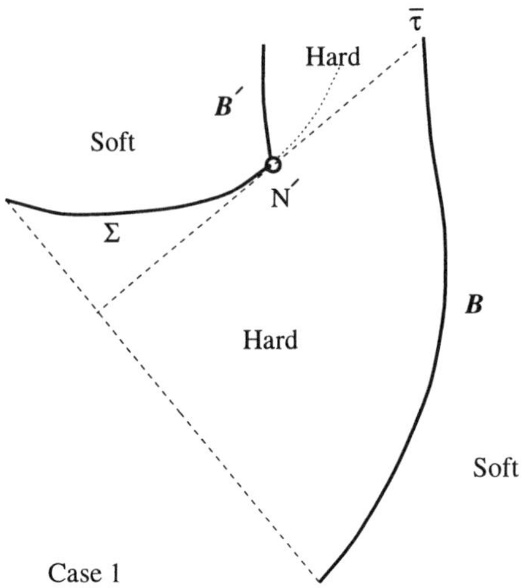

Case 2: $\overline{\rho} = 1, \overline{\gamma} < 1$; a point of intersection with a spacelike transition from soft to hard, from which point the discontinuity propagates in the hard phase as a null shock.

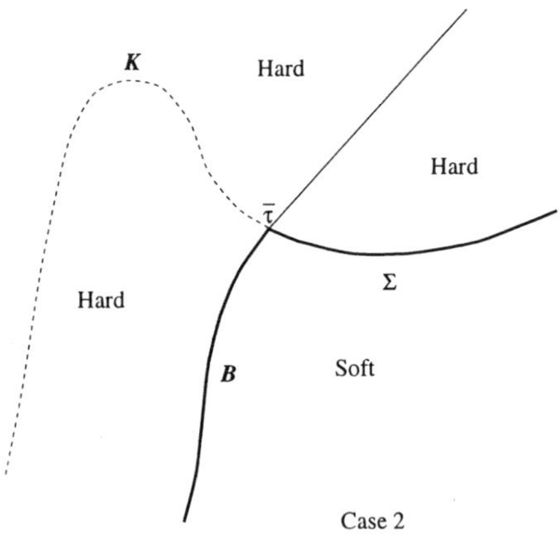

Case 3: $\overline{\rho} < 1, \overline{\gamma} = 1$; a point of intersection with a spacelike transition from hard to soft, from which point the discontinuity propagates as a contact discontinuity in the soft phase.

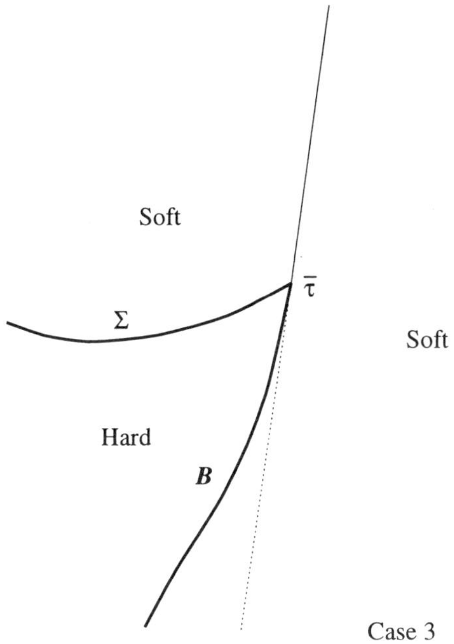

Case 3

(The case $\overline{\rho} = \overline{\gamma} = 1$ does not occur.)

The results presented here are to be found in the papers:

D.C. "Self-gravitating relativistic fluids: a two-phase model", submitted to *Arch. Rat. Mech. Anal.*

D.C. "Self-gravitating relativistic fluids: The Continuation and Termination of A Free Phase Boundary", submitted to *Arch. Rat. Mech. Anal.*

D.C. "Self-gravitating relativistic fluids: the Formation of a Free Phase boundary in the Phase transition from Soft to Hard", submitted to *Arch.Rat.Mech.Anal.*

D.C. "Self-gravitating relativistic fluids: the Formation of a Free Phase boundary in the Phase transition from Hard to Soft", submitted to *Arch.Rat.Mech.Anal.*

On the Courant-Friedrichs-Lewy Condition Equipped with Order for Hyperbolic Differential Equations

Rolf Jeltsch[1], Rosemary A. Renaut[2], J.-H. Smit[3]

Abstract

In 1928 Courant, Friedrichs and Lewy showed a necessary condition for convergence of a difference scheme for solving a hyperbolic differential equation. This condition states that the numerical domain of dependence has to include the exact domain of dependence. Applying this argument to the linear advection equation, where the exact domain of dependence is given by the characteristic line, says that an explicit difference stencil must have at least one stencil point on both sides of the characteristic line through the stencil point on the newest time level. In 1985 this result was generalized to explicit and implicit, normalized, two time level difference schemes in the form that a stable scheme with local error order p must have at least $\lceil p/2 \rceil$ stencil points on each side of the characteristic line. It is conjectured that this statement is correct for multistep difference schemes. In the present paper we consider three time level schemes and can show the correctness of the conjecture for schemes of optimal order $p =$ number of stencil points -2 where the Courant number μ satisfies $-1/2 < \mu < 0$. Moreover, the conjecture can be shown for all schemes with a convex increasing stencil using a conjecture on the so-called order star of the scheme. In particular this result applies to implicit schemes.

1 Introduction

When using linear difference schemes for initial boundary value problems of linear hyperbolic partial differential equations it has been shown that it is essential to study the schemes when applied to the pure Cauchy problem for the scalar advection equation

$$
\begin{aligned}
u_t(t,x) &= c\, u_x(t,x),\ x \in \mathbb{R},\ t \geq 0 \\
u(0,x) &= u_0(x) \text{ given },
\end{aligned} \tag{1}
$$

see e.g. [4], [2]. As indicated in the lecture of VanLeer in this conference conditioning of the Euler equations of gas dynamics is a way of preparing these nonlinear equations such that they behave more like linear equations and hence our linear analysis sheds some light even on this nonlinear case. However it should be noted that the schemes we are applying to (1) in the present article are linear only. We consider the following class of multistep ($(k+1)$-time-level) difference schemes of the form

$$
\sum_{i=0}^{k} \sum_{j=-r_i}^{s_i} a_{ij}\, u_{n+i,m+j} = 0\ ,\ n = 0, 1, 2, ..., m \in \mathbb{Z}\ . \tag{2}
$$

Here $u_{n,m}$ is supposed to approximate the exact solution of (1) $u(n\Delta t, m\Delta x)$ where Δt and Δx are the uniform stepsizes in time and space. Schemes of form (2) can be used for more general systems of partial differential equations either as an interior scheme or as a boundary scheme. As an interior scheme one always assumes that such a scheme is stable in the von Neumann sense when applied to the pure Cauchy problem (1), see e.g. [4]. Observe that Goldberg and Tadmor

[1]Seminar für Angewandte Mathematik, ETH-Zürich, 8092 Zürich, Switzerland
[2]Department of Mathematics, Arizona State University, Tempe, AZ 85287, USA
[3]Department of Mathematics, University of Stellenbosch, 7600 Stellenbosch, South-Africa

[2], [3] gave practical sufficient conditions for stability of global schemes for initial boundary value problems. These conditions entail, among others, that the boundary scheme also has to be stable in the sense mentioned above for the interior scheme. Whether the scheme is used in the interior or at the boundary, it has to represent the equation (1) with a certain accuracy; that is a smooth solution of (1) has to satisfy the difference formula (2) up to terms of the form $O(\Delta x^{p+1})$. Here we have assumed that $c\, \Delta x/\Delta t$ is kept constant. p is called the order of accuracy. In this paper we investigate the interplay between the accuracy and the stability of three-time-level schemes. The stability requirement gives a bound for the highest possible order a scheme with a prescribed stencil can have.

The above mentioned interplay between the error order and the stability of the scheme is well understood for one-step formulas ($k = 1$), [10]. For multistep schemes partial results have been given in [6], [8], [12], [13]. These are in agreement with the conjecture which states that the order of accuracy p of a stable scheme satisfies the inequality

$$p \leq 2\ \min\{R, S\}\ . \tag{3}$$

Here R and S denote the number of stencil points to the left and right, respectively, of the characteristic line through the point $(n\,\Delta t, m\,\Delta x)$. For explicit schemes, $r_k = s_k = 0$, this characteristic line represents the exact domain of dependence. If the scheme is implicit, i.e. at least two coefficients of a_{kj} are nonzero, this line has a similar function if the scheme is normalized in a fashion to be in agreement with the Wiener-Hopf theory on solutions of doubly infinite linear systems, see Section 2 or [5], [10]. The conjecture (3) says that a stable scheme of order p needs at least $\lceil p/2 \rceil$ stencil points on each side of the characteristic line described above. ($\lceil \alpha \rceil$ denotes the smallest integer which is not smaller than α). Observe that for explicit schemes of order $p \geq 1$ the conjecture says that on either side of the characteristic line one needs at least one stencil point. This can be interpreted as the statement of Courant-Friedrichs-Lewy that the numerical domain of dependence has to include the exact domain of dependence in order for a scheme to be convergent, [1].

The conjecture (3) has been proved in [10] for one-step schemes. In [6] and [8] many examples in support of the conjecture for multistep schemes, $k > 1$, were given. In [7] a stable $(k+1)$-time level scheme has been given which shows that if the barrier (3) were correct, it would be a sharp bound. In the present article we follow the lines of [12] where the conjecture (3) has been proved for a small class of explicit three-time-level schemes. In order to prove the result for all explicit and implicit three-time-level schemes we have to resort to another conjecture, Conjecture 5.1. In the important case of schemes of maximal order this conjecture and hence (3) can be proved at least for **convex schemes** with an increasing stencil. This proof could only be done for $-1/2 < \mu < 0$. However we do believe that we shall be able to extend it to $0 < \mu < 1/2$.

In this article we can, due to space limitations, only highlight the ideas of the proofs. For details we refer to [9]. In Section 2 we give the basic definitions and formulate the main theorem in Section 3. To start explaining the proof we describe the order star in Section 4 and give in Section 5 the important bounds for the multiplicities of the order star components. There we also explain Conjecture 5.1. In the last section we outline the proof and give an interpretation of the result.

2 Order, stability and normalization of schemes

In this article we shall restrict ourselves to three-time-level schemes, i.e. schemes of form (2) with $k = 2$ for solving the Cauchy-problem (1). We assume that while Δx and Δt tend to zero the Courant number $\mu = c\Delta t/\Delta x$ is kept constant. The coefficients a_{ij} are real and depend in general on μ. Further r_i, $s_i \in \mathbb{Z}$ with $r_2 \geq 0$, $s_2 \geq 0$ and $-r_i \leq s_i$, $i = 0,1$ and $a_{i,-r_i} \neq 0$, $a_{i,s_i} \neq 0$ for $i = 0,1,2$. If $r_2 = s_2 = 0$ a scheme is said to be **explicit**. Otherwise it is called **implicit**.

The interplay between accuracy and stability can best be handled mathematically by making a discrete Fourier Transform in space of $u_{n,m}$ in (2). One obtains

$$\sum_{i=0}^{2} a_i(e^{i\Delta x\xi})\, \tilde{u}_{n+i}(\xi) = 0,\ n = 0,1,2,... \tag{4}$$

where $\tilde{u}_n(\xi)$ is the Fourier Transform of $u_{n,m}$ at $t = \Delta t \cdot n$ and

$$a_i(z) = \sum_{j=-r_i}^{s_i} a_{ij}\, z^j,\ i = 0,1,2,\ . \tag{5}$$

In order to be able to solve (2) or (4) for the values on the new time-level $t = (n+2)\Delta t$ we need

$$a_2(z) \neq 0 \ \text{ for } \ |z| = 1\ . \tag{6}$$

Given a scheme with a stencil the numbers r_i, s_i are *a priori* not uniquely defined since one could replace these by $r_i' = r_i + 1$, $s_i' = s_i - 1$ without changing formula (2). In order to make the definition unique we observe that $z^{r_2} a_2(z)$ has exactly $r_2 + s_2$ zeros, none of which are on the unit circle. Hence we define r_2 and s_2 according to the following **normalization condition**.

Normalization condition

$$\begin{aligned} r_2 &= \text{number of zeros of } a_2(z) \text{ with } |z| < 1 \\ s_2 &= \text{number of zeros of } a_2(z) \text{ with } |z| > 1\ . \end{aligned} \tag{7}$$

Observe that while this normalization seems arbitrary for the pure Cauchy problem, it is by the Wiener-Hopf theory a necessary requirement when solving initial boundary value problems, see [5]. Furthermore by the normalization condition (7) the characteristic line mentioned in the introduction is uniquely defined even for implicit schemes.

Equation (4) is a three-term-recurrence relation. Hence its solutions are bounded if the roots of the polynomial

$$a_2(e^{i\Delta x\xi})\, w^2 + a_1(e^{i\Delta x\xi})\, w + a_0(e^{i\Delta x\xi})$$

have modulus not exceeding 1 and those of modulus 1 are simple. Therefore we introduce the **characteristic function**

$$\Phi(z,w) := a_2(z)\, w^2 + a_1(z)\, w + a_0(z) \tag{8}$$

and say that the scheme (2) is **stable** if

$$\left.\begin{array}{r} \Phi(z,w) = 0 \\ |z| = 1 \end{array}\right\} \Longrightarrow \left\{\begin{array}{l} |w| \leq 1 \text{ and if } |w| = 1 \\ \text{then } w \text{ is a simple root}\ . \end{array}\right. \tag{9}$$

As scheme has **error order** p if for any smooth solution $u(t,x)$ of (1) we have

$$\begin{aligned} \sum_{i=0}^{2} \sum_{j=-r_i}^{s_i} a_{ij}\, u(t+i\Delta t, x+j\Delta x) &= C\, \frac{\partial^{p+1}}{\partial x^{p+1}} u(t,x)(\Delta x)^{p+1} + O((\Delta x)^{p+2}) \\ &\quad \text{if } \Delta x \to 0 \text{ and } \mu = \text{constant}\ . \end{aligned} \tag{10}$$

We are interested in schemes with a positive order, hence we always assume that

$$\Phi(1,1) = \sum_{i=0}^{2} \sum_{j=-r_i}^{s_i} a_{ij} = 0 . \tag{11}$$

Hence $\Phi(1, w)$ has a root which is 1. If the scheme is stable then by (9) this root is a simple one. This implies that the **algebraic function** $w(z)$ defined implicitly by

$$\Phi(z, w(z)) \equiv 0 , \tag{12}$$

has a branch $w_1(z)$ which is analytic in a neighborhood of $z = 1$ and satisfies $w_1(1) = 1$. One can easily show that a stable scheme has order p if and only if this branch satisfies

$$z^{\mu} - w_1(z) = O((z-1)^{p+1}) \text{ as } z \to 1 , \tag{13}$$

see [8], [13]. This implies that one can assume without loss of generality that Φ is irreducible.

Observe that (10) represents linear conditions for the coefficients a_{ij}. Let I be the set of indices

$$I = \{(i,j) \in \mathbb{Z} \times \mathbb{Z} \mid 0 \le i \le 2, \ -r_i \le j \le s_i\} .$$

A scheme is said to be **regular** if no characteristic line passes through two stencil points. For regular stencils the system of equations for the a_{ij} is regular for the highest possible order $p = |I| - 2$ and hence we always have a unique scheme of highest error order for a given regular stencil. Of course, such a scheme need not be stable.

Due to the techniques used in the proofs one can only hope to prove the conjecture (3) for **convex** schemes. A scheme is called **convex** if the convex hull of the stencil contains all extreme points $(-r_i, i)$, (s_i, i), $i = 0, 1, 2$ of the stencil. For illustration the stencils in Fig. 1a),b) are not convex since the grid point $(-1, 1)$ does not lie on the convex hull. The scheme of Fig. 1c) however is convex. In the present paper we restrict ourselves to schemes with a convex and increasing stencil. A scheme is said to have an **increasing stencil** if r_i and s_i are non-decreasing with respect to i. Hence the stencil of Fig. 1c) is not increasing but convex. The stencil of Fig. 1d) however is convex and increasing. For two step-schemes the condition to be convex and have an increasing stencil is equivalent to the condition

$$\left\{ \begin{array}{lllll} 0 & \le & r_0 - r_1 & \le & r_1 - r_2 \\ 0 & \le & s_0 - s_1 & \le & s_1 - s_2 . \end{array} \right. \tag{14}$$

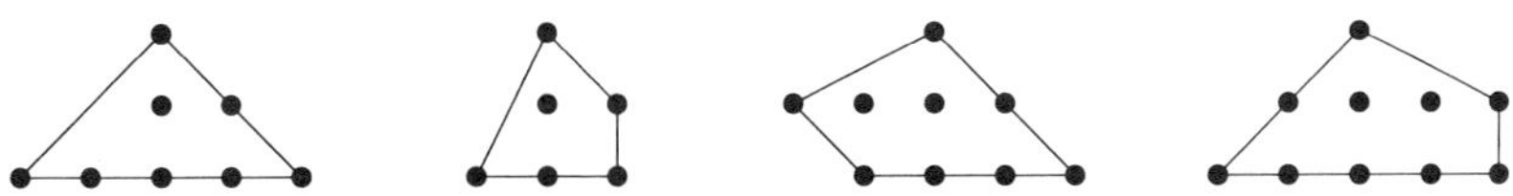

a) b) c) d)

Fig. 1: Stencils and their convex hull

From a computational point of view these schemes seem the most interesting class of schemes to consider. While we believe the conjecture (3) can be proved for convex schemes with an increasing stencil the actual proof in [9] is only done for a class of schemes which is slightly smaller. We shall indicate this difference in Section 4, Remark 4.2.

3 Main results

For two step schemes we define exactly the numbers R and S by

$$R = R_0 + R_1 + r_2, \; S = S_0 + S_1 + s_2 \; , \tag{15a}$$

where R_i, S_i are defined as follows

$$R_1 = \begin{cases} 0 & \text{if} \quad \mu < -r_1 \\ \lfloor r_1 + \mu \rfloor + 1 & \text{if} \quad -r_1 < \mu < s_1 \\ r_1 + s_1 + 1 & \text{if} \quad \mu > s_1 \end{cases} , \; S_1 = r_1 + s_1 + 1 - R_1 \tag{15b}$$

$$R_0 = \begin{cases} 0 & \text{if} \quad \mu < -r_0 \\ \lfloor r_0 + 2\mu \rfloor + 1 & \text{if} \quad -r_0 < 2\mu < s_0 \\ r_0 + s_0 + 1 & \text{if} \quad 2\mu > s_0 \end{cases} , \; S_0 = r_0 + s_0 + 1 - R_0 \tag{15c}$$

Here $\lfloor \alpha \rfloor$ denotes the largest integer not exceeding α.

Theorem 3.1 *Let a convex scheme (4) with an increasing stencil be normalized and have a Courant number μ satisfying $0 < |\mu| < 1/2$. If the scheme is stable, then the order p of the scheme is bounded by*

$$p \leq 2 \min\{R, S\} \, . \tag{16}$$

□

Remark 3.2

a) In [12] the bound (16) was proved for a small subclass of explicit schemes of type (4). In the present paper we generalize it, making use of a conjecture introduced in Section 5, for a class of explicit and implicit schemes of type (4) which are convex and have an increasing stencil.

b) In [9] we prove the conjecture for schemes of maximal order $p = |I| - 2$ and $-1/2 < \mu < 0$ and give here a brief hint of the proof in Remark 5.2.

c) As indicated the current proofs are valid only for a slightly smaller class than the convex schemes with increasing stencil, see Remark 4.2. However this smaller class includes all schemes with

$$\begin{cases} 0 \leq r_0 - r_1 < r_1 - r_2 \\ 0 \leq s_0 - s_1 < s_1 - s_2 \end{cases} . \tag{17}$$

d) The result (16) can be extended to $|\mu| > \frac{1}{2}$ by making use of the following transformation. Assume that a stable scheme is represented by $\Phi(z, w)$, where $w(z)$ approximates z^μ in a neighbourhood of the point $z = 1, w = 1$. Then we consider the scheme represented by the characteristic function

$$\widetilde{\Phi}(z, u) = z^2 \, \Phi(z, \tfrac{u}{z}) \, .$$

Since $u = zw$, the new scheme is stable and approximates $z^{\widetilde{\mu}} = z^{\mu+1}$ with the same order as the original scheme. The stencil undergoes the following transformations:

$$\widetilde{r}_1 = r_1 - 1, \quad \widetilde{s}_1 = s_1 + 1, \quad \widetilde{r}_0 = r_0 - 2, \quad \widetilde{s}_0 = s_0 + 2 \, .$$

4 The algebraic function w and order stars

The algebraic function $w(z)$ defined by Φ as a solution of (12) can either be viewed as a double valued function, consisting in general for each z of two values $w_1(z)$ and $w_2(z)$ or we can interpret $w(z)$ as a single valued function on the compact Riemann surface,

$$M = \{(z,w) \in \bar{\mathbb{C}} \times \bar{\mathbb{C}} \mid \Phi(z,w) = 0\} .$$

M is a connected set on which w is analytic except for points (z_i, ∞) where w has a pole. Since in the order star technique poles play an important role we need to know where these occur.

Proposition 4.1 *The function w has a pole at every point where $a_2(z) = 0$. In view of normalization (7), there will be exactly r_2 poles in $|z| < 1$ and s_2 in $|z| > 1$. Moreover if there is no branch point of $w(z)$ at $z = 0$ then the branches $w_i(z)$ have the following expansions*

$$\text{(18)} \qquad \begin{aligned} w_1(z) &= z^{-(r_1-r_2)}(c_0 + c_1\, z + c_2\, z^2 + ...) \\ w_2(z) &= z^{-(r_0-r_1)}(d_0 + d_1\, z + d_2\, z^2 + ...) \end{aligned}$$

at $z = 0$.

Remark 4.2

At this point we can explain the small set for which our proof has not been carried through completely. In the proof we assume that there is no branch point at $z = 0$. Branch points are not really special points on a Riemann surface; however, in the order star technique used here, $z = 0$ is a very special point. If $r_1 - r_0 = r_2 - r_1$ it can in rare cases happen that there is a branch point at $z = 0$. However then a slight perturbation of the coefficients a_{ij} will move the branch point away from zero. However a stencil of form given in Fig. 1b) always has a branch point at $z = 0$ even if one perturbs the coefficients a_{ij} slightly. This is why we have excluded these type of stencils by requesting convexity. Observe that condition (17) automatically ensures that there is no branch point at zero. In the following we shall always assume that $z = 0$ and $z = \infty$ are not branch points of w.

At this point it is possible to explain why one requests that the stencil is increasing. The zeros of w are at points $(z_i^\circ, 0)$ with $a_0(z_i^\circ) = 0$ and at $z = 0$ or $z = \infty$ if the stencil is not increasing. For example from (18) we see that a stencil of the form given in Fig. 1c) has a zero at $z = 0$. Since we currently have no definite proofs to locate the zeros we cannot adapt our proofs to work with zeros. Hence the shape of the stencil which is related to zeros cannot be efficiently treated and therefore we have to restrict ourselves to increasing stencils.

Definition 4.3 The set

$$\Omega = \{(z,w) \in M \,\Big|\, |z^{-\mu} w| > 1\}$$

is said to be the **order star**. □

We shall frequently work with the function

$$\varphi(z,w) = z^{-\mu} w, \ (z,w) \in M$$

which is multivalued on M due to the factor $z^{-\mu}$. Ω^c denotes the complement of Ω, i.e. $\Omega^c = M \backslash \Omega$. Because the coefficients a_{ij} are real Ω is symmetric with respect to the real axis. As in the classical order star technique introduced in [14] the following two lemmata express stability and accuracy in terms of the order star.

Lemma 4.4 *(Stability)*

If a scheme is stable then

$$\Omega \cap \{(z,w) \in M \,\Big|\, |z| = 1\} = \emptyset \,. \qquad \square$$

Lemma 4.5 *(Order)*

A stable scheme of form (2) has order p if and only if at the point $z = 1$ the point of the order star induced by $z^{-\mu}w_1(z)$ locally consists of $p+1$ sectors of an angle $\frac{\pi}{p+1}$, separated by $p+1$ sectors of Ω^c, each with the same angle. $\square$

Observe that Ω usually consists of several disconnected components. We say a component has **multiplicity** m if it contains m of the sectors described in Lemma 4.5.

The basic philosophy of the proof is now as follows. By Lemma 4.4 one can distinguish between the components of Ω inside the unit disk $\Delta = \{(z,w) \in M \,\Big|\, |z| \leq 1\}$ and those outside. At this point one wants to apply the argument principle to $f(z)$. The question arises whether one should work on M or on the noncompact Riemann surface induced by $z^{-\mu}w(z)$. We will work on M and therefore we have to be able to define $z^{-\mu}$ uniquely in the component of Ω on M with which we want to work. On a technical level this is done as follows. M is represented over z as two sheets which are connected through branch cuts. There are only finitely many of these cuts. They connect the branch points. There is not a unique way of choosing these but the connectivity of the Ω components are independent of this choice since it is a property related to M and not to a particular representation over z. Once the cuts have been chosen we call the sheet containing $(1,1)$ the principal sheet and the other one the secondary sheet. To be able to define $z^{-\mu}$ uniquely on M we cut both sheets with two cuts L_1, L_2 which connect $(0, w_i^0)$ with (∞, w_i^∞) such that the projections of the cuts onto the z-plane are identical and do not intersect the cuts used in the representation of M. These cuts L_1, L_2 can always be chosen. We can now define $\log z$ and hence $z^{-\mu} = e^{-\mu \log z}$ uniquely on $M \backslash \{L_1, L_2\}$ and the choice is done such that in a neighborhood of $(1,1) \in M$ $\log z$ is real for z real. When we apply the argument principle on a component, we are not allowed to cross either of the cuts L_1 or L_2. In both cases we move along the cut to $z = 0$, encircle this point to get to the other side of the cut and move back along the other side of the cut until we reach the same location on the other side. Such integration paths are illustrated in Fig. 2.

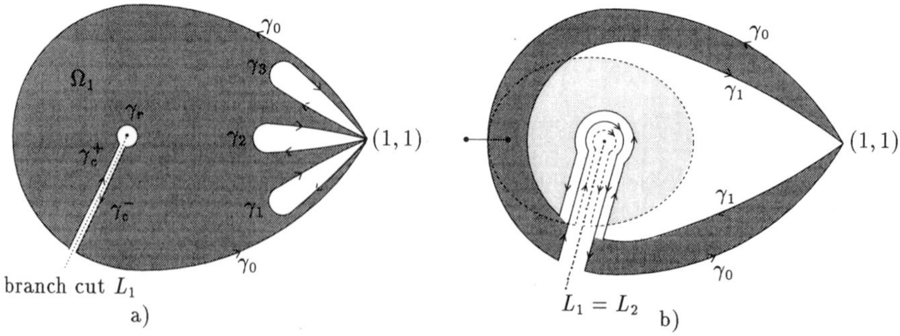

Fig. 2: Ω components Ω_1 with integration paths

Using the argument principle one can show that each component of Ω in Δ has either to contain one of the zero points $(0, w_i^0) \in M$, $i = 1,2$; or a pole (z_i, ∞), $i = 1, 2, ..., r_2$, where $a_2(z_i) = 0$. The order star technique uses the argument principle to relate the order of these poles, i.e. the leading exponent of the expansion of these points, to the multiplicity of the component. This will be done in Section 5 and we shall use this relation to prove the main theorem 3.1 in Section 6.

5 Bounds for the multiplicity of Ω components

We need to look at the expansion of φ at the zeropoints $(0, w_i^0)$ and at the poles of $w(z)$. Let $-\alpha_i$ be the leading exponent of φ at $(0, w_i^0)$.

Conjecture 5.1 The multiplicity m for the Ω-component Ω_1 in Δ containing both zero points $(0, w_i^0)$, $i = 1, 2$, and no poles satisfies

$$m \leq \lfloor \alpha_1 \rfloor + 1 + 2\lfloor \alpha_2 \rfloor + \max\{1, 2\lfloor \delta_1 + \delta_2 \rfloor\} ,$$

where $\alpha_i = \lfloor \alpha_i \rfloor + \delta_i$, $i = 1, 2$. □

Remark 5.2

The conjecture has been proved in [9] for schemes of optimal order i.e. $p = |I| - 2$ and $-1/2 < \mu < 0$. Due to its length we have to omit this proof here. It works by also counting the number of branch points of the algebraic function. To do this one also has to work with the part of Ω outside Δ.

Note that in general, i.e. if $p < |I| - 2$ or $p = |I| - 2$ and $0 < \mu < 1/2$, we can only prove (19) with a factor 2 in front of the term $\lfloor \alpha_1 \rfloor$. In the situation where the two zero points lie in different components we had shown already in [12] the following proposition

Proposition 5.3 *The multiplicity m of an Ω-component Ω_1 in Δ containing exactly one zero point and no poles satisfies either*

$$m \leq \lfloor \alpha_1 \rfloor + 1, \ (0, w_1^0) \in \Omega_1 \tag{19}$$

or

$$m \leq 2\lfloor \alpha_2 \rfloor + \max\{1, 2\lfloor \delta_1 + \delta_2 \rfloor\}, \ (0, w_2^0) \in \Omega_1 \tag{20}$$

where $\alpha_i = \lfloor \alpha_i \rfloor + \delta_i$, $i = 1, 2$. The Ω_1 component which allows bound (19) is called **simple** *while the one allowing bound (20) is called* **binary**. *Two disconnected binary components cannot exist simultaneously.* □

At this point we can indicate where our proving technique to show Conjecture 5.1 fails. Consider the order stars depicted in Fig. 3.

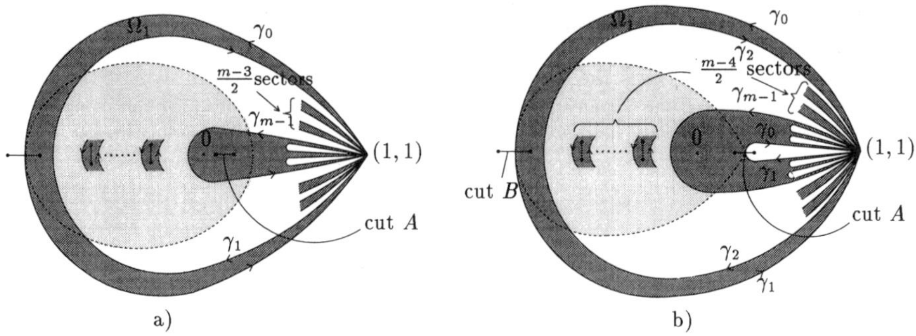

Fig. 3: Example of Ω-component where bound (19) cannot be shown

If in Fig. 3a) the cut A is missing we can apply Proposition 5.3 and get as bound for the total multiplicity the sum of the two bounds (19) and (20) and hence (19) holds. However if the cut A is present our techniques give a factor two in front of $\lfloor\alpha_1\rfloor$. Moreover in Fig. 3b) it is not clear how one could remove the cut A.

In Conjecture 5.1 and Proposition 5.3 the poles of w away from $z = 0$ have been excluded for the following reason. Since we want to bound m for all possible order stars we have to consider those where this bound becomes as large as possible. With proofs similar to the one used to show Proposition 5.3 one sees that the poles give rise to the highest bound when the components of Ω with a zero point do not contain a pole away from zero and each of these components contains exactly one pole and this pole has order 1.

Proposition 5.4 *Assume that φ has q poles away from zero in Ω-components in Δ which don't contain the zero points. Hence the sum m of all the multiplicities of these components satisfies the inequality*

$$m \le 3q\,. \tag{21}$$

Equality in (21) occurs if each of these poles is simple and each component contains exactly one of these poles. □

We collect Conjecture 5.1, Proposition 5.3 and 5.4 in the following

Corollary 5.5 *Let m be the total multiplicity of Ω in Δ of a stable normalized scheme with a convex and increasing stencil. Then we have the following three bounds:*

i) If Ω contains no zero point $(0, w_i^0)$ then

$$m \le 3r_2\,. \tag{22}$$

ii) If Ω contains exactly one zero point $(0, w_i^0)$ then

$$m \le 2\lfloor\max\{\alpha_1, \alpha_2\}\rfloor + \max\{1, 2\lfloor\delta_1 + \delta_2\rfloor\} + 3r_2\,. \tag{23}$$

iii) If Ω contains both zero points $(0, w_i^0)$ then

$$m \leq \lfloor \min\{\alpha_1, \alpha_2\} \rfloor + 1 + 2\lfloor \max\{\alpha_1, \alpha_2\} \rfloor + \max\{1, 2\lfloor \delta_1 + \delta_2 \rfloor\} + 3r_2 \,. \tag{24}$$

□

Remark 5.6

Observe that Conjecture 5.1 is only used in (24) if $p < |I| - 2$ or if $p = |I| - 2$ and $0 < \mu < 1/2$. Moreover as explained in Remark 4.2 the actual proofs have been carried through for the cases where $z = 0$ is not a branch point of w.

6 Proof of the main theorem

We shall indicate in the proof where we use Conjecture 5.1. Observe, however, that for $p = |I| - 2$ and $-1/2 < \mu < 0$ the conjecture has been proved in [9]. The proof of Theorem 3.1 will be divided into two parts. First we show that

$$p \leq 2R \,, \tag{25}$$

where R is the number of downwind points defined in (15). Since the scheme is stable we know that the order star Ω does not intersect the unit circle $\{(z, w) \in M \,\big|\, |z| = 1\}$ and hence all components of Ω in Δ are bounded and the bounds for the multiplicities of m given in Section 5 apply. Since the scheme is normalized we have exactly r_2 poles in Ω in Δ. From Proposition 4.1 we have the following expansions of $\varphi(z)$ at $z = 0$, w_i

$$\begin{aligned} \varphi(z) &= z^{-(r_1 - r_2) - \mu}(c_0 + c_1\, z + c_2\, z^2 + \ldots) \text{ at } (0, w_1^0) \\ \varphi(z) &= z^{-(r_0 - r_1) - \mu}(d_0 + d_1\, z + d_2\, z^2 + \ldots) \text{ at } (0, w_2^0) \,. \end{aligned} \tag{26}$$

It should be noted here that we have no means of associating a certain expansion with the zero point on a specific sheet. Hence the expansions will be associated with the zero points in the way which leads to the highest possible multiplicity.

In the remainder of the proof we have to work separately with the cases where $\mu < 0$ and where $\mu > 0$.

a) We first assume $-\frac{1}{2} < \mu < 0$. Then the following choices of the indices r_0, r_1, r_2 lead to different combinations of Ω-components inside Δ.

   (i) $r_0 = r_1 = r_2 = 0$. Then also $R = 0$. According to the Courant-Friedrichs-Lewy condition the scheme cannot be convergent, i.e. it is impossible to have simultaneously order $p \geq 1$ and stability.

   (ii) $r_0 = r_1 = r_2 > 0$. There are r_2 poles away from $z = 0$ inside Δ, while $\varphi(z)$ has positive leading exponents of $-\alpha_i = -\mu$ at $z = 0$ on both sheets of M, implying that both zero points belong to Ω^c. We apply Proposition 5.4 or i) of Corollary 5.5 to obtain

$$m \leq 3r_2 = r_0 + r_1 + r_2 = R \,.$$

   (iii) $0 = r_0 - r_1 < r_1 - r_2$. Then $-\alpha_1 = -\mu > 0$, implying that $(0, w_1^0) \in \Omega^c$, and $-\alpha_2 = -(r_1 - r_2) - \mu < 0$, implying that $(0, w_2^0) \in \Omega$. We apply ii) of Corollary 5.5. Then we have

$$\begin{aligned} m &\leq 3r_2 + 2\lfloor r_1 - r_2 + \mu \rfloor + 2\lfloor 2 + 2\mu \rfloor \\ &= 3r_2 + 2(r_1 - r_2 - 1) + 2 \\ &= r_2 + 2r_1 = r_2 + r_1 + r_0 = R \,. \end{aligned}$$

(iv) $0 < r_0 - r_1 < r_1 - r_2$. Then $-\alpha_1 = -(r_0 - r_1) - \mu < 0$ and $-\alpha_2 = -(r_1 - r_2) - \mu < 0$, implying that both $(0, w_1^0)$ and $(0, w_2^0)$ belong to Ω. We apply iii) of Corollary 5.5. Then the total multiplicity m inside Δ is bounded by

$$\begin{aligned} m &\leq 3r_2 + \{2\lfloor r_1 - r_2 + \mu\rfloor + 2\lfloor 2 + 2\mu\rfloor\} + \{\lfloor r_0 - r_1 + \mu\rfloor + 1\} \\ &= 3r_2 + \{2(r_1 - r_2 - 1) + 2\} + \{(r_0 - r_1 - 1) + 1\} \\ &= r_0 + r_1 + r_2 = R\,. \end{aligned}$$

b) Assume $0 < \mu < \frac{1}{2}$. Then we have $-\alpha_1 = -(r_0 - r_1) - \mu < 0$ and $-\alpha_2 = -(r_1 - r_2) - \mu < 0$, implying that both $(0, w_1^0)$ and $(0, w_2^0)$ belong to Ω. We apply iii) of Corollary 5.5. This leads to the bound

$$\begin{aligned} m &\leq 3r_2 + \{2\lfloor r_1 - r_2 + \mu\rfloor + 1\} + \{\lfloor r_0 - r_1 + \mu\rfloor + 1\} \\ &= 3r_2 + \{2(r_1 - r_2) + 1\} + \{(r_0 - r_1) + 1\} \\ &= r_0 + r_1 + r_2 + 2 = R\,. \end{aligned}$$

Observe that in the case b) and iv) of a) we have used Conjecture 5.1 if $p < |I| - 2$ and if $p = |I| - 2$ and $0 < \mu < 1/2$. Moreover in both cases it could happen that w has a branch point at $z = 0$, see Remark 4.2.

In all cases we obtain

$$m \leq R\,.$$

The remainder of the proof makes use of Lemma 4.5. Observe that by this lemma and stability the multiplicity outside Δ can be at most $m + 1$. Hence we find

$$p + 1 \leq m + (m + 1) \leq 2R + 1$$

which leads to the bound (25) $p \leq 2R$.

Concerning the bound $p \leq 2S$ we could work in exactly the same fashion but outside Δ. However by using the mapping $z \to 1/z$ and $\mu \to -\mu$ we can apply the bound just shown. To do this our assumptions on the scheme and the bound just proved have to be invariant with respect to this mapping. This completes the proof of Theorem 3.1. □

Remark 6.1

It is interesting to observe the following since the order condition represents a regular system of linear equations. The presence of a stencil point raises the highest reachable order by 1. To each stencil point $(2, j)$ on the newest time-level belong, due to our assumption of an increasing stencil, two stencil points $(1, j)$ and $(0, j)$, which means, by collecting these, that for our schemes a stencil point $(2, j)$ raises the highest reachable order by 3. This factor can be found in all three bounds of Corollary 5.5 associated with r_2. In a certain sense we can say the stencil points $(0, j)$, $(1, j)$ with $j \in \{-r_2, -r_2 + 1, ..., -1\}$ are used up by the implicit stencil points $(2, j)$, $j \in \{-r_2, -r_2 + 1, ..., -1\}$. For simplicity assume that (16) is satisfied, i.e. $0 \leq r_0 - r_1 < r_1 - r_2$. Hence $\alpha_2 = r_1 - r_2 + \mu > \alpha_1 = r_0 - r_1 + \mu$. Again to each stencil point $(1, j)$ which has not been used by the implicit part, i.e. $j \in \{-r_1, -r_1 + 1, ..., -r_2 - 1\}$ we can associate one stencil point $(0, j)$. Again collecting these, one has that each $(1, j)$, $j \in (-r_1, ..., -r_2 - 1\}$ raises the highest reachable order by 2. This factor can be found in the bounds (23) and (24) in front of the term with the larger α_i which, according to our assumptions, is the α_i associated with the stencil points on the time level t_{n+1} which have not been used by the implicit part. Similarly one sees that the stencil points $(0, j)$ which have not been used up by the other time levels i.e., the ones with $j \in \{-r_0, -r_0+1, ..., -r_1-1\}$, are associated with the factor 1 in front of smaller of the two α_i in bound (24). In further investigations to more than 3 time level schemes we have already discovered that this pattern continues.

Acknowledgements:

The work of the second author was supported by an NSF US-Switzerland cooperative research grant INT9123314 and funding from the Forschungsinstitut für Mathematik, ETH Zürich. The work of the first and third authors was supported under project Nr. 21-33551.92 of the Schweizerische Nationalfonds. Travel funds for the third author were provided by the University of Stellenbosch.

References

[1] Courant R., Friedrichs K.O., Lewy H., Über die partiellen Differenzengleichungen der mathematischen Physik, Math. An., **100**, 32-74, 1928.

[2] Goldberg M., Simple stability criteria for difference approximations of hyperbolic initial-boundary value problems II, in: Proceedings of the Third International Conference on Hyperbolic Problems, Uppsala, June 11-15, 1990, ed. Engquist B. and Gustafsson B., Chartwell-Bratt, 519-527, (1991).

[3] Goldberg M., Tadmor E., Convenient stability criteria for difference approximations of hyperbolic initial-boundary value problems II, Math. Comp. **48**, 503-520, (1987).

[4] Gustafsson B., Kreiss H.-O., Sundström A., Stability theory of difference approximations for mixed initial boundary value problems II, Math. Comp. **26**, 649-686, (1972).

[5] Iserles A., Strang G., The optimal accuracy of difference schemes, Trans. Amer. Math. Soc. **277**, 779-803, (1983).

[6] Jeltsch R., Order barriers for difference schemes for linear and non-linear hyperbolic problems, in: Numerical Analysis, Proceedings of the 1987 Dundee conference, ed. D. F. Griffiths, D. A. Watson, Pitman, (1988).

[7] Jeltsch R., Kiani P., Stability of a family of multi-time-level difference schemes for the advection equation, Numer. Math. **60**, 77-95 (1991).

[8] Jeltsch R., Kiani P., Raczek K., Counterexamples to a stability barrier, Numer. Math. **52**, 301-316, (1988).

[9] Jeltsch R., Renaut R., Smit J., An accuracy barrier for stable three-time-level difference schemes for hyperbolic equations, Research Report No. 95-01, Seminar für Angewandte Mathematik, ETH Zürich, 1995.

[10] Jeltsch R., Smit J. H., Accuracy barriers of difference schemes for hyperbolic equations, SIAM J. Numer. Anal. **24**, 1-11, (1987).

[11] Jeltsch R., Smit J. H., Accuracy barriers of three-time-level difference schemes for hyperbolic equations, in: Proceedings of the Third International Conference on Hyperbolic

Problems, Uppsala, June 11-15, 1990, ed. Engquist B. and Gustafsson B., Chartwell-Bratt, 611-627, (1991).

[12] Jeltsch R., Smit J. H., Accuracy barriers of three-time-level difference schemes for hyperbolic equations, Ann. University of Stellenbosch, 1992/2, 1-34, (1992).

[13] Strang G., Iserles A., Barriers to stability, SIAM J. Numer. Anal. **20**, 1251-1257, (1983).

[14] Wanner G., Hairer E., Nørsett S. P., Order stars and stability theorems, BIT 18, 475-489, (1978).

NONUNIQUENESS OF SOLUTIONS OF RIEMANN PROBLEMS CAUSED BY 2-CYCLES OF SHOCK WAVES

ARTHUR AZEVEDO, DAN MARCHESIN, BRADLEY J. PLOHR, AND KEVIN ZUMBRUN

ABSTRACT. We investigate a general mechanism, based upon transitional waves, for nonuniqueness of solutions of Riemann initial-value problems for systems of two conservation laws. This nonuniqueness occurs whenever there exists a pair of viscous shock waves forming a 2-cycle, *i.e.*, two states U_1 and U_2 such that a traveling wave leads from U_1 to U_2 and another leads from U_2 to U_1. We show that a 2-cycle gives rise to an open region of Riemann data for which there exist multiple solutions of the Riemann problem, and we describe all such solutions.

1. INTRODUCTION

1.1. **Overview.** Recent work in Refs. [1, 2] describes a system of two conservation laws of mixed elliptic-hyperbolic type which, for an open set of choices of Riemann initial data, exhibits multiple solutions. This phenomenon supports a point of view advocated by Glimm and Sharp [9], that the basic building block in the long-time behavior of conservation laws is not the solution operator for the Riemann problem, which can be thought of as a map associating Riemann data at $t = 0$ to the $t \to \infty$ asymptotic solution, but rather the so-called scattering operator, mapping incoming waves from $t \to -\infty$ to outgoing waves at $t \to \infty$. The solution operator is, of course, ill-defined when there is nonuniqueness. The distinction between these operators, important from a conceptual point of view, also has practical implications, for example in the design of numerical schemes.

Nonuniqueness deserves study. It is important, first, to understand the mechanisms that give rise to multiple solutions; and, second, to verify the stability of these multiple solutions under small changes both in the initial data and in the governing partial differential equations. In the present paper, we investigate a new mechanism for nonuniqueness.

The multiple solutions described in Refs. [1, 2] result from projection effects: even for two equations, a nonclassical solution can have a rich structure, involving several intermediate states, in contrast to the classical solution constructed by Lax, which has a single middle state; the manifold of solutions thus resides in a higher dimensional space, making more likely the possibility of folds when it is projected onto the space of endstates specifying Riemann problems. In the present paper, we investigate a different cause for multiple solutions, involving global bifurcations. We isolate a general mechanism, occurring for an open set of models and data, by which successions of nonclassical waves, associated with saddle-to-saddle connections, lead to multiple solutions for the Riemann problem. Some of the solutions contain solitary waves, which, in the zero-viscosity limit, are indistinguishable from constant solutions. However, the other solutions remain distinct in the hyperbolic limit, a behavior that is in marked contrast to that observed in some other examples of nonuniqueness [5, 6, 7, 8].

1991 *Mathematics Subject Classification.* 35L65, 35L67.

Key words and phrases. conservation law, Riemann problem, viscous profile, nonuniqueness.

This work was supported in part by: the Conselho Nacional de Desenvolvimento Tecnológico e Científico under Grant CNPq/NSF 910087/92-0; the Conselho Nacional de Desenvolvimento Tecnológico e Científico under Grant 530054/93-0; the Financiadora de Estudos e Projetos under Grant 65920311-00; the National Science Foundation under Grant INT-9216357; the U. S. Army Research Office under Grant DAAL03-91-C-0027 to the Mathematical Sciences Institute of Cornell University, through subcontract to Stony Brook; the Applied Mathematics Subprogram of the U. S. Department of Energy under Grant DE-FG02-90ER25084; the Office of Naval Research under Grant N00014-94-1-0456; and the National Science Foundation under Grant DMS-9107990.

1.2. **Results.** Recall that a Riemann solution U of a system of two conservation laws

$$U_t + F(U)_x = 0 \tag{1.1}$$

satisfies Riemann initial data, of form

$$U(x,0) = \begin{cases} U_L & \text{for } x < 0, \\ U_R & \text{for } x > 0. \end{cases} \tag{1.2}$$

Each shock wave appearing in a solution is required to be *admissible*, *i.e.*, the limit, as $\epsilon \to 0$, of a traveling wave for an augmented system

$$U_t + F(U)_x = \epsilon \left[B(U)U_x\right]_x, \tag{1.3}$$

where the viscosity matrix $B(U)$ has positive eigenvalues. In other words, for a shock wave with left state U_-, right state U_+, and speed s to be admissible, there must be an orbit Γ for the dynamical system

$$\dot{U} = B(U)^{-1}\left[F(U) - F(U_-) - s(U - U_-)\right] \tag{1.4}$$

leading from U_- to U_+. In this work, a shock wave is of one of three types:

(1) a *classical shock wave*, either of the 1-family, when U_- is a repelling equilibrium point and U_+ is a saddle point, or else of the 2-family, when U_- is a saddle point and U_+ is an attracting equilibrium point;
(2) a *transitional shock wave*, for which both U_- and U_+ are saddle points;
(3) a *homoclinic shock wave*, for which $U_- = U_+$ and a homoclinic orbit joins this saddle point to itself.

Of course, a homoclinic shock wave, in the inviscid limit, is indistinguishable from the constant solution; nevertheless, we maintain a distinction since a homoclinic wave represents possible asymptotic behavior of parabolic solutions of system (1.3).

Nonuniqueness, of the type we study here, arises when there exists a *2-cycle of shock waves*, meaning that there is an orbit for Eq. (1.4) leading from U_+ to U_-, in addition to an orbit from U_- to U_+, so that the reversed shock wave, with left state U_+ and right state U_-, is also admissible. More precisely, we suppose that there are states U_1^* and U_2^* and speed s^* such that, for the dynamical system

$$\dot{U} = B(U)^{-1}\left[F(U) - F(U_1^*) - s^*(U - U_1^*)\right], \tag{1.5}$$

there is an orbit Γ_a^* leading from U_1^* to U_2^* and an orbit Γ_b^* leading from U_2^* to U_1^*. In other words, the dynamical system (1.5) has a 2-cycle, as illustrated in Fig. 1.1. We note that 2-cycles are nonclassical phenomena, incompatible with the existence of a convex entropy for system (1.1); they do not occur, for example, in gas dynamics.

In the following, we shall make several assumptions about the family (1.4) of dynamical systems, all of which will be "genericity" conditions holding in an open dense set in the space of models (1.3). The first of these is the assumption that

$$U_1^* \text{ and } U_2^* \text{ are hyperbolic saddle points.} \tag{G0}$$

Then the two shock waves corresponding to the orbits from U_1^* to U_2^* and *vice versa* are transitional shock waves.

We shall consider Riemann problems for which the states U_L and U_R lie near U_2^*. For these data, we will construct solutions of Eqs. (1.1)–(1.2) with the following wave patterns:

(a) *Lax solution* — a 1-family shock or rarefaction wave from U_L to a nearby state U_M, followed by a 2-family shock or rarefaction wave from U_M to U_R;

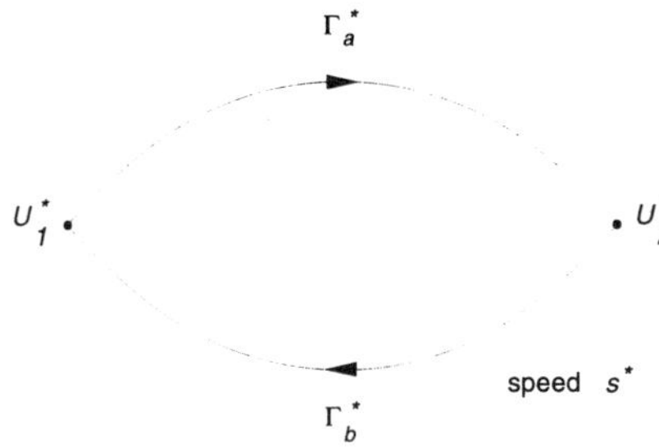

FIGURE 1.1. The phase portrait for a 2-cycle of shock waves.

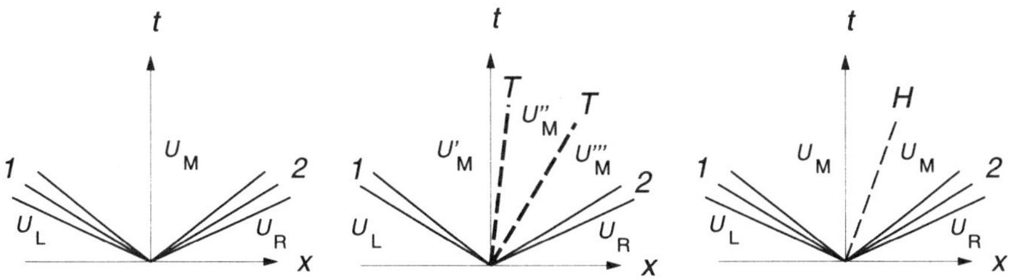

(a) Lax (b) two-transitional (c) homoclinic

FIGURE 1.2. The three types of wave patterns, depicted in space-time.

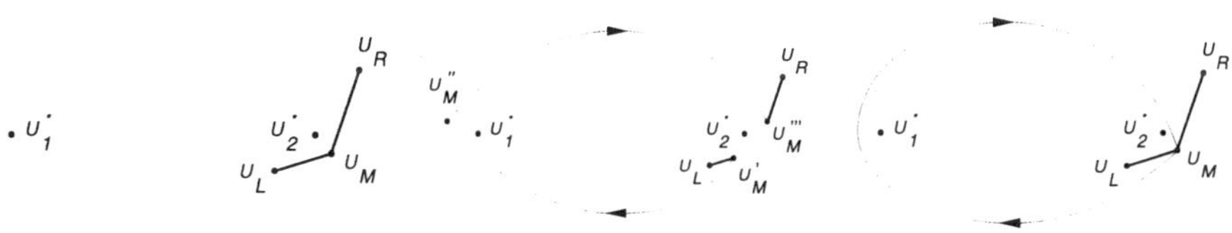

(a) Lax (b) two-transitional (c) homoclinic

FIGURE 1.3. The three types of wave patterns, depicted in state space.

(b) *two-transitional solution* — a 1-family shock or rarefaction wave from U_L to a nearby state U'_M, followed by a transitional shock wave from U'_M to a state U''_M near U_1^*, a transitional shock wave from U''_M to a state U'''_M near U_2^*, and a 2-family shock or rarefaction wave from U'''_M to U_R;

(c) *homoclinic solution* — a 1-family shock or rarefaction wave from U_L to a nearby state U_M (the same as in the Lax solution), followed by a homoclinic shock wave from U_M to U_M, followed by a 2-family shock or rarefaction wave from U_M to U_R.

These wave patterns are illustrated in Figs. 1.2 and 1.3.

The principal result of this paper, stated roughly, is the following. Assuming certain nondegeneracy conditions, there is a smooth curve $\mathcal{C}(U_L)$ near U_2^*, which depends smoothly on U_L, such that for U_R near U_2^*:

- there is a Lax solution;
- there is a homoclinic solution precisely when U_R lies on a certain side of $\mathcal{C}(U_L)$ (the "homoclinic" side);
- and there is a two-transitional solution precisely when U_R lies on a certain side of $\mathcal{C}(U_L)$ (the "transitional" side).

The homoclinic and transitional sides may be coincident or opposite. If they are coincident, all three types of solution occur for U_R on the joint homoclinic-transitional side, and only the Lax solution occurs for U_R on the other side; thus, there is three-fold nonuniqueness when U_R lies on one side and uniqueness when U_R lies on the other side. If the homoclinic and transitional sides are opposite, Lax and homoclinic solutions occur for U_R on the homoclinic side, whereas Lax and transitional solutions occur for U_R on the transitional side. In other words, there is two-fold nonuniqueness for U_R anywhere except on $\mathcal{C}(U_L)$. Depending on the nature of the 2-cycle, there can also occur additional solutions, which involve sequences of transitional and homoclinic waves, when U_R lies on the transitional side of $\mathcal{C}(U_L)$.

1.3. **Outline.** The plan of the paper is as follows. In section 2, we describe the bifurcation of the phase portrait of Eq. (2.1) as the parameters U_1 and s vary near U_1^* and s^*. In sections 2.1–2.2, we describe the transitional and homoclinic surfaces that make up the various boundaries in the associated bifurcation diagram. These surfaces implicitly define pairs of transitional and homoclinic maps, relating points U_1 to corresponding points U_2 for which there exist certain saddle-to-saddle or homoclinic connections. In sections 2.3–2.4, we describe these maps, along with the associated speed maps relating U_1 and U_2 to the shock speed s. In section 2.5 we describe the relationships among the speeds of three successive transitional waves. In section 3, we analyze all possible solutions of Riemann problems with data given near U_2^*. Finally, in section 4, we discuss our results and related open problems.

2. Bifurcation analysis

The bifurcation analysis of the two-saddle cycle shown in Fig. 1.1 is well known; see, *e.g.*, [4]. In this section, we employ this analysis to establish the existence of transitional and homoclinic shock waves near the 2-cycle of shock waves defined by U_1^*, U_2^*, and s^*.

2.1. **Transitional surfaces.** Consider the differential equation

$$\dot{U} = B(U)^{-1}\left[F(U) - F(U_1) - s(U - U_1)\right] \tag{2.1}$$

for U_1 near U_1^* and s near s^*. When $U_1 = U_1^*$ and $s = s^*$, Eq. (2.1) reduces to Eq. (1.5), which has orbits Γ_a^* and Γ_b^*, each a separatrix for both U_1^* and U_2^*. When $U_1 \neq U_1^*$ or $s \neq s^*$, a second equilibrium U_2 persists near U_2^*; in general, however, the separatrix connections are broken. One possible phase portrait is shown in Fig. 2.1.

To quantify the breaking of connections, choose lines transverse to Γ_a^* and Γ_b^* and orient them as indicated in Fig. 2.1. Take the origin on the transversal for Γ_a^* to be the point where the stable manifold of U_2 meets it; then the unstable manifold of U_1 meets this transversal at the point $d_a(U_1, s)$. We can also define $d_b(U_1, s)$ as in Fig. 2.1. The *separation functions* d_a and d_b are smooth, and their derivatives at (U_1^*, s^*) can be expressed as Melnikov integrals [10].

We make the following genericity assumption:

$$\frac{\partial d_a}{\partial s}(U_1^*, s^*) \neq 0 \quad \text{and} \quad \frac{\partial d_b}{\partial s}(U_1^*, s^*) \neq 0. \tag{G1}$$

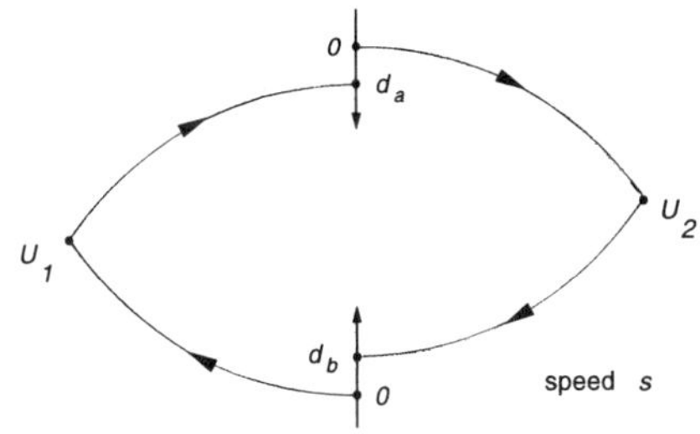

FIGURE 2.1. A possible phase portrait for Eq. (2.1).

Consequently, the equation $d_a(U_1, s) = 0$ (respectively, $d_b(U_1, s) = 0$) can be solved locally for s in terms of U_1: $s = s_a(U_1)$ (respectively, $s = s_b(U_1)$).

Geometrically, the equation $d_a(U_1, s) = 0$ defines a smooth surface $\mathcal{T}_a$ in (U_1, s)-space. If (U_1, s) lies on $\mathcal{T}_a$, then Eq. (2.1) has an orbit Γ_a leading from U_1 to U_2. This orbit is a separatrix for both U_1 and U_2. Similarly, $d_b(U_1, s) = 0$ defines $\mathcal{T}_b$. We refer to $\mathcal{T}_a$ and $\mathcal{T}_b$ as *transitional surfaces.* By virtue of assumption (G1), the transitional surfaces $\mathcal{T}_a$ and $\mathcal{T}_b$ are the graphs of s_a and s_b.

Another genericity assumption is that $\mathcal{T}_a$ and $\mathcal{T}_b$ meet transversally. Because of assumption (G1), transversality can be phrased in terms of s_a and s_b:

$$\frac{\partial s_a}{\partial U_1}(U_1^*) \neq \frac{\partial s_b}{\partial U_1}(U_1^*). \tag{G2}$$

It follows that the equation $s_a(U_1) = s_b(U_1)$ defines a smooth curve $\mathcal{B}_1$ through U_1^*. If $U_1 \in \mathcal{B}_1$ and s is the common value of $s_a(U_1)$ and $s_b(U_1)$, then Eq. (2.1) has a two-saddle cycle. Thus there exists a one-parameter family of two-saddle cycles connecting pairs of equilibria near U_1^* and U_2^*.

2.2. **Homoclinic surfaces.** The bifurcation analysis of the two-saddle cycle shown in Fig. 1.1 is contingent upon further genericity assumptions. Let $\mu_1^* > 0 > -\nu_1^*$ denote the eigenvalues at the equilibrium U_1^* for Eq. (1.5), and let $r_1^* = \nu_1^*/\mu_1^*$ denote the *hyperbolicity ratio.* Similarly, define r_2^* for the equilibrium U_2^*. We assume that

$$r_1^* \neq 1, \quad r_2^* \neq 1, \quad \text{and} \quad r_1^* r_2^* \neq 1. \tag{G3}$$

Under assumption (G3), the bifurcation analysis of Ref. [4] establishes the existence of two *homoclinic surfaces* $\mathcal{H}_1$ and $\mathcal{H}_2$ embedded in (U_1, s)-space. If (U_1, s) lies on $\mathcal{H}_1$, then Eq. (2.1) has a homoclinic orbit Γ_1 leading from U_1 to itself, while if (U_1, s) lies on $\mathcal{H}_2$, then Eq. (2.1) has a homoclinic orbit Γ_2 leading from U_2 to itself. Moreover:

(H1) $\mathcal{H}_1$ lies in the sector where $d_a > 0$ and $d_b < 0$;
(H2) $\mathcal{H}_2$ lies in the sector where $d_a < 0$ and $d_b > 0$;
(H3) the closure $\overline{\mathcal{H}_1}$ is tangent to $\mathcal{T}_a$ along $\partial\mathcal{H}_1 = \mathcal{T}_a \cap \mathcal{T}_b$ if $r_2^* < 1$, and otherwise is tangent to $\mathcal{T}_b$;
(H4) the closure $\overline{\mathcal{H}_2}$ is tangent to $\mathcal{T}_a$ along $\partial\mathcal{H}_2 = \mathcal{T}_a \cap \mathcal{T}_b$ if $r_1^* > 1$, and otherwise is tangent to $\mathcal{T}_b$.

The first two statements are made clear by Fig. 2.1. The last two statements reflect the facts that $\mathcal{H}_1$ is defined, roughly speaking, by an equation of the form $d_a = \alpha(-d_b)^{1/r_2^*}$ with $\alpha > 0$ and that $\mathcal{H}_2$ is defined by $d_a = -\beta(d_b)^{r_1^*}$ with $\beta > 0$ [4]. These formulae originate from the behavior of orbits near the saddle points; a precise analysis, using transport by the flow, relates the local picture to the behavior near the transversals. Statements (H3)–(H4) follow from the observation that $\mathcal{T}_a$ is the locus where $d_a = 0$ and $\mathcal{T}_b$ is the locus where $d_b = 0$.

2.3. **Transitional maps.** For each U_1 near U_1^* and s near s^*, there is a unique equilibrium U_2 near U_2^* for Eq. (2.1). This follows from the implicit function theorem and assumption (G0). We denote this equilibrium by $U_2 = \Phi(U_1, s)$. Thus the *Hugoniot map* Φ satisfies the Rankine-Hugoniot condition

$$F(\Phi(U_1,s)) - F(U_1) - s\,[\Phi(U_1,s) - U_1] = 0 \tag{2.2}$$

identically. Analogously, we can write $U_1 = \tilde{\Phi}(U_2, s)$ to define the backward Hugoniot map $\tilde{\Phi}$, which satisfies

$$F(U_2) - F(\tilde{\Phi}(U_2,s)) - s\left[U_2 - \tilde{\Phi}(U_2,s)\right] = 0. \tag{2.3}$$

In particular, if we take $s = s_a(U_1)$, U_1 and $\Phi(U_1, s_a(U_1))$ are the left and right states, respectively, of a transitional shock wave with speed $s_a(U_1)$. This association defines the *transitional map* T_a with $T_a(U_1) = \Phi(U_1, s_a(U_1))$. We define the (backward) transitional map $\check{T}_b$ so that $\check{T}_b(U_1) = \Phi(U_1, s_b(U_1))$ and U_1 are the left and right states, respectively, of a transitional shock wave with speed $s_b(U_1)$.

We shall be interested in the generic situation where the transitional maps T_a and $\check{T}_b$ are invertible, *i.e.*,

$$\det DT_a(U_1^*) \neq 0 \quad \text{and} \quad \det D\check{T}_b(U_1^*) \neq 0. \tag{G4}$$

As a result of this condition, we can define the (forward) transitional map T_b as the inverse of $\check{T}_b$. Then U_2 and $T_b(U_2)$ are the left and right states, respectively, of a transitional shock wave with speed $s_b(T_b(U_2))$.

2.4. **Homoclinic maps.** Since $\mathcal{T}_a$ and $\mathcal{T}_b$ are graphs parameterized by U_1, so are the homoclinic surfaces $\mathcal{H}_1$ and $\mathcal{H}_2$. These surfaces will be parameterized in the form $s = s_1(U_1)$ and $s = s_2(U_1)$, respectively. Here s_1 is defined for states U_1 that lie on one particular side of the two-saddle cycle curve $\mathcal{B}_1$, and similarly for s_2.

The map taking U_1 to the point $U_2 = \Phi(U_1, s_2(U_1))$, which is joined to itself with a homoclinic connection, defines the *homoclinic map* H_2. Analogously, on the side where s_1 is defined, the (backward) homoclinic map $\check{H}_1$ is defined by $\check{H}_1(U_1) = \Phi(U_1, s_1(U_1))$. Since $\mathcal{H}_1$ is tangent to a transitional surface, assumption (G4) guarantees that $\check{H}_1$ is invertible; therefore we can define $H_1 = \check{H}_1^{-1}$.

2.5. **Speed relationships.** To construct a Riemann solution by a succession of waves, it is essential to guarantee that the waves are concatenated in order of increasing speed. If U_1 is sufficiently near to U_1^*, then there is a transitional shock wave on $\mathcal{T}_b$, with left state $T_b^{-1}(U_1)$, right state U_1, and speed $s_b(U_1)$. Similarly there is a transitional shock wave on $\mathcal{T}_a$, with left state U_1, right state $T_a(U_1)$, and speed $s_a(U_1)$. Clearly, the first wave can be followed by the second if and only if $s_b(U_1) < s_a(U_1)$. In other words, U_1 must lie on the side of the two-saddle cycle curve $\mathcal{B}_1$ where $s_b < s_a$.

Can a pair of transitional waves be followed by a third? This is possible if and only if either

$$s_b(U_1) < s_a(U_1) < s_b(T_b \circ T_a(U_1)) \tag{2.4}$$

or

$$s_a(T_a^{-1} \circ T_b^{-1}(U_1)) < s_b(U_1) < s_a(U_1). \tag{2.5}$$

It is possible to show, using conditions (G0)–(G2), that a succession $T_b - T_a - T_b$ of transitional waves corresponding to inequalities (2.4) can occur if and only if $\det DT_b(U_2^*) < 0$, and a succession $T_a - T_b - T_a$ corresponding to inequalities (2.5) can occur if and only if $\det DT_a(U_1^*) < 0$. In particular, a succession of four or more transitional waves can occur if and only if both $\det DT_a(U_1^*)$ and $\det DT_b(U_2^*)$ are negative.

3. Riemann solutions

We can now give a description of the multiple solutions of the Riemann problem for data U_L and U_R close to U_2^*. Following Lax [11], let $\mathcal{W}_1^+(U_L)$ denote the forward 1-family wave curve through U_L, consisting of points U_M (in the vicinity of U_2^*) that can be reached from U_L by a 1-family shock or rarefaction wave; similarly, let $\mathcal{W}_2^-(U_R)$ denote the backward 2-family wave curve through U_R, consisting of points U_M from which U_R can be reached by a 2-family shock or rarefaction wave. As a consequence of assumption (G0), the 1- and 2-family wave curves through U_2^*, $\mathcal{W}_1^+(U_2^*)$ and $\mathcal{W}_2^-(U_2^*)$, are transverse at U_2^*.

Recall from section 2.1 that there exist a curve $\mathcal{B}_1$ of two-saddle cycles passing through U_1^*. Each point $U_1 \in \mathcal{B}_1$ is associated with a unique point $U_2 = T_a(U_1) = T_b^{-1}(U_1)$ such that there is one transitional shock wave connecting U_1 to U_2 and another connecting U_2 to U_1. Such points U_2 constitute the two-saddle cycle curve $\mathcal{B}_2$ passing through U_2^*.

We make the following genericity assumptions concerning the point U_2^*, in addition to the assumptions (G0)–(G4) already described:

$$\mathcal{B}_2 \text{ is transverse to } \mathcal{W}_1^+(U_2^*) \text{ at } U_2^*; \tag{G5}$$

$$T_a \circ T_b \left[\mathcal{W}_1^+(U_2^*)\right] \text{ is transverse to } \mathcal{W}_2^-(U_2^*) \text{ at } U_2^*. \tag{G6}$$

For any U_L sufficiently close to U_2^*, the 1-family wave curve $\mathcal{W}_1^+(U_L)$ through U_L meets $\mathcal{B}_2$ at a (locally) unique point, $U_{\mathcal{B}_2}(U_L)$. We define the curve $\mathcal{C}(U_L)$ to be $\mathcal{W}_2^+(U_{\mathcal{B}_2}(U_L))$, the forward 2-family wave curve through $U_{\mathcal{B}_2}(U_L)$. It proves convenient to distinguish certain sides of $\mathcal{C}(U_L)$, which are defined as follows. Refer to Fig. 3.1.

Notice that the function $s_b \circ T_b - s_a \circ T_b$ vanishes on $\mathcal{B}_2$ and has a nonzero gradient by assumptions (G2) and (G4). By assumption (G5), its derivative along $\mathcal{W}_1^+(U_L)$ is nonzero. Therefore we can define the *transitional half of the wave curve* $\mathcal{W}_1^+(U_L)$, containing states $U_2 \in \mathcal{W}_1^+(U_L)$ such that $s_b(T_b(U_2)) < s_a(T_b(U_2))$. By assumption (G6), the image of this half curve under $T_a \circ T_b$ lies on one side of $\mathcal{C}(U_L)$, which we call the *transitional side of* $\mathcal{C}(U_L)$. If U_M' belongs to the transitional half of $\mathcal{W}_1^+(U_L)$, then the 1-family wave from U_L to U_M' can be followed by a transitional wave from U_M' to $U_M'' = T_b(U_M')$, which in turn can be followed a second, faster, transitional wave from U_M'' to $U_M''' = T_a(U_M'')$; finally, these two transitional waves can be followed by a 2-family wave from U_M''' to any point $U_R \in \mathcal{W}_2^+(U_M''')$, which lies on the transitional side of $\mathcal{C}(U_L)$.

Similarly, we define the *homoclinic half of* $\mathcal{W}_1^+(U_L)$ to consist of states $U_2 \in \mathcal{W}_1^+(U_L)$ such that $\tilde{s}_2$ is defined at U_2; this half lies in what we define to be the *homoclinic side of* $\mathcal{C}(U_L)$. If U_M lies on the homoclinic side, then the 1-family wave from U_L to U_M can be followed by a homoclinic wave from U_M to itself, which then can be followed by a 2-family wave from U_M to any point $U_R \in \mathcal{W}_2^+(U_M)$, which lies on the homoclinic side of $\mathcal{C}(U_L)$.

It is to be noted that the relationship between the transitional and homoclinic sides of $\mathcal{C}(U_L)$, as opposed to the transitional and homoclinic halves of $\mathcal{W}_1^+(U_L)$, depends on further information, namely how $T_a \circ T_b$ places the transitional half of $\mathcal{W}_1^+(U_L)$ relative to the 2-family wave curve $\mathcal{C}(U_L)$. See Fig. 3.1, which depicts the situation when the transitional half of $\mathcal{W}_1^+(U_L)$ lies on the transitional side of $\mathcal{C}(U_L)$.

Our main result is

Theorem 3.1. *Suppose that the genericity conditions* (G0)–(G6) *hold. Then for* U_L *and* U_R *sufficiently close to* U_2^*, *the Riemann problem* (1.1)–(1.2) *admits the following solutions:*

(i) *a Lax solution, consisting of weak* 1- *and* 2-*family waves, for any* U_R*;*

(ii) *a homoclinic solution, of the form* $1 - H_2 - 2$, *for* U_R *on the homoclinic side of* $\mathcal{C}(U_L)$*;*

(iii) *a two-transitional solution, of the form* $1 - T_b - T_a - 2$, *for* U_R *on the transitional side of* $\mathcal{C}(U_L)$*;*

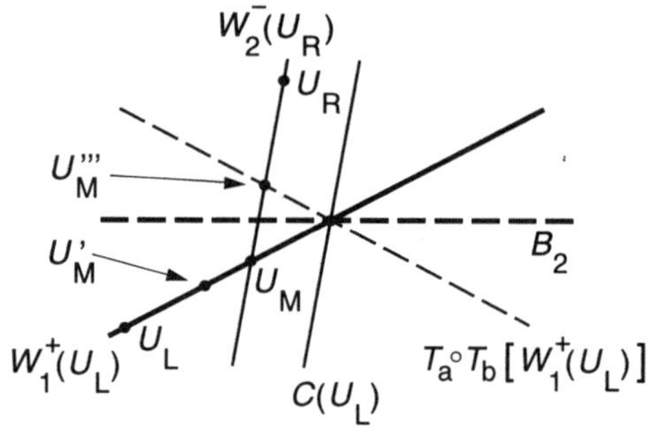

FIGURE 3.1. State space in the vicinity of U_2^*. The wave curve $\mathcal{W}_1^+(U_L)$ meets the 2-cycle curve $\mathcal{B}_2$ at $U_{\mathcal{B}_2}(U_L)$, through which point is drawn the forward 2-family wave curve $\mathcal{C}(U_L) = \mathcal{W}_2^+(U_{\mathcal{B}_2}(U_L))$. The transitional half of $\mathcal{W}_1^+(U_L)$ contains U'_M. This half-curve is mapped to the half of $T_a \circ T_b\left[\mathcal{W}_1^+(U_L)\right]$ containing U'''_M, which lies in the transitional side of $\mathcal{C}(U_L)$.

The following are also possible in certain cases:

(iv) *a transitional/homoclinic/transitional solution, of the form* $1 - T_b - H_1 - T_a - 2$, *for* U_R *on the transitional side of* $\mathcal{C}(U_L)$;

(v) *a homoclinic/two-transitional solution, of the form* $1 - H_2 - T_b - T_a - 2$, *for* U_R *on the transitional side of* $\mathcal{C}(U_L)$;

(vi) *a two-transitional/homoclinic solution, of the form* $1 - T_b - T_a - H_2 - 2$, *for* U_R *on the transitional side of* $\mathcal{C}(U_L)$;

(vii) *a two-homoclinic/two-transitional solution, of the form* $1 - H_2 - T_b - H_1 - T_a - 2$, *for* U_R *on the transitional side of* $\mathcal{C}(U_L)$;

(viii) *a two-homoclinic/two-transitional solution, of the form* $1 - T_b - H_1 - T_a - H_2 - 2$, *for* U_R *on the transitional side of* $\mathcal{C}(U_L)$;

(ix) *a two-homoclinic/two-transitional solution, of the form* $1 - H_2 - T_b - T_a - H_2 - 2$, *for* U_R *on the transitional side of* $\mathcal{C}(U_L)$;

(x) *a three-homoclinic/two-transitional solution, of the form* $1 - H_2 - T_b - H_1 - T_a - H_2 - 2$, *for* U_R *on the transitional side of* $\mathcal{C}(U_L)$;

Moreover, among solutions composed of weak 1- *and* 2-*family waves together with the transitional and homoclinic waves described above, these are the only solutions with at most two transitional waves. Solutions with more than two transitional waves are possible if and only if* $\det DT_a(U_1^*) < 0$ *and* $\det DT_b(U_2^*) < 0$.

4. DISCUSSION

Assuming the existence of a 2-cycle of shock waves, we have shown that a multiplicity of solutions necessarily occurs for Riemann data near an endstate of the shock waves. We have presented a complete description of the mechanism for this nonuniqueness and a classification of possible solutions in terms of the location of U_R relative to U_L.

A 2-cycle configuration is generic, in the sense that it occurs for an open set of model parameters. It is an open problem to determine conditions under which a 2-cycle occurs. In ongoing work, we have shown that 2-cycles of shock waves occur generically for quadratic flux models in case I [13].

This suggests that 2-cycles arise in Stone's model of three-phase flow in a porous medium for certain values of model parameters (see, *e.g.*, Refs. [3, 12, 14]). In fact, more appears to be true: there is a one-parameter family of 2-cycles that shrinks to a single point, and this point, in general, lies in the strictly hyperbolic domain. Therefore it seems possible to find strictly hyperbolic systems and arbitrarily small neighborhoods in state space such that Riemann problems have multiple solutions involving only states within these neighborhoods.

REFERENCES

1. A. Azevedo and D. Marchesin, *Multiple viscous profile Riemann solutions in mixed elliptic-hyperbolic models for flow in porous media*, Hyperbolic Equations that Change Type (B. Keyfitz and M. Shearer, eds.), IMA Volumes in Mathematics and its Applications, vol. 27, Springer-Verlag, New York–Heidelberg–Berlin, 1990, pp. 1–17.
2. ———, *Multiple viscous solutions for systems of conservation laws*, Trans. Amer. Math. Soc. (1995), to appear.
3. J. Bell, J. Trangenstein, and G. Shubin, *Conservation laws of mixed type describing three-phase flow in porous media*, SIAM J. Appl. Math. **46** (1986), 1000–1017.
4. F. Dumortier, R. Roussarie, J. Sotomayor, and H. Żoładek, *Nilpotent singularities and Abelian integrals*, Lect. Notes in Math., vol. 1480, Springer-Verlag, New York–Heidelberg–Berlin, 1991.
5. H. Freistühler, *Instability of vanishing viscosity approximation to hyperbolic systems of conservation laws with rotational invariance*, J. Differential Equations **87** (1990), 205–226.
6. ———, *Rotational degeneracy of hyperbolic systems of conservation laws*, Arch. Rat. Mech. Anal. **113** (1991), 39–64.
7. ———, *Dynamical stability and vanishing viscosity: A case study of a nonstrictly hyperbolic system of conservation laws*, Comm. Pure Appl. Math. **45** (1992), 561–582.
8. H. Freistühler and T.-P. Liu, *Nonlinear stability of overcompressive shock waves in a rotationally invariant system of viscous conservation laws*, Comm. Math. Phys. **153** (1993), 147–158.
9. J. Glimm and D. Sharp, *An S-matrix theory for classical nonlinear physics*, Found. Phys. **16** (1986), 125–141.
10. J. Guckenheimer and P. Holmes, *Nonlinear oscillations, dynamical systems, and bifurcations of vector fields*, Springer-Verlag, New York–Heidelberg–Berlin, 1986.
11. P. Lax, *Hyperbolic systems of conservation laws II*, Comm. Pure Appl. Math. **10** (1957), 537–566.
12. D. Marchesin and H. Medeiros, *A note on the stability of eigenvalue degeneracy in nonlinear conservation laws of multiphase flow*, Current Progress in Hyperbolic Systems: Riemann Problems and Computations (Bowdoin, 1988) (B. Lindquist, ed.), Contemporary Mathematics, vol. 100, American Mathematics Society, Providence, RI, 1989, pp. 215–223.
13. D. Schaeffer and M. Shearer, *The classification of* 2×2 *systems of non-strictly hyperbolic conservation laws, with application to oil recovery*, Comm. Pure Appl. Math. **40** (1987), 141–178.
14. J. Trangenstein, *Three-phase flow with gravity*, Current Progress in Hyperbolic Systems: Riemann Problems and Computations (Bowdoin, 1988) (B. Lindquist, ed.), Contemporary Mathematics, vol. 100, American Mathematics Society, Providence, RI, 1989, pp. 147–159.

DEPARTAMENTO DE MATHEMÁTICA, UNIVERSIDADE DE BRASILIA, 70910 BRASILIA, DF, BRAZIL
E-mail address: arthur@mat.unb.br

INSTITUTO DE MATEMÁTICA PURA E APLICADA, ESTRADA DONA CASTORINA 110, 22460 RIO DE JANEIRO, RJ, BRAZIL
E-mail address: marchesi@impa.br

DEPARTMENTS OF MATHEMATICS AND OF APPLIED MATHEMATICS AND STATISTICS, STATE UNIVERSITY OF NEW YORK AT STONY BROOK, STONY BROOK, NY 11794-3651
E-mail address: plohr@ams.sunysb.edu

DEPARTMENT OF MATHEMATICS, INDIANA UNIVERSITY, BLOOMINGTON, IN 47405
E-mail address: kzumbrun@ucs.indiana.edu

RESERVOIR SIMULATION BY FRONT TRACKING

T. GIMSE[1], H. HOLDEN[2], AND N. H. RISEBRO[3]

1. Introduction.

We will in this paper give a review of part of the work we have been involved with over the last decade in connection with the study of flow in porous media. The project was started with the aim of developing a computer code that should be able able to simulate the complicated flow of hydrocarbons in a porous medium, the impetus coming from the existence of the huge oil fields in the North Sea. It became clear rather soon that this was a formidable challenge, and that the only possible way of accomplishing this task was to start a fundamental analysis of the equations.

There are many scales on which the very complicated processes that take place in a porous medium can be described. With the aim of developing a computer simulator that describes the fluid flow on the scale of a full reservoir, the standard model used by the petroleum industry consists of a system of nonlinear partial differential equations. The traditional approach is to split these equations into two classes of equations with rather different properties, and then to solve the system numerically by iterating between solving these two classes. This method is called the IMPES method, and will be discussed in more detail in Section 2. Here it suffices to say that one equation is a system of multidimensional conservation laws, and one is an elliptic or parabolic second order equation. Our research has almost exclusively been concentrated on the hyperbolic equations.

At the beginning of the project, we decided to concentrate on *the front tracking* concept developed by J. Glimm and co-workers, as that seemed to be the most promising approach both from a theoretical point of view as well as from the point of view of computations. Furthermore, it was being used for precisely the simulation of flow in porous media. For some of the early papers on front tracking we refer to [11] and [12].

Over the years, our experience been that that it is easier said than done to develop a full scale field reservoir simulator according to the demands of the petroleum industry. We will present some examples that reflect the current status of the simulator in Section 3.

Let us now turn to a more detailed discussion of what we mean by *front tracking*. As is well-known in this forum, conservation laws have intrinsically discontinuous solutions. Traditional methods, e.g., finite differences schemes, have problems resolving this phenomenon, and have a tendency to smear out discontinuities, thus obtaining a non-physical solution. Front tracking, on the other hand, adds the discontinuities as additional degrees of freedom which are treated independently of a fixed grid structure which may be required in e.g., the solution of the pressure equation. Thus front tracking preserves the sharp fluid interfaces, the fronts, and they are tracked in the computations. Another important issue in the simulation of full field petroleum reservoirs, is that due to the enormous dimensions of the reservoir, the actual numerical grid that one has to use will be several magnitudes larger than the size of the fluid interface which will thus be a sub-grid phenomena. A resolution of a fluid interface over several grid blocks will not be acceptable in a reservoir engineering context. Although the basic philosophy of front tracking is dimension independent, the fact that fronts are curves in two dimensions and surfaces in three dimensions has drastic consequences for the actual development of numerical algorithms based on front tracking in three dimensions.

Before we discuss the multidimensional situation, we will explain the one-dimensional algorithm in some detail. Consider the one-dimensional scalar conservation law

$$u_t + f(u)_x = 0, \quad u(x,0) = u_0(x). \tag{1.1}$$

By approximating the initial data by a step function, i.e., a piecewise constant function, we obtain an initial value problem that consists of multiple Riemann problems each of which can solved independently. However, as the solution evolves, the waves coming from different Riemann problems will interact. In order to be able to resolve these interactions one can approximate the flux function f by a polygon, i.e., a continuous, piecewise linear function. By this approximation, rarefaction waves are replaced by (many minor) discontinuities, and their interaction with shocks can be tracked. This idea is due to Dafermos [9], and was further analyzed both as a theoretical and numerical method by Holden, Holden and Høegh-Krohn

[15], [16]. This method has proved to be very effective in numerical computations. The approximation parameters in this approach are the fineness of the polygon approximation of f, as well as the step function approximation of the initial data. The precise statement of this is the following error estimate due to Lucier [22]

$$||u(\cdot,t) - v(\cdot,t)||_1 \leq ||u_0 - v_0||_1 + tC||f - g||_{\text{Lip}} \tag{1.2}$$

where u and v are solutions of (1.1) and

$$v_t + g(v)_x = 0, \quad v(x,0) = v_0(x) \tag{1.3}$$

respectively, and the constant C depends on the total variation of the initial data.

At this point we would also like to stress that the front tracking approach has proved to be very useful also for theoretical studies of one-dimensional systems of conservation laws, cf. [24], and the very recent results of Bressan and Columbo [6], [7].

The extension of this idea to more than one dimension has proved to be difficult in practice. Although one can approximate the flux functions as before, the solution of even one single Riemann problem in this approximation turned out to be very difficult indeed, see [25]. However, another approach which preserves many of the advantages of the one dimensional case, that has proved to be useful in two and three spatial dimensions, is dimensional splitting, see [17]. For this approach consider the two dimensional scalar conservation law

$$u_t + f(u)_x + g(u)_y = 0, \quad u(x,y,0) = u_0(x,y). \tag{1.4}$$

Let $S^{f,x}(t)u_0(\cdot,y)$ denote the entropy solution of

$$u_t + f(u)_x = 0, \quad u(x,y,0) = u_0(x,y) \tag{1.5}$$

with y as a parameter. Define $S^{g,y}(t)u_0(x,\cdot)$ similarly. The idea of dimensional splitting (or fractional steps or operator splitting) is that

$$u(x,y,t) \approx [S^{f,x}(\Delta t)S^{g,y}(\Delta t)]^n u_0(x,y), \quad t = n\Delta t. \tag{1.6}$$

In our case we use an approximate solution based on front tracking as described above to solve the one-dimensional conservation law numerically. One can prove that as $n \to \infty$ with t fixed, the approximation constructed by operator splitting using the front tracking approximation as the one-dimensional method converges to the correct entropy solution, see [17], and one can also obtain the order of convergence rigorously, [28], [19]. One important advantage with this method is that it is not restricted by the CFL-condition. The precise results will be described in Section 2.

This approach has been implemented as a part of a full field reservoir simulator: FRONTSIM. Some examples from this and comparisons with other simulators will be presented in Section 3. FRONTSIM is able to simulate full field data for two-phase black oil reservoirs in three space dimensions.

Another approach that has proved to be more effective than FRONTSIM, albeit for a more restrictive class of problems, is based on a streamline method. If we take the initial saturation values and solve the pressure equation, we obtain a set of streamlines that cover the reservoir. Each of these streamlines can be considered as a one-dimensional flow problem, and along each streamline we may solve the conservation law by the one dimensional front tracking method. One may then use these updated saturation values to compute new streamlines if necessary. Due to the streamline approach, one cannot include gravitational effects in this model. The full field simulator based on the streamline method as explained here is called FRONTLINE, and comparisons between FRONTSIM and FRONTLINE will be presented in Section 3.

Both FRONTSIM and FRONTLINE are currently being used by several Norwegian and international oil companies. A nontechnical summary of our experience with the front tracking simulators in connection with reservoir simulators is that the method is demanding and time consuming to implement, but once implemented offers high resolution and physically plausible simulations that are very CPU-efficient.

Some of our results have been summarized in [1], [2], [3], and [4], and these papers may also be consulted for more complete references to further literature.

The rest of this report is organized as follows. In Section 2 we give more thorough explanation of the streamline front tracking, and of the dimensional splitting approach, and of the implementation of these algorithms in FRONTLINE and FRONTSIM. In Section 3 we present some examples of simulations and some comparisons between the two simulators. In Section 4 we discuss some extensions and possible future applications of the front tracking technique.

2. Front tracking.

In this section we explain the implementation of a reservoir simulator for two dimensional two phase reservoir models. This implementation is based on a streamline approach, thus it does not allow for gravitational effects. We then explain how such effects may be incorporated into another code, also based on front tracking, by using dimensional splitting. This results in a reservoir simulator that can simulate on a wider variety of reservoir models, but gives less accurate results in a given CPU time.

In order to explain the streamline model, we will first briefly recapitulate Dafermos' method [9] for one dimensional conservation laws

$$s_t + f(s)_x = 0. \tag{2.1}$$

This method consists in approximating the flux function $f(s)$ with a piecewise linear continuous function f_δ with a finite number of linear segments. The solution of the Riemann problem for the equation with f_δ replacing f consists of a finite number of constant values separated by discontinuities moving apart. This property can now be utilized to solve the initial value problem if the initial function $s(x, 0) = s_0(x)$ is piecewise constant. To do this one solves the Riemann problems associated with each discontinuity in s_0. These solutions will give discontinuities which propagate linearly in (x, t) space from each initial location, and the solution is constant between discontinuities. When two discontinuities collide, one solves the Riemann problem defined by the constant states to the left and right of the collision. It was shown in [15], [16] that this gives a well defined approximation, furthermore the approximation converges to the unique weak solution to (2.1) as the approximation f_δ converges to f. This algorithm we call *front tracking*. The one dimensional front tracking algorithm has also been extended to various systems of conservation laws, both for theoretical, [6], [7], [25], and for computational [26], [27], [20], [21] purposes. For one dimensional conservation laws, front tracking was found to be a very good numerical method in the sense that it generates accurate solutions using very little computer time.

It is not straightforward to generalize the one dimensional front tracking approach to scalar equations in more dimensions. To do this one would have to solve the Riemann problem in two or three dimensions. The solution to this is only partially known [14], [24], and the solution is considerably more complicated than in the one dimensional case. There is however, one class of two dimensional conservation laws where a straightforward adaptation of one dimensional front tracking is possible. This is the so-called *streamline model*

$$s_t + \nabla(\vec{v} f(s)) = 0. \tag{2.2}$$

Here $\vec{v}$ denotes some vector field, assumed to be independent of the unknown s. In (2.2) one can change variables, at least locally, to have one coordinate ξ parallel to $\vec{v}$ and one coordinate orthogonal to $\vec{v}$. In these coordinates, (2.2) reads

$$s_t + f(s)_\xi = 0, \tag{2.3}$$

which is the one dimensional conservation law.

We can now use this to construct a numerical method. First we construct a finite set of integral curves of $\vec{v}$. These are called ***streamlines***, hence the name "streamline method". This set should be chosen so that it in some sense "covers" our computational domain up to the level of accuracy we are interested in. Then we can map the initial function $s_0(x, y)$ to a set of initial functions $s_0^i(\xi_i)$, one for each streamline. Then one solves the finite set of one dimensional conservation laws on each streamline. In order to obtain the solution in the whole computational domain, not only on the set of streamlines, one must map the s values defined on streamlines back to the two dimensional computational domain. There are, of course, many ways to do this.

The above algorithm might be called a "pure streamline method", and can be expected to give reasonable results if the streamlines are independent of the unknown s. This is not the case for reservoir models.

In slightly simplified form, the equations modeling two dimensional two phase incompressible flow in a porous medium are

$$s_t + \nabla(v_D f(s)) = 0, \tag{2.4}$$

$$\nabla \vec{v} = q. \tag{2.5}$$

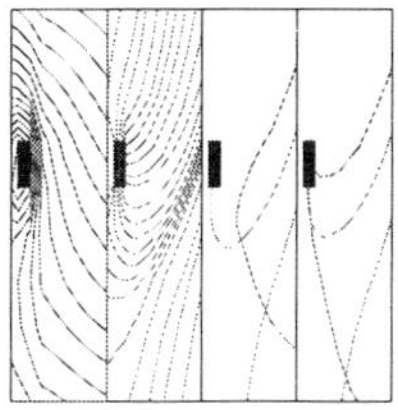

FIGURE 1. On the left: Isobars of the pressure field, corresponding streamlines, one initial front (dotted) with two streamlines, and on the right: final position of the front as the points are propagated along the streamlines.

Here the unknown $s = s(x, y, t)$ denotes the saturation of one of the two phases, usually oil or water. The saturation of a phase is defined to be the fraction of the available pore volume occupied by that phase; consequently saturations are always in the interval $[0, 1]$. The equations, (2.4) and (2.5), express Darcy's law and conservation of mass, and are referred to as the *saturation equation* and the *pressure equation* respectively.

If s denotes the water saturation, and subscripts "w" or "o" denote quantities pertaining to water or oil respectively, the function f is usually given by

$$f(s) = \frac{\lambda_{\mathrm{w}}(s)}{\lambda_o(s) + \lambda_{\mathrm{w}}(s)}. \tag{2.6}$$

The functions λ_{i} are called mobilities. The mobilities are computed from

$$\lambda_{\mathrm{i}}(s) = \frac{k_{rock} k_{\mathrm{i},rel}(s)}{\mu_{\mathrm{i}}}. \tag{2.7}$$

Here k_{rock} denotes the *rock permeability*. This is a rock property, hence it depends on the location. The rock permeability is always non-negative. The parameter μ_{i} denotes the viscosity of phase i, and $k_{\mathrm{i},rel}$ is called the *relative permeability* of phase i. The relative permeabilities are assumed to be known convex positive functions of the saturation. The function (2.6) is often referred to as the fractional flow function.

The vector v_D is called the *total Darcy velocity* and is given as the sum of the *water Darcy velocity* v_{w} and the *oil Darcy velocity* v_{o}. These are in turn determined by the pressure from

$$v_{\mathrm{i}} = -\lambda_{\mathrm{i}}(s)\nabla p. \tag{2.8}$$

This equation is called Darcy's law. Here $p = p(x, y, t)$ denotes the pressure. This model assumes the capillary pressure to be zero, therefore we have one common pressure for both phases.

Finally, q (in (2.5)) denotes sources and contributions from boundaries.

When solving this set of equations, (2.4) and (2.5), we use an operator splitting technique. First, the pressure equation (2.5), is solved, with a fixed saturation. The pressure equation is an elliptic, and for compressible flow parabolic, second order equation. Therefore it is natural to use an element method to obtain numerical solutions. Element methods have the advantage that it is easy to adapt the grid to the coefficients of the equation, which depend on the location through the geology and the saturation. Both the saturation and the rock permeability may change discontinuously. It increases the accuracy of the numerical solution if these discontinuities can be represented in the coefficients used in the numerical solution. In the implementations presented here, we have used linear and bilinear elements.

Once the pressure equation has been solved, the saturation equation (2.4) can be solved for some time interval Δt, assuming that the pressure does not change in this interval. Next, using the saturation obtained at $t = \Delta t$, the pressure equation (2.5) is solved, and the procedure is repeated.

In reservoir simulation this operator splitting is common, and is usually referred to as an IMPES (Implicit Pressure Explicit Saturation) method, see [23]. Nevertheless, to the authors' knowledge, there exists no rigorous justification for this procedure.

2.1. Streamline front tracking. Using an IMPES method, one can of course use any method to solve the saturation equation. We use a slight modification of the streamline front tracking method described above.

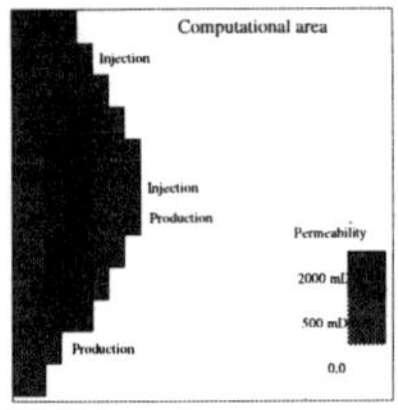

FIGURE 2. Computational area. The grid is uniform 24×36.

First we approximate the fractional flow function (2.6) by a piecewise continuous linear function, made by interpolating from a table of relative permeabilities. Then we discretize the initial saturation in the computational domain, so that the saturation is piecewise constant, and takes values from the set where the derivative of the approximate fractional flow function has discontinuities. The reason for this is that these s values are the only ones that will appear in the solutions to the Riemann problem with such values as initial values.

This modified streamline method also uses the concept of *fronts*. Here, a front is defined to be a doubly linked list of points. Each point has a left and right neighbor, or is an endpoint. In addition each front has a left and right saturation; this is analogous to the discontinuities in the solution of the one dimensional Riemann problem described above. The initial position of the fronts coincides with the discontinuities in the approximation to the initial saturation. The fronts have one additional discretization parameter which is analogous to the number of streamlines generated in a pure streamline approach. This is the average distance between neighboring points on a front. It is clear that complicated saturation contours will be more accurately represented if this distance is small.

When solving the saturation equation using the modified streamline method, we calculate local streamlines through each point on all fronts. We can then solve the one dimensional Riemann problems at each point. The solutions will consist of discontinuities separated by constant states; these states are also breakpoints of the approximate flux function. These discontinuities will propagate along the local streamline. If the solution of a Riemann problem consists of several discontinuities, then a new front is created corresponding to each new discontinuity.

The length of a local streamline is chosen so that a point will move along the whole local streamline during the interval Δt. Therefore we refer to the process of solving all the Riemann problems at every point as "moving the fronts".

Since we use linear or bilinear elements, the streamlines will be piecewise linear or piecewise quadratic. In both cases the trajectories of a point can be found analytically.

Generally, interactions between different fronts will not occur along the same local streamline. This would correspond to two points lying on the same integral curve of ∇p. Therefore interactions, i.e., overlaps, between fronts are resolved after all fronts have been moved. Similarly, if two points have become too far apart after the fronts have been moved, additional points are inserted between the original locations of these points until the final distance is acceptable. Also, if a pair of neighboring points are too close after moving the fronts, they are replaced by a single point. This postprocessing, or "untangling" of the fronts is rather *ad hoc*, but is observed to work well. It is also necessary to keep the code robust.

After the saturation is updated, the saturation is mapped to the grid used to calculate the pressure equation. This is done by adapting the grid used for the elements to the location of the fronts. Although the size of this grid is independent of the spacing between the points that make up the fronts, to get the best results the size of the elements should be roughly comparable to this spacing.

Note that the modified streamline method, or *front tracking*, involves several independent parameters that (presumably) determine the accuracy of the numerical solution:

(i) The time interval for the operator splitting; Δt.
(ii) The number of linear segments in the approximation to the fractional flow function (2.6).
(iii) The spacing between points on a front.
(iv) The grid size used to solve the pressure equation (2.5).

One hopes that as parameters (i), (iii) and (iv) tend to zero, and (ii) tends to infinity, the numerical solution converges to a solution of the system (2.4) and (2.5). Presently, there exits no rigorous demonstration of this, but computational investigations indicate that this is the case. Here, we will present a few examples

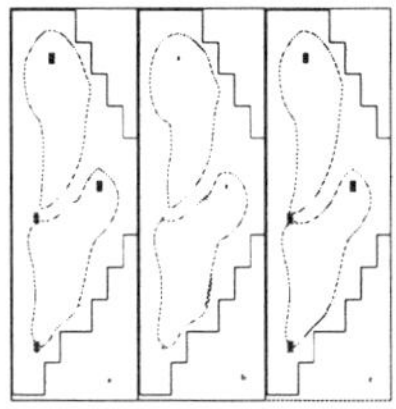

FIGURE 3. Buckley-Leverett front at 8 years for (a) 24×36 grid, (b) 72×108 grid, and for four times as many pressure updates (c), 24×36 grid. The CPU times are 45, 217, and 56 seconds, respectively. The recovery is within the limits of material balance.

which demonstrate how a change in these parameters influence the numerical solution. A more thorough "numerical convergence" study is in preparation.

2.2. Dimensional splitting. While the modified streamline front tracking method works very well on two dimensional reservoir models, some difficulties immediately come to mind.

Firstly the front tracking approach becomes vastly more difficult in three dimensions than in two. A global front object in three dimensions will be a surface. The front moving algorithm would presumably not be much more difficult in three than in two dimensions, but the untangling would be much more complicated. Making such an untangling algorithm robust, and at the same time flexible, is very difficult. Despite the recent progress by the front tracking group at Stony Brook, the authors feel (and fear) that any attempt at writing such a piece of code is incommensurable with commercial code development.

Another difficulty is *gravitational effects*. If there is a density difference between the phases the saturation equation reads

$$\frac{\partial s}{\partial t} + \nabla \left(f(s)(v_D + (\rho_{\rm o} - \rho_{\rm w})\lambda_{\rm o}(s)k_{rock}g) \right) = 0. \tag{2.9}$$

Here $\rho_{\rm i}$ denotes the density difference between the phases and g the gravitational acceleration. It is not possible to determine a direction in which (2.9) becomes one dimensional. Consequently, a streamline algorithm cannot be formulated to solve (2.9). Another complication is that in one space dimension, (2.9) is a so-called resonant conservation law, see [18, 10]. The two dimensional Riemann problem for resonant conservation laws have not been solved.

Because of these difficulties, we use another numerical method when solving (2.5) and (2.9), or for problems where gravitational effects are important: dimensional splitting. Dimensional splitting for conservation law was introduced by Godunov in connection with gas dynamics [13]. Crandall and Majda showed that dimensional splitting, for a scalar conservation law in several space dimensions, converged if a so called monotone difference method was used to solve the one dimensional equations, [8]. See also [17] and [28], [19] for theoretical work in connection with scalar conservation laws and front tracking.

In this context, dimensional splitting means that the conservation law (2.9) is solved numerically by Dafermos' method in one direction at a time. More precisely, we define the one dimensional solution operator S_x by denoting the solution $s(x, y, t)$ of

$$\begin{gathered} \frac{\partial s}{\partial t} + \frac{\partial}{\partial x}(f(s)(v_D + (\rho_{\rm o} - \rho_{\rm w})\lambda_{\rm o}(s)k_{rock}g)) = 0, \\ s(x, y, 0) = s_0(x, y), \end{gathered} \tag{2.10}$$

by $S_x(t)s_0(x, y)$. Here y is only a parameter. The solution operator $S_y(t)$ is similarly defined. In order to apply Dafermos' method, we approximate the flux function

$$f(s)(v_D + (\rho_{\rm o} - \rho_{\rm w})\lambda_{\rm o}(s)k_{rock}g) \tag{2.11}$$

by a piecewise linear continuous function. Then we define the approximate solution operators S_x^δ and S_y^δ to be solution operators for the approximate flux functions in the x and y directions respectively. Here δ is the length of the linear segments used when approximating the flux function (2.11). We also introduce an additional grid on the computational domain. This grid divides the domain into strips $[i\Delta x, (i+1)\Delta x]$ and $[j\Delta y, (j+1)\Delta y]$. To apply $S_x^\delta(\tau)$, the initial saturation is mapped to a function that is constant on each grid block $[i\Delta x, (i+1)\Delta x] \times [j\Delta y, (j+1)\Delta y]$, then we apply Dafermos' method in each strip

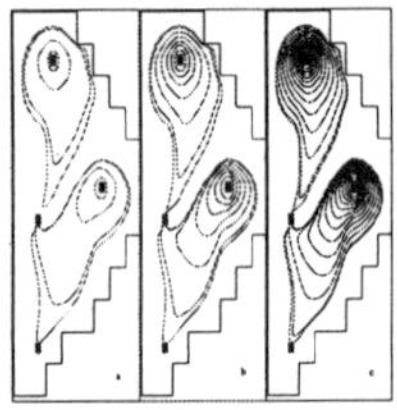

FIGURE 4. Fronts using (a) 3, (b) 9, and (c) 19 linear segments in the fractional flow function.

$[j\Delta y, (j+1)\Delta y]$ for $j = 1, \ldots, N_y$ to obtain N_y solutions at a time τ. The solution is then again mapped to a function which is constant on grid blocks, and the operator S_y^δ is applied on each strip. This grid is called the ***saturation grid*** and is independent of the grid used to solve the pressure equation (2.5).

If $n\tau = \Delta t$, this process can formally be written

$$s(x, y, t + \Delta t) = \pi[S_y^\delta \circ \pi \circ S_x^\delta]^n \pi s(x, y, t). \tag{2.12}$$

The pressure equation (2.5) is solved using an element method as before, and the elements are adapted to follow saturation contours from the saturation grid.

The saturation grid does not need to be uniform. It can be adapted dynamically so that the discontinuities in the one dimensional solutions can be more accurately represented. In the implementation we present here, the algorithm used to construct the saturation grid is fairly complicated, representing a tradeoff between the wish to represent discontinuities in the saturation and to minimize the total number of grid blocks.

This dimensional splitting algorithm also has many independent parameters controlling the accuracy of the solution.

(i) The time interval for the operator splitting; Δt.
(ii) The time interval for the dimensional splitting; τ.
(iii) The number of linear segments in the approximation to the flux function (2.11).
(iv) The size of the saturation grid; Δx, Δy.
(v) The grid size used to solve the pressure equation (2.5).

In the next section we present computations which show the effect on the computed solution when changing some of these parameters.

3. Examples.

The purpose of this section is to present some of the numerical aspects that are specific to front tracking. The discussion will be focused on discretization parameters, and we will provide a few examples which show how these parameters influence the computed solutions. For a more detailed study of the performance of the method with respect to numerical diffusion, grid orientation effects, and computational efficiency, see, e.g., [1, 2, 3, 4].

Although the these features may be illuminated by simple, "academic" test cases, we will provide more realistic, full-field-type examples. This is done to emphasize that the front tracking algorithms as presented in the previous sections have been implemented and tested thoroughly in full scale reservoir simulators.

Presently, two simulator codes using front tracking have been developed by Technical Software Consultants a/s, Oslo, Norway. The two codes are based on streamline computations and dimensional splitting respectively.

The streamline-based code, FRONTLINE, solves Riemann problems locally along streamlines. Note that FRONTLINE is a two-dimensional code, ignoring gravity in its present version. As described in the previous section, the saturation is represented by level set curves. Along the curves a certain number of points are placed, and between these points the curves are piecewise linear, i.e., the level set curves are polygons in the plane. Each point then represents a Riemann problem along the streamline through this point. Because of unfavorable mobility ratio or changed injection rates, it is usually necessary to update the pressure field several times throughout a simulation. Consequently, it suffices to determine the solution of a Riemann problem locally. Therefore, the streamlines are not computed globally, rather on one block at a time along the streamline through each point.

In Figure 1 we show the steps within each global timestep:

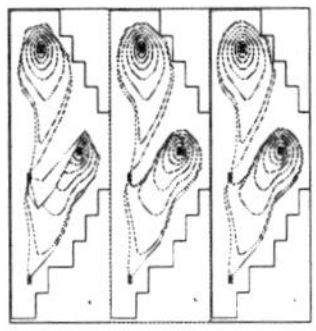

FIGURE 5. Simulations using increased (from a to c) density of the points representing the fronts.

(i) Given a pressure field (a), the streamlines are computed (b).
(ii) Given the initial position of a front (c), the Riemann problems defined by the two states are solved.
(iii) The one dimensional fronts are propagated along the streamlines to obtain new front positions.

Within FRONTLINE there are several options for determining the velocity field, and thus the streamlines and point paths. The pressure equation is solved by a finite element method, on triangular or quadrangular elements. On each element we use linear or bilinear basis functions. Accordingly, for triangular elements, numerical differentiation of the pressure yields a blockwise constant velocity field. In this case the streamlines are linear on each element. For some cases, this gives enough accuracy, but particularly along boundaries with a no-flow boundary condition, and in places where the velocity field converges or diverges strongly, we improve the accuracy of the streamlines by using bilinear elements or grid refinement.

A more straightforward way of improving the accuracy of the streamlines is to use *mixed methods*, thereby computing the pressure and the velocity independently. The idea is to formulate the pressure equation (2.5) as two equations:

$$\nabla v_D = q \tag{2.13}$$

and

$$v_D = -\lambda \nabla p. \tag{2.14}$$

Then we solve both (2.13) and (2.14) simultaneously, with both the pressure p and total Darcy velocity v_D as unknowns. Such a method yields a larger number of unknowns: the (scalar) pressure and the (two components) velocity. Consequently, solving the pressure equation takes more time. On the other hand, since the velocity is the important variable for the saturation equation, we may use a low order approximation (piecewise constant) for the pressure itself, and linear approximation for the velocity.

In the sample cases below, we will focus on the performance of front tracking, and only briefly comment on the coupling to the pressure/velocity. However, in general, it turns out that local refinement during the differentiation of the pressure field is sufficient to achieve an adequate saturation solution.

3.1. Examples — FRONTLINE. In this section we focus on the parameters that determine the accuracy of FRONTLINE. First we display a comparison between a coarse and a fine grid pressure computation. The case that will be used throughout this section is a horizontal reservoir with two injection wells and two production wells. The reservoir is homogeneous with porosity 0.3 and permeability $500\text{m}D$, except for a region with $2000\text{m}D$. In Figure 2 we show the region. The original grid size is 24×36, and the overall dimensions of the field are $800 \times 2160\text{m}^2$.

The production rates are 1500m^3 and 1250m^3 per day for the two wells, and the ratios of injection is 0.4546 and 0.5454, wells numbered from the bottom of Figure 1. The two phases, water and oil, are assumed to be incompressible. The densities and viscosities respectively are $780\text{k}g/\text{m}^3$ and $4.0cp$ for oil, and $1014\text{k}g/\text{m}^3$ and $0.42cp$ for water. For the relative permeability we have used an s^2 type relation, and we make the fractional flow function piecewise linear on 20 intervals. This corresponds to a very fine resolution compared with most realistic data. To the left (a) in Figure 2 we show the position of the leading Buckley-Leverett front after 8 years of constant injection for the original 24×36 grid, and in the middle we show the leading front using a 72×108 grid for the pressure. In both simulations we have used timesteps of length 4 months. The rightmost figure shows the position of the same front using four times as many timesteps on a 24×36 grid. Thus, the velocity field has been updated 96 times during this simulation. Nevertheless, save for a slightly better resolution of the front in the vicinity of the heterogeneity, one cannot see any important differences among the three simulations.

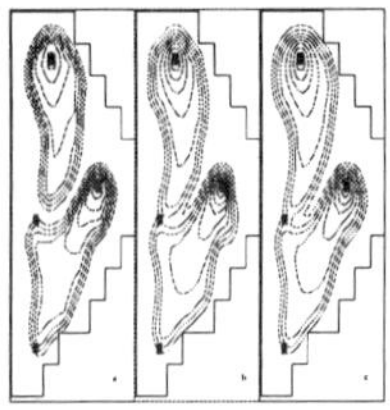

FIGURE 6. Saturation contours computed by using different grids for the dimensional splitting. A refinement of the pressure grid of: (a) 1, (b) 5, and (c) 15. The CPU times for the saturation equation are (a) 6, (b) 37, and (c) 64, respectively.

We now want to investigate the effects of the initial discretization of the flow function. In Figure 4 we show the results of the same simulation using (from the left) 3, 9, and 19 linear segments in the approximation to the fractional flow function. One should note that the number of fronts visible corresponds to these numbers. Furthermore, when using few linear segments, the maximum front speed will be smaller, since this gives a less steep approximation of the flow function.

Finally, we will demonstrate the effects of the density of points chosen to represent the two-dimensional fronts. The importance of using many points, of course, is greater as the flow pictures are complicated. Particularly at narrow passages, or if the reservoir is highly heterogeneous, the fronts may have high curvature, and thus require a high point density to be represented accurately. The density is given as the average number of points compared to the average pressure grid block diameter. In Figure 5, we have used (from the left) a point density of 0.2, 1.0, and 5.0 respectively. Note that a high density represents the finer structure of the front very close to the production well, and a gives smoother overall picture. The density used in Figure 1, which shows a zoom-in close to the upper production well, is 1.4.

3.2. Examples — FRONTSIM. First we will demonstrate the effects of gridsize used in the dimensional splitting. In this example, we have used a grid for the dimensional splitting that is a refinement of the pressure grid. In Figure 6, the refinement factor used is 1, 5, and 15 tubes in each grid block respectively. Note that the CPU time increases less than linearly. To display the saturation when using FRONTSIM, we draw contour lines based on the average saturation in each block in the pressure grid.

Finally, we show the effects of refining the timestep used in the dimensional splitting; τ. The effect of this is most significant where we expect flow around corners. In Figure 7 we display the solution computed using $\tau = \Delta t$, $3\tau = \Delta t$, and $8\tau = \Delta t$.

We conclude this section by stating that the above examples are chosen as representative cases, with typical behavior. Of course, extreme test cases can be constructed, to focus on specific features of the methods, but the purpose of this section was to demonstrate how an algorithm based on front tracking may behave on realistic problems in several space dimensions.

4. Future work.

We have in this review paper concentrated on the application of the front tracking technology to two phase flow of hydrocarbons in a porous medium. Within this area the methods are well understood, and the simulators have been extensively tested on full scale reservoirs and on more theoretical test cases, and one has a good understanding of the behavior of the simulators.

However, it is quite clear that the methods can, and should, be extended to more general situations. Within the context of petroleum reservoirs a very important extension is to develop a front tracking simulator that simulates three phase flow that, e.g., includes simple phase transitions. Another interesting extension would be to include polymers in the simulators.

Front tracking provides a method to handle intrinsic discontinuities in the solutions of hyperbolic conservation laws, and should in principle be applicable to other phenomena governed by such equations. We have made some applications of front tracking to the simulation of ground water flow, where the general setting is close to the one encountered in petroleum reservoirs, see [5].

Finally, applications to gas dynamics provide another challenge for the front tracking techniques, see e.g. [20], [21].

Acknowledgments. NHR is grateful for the invitation to present these results at the conference. HH and NHR were supported in part by VISTA, a research cooperation between the Norwegian Academy of

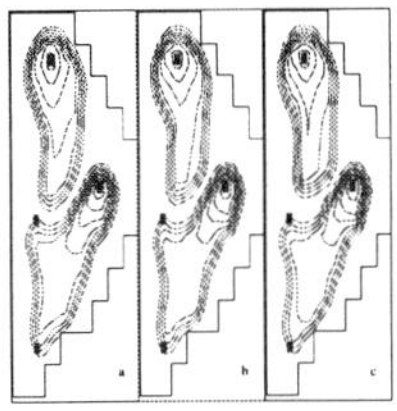

FIGURE 7. Saturation contours obtained by dimensional splitting using (a) 1, (b) 3, and (c) 8 dimensional splitting timesteps between each pressure solution. The CPU times are 6, 14, and 36, respectively.

Science and Letters and Den norske stats oljeselskap (Statoil), as well as Norges forskningsråd, and the support is gratefully acknowledged. We would like to thank Frode Bratvedt, Kyrre Bratvedt, Christian Buchholz, Lars Holden, and Rudi Olufsen for collaboration on the front tracking project.

References

1. F. Bratvedt, K. Bratvedt, C. F. Buchholz, H. Holden, L. Holden, N. H. Risebro, A new front tracking method for reservoir simulation *SPE Reservoir Engineering, Feb. 1992*: 107–116.
2. F. Bratvedt, K. Bratvedt, C. F. Buchholz, T. Gimse, H. Holden, L. Holden, R. Olufsen, N. H. Risebro, Three-dimensional reservoir simulation based on front tracking, in *North Sea Oil and Gas Reservoirs - III*, J. O. Aasen et al. (eds.), Kluwer, Dordrecht, 1994, pp. 247–257.
3. F. Bratvedt, K. Bratvedt, C. F. Buchholz, T. Gimse, H. Holden, L. Holden, N. H. Risebro, Frontline and Frontsim: two full scale, two-phase, black oil reservoir simulators based on front tracking, *Surv. Math. Ind., 3*: 185–215 (1993).
4. F. Bratvedt, K. Bratvedt, C. F. Buchholz, T. Gimse, H. Holden, L. Holden, N. H. Risebro, Front tracking for petroleum reservoirs, in *Ideas and Methods in Mathematical Analysis, Stochastics, and Applications*, S. Albeverio, J. E. Fenstad, H. Holden, T. Lindstrøm (eds.), Cambridge UP, Cambridge, UK 1992, pp. 409–427.
5. F. Bratvedt, K. Bratvedt, C. F. Buchholz, T. Gimse, H. Holden, L. Holden, N. H. Risebro, Front tracking for ground water simulations, in *Computational Methods in Water Resources, IX*, T. F. Russel, R. E. Ewing, C. A. Brebbia, W. G. Grey, G. F. Pinder (eds.) Elsevier, 1992, pp. 97–106.
6. A. Bressan, Global solutions of systems of conservation laws by wave-front tracking, *J. Math. Anal. Appl., 170*:414–432, (1992).
7. A. Bressan, R. Columbo, The semigroup generated by 2×2 conservation laws, *SISSA preprint, 102*, (1994).
8. M. Crandall, A. Majda, The method of fractional steps for conservation laws, *Numer. Math., 34*: 285–414 (1980).
9. C. Dafermos, Polygonal approximation of solutions of the initial value problem for a conservation law, *J. Math. Anal. Appl., 38*: 33–41 (1972).
10. T. Gimse, N. H. Risebro, Solution of the Cauchy problem for a conservation law with a discontinuous flux function, *SIAM J. Math. Anal., 23*:635–648,(1992).
11. J. Glimm, E. Isaacson, D. Marchesin, O. A. McBryan, Front tracking for petroleum reservoir simulations, *SPE preprint 12238*: 41–49 (1981).
12. J. Glimm, B. Lindquist, O. A. McBryan, B. Plohr, S. Yaniv, Front tracking for hyperbolic systems, *Adv. Appl. Math., 2*: 91–119 (1981).
13. S. K. Godunov, Finite difference methods for numerical computations of discontinuous solutions of the equations for fluid dynamics, *Mat. Sb., 47*: 271–306 1959 (Russian).
14. J. Guckenheimer, Shocks and rarefactions in two space dimensions *Arch. Rat. Mech. Anal.,59*:281–291 1975.
15. H. Holden, L. Holden, R. Høegh-Krohn, A numerical method for first order nonlinear scalar conservation laws in one dimension, *Comput. Math. Anal., 15*: 595–602 (1988).

16. H. Holden, L. Holden, On scalar conservation laws in one dimension, in *Ideas and Methods in Mathematical Analysis, Stochastics, and Applications*, S. Albeverio, J. E. Fenstad, H. Holden, T. Lindstrøm (eds.), Cambridge UP, Cambridge, UK 1992, pp. 480–509.
17. H. Holden, N. H. Risebro, A method of fractional stepes for scalar conservation laws without the CFL condition, *Math. Comp., 60*: 221–232 (1993).
18. E. Isaacson, B. Temple, Examples and classification of non-strictly hyperbolic systems of conservation laws, *Abstracts of AMS*, January 1985.
19. K. H. Karlsen, On the accuracy of a dimensional splitting method for scalar conservation laws, *Cand. Scient. thesis, UiO* (1994).
20. J. O. Langseth, N. H. Risebro, A. Tveito, A conservative front tracking scheme for 1D hyperbolic conservation laws, *Notes Num. Fl. Dyn., 43*:385–392, (1993).
21. J. O. Langseth, On an implementation of a front tracking method for hyperbolic conservation laws, *Preprint, UiO, Informatics, no 6* (1994).
22. B. Lucier, A moving mesh numerical method for hyperbolic conservation laws, *Math. Comp., 46*: 59–69 (1986).
23. D. W. Peaceman, *Fundamentals of numerical reservoir simulation*, Elesevier, Amsterdam (1977).
24. N. H. Risebro, The partial differential equation $u_t + \sum_i f_i(u)_{x_i} = 0$, a numerical method, *Cand. scient. thesis*, University of Oslo, 1987.
25. N. H. Risebro, A front tracking alternative to the random choice method, *Proc. Amer. Math. Soc.,117*: 1125–1139 (1993).
26. N. H. Risebro, A. Tveito, A front tracking method for conservation laws in one dimension, *J. Comput. Phys., 101*:130–139, (1992)
27. N. H. Risebro, A. Tveito,Front tracking applied to a nonstrictly hyperbolic system of conservation laws, *SIAM J. Sci. Stat. Comp., 12*:1401–1419, (1991).
28. Z. H. Teng, On the accuracy of fractional steps methods for conservation laws in two dimensions, *SIAM J. Numer. Anal., 31*:43–63, (1994).

[1] Technical Software Consultants, Gaustadalléen 21, 0371 Oslo, Norway
E-mail address: gimse@ifi.uio.no

[2] Department of Mathematical Sciences, Norwegian Institute of Technology, University of Trondheim, N–7034 Trondheim, Norway
E-mail address: holden@imf.unit.no

[3] Department of Mathematics, University of Oslo, P.O. Box 1053, 0314 Oslo, Norway
E-mail address: nilshr@math.uio.no

A Survey of Granular Flow

David G. Schaeffer[1,2]

INTRODUCTION

This paper reports on an integrated program of analysis, experiment, and computation to better characterize and understand granular flows. Other participants in the program include R. Behringer, P. Thompson, and J. Trangenstein of Duke University; F.X. Garaizar and M. Shearer of N.C. State University; and E.B. Pitman of SUNY Buffalo. Our three-pronged approach has been more fruitful than separate efforts. For instance, an experimental search for instabilities predicted by analysis led to the discovery of density waves and of a self-affine time-dependence in hopper flows, which in turn influenced analysis by focusing attention on the flutter instability and raising the issue of fluctuations for granular flows. Similarly, analytical understanding of a model problem for shear bands cleared the way for progress on computations for shear bands, which in turn brings within our grasp the goal of simulating both the experiments in this project and industrially relevant flows.

Three broad scientific issues regarding the dynamics of granular materials arise in this work:

(i) Instability, including pattern formation in the post-instability regime;
(ii) The continuum limit in granular flow; and
(iii) Reliable computation of granular flows.

The analogy with fluid mechanics and its fascinating instabilities highlights the obvious intellectual interest of the first issue. At the same time, this issue has great practical significance. For example, porosity waves in a hopper in industrial use can affect its performance. In extreme cases, such waves can lead to "thumping" which may threaten the structure. Similar issues arise in metal plasticity where nonuniform deformations can compromise the quality of a formed product.

Regarding the second issue, the continuum limit in granular flow is more problematic than in fluid mechanics. In particular, it seems that fluctuations are far more important in granular flow. Of course, there is the obvious point that the "molecules" in granular flow are much larger than in fluid mechanics. More subtly, however, the assumption of molecular chaos breaks down because stresses in a granular material are carried by *stress chains*, i.e., sequences of individual grains with strong contact forces between them. The exact structure of such chains may vary greatly in repetitions of an experiment under apparently identical circumstances, and this gives rise to fluctuations that may be of the same order of magnitude as the mean stresses. Moreover, such stress chains typically extend for many grain diameters; indeed, it is not known whether there is a characteristic length above which the motion of grains becomes uncorrelated. In other words, it is conceivable that fluctuations are important no matter how large the system. In particular, fluctuations of this type may be the cause of

[1]Research supported by NSF Grant 9201034, which includes funds from AFOSR.
[2]Department of Mathematics, Duke University, Durham, NC 27708

the frequent, mysterious collapse of silos.

Regarding the third issue, computations with plasticity models for granular flow pose the following challenges: (i) solving highly nonlinear time–dependent equations, (ii) switching between elastic and plastic behavior as called for by the solution, (iii) tracking the formation and propagation of shear bands, and (iv) allowing for the multiple time scales resulting from the extreme smallness of a typical velocity as compared to the elastic wave speed. Computations with molecular-dynamics models for granular flow pose a different set of challenges. We plan to develop a hybrid code which uses a plasticity model in regions where the solution varies smoothly and molecular dynamics near shear bands. In our opinion, such a code will represent a major advance in simulation capability.

A better understanding of granular flow will be useful in a variety of problems related to materials handling in manufacturing industries. Granular flow is also important in other engineering contexts (e.g., soil mechanics) and in several geological contexts (e.g., beaches, deserts, earthquakes, avalanches).

1 Experiment

1.1 Convection in shaken sand.

In shaker experiments we study the convection induced in granular material by vertical vibration. (Already in 1831, Michael Faraday [8] observed the convection of fine powders sprinkled on vibrating plates.) Scientifically, this is interesting because the energy that is available from shaking makes the instabilities of the system more apparent. Practically, this is interesting because shaking is used in industrial settings to break up flow blockages in silos and to transport material on conveyors. In our experiments, an annular container is filled to a height h with grains of diameter d, and the container is shaken vertically (i.e., along its axis), say with displacement $A\sin\omega t$. The basic instability typically occurs at $\Gamma \approx 1.2$, where $\Gamma = A\omega^2/g$ is a bifurcation parameter, and g is the acceleration of gravity. Initially the top surface of the grains is flat. As a result of the instability, this surface eventually stabilizes to a non–axisymmetric shape with one peak and one valley. (See Figure 1.) This shape is supported by internal convection moving sand upward, in competition with grains falling down the top surface of the heap.

Rather surprisingly, this basic instability is suppressed [20] if the pressure P of the ambient gas is sufficiently low. More quantitatively, we show that the heap height, H, has a universal variation with pressure H proportional to $\tanh(kP)$, where k is a very large parameter depending on the gas as well as the size and structure of the particles. The asymptotic form of $\tanh z$ as $z \to \infty$ reproduces the astonishing behavior observed experimentally; i.e., *H is virtually independent of P until P is reduced to the order of* 0.02 *atm!* We have also developed a theoretical model which gives insight into this behavior.

At larger values of Γ, the basic instability is followed by various secondary instabilities [17], [18], [19]. With a complexity rivaling that of fluid mechanics, these depend on several auxiliary parameters, in particular an energy parameter $A^2\omega^2/gd$, a grain-size parameter

h/d, and P. In different regimes, we show that the large-Γ behavior of the system consists of:

(a) Traveling waves. In this state, the top surface of the grains is unsteady, with waves traveling against gravity from valley to peak.

(b) Bubbling. This chaotic state resembles qualitatively the behavior of fluidized beds.

(c) Subharmonic standing waves. This state [6,16] occurs at low gas pressure or moderate h/d, where the primary heaping bifurcation is suppressed. Different regions of material have 180° phase difference in their vertical motion; the resulting pattern typically has a well defined wavelength.

1.2 Hopper flow.

(a) Stress fluctuations at the wall. High precision, high speed measurements have shown that there are fluctuations in the normal stress on the hopper wall during a typical flow. These fluctuations, which occur on a broad range of time scales, resemble $1/f$ noise; they can be a substantial fraction of the mean stress. Their presence suggests that models which lack fluctuational effects may be inadequate. Theoretically, these fluctuations are also interesting because they are, at least approximately, Brownian-like (actually fractional Brownian) with power spectra characterized by power laws in the frequency: i.e., $P = A\omega^{-\alpha}$, where $1 < \alpha < 2$.

(b) Porosity waves. In an effort to understand stress fluctuations, we examined the internal structure of hopper flow with x-rays. This led to the surprising discovery of porosity waves, i.e., alternating regions of dense and diffuse material which propagate relative to the flow (see Figure 2). These waves occur for rough grains but not for smooth ones, behavior which is not yet understood. In subsequent experiments, using our unique capability for simultaneous x-ray and velocimetry experiments, we have shown conclusively [21] that there is a steady flow of material through the low-density propagating regions. Hence, the porosity waves cannot be shear bands in the conventional sense.

1.3 Couette experiments.

Two Couette shear experiments are in progress and have yielded preliminary results.

(a) Temporal stress fluctuations. Figure 3 depicts an apparatus for measuring stress fluctuations. Granular material is held in an annular ring and subjected to continuous shearing. The variations in normal and shear stresses seen at the bottom of the material are detected and recorded as a function of time using high-speed,[1] very sensitive strain gauges. Time-dependent data for the normal stress is recorded by a laboratory computer and then shipped to a workstation for analysis. We have already found that the power spectra, $P(\omega)$ (i.e., the absolute square of the Fourier transform), have striking power-law dependence on the frequency $\omega : P \propto \omega^{-\alpha}$, where $\alpha \simeq 1$ at lower frequencies, and $\alpha \simeq 2$ at higher frequencies.

[1]Shearing is performed under low-speed, quasistatic conditions. Even under these conditions, fluctuations occur on many time scales, including some much faster than the underlying flow. The high-speed gauges allow us to track these fluctuations.

We will also use shear-stress transducers so that we can record both the normal and shear stresses. From the experiments we hope, in particular, to discover the relation between the relative sizes of the fluctuations and particle to gap size.[2] In later stages of this experiment, we plan to make multipoint measurements of the stress in order to look for spatial and temporal correlations.[3]

(b) Stress chains. Figure 4 sketches an apparatus for measuring stress chains in granular flow. Photoelastic disks are sheared between a rotating inner wheel and an outer boundary, and the whole assembly is subject to normal stress from the outer boundary. We determine the local internal stress state of the grains by viewing the apparatus through crossed polarizers and by exploiting the photoelasticity of the grains: grains experiencing a high stress are brighter than those experiencing a low stress.

In our preliminary experiments so far, we see that after moderate shearing, a shear band forms. This separates an active inner region, in which ordered regions continually form and break up, from a more stable outer region, in which particles tend to maintain their relative positions [33]. Although relative positions in the outer region remain fairly constant, stresses in that region vary greatly, in response to structural changes in the inner region; moreover, stresses in the outer region are very nonuniform, with discrete stress chains carrying the bulk of the load. In our experiments we gather careful statistics to quantify the extent to which these phenomena can or cannot be described by homogenized dynamics.

It is interesting to note that, fluctuations aside, the *mean* stress seen at the outer boundary evolves very slowly with time and only seems to approach an asymptotic value after several hours, even at rotation rates of the inner cylinder of $\approx 1Hz$. This point also bears on the question of when a homogenized theory can be applied: i.e., if hours are required to obtain a steady state, many technically relevant granular flows may never reach such a steady state before they are terminated.

1.4 Biaxial constitutive tests.

Our recently constructed biaxial tester, a schematic of which is shown in Figure 5, is similar to that of Drescher et al. [7], with two additional innovative features: acoustic transducers (to measure the speed of sound) and continuous monitoring by x-rays at 30 frames per second (to measure density variations and to follow particle paths). A prismatic sample is confined inside an elastic lubricated latex membrane. The material is deformed quasi-statically in the z-direction. Side plates restrain motion in the y-direction, while hydrostatic pressure is applied from the x-direction. The base plate rides on a bearing-supported sled which can move in the x-direction once a shear band starts to form, allowing

[2]Numerical calculations by Savage [24] suggest that the relative fluctuations should be largest when the grain size to gap width ratio d/w is largest. However, to our knowledge, there is no information on the size of these fluctuations as d/w becomes large, i.e., when one would hope to reach the continuum limit.

[3]In kinetic theory models of granular materials, which assume molecular chaos, the correlations between stress fluctuations between two spatial points should be delta function correlated. This means that a fluctuation of the stress from the mean value at point $\mathbf{r}$ and time t and at $\mathbf{r}'$ and t' should be correlated only if $\mathbf{r} = \mathbf{r}'$ and $t = t'$. Although this assumption is often made for many granular flows, we are unaware of tests which would establish the actual correlations.

us to capture one shear band at a time.

Experiments with this apparatus will generate a wealth of valuable data. For example, the traditional biaxial constitutive test can be supplemented with quantitative x-ray data on the density and on trajectories of individual trace particles. Resonance measurements from sinusoidal driving or time-of-flight measurements from impulse driving (either loading or unloading) are both possible. We can also study the softening which precedes shear-band formation and even measure how sound propagation is affected by the presence of shear bands.

2 Analysis

2.1 Shear bands.

Analytically the most challenging phenomenon in granular flow is the formation of shear bands: i.e., narrow regions in which intense deformation occurs while material outside the band(s) may even unload. Much of our effort concerns a model problem [25] roughly based on anti-plane shear. In this model, which we believe captures the essence of shear banding, only the x_3-component of the velocity is nonzero (so the velocity is a scalar v), and only the T_{31} and T_{32} components of the stress influence the dynamics (so the stress may be represented by a 2-component vector τ). These quantities, which depend on x_1, x_2, and t, obey the equations

$$\begin{array}{lrl} (a) & \partial_t v & = \operatorname{div} \tau \\ (b) & \partial_t + h^{-1}(\gamma)(R\tau)\partial_t\gamma & = \operatorname{grad} v \\ (c) & \partial_t\gamma & = \begin{cases} (\partial_t|\tau|)_+ & \text{if } |\tau| = \gamma \\ 0 & \text{if } |\tau| < \gamma, \end{cases} \end{array} \tag{2.1}$$

where $\gamma(x,t)$ measures hardening, h is a positive decreasing function, Q_+ denotes a quantity Q if $Q > 0$, or zero if $Q < 0$, and R is the 2×2 rotation matrix through a constant angle α that specifies the degree of nonassociativity in the model.

References [28] and [30] solve the Riemann problem for system (2.1), for cases in which strong loading forces a shear band to form. Mathematically the solution with a shear band is interesting because (i) the Riemann problem does not admit a scale-invariant solution and (ii) the generalized solution leads to a natural free boundary problem for the wave equation. Reference [30] analyzed the small-time asymptotics of the problem; [28] proves a rigorous existence theorem. The latter was surprisingly tricky, apparently requiring the Nash-Moser theorem. These analytic results played a major role in guiding the computations with shear bands reported below.

Currently we are working on extending results for the model problem to the equations for full two-dimensional motion. Preliminary calculations indicate that, even in the regime before shear banding, the wave speeds have the following interesting properties: (i) The plastic wave speed for plane waves in some directions is greater than the elastic wave speed;

(ii) Two plastic wave speeds may coincide, warning of possible flutter instability in perturbed models (see [1,27]).

After shear banding begins (i.e., when hyperbolicity is lost), Riemann problems fail to have scale-invariant solutions; we plan to show that the problem admits a generalized solution containing a shear band (as was done in [30] for the model problem). In two dimensional motion there are two directions preferred by a shear band; we will investigate the mechanism by which one direction is selected. [4] Finally we need to analyze the propagation of the tip of the shear band as the shear band expands across the material. The objective will be to find traveling wave solutions of the equations in two dimensions, including the dynamic jump conditions at the shear band. Understanding the structure of these solutions is crucial to designing the two-dimensional code. Moreover, these truly two-dimensional analytic solutions will be invaluable in testing the code. A similar problem in metal plasticity is being investigated by Wright [37].

2.2 Viscoplasticity.

Viscoplasticity eliminates the choice between elastic and plastic behavior in (rate-independent) elastoplasticity by adding a stiff relaxation term. In [23], Pitman derived an energy bound for the solution of a one-dimensional viscoplasticity model, and under the assumption of loading, proved that the stress converges (in L^2) to the plastic yield surface. Results from numerical experiments were also presented as evidence of the broader claim that the solution of the visco-plastic equations converges to the solution of the underlying elasto-plastic equations as the relaxation time tends to zero.

As the stiffness parameter tends to infinity, the viscoplasticity equations formally reduce to the rate-independent equations. Preliminary calculations suggest that, before the formation of shear bands, this reduction may be made rigorous with the techniques of [4], although there are two technical difficulties: (i) at least in the theory most accurate for metals, stiffness comes from a large *exponent*, not a large coefficient and (ii) the reduced equations are not in conservation form.

The main thrust of our effort is to understand the effect of rate dependence on the formation and development of shear bands, a new development in the analysis of equations with relaxation. Specifically, although the viscoplasticity equations are well-posed under all circumstances, the reduced equations (assuming plastic deformation everywhere) become ill-posed at moderately large total deformation (on the order of 6%). As discussed above, we have dealt with this behavior by allowing the plastic deformation to localize into a shear band which we represented by an ad hoc jump condition. Viscoplasticity offers the promise of rationalizing this process; in particular, of providing a shock width and shock structure.

Another interesting problem related to viscoplasticity concerns the propagation speed of wave trains generated in the acoustic biaxial tester when there is unloading in some part of the cycle. In elastoplasticity, the elastic and plastic wave speeds are different, and hence, it is a challenging problem to analyze the interactions and extract the net speed of propagation of such a wave train. By contrast, in viscoplasticity, there is a unique wave

[4]We gave one such treatment in [26] for yield vertex models.

speed but interactions occur via the relaxation term. We propose to study propagation speeds of a wave train in both theories and compare the results with experiment. In this way, quantitatively new information about rate dependence in granular materials may be extracted.

2.3 Other Work.

(a) Incrementally nonlinar constitutive laws. References [29] and [31] begin the theoretical analysis of wave propagation with rate-independent, incrementally nonlinear constitutive laws, a class which contains both yield-vertex models and hypoplastic models. Physically these are important in improving the accuracy of modeling, especially in the presence of large stress rotations, and mathematically they are interesting because the resulting partial differential equations are fully nonlinear. These papers demonstrate that, for fully nonlinear equations, there is a subtle relationship between linear well-posedness (i.e., hyperbolicity) and nonlinear well-posedness (existence of a unique solution of initial value problems and continuous dependence on the data). In reference [31], we analyze the structure of discontinuous solutions and identify a class of 2×2 model systems, for which well-posedness of the Riemann problem can be identified with property of linear hyperbolicity. Reference [16] studies numerically the blow-up of the solution in an ill-posed case. In recent work, we are extending the analysis to the full 5×5 system for hypoplastic plane waves in two dimensions.
(b) Modeling of flows of granular materials. We seek to understand the origin of porosity waves (cf. Section 1.2) in terms of the flutter instability. An and Schaeffer [1] showed that: (i) With even the simplest constitutive models, for small Poisson's ratio, the equations of granular flow lose strict hyperbolicity (in the terminology of [27], have an *umbilic point*), and (ii) Under a generic perturbation of the constitutive relation, the umbilic point broadens to a small region in state space where the equations are ill-posed. This approach suggests an explanation of the experimental fact that porosity waves occur in rough-grained materials but not in smooth-grained materials, even though both kinds of materials have the same angle of repose: specifically, the different behavior may result from different *elastic* properties in the two cases, in particular from different Poisson ratios.

Another project is to study the equations of two phase flow. This project derives its importance from the fact that industrial applications often involve the coupled flow of solid particles and a fluid [36,14]. Moreover, our shaker experiments (cf. Section 1.1) indicate that interstitial fluid can play a crucial role in granular deformation.

3 Computation

Our goal is to develop a flexible code that can be used to simulate a variety of experimental and industrial configurations. We are working with both continuum models and molecular-dynamics models. We are writing these codes with the goal of parallelization in view.

3.1 Continuum models.

In paper [9] we carry out one-dimensional simulations of the antiplane-shear model (2.1) before the formation of a shear band. The Riemann problem is solved for initial data that do not lead to a shear band. Part of the analysis of this paper is the identification of initial data for which the solution involves only classical shock waves and rarefaction waves. A one-dimensional algorithm, based on Godunov methods, is designed and implemented; it incorporates the elastic-plastic transition of the material during a single timestep integration.

In paper [10], we carry out one-dimensional simulations of the antiplane-shear model after the formation of a shear band. In this paper we also use a Godunov-type method, supplemented by 1D-front-tracking techniques to resolve the shear band. These techniques achieve the proper localization of the shear band and the evolution of the material near it. We also validate the analytical results concerning the existence of unloading fronts originating from the shear band. An appendix to this paper extends the results of [28] to a large-time existence result and derives the appropriate asymptotic behavior.

In paper [32], an interim report, we continue this program by incorporating adaptive mesh refinement (AMR) into the one-dimensional code. AMR introduces a finer grid in the region surrounding the shear band, effectively concentrating the computational efforts on those regions. Without this refinement, especially in two or more dimensions, the CFL condition would combine with the multiple-time-scale problem to make realistic computations prohibitively expensive. Incidentally, this paper also contains an improvement over [9] and [10] in that in [32] the full, circular yield surface is used rather than merely a linear approximation of it. Paper [22] presents a simple method for computing shear band formation in a one-dimensional model. This paper serves as a test for extending the shear band tracking idea to two dimensions.

All the above papers were preparatory for computations on shear bands in two dimensions. In paper [11] we present a numerical code for 2D antiplane shear based on Godunov methods, to which we have incorporated 2D front-tracking for the localization and growth of the shear band (see Figure 6–the originals are in color). One of the main difficulties addressed in this paper is that the shear band is not a front in the traditional sense, for two reasons: (i) it behaves as an internal boundary in the material (i.e., the displacement, not merely the velocity, is discontinuous across a band) and (ii) it does not divide the computational domain into separate regions (i.e., at intermediate stages of the computation, the shear band does not extend all the way across the computational domain). Concerning the latter, the code must track the lengthwise growth of the shear band, which is complicated by the fact that this growth is not dominated by the acoustic speed. This means that the shear band, unlike a hyperbolic wave, can grow by many cells in one time step. The code accommodates this behavior in two ways: by placing shear-band interfaces according to a stress-loading criterion, and by updating the shear band in a sequence of partial time steps dictated by the band itself.

The next steps towards our ambitious goal are as follows:

(a) Completion of the antiplane-shearing code. The main remaining task is to incorporate AMR into the existing 2D-code for the antiplane shear model (2.1). Note that a shear band does not divide the computational domain into separate regions. Also, shear bands can grow

across many cells in the time it would take for an acoustic wave to cross one cell. This difficulty has been resolved by imbedding the AMR regridding process in the algorithm used to model shear band growth. This achieves the twin objectives of enabling the refinement to contain the shear band and providing maximum resolution.

(b) Extension to two-dimensional motion. New challenges to be faced include: (i) For two dimensional motion, when ill-posedness arrives there are two crossed directions for a possible shear band, and the code must follow both of these at least until one of them dies out. (ii) Tracking a shear band is complicated by the fact that, unlike for the antiplane–shearing model, a shear band also *moves* (in Eulerian coordinates) while it grows. (iii) Accepted plasticity models for soils have some nonassociativity, which means that for certain directions the plastic wave speed exceeds the elastic wave speed.

(c) The quasistatic limit. Many problems in granular flow suffer from a multiple–time–scale difficulty: specifically, that the elastic wave speed c_e is enormously greater than a typical flow velocity. In consequence, the CFL condition for an explicit numerical scheme poses severe limitations on the time step. To avoid these restrictions, we are considering both (i) implicit methods and (ii) direct simulation of the mixed elliptic–hyperbolic system resulting from the quasistatic limit $c_e \longrightarrow \infty$. The latter approach would make a connection with other work in plasticity.

(d) Three dimensions with cylindrical symmetry. In the long range we plan to extend the code for 2D motion to one that simulates (time-dependent) axisymmetric motion in 3D. Of course, in terms of computational effort, this is a 2D problem.

3.2 Molecular Dynamics.

In a molecular-dynamics (MD) simulation of granular flow, one integrates Newton's equations of motion for a many-particle system, assuming an appropriate force rule for interacting grains. Of course, since there are so many grains, great effort and clever numerics are needed to integrate the equations. Even interpreting the wealth of data so generated is a challenge. This approach has been developed by [3, 5, 13, 33, 34, 36].

(a) Fundamental questions regarding the continuum limit. We have seen from our experiments that, at least in laboratory-scale granular flow, the stress fluctuates over a substantial fraction of the mean stress. [5] Moreover, intuitive ideas about stress chains make this behavior seem inevitable. However, this intuition does not answer pressing questions about change of scale. Specifically, in a large granular system, *is there an intrinsic upper bound on the length of stress chains, or do stress chains occur on all length scales up to the dimensions of the flow vessel?* [6] In the former case presumably one can obtain an accurate description of the flow using only mean quantities in some appropriate continuum theory.

We are studying this question by means of MD simulations. Consider a granular assembly of N particles. For each grain, let $F_i(t) = \sum_{\alpha=1}^{n_i} \vec{\mathbf{F}}_{i\alpha} \cdot \vec{\mathbf{N}}_{i\alpha}$ where $\{\vec{\mathbf{F}}_{i\alpha} : \alpha = 1, \ldots, n_i\}$

[5] In this subsection we focus on stress fluctuations; in most circumstances velocity fluctuations seem to be much less significant.

[6] Although the precise mathematical issues are different, this question has much in common with questions about length scales in dispersion in random media (e.g., see [12]).

are the forces exerted on the ith grain by the n_i other grains in contact with it and $\vec{\mathbf{N}}_{i\alpha}$ is the unit inward normal at the point of contact. This time-dependent, scalar quantity provides a measure of the total compressive force acting on the ith grain. Define the average $\bar{F}(t) = N^{-1}\sum_{i=1}^{N} F_i$ and the correlation function[7]

$$G(\vec{x},t) = N^{-1}\sum_{i=1}^{N}(F_i - \bar{F})(F_{I(\vec{x},i)} - \bar{F})$$

where $I(\vec{x},i)$ denotes the particle whose center at time t is closest to $\vec{r}_i(t) + \vec{x}$, with $\vec{r}_i(t)$ being the position of the ith particle. Information about long-range stress correlations is contained in the decay or nondecay of $G(\vec{x},t)$ as $|\vec{x}|$ gets large.

Consider a two-dimensional granular assembly in a rectangular box, the bottom face being at rest, the load-carrying top face moving at constant velocity, and the side faces being subject to periodic boundary conditions.[8] In this configuration, the primary loading is vertical; thus stress chains should tend to be approximately vertical, perhaps tilted somewhat by the underlying shearing motion. One expects $G(\vec{x},t)$ to decay rapidly if $\vec{x}$ varies perpendicular to the stress chains, less rapidly (if at all) if $\vec{x}$ varies parallel to the stress chains. If for all times t, in repeated simulations, we find that $G(\vec{x},t)$ decays with a consistent length scale, we shall regard that length as a characteristic length for stress chains.

If it turns out that the length of stress chains is not bounded, we shall gather careful statistics on stress fluctuations in multiple runs of MD simulations. We shall then investigate whether introduction of some controlled amount of randomness in material properties in a continuum computation can reproduce the MD results at greatly reduced effort.

(b) MD simulation of shear bands. MD offers an exciting possibility for improving the numerical treatment of shear bands. To date, in our work shear bands are treated as discontinuities in some fields (e.g., the velocity) in a continuum formulation, subject to ad hoc jump conditions. Such a treatment neglects the detailed granular mechanics within the shear band, which typically has a thickness of about ten grain diameters. We propose to develop a hybrid code that uses continuum variables in regions where the solution varies slowly and uses MD simulations in narrow regions around shear bands, the latter being selected by the AMR feature of the code.

This approach provides an appealing resolution of a multiple-length-scale problem in that the intensive MD computations are restricted to the small regions where micromechanics dominates the flow. However, it must be acknowledged that this is a long-term project. In particular, it will require a much more flexible treatment of boundary conditions in MD simulations than is currently available to smoothly make the transition between MD cells and continuum cells. To help develop the requisite understanding, we are first studying lubrication problems with the goal of combining continuum models for the lubricating fluid with MD calculations near the shearing surface.

---

[7] Strictly speaking, one must restrict the sum defining $G(\vec{x},t)$ to exclude particles for which $\vec{r}_i(t)+\vec{x}$ would lie outside the domain under study and reduce N accordingly.

[8] Apart from the neglect of curvature, this corresponds to the Couette geometry of Section 1.3a.

References

[1] An, L.-J. and D. Schaeffer, *The flutter instability in granular flow*, J. Mech. Phys. Solids **40** (1992), 683–698.

[2] G. W. Baxter, R. Leone and R. P. Behringer, *Experimental Tests of Time Scales in Flowing Sand*, Europhys. Lett. **21** (1993), 569-574.

[3] C.S. Campbell and L.E. Brennen, *Computer simulation of granular shear flow.* J. Fluid Mech. **151** (1985), 167.

[4] G.-Q. Chen, C.D. Levermore, T.-P. Liu, Hyperbolic conservation laws with stiff relaxation terms and entropy. MSRI preprint #30, 1992.

[5] P. Cundall and O. Strack, *A discrete numerical model for granular assemblies.* Geotech. (1979).

[6] S. Douady, S. Fauve and C. Laroche, *Subharmonic instabilities and defects in a granular layer under vertical vibrations,* Europhys. Lett. **8** (1989), 621-627.

[7] A. Drescher, I. Vardoulakis, and C. Han, *A biaxial apparatus for testing soils*, Geotechnical Journal ASTM **13** (1990), 226–234.

[8] M. Faraday, Phil. Trans. Roy. Soc. London **121** (1831), 299–340.

[9] X. Garaizar, *Numerical computations for antiplane shear in a granular flow model.* Q. Appl. Math., **52** (1994), 289–309.

[10] F.X. Garaizar and D.G. Schaeffer, *Numerical computations for shear bands in an antiplane shear model.* J. Mech. Phys. Solids, to appear.

[11] F.X. Garaizar and J. Trangenstein, *Tracking of a shear band in 2D antiplane shear flow.* In preparation.

[12] J. Glimm, W.B. Linquist, F. Pereira, and Q. Zhang, *A theory of macrodispersion for the scale up problem*, preprint (SUNYSB publication number AMS-92-19).

[13] D.C. Hong and J.A. McLennan, *Molecular dynamics simulations of hard sphere granular particles*, Physica A **187** 159–171 (1992).

[14] A. W. Jenike, *Analysis of solids densification during the pressurization of lock hoppers*, in International Symposium on the Role of Particle Interactions in Powder Mechanics, Aug 1983.

[15] F. Melo, P. Unbarihawu and H.L. Swinney, *Transition to parametric wave patterns in a vertically oscillated granular layer,* Phys. Rev. Lett. **72** (1994), 172-175.

[16] S. Myagchilov, *A class of model equations related to hypoplasticity.* M.Sc. project report, N.C. State University.

[17] H. K. Pak and R. P. Behringer, *Surface Waves in Vertically Vibrated Granular Materials*, Phys. Rev. Lett. **71** (1993), 1832-1835.

[18] H. K. Pak and R. P. Behringer, *Pattern Formation of Vertically Vibrated Granular Materials*, Proceedings of the First International Workshop on Nonlinear Dynamics and Chaos, Pohang, Korea, pp. 91-113, 1993.

[19] H. K. Pak and R. P. Behringer, *Bubbling in Vertically Vibrated Granular Materials*, submitted to Nature, 1994.

[20] H. K. Pak, E. Van Doorn, and R. P. Behringer, *Effect of Ambient Gas on Vertically Vibrated Granular Materials*, submitted to Phys. Rev. Lett., 1994.

[21] H. K. Pak, E. Van Doorn, and R. P. Behringer, *Velocity and Density profiles for Density waves in Hopper Flows*, to appear.

[22] E.B. Pitman. *A Godunov Method for Localization in Elasto-Plastic Granular Flow.* Int. J. Num. Analy. Methods in Geomech. **17** (1993), 385–400.

[23] E.B. Pitman *Viscoelastic relaxation: convergence and localization*, preprint.

[24] S.B. Savage, *Numerical simulations of Couette flow of granular materials: spatio-temporal cohenence of* $1/f$ *noise,* Physics of Granular Media, Ed. D. Bideau and J. Dodds. Nova Science, 1992.

[25] D.G. Schaeffer, *A mathematical model for localization in granular flow: postcritical behavior*, Proc. Royal Soc. London **A436** (1992), 217–250.

[26] D.S. Schaeffer and M. Shearer, *Loss of hyperbolicity in yield vertex plasticity models under nonproportional loading* in *Nonlinear evolution equations that change type*, (B. Keyfitz and M. Shearer, eds.) Springer, 1990.

[27] D.G. Schaeffer and M. Shearer, *Riemann problems for nonstrictly hyperbolic* 2×2 *systems of conservation laws*, Trans. AMS **304** (1987), 267–306.

[28] D.G. Schaeffer and M. Shearer, *Unloading near a shear band: A free boundary problem for the wave equation.* Comm. P.D.E., **18** (1993), 1271-1298.

[29] D.G. Schaeffer and M. Shearer, *Fully nonlinear hyperbolic systems of partial differential equations related to plasticity.* Preprint.

[30] M. Shearer and D.G. Schaeffer, *Unloading near a shear band in granular material.* Quart. Appl. Math., to appear.

[31] M. Shearer and D.G. Schaeffer, *A class of fully nonlinear* 2×2 *systems of partial differential equations.* Submitted to Comm. P.D.E.

[32] M. Shearer, F.X. Garaizar, D.G. Schaeffer and J. Trangenstein, *Formation and development of shear bands in granular material. Proceedings,* Annual ARO Conference on Applied Mathematics and Computing, 1993.

[33] P.A. Thompson and G.S. Grest, *Granular flow, friction and dilatancy transition*, Phys. Rev. Lett. **67** 1751–1754 (1991).

[34] P.A. Thompson, *Vibrated granular assemblies: Fluidization and convection,* in *Computer Simulations in Condensed Matter Physics VI*, D.P. Landau, K.K. Mond and H.B. Schuttler, eds. Springer-Verlag, New York, 1993.

[35] I. Vardoulakis, *Dynamic stability of undrained simple shear on water saturated granular soils*, Int. J. Num. Analy. Methods in Geomechanics **10**, 1986, p177-190.

[36] O. Walton and R.L. Braun, *Viscosity, granular temperature, and stress calculations for shearing assemblies of inelastic frictional disks.* J. Rheol. **30** (1986), 949-980.

[37] T. Wright, Presentation at conference on Hyperbolic Problems, SUNY Stony Brook, June 1994.

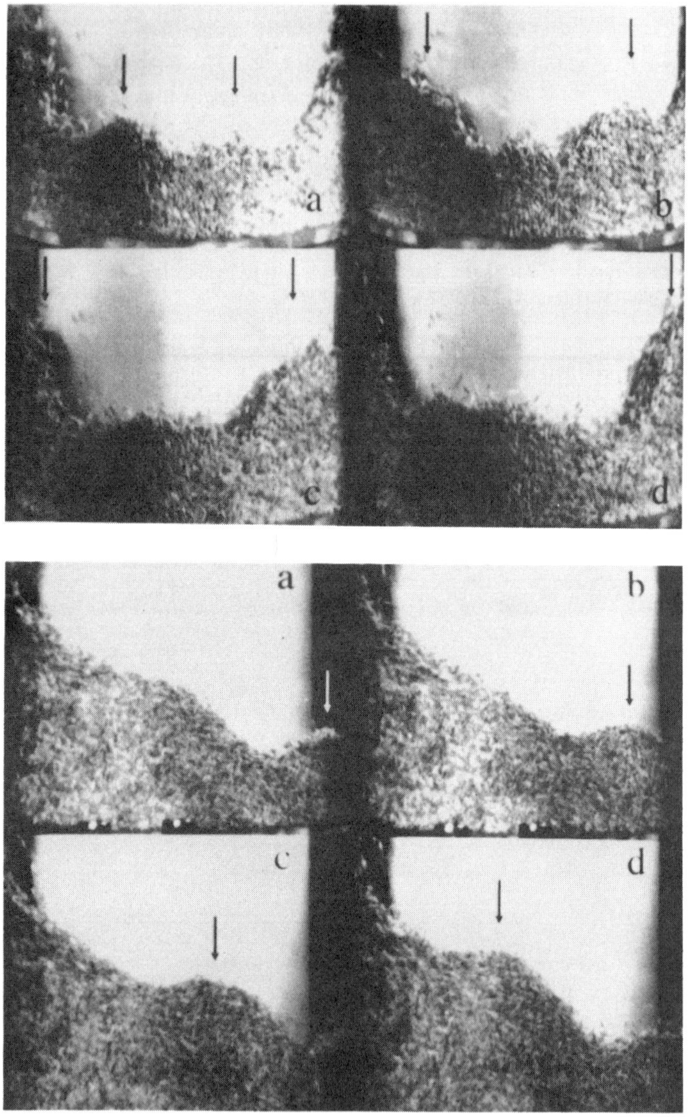

FIGURE 1 Traveling waves for a granular material in an annulus subject to vertical vibration The top 4 images show a view opposite the heap. The bottom 4 views show the heap from the side. Arrows indicate the wave pulse location. The time between successive views is about 4s.

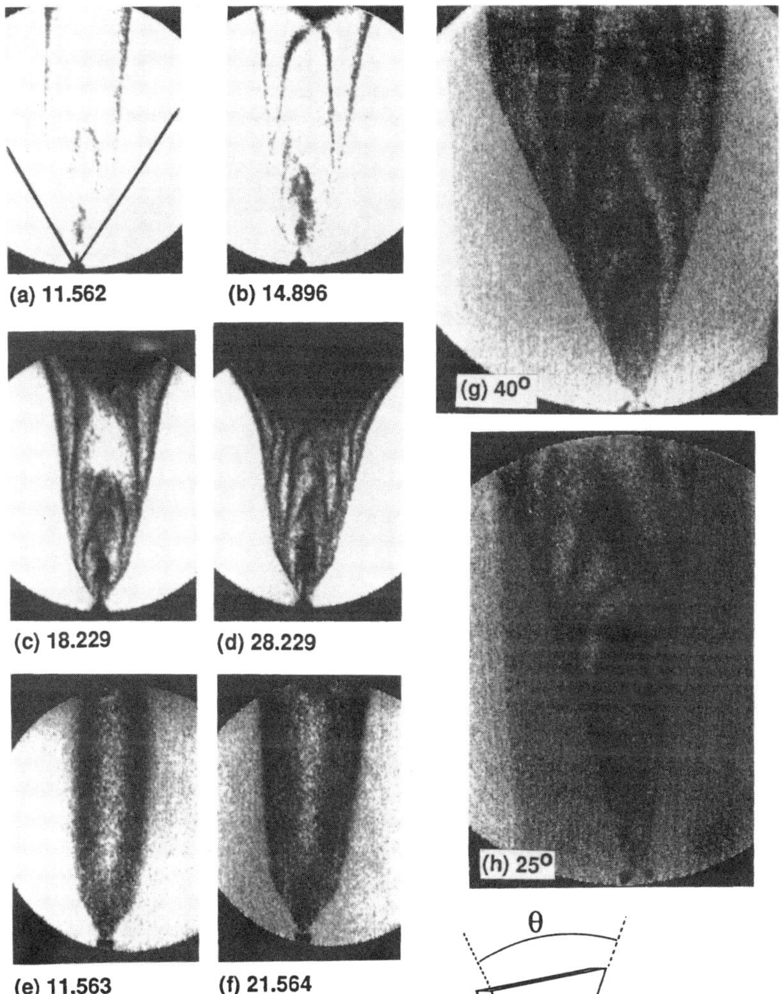

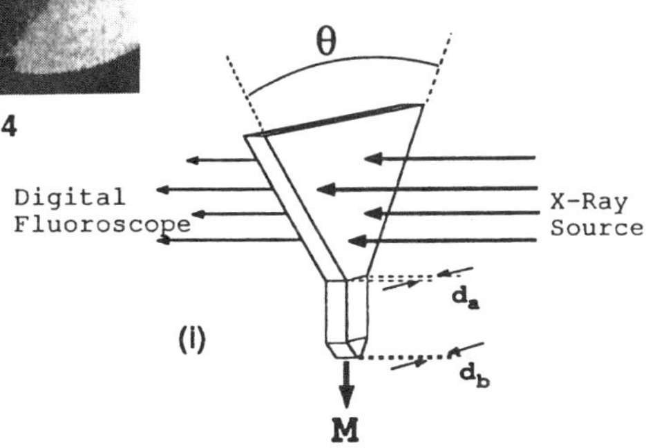

FIGURE 2 X-ray images of rough and smooth sand. (a-d) images obtained with rough sand and $\theta = 66.8$ deg. (e,f) images obtained with smooth sand and $\theta = 66.8$ deg. (Numbers indicate elapsed time.) (g,h) images with rough sand and $\theta = 40$ deg and 25 deg respectively. (h) Schematic of the apparatus.

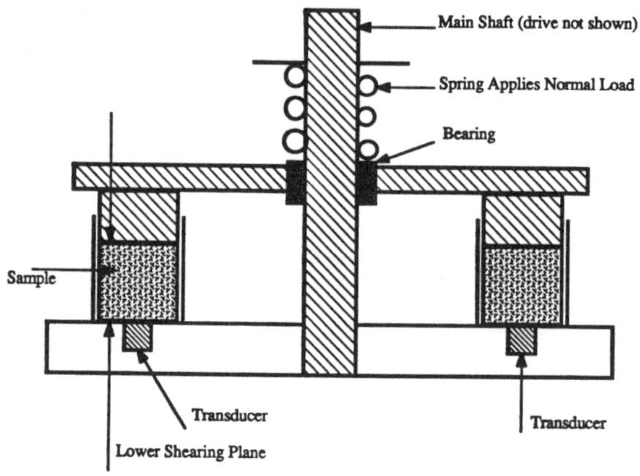

FIGURE 3

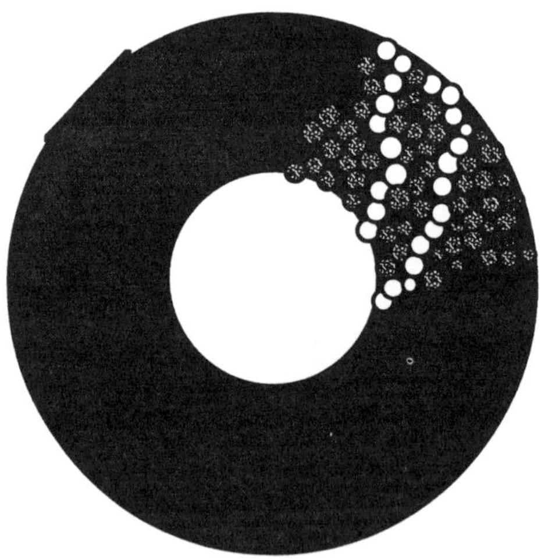

FIGURE 4

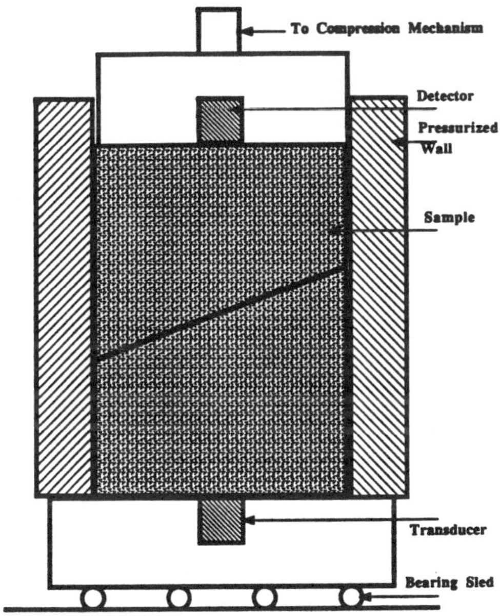
To Compression Mechanism
Detector
Pressurized Wall
Sample
Transducer
Bearing Sled

FIGURE 5

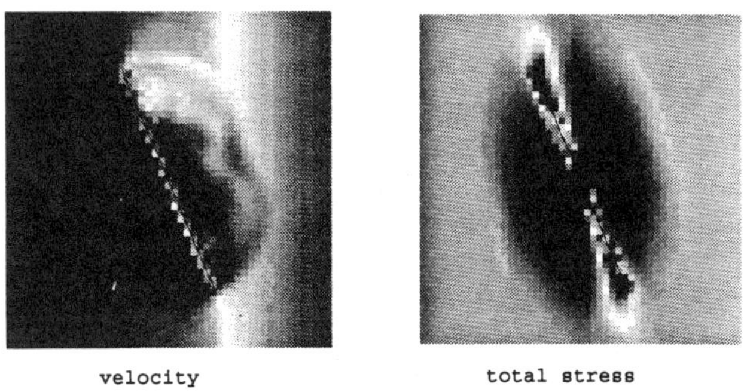

FIGURE 6

Shock-Waves and Irreversibility in Einstein's Theory of Gravity

Joel Smoller[1] *Blake Temple*[2]

1 Introduction

In this article we discuss the issues, and summarize our recent work on the construction of a class of spherically symmetric shock-wave solutions of the Einstein gravitational field equations for a perfect fluid. Our construction entails matching Friedmann-Robertson-Walker (which we refer to as (R-W), c.f. [18], page 412) to Oppenheimer-Tolman[3] (O-T) type metrics at shock-wave interfaces across which the gravitational metric is Lipschitz continuous. In our dynamically matched solution, we imagine the R-W metric as an exploding inner core (of a star or the universe as a whole), and the boundary of this inner core is a shock surface that is driven by the expansion behind the shock into the static O-T solution, which we imagine as the outer layers of a star, or the outer regions of the universe. Our solution addresses the problem first posed by Oppenheimer-Synder (O-S) in 1939 of extending their solution to the case of non-zero pressure. The O-S model is obtained by matching the R-W metric to the *empty* space Schwarzschild metric, and since in this case mass and momentum cannot cross the interface, (which in this case models the surface of a star), and keep the outer solution empty, O-S must make their well known (unphysical) assumption that the pressure be identically zero inside the star. In the classical theory of shock-waves, the interface in the O-S solution is a contact discontinuity, and this means that the solution is time-reversible. In contrast, our shock-wave solution in which $p \neq 0$, is an irreversible solution of the Einstein gravitational field equations in which the irreversibility, loss of information, and increase of entropy in the fluids puts time-irreversibility into the dynamics of the gravitational field.

The shock-waves we identify are described by an autonomous system of ODE's that determine the shock position and the R-W pressure for any given O-T solution. In this article we discuss the derivation of this system of ODE's, as well as an explicit solution of the ODE's that models a shock-wave exploding into a *static, singular, isothermal sphere*. The outer static solution for an isothermal sphere has an inverse-square density profile, and this solution is important in theories of how stars form from gaseous clouds, [1]. The idea is that a star begins as a diffuse gas cloud which slowly contracts under its own gravitational force, until it reaches the point where the mean free path for transmission of light is small enough that the scattering of light within the cloud has the effect of equalizing the temperature

[1]Department of Mathematics, University of Michigan, Ann Arbor, MI 48109; Supported in part by NSF Applied Mathematics Grant Number DMS-89-05205, in part by ONR, US NAVY grant number N00014-94-1-0691, and by the Institute of Theoretical Dynamics, UC-Davis.

[2]Department of Mathematics, University of California, Davis, Davis CA 95616; Supported in part by NSF Applied Mathematics Grant Number DMS-86-13450, in part by ONR, US NAVY grant number N00014-94-1-0691, and by the Institute of Theoretical Dynamics, UC-Davis.

[3]We choose this as a name for static, spherically symmetric metrics that solve the Einstein equations for a perfect fluid. It appears that O-T solutions have not been given a name in the literature, and we consider this name to be appropriate, c.f., [17]. In the special case when the density is constant, this metric is commonly referred to as the *Interior Schwarzschild* metric.

within the cloud. The gas then drifts toward the static solution for an isothermal equation of state, namely, a static, singular, isothermal sphere. Since this solution is inverse square in the density and pressure, the density tends to infinity at the center of the sphere, and this ignites the thermonuclear reactions. The result is a shock-wave explosion emanating from the center of the sphere. Our explicit solution is an exact, general relativistic model for this shock-wave explosion. In this solution, both the inner R-W sound speed $\sqrt{\sigma}$ and the outer O-T sound speed $\sqrt{\bar{\sigma}}$ are assumed to be constant, and are related by an algebraic relation of the form $\bar{\sigma} = H(\sigma) < \sigma$, which we interpret as describing isothermal fluids at different temperatures. (We assume the speed of light $c = 1$.) In this model, the shock-wave emerges from a singularity that represents the initial *Big Bang* in the R-W metric, and so this model provides a scenario by which the universe begins with a shock-wave explosion. We observe that in this explicit solution, the shock speed remains everywhere less than the speed of light so long as the R-W sound speed $\sqrt{\sigma}$ satisfies $\sigma < \sigma_2 \equiv \sqrt{5}/3 \approx .745$; and the Lax characteristic condition holds so long as $\sigma < \sigma_1 \approx .458$. When $\sigma_1 < \sigma < \sigma_2$, the shock speed becomes supersonic with respect to the fluid on both sides of the shock, and tends to the speed of light as $\sigma \to \sigma_2$.

2 General Discussion

The Einstein gravitational field equations can be written in the compact form

$$G = \kappa T, \tag{1}$$

where G denotes the Einstein curvature tensor, T denotes the stress energy tensor, $\kappa = 8\pi/c^2$, and c denotes the speed of light. In Einstein's theory, all properties of the gravitational field are determined by a Lorentzian metric g, (a non-degenerate symmetric 2-tensor of signature $(-1, 1, 1, 1)$), and equation (1) describes the time evolution of g, together with the sources in T, *simultaneously.*

In a given coordinate system $x = (x^0, ..., x^3)$ defined on spacetime, the components of G are given by

$$G_{ij} = R_{ij} - \frac{1}{2} R g_{ij}, \tag{2}$$

where the Ricci tensor R_{ij} and the scalar curvature R are obtained from the Riemann curvature tensor R^i_{jkl}. In the important case of a perfect fluid, the stress energy tensor takes the form

$$T_{ij} = (p + \rho c^2) u_i u_j + p g_{ij}, \tag{3}$$

where ρ is the mass-energy density of the fluid, (as measured in a frame moving with the fluid), p is the pressure and u^i denote the components of the unit 4-velocity vector u of the fluid. When an equation of state for the fluid $p = p(\rho)$ is specified, system (1) imposes ten equations in 14 unknowns: the ten unknown metric components g_{ij}, the density ρ, and three of the four components of the unit vector u. Four remaining equations can then be imposed to specify a coordinate system.

It is well known that G satisfies the fundamental identity

$$divG = 0, \tag{4}$$

where *div* denotes the covariant divergence for the unique symmetric connection Γ which is compatible with the metric g. The Einstein theory then has the remarkable property that conservation of energy and momentum hold *automatically* on solutions of the field equations (1). For example, in the case of a perfect fluid, if g, ρ, u solve (1), then the conservation law

$$divT = 0 \tag{5}$$

must hold as a consequence of (4). In the case when g is identically equal to the flat Minkowski metric η, div reduces to the ordinary divergence, and (3), (5) reduce to the relativistic Euler equations, [14], which in turn reduce to the classical compressible Euler equations for an inviscid perfect fluid in the limit of small velocities.

One important issue in the subject of general relativity is the problem of explaining how time irreversibility should enter the theory, [12]. Indeed, all of the exact solutions of the Einstein equations (1) for classical sources assume the stress tensor for a perfect fluid, or for electro-magnetism, and it is well known that smooth solutions of the Einstein equations with these sources are time reversible. However, *it is our thesis here that time-irreversibility has been built into the theory of general relativity from the very start by way of the stress tensor for a perfect fluid, and that the irreversibility, loss of information and increase of entropy in solutions of (1) is encoded in the properties of the shock-waves.*

The idea is that, in the case of a perfect fluid (3), the relativistic compressible Euler equations appear as a subsystem of the Einstein gravitational field equations (1) through the identity (5), and the relativistic Euler equations form a system of nonlinear hyperbolic conservation laws, the setting for the mathematical theory of shock-waves. Therefore, like the classical compressible Euler equations, solutions of the Einstein equations should be time reversible only up until the formation of shock-waves. For the classical Euler equations, the dissipation and increase of entropy at shock-waves can be interpreted physically as the non-negligible effects of viscosity that must appear because the inviscid Euler equations represent a *singular limit* of the Navier-Stokes equations as the viscosity is taken to zero: however small the viscosity parameters, the dissipative effects of viscosity are not negligible in the surface layer of a shock, and this dissipation is fully accounted for in the shock-waves that appear in the zero viscosity limit.

Unfortunately, shock-waves are exceedingly complicated in the setting of general relativity. For one thing, there is no simple model like the classical Navier-Stokes equation that plays a similar role in relativistic theories, partly because the introduction of second order parabolic terms to model dissipation introduces an infinite speed of propagation similar to that observed in the linear heat equation. Thus the classical interpretation of the Euler equations as the *zero viscosity limit* of the Navier-Stokes equations is not so easily made for the Einstein equations. In the setting of 4-dimensional spacetime, the mathematical theory of the classical Euler equations is woefully incomplete. The theory of conservation laws and shock-waves is well understood mathematically only in the case of one space-one time dimension, but we believe that this theory does set out the *role* we expect shock-waves to play in the theory of the Einstein equations for a perfect fluid. For a classical system of conservation laws in one space-one time dimension, it was shown by Lax [6] that shock-waves always form when the solution is compressive. Lax also made the point that irreversibility, loss of information and increase of entropy could be attributed to a shock-wave in a general system of conservation laws based on the fact that characteristics always impinge on shock- waves. In this setting a celebrated existence theory has been given by Glimm [3], and a rather complete analysis of the irreversibility of shock-waves is described in the work of Glimm, Lax, Diperna and Liu [13] in which the time asymptotic decay of solutions to shock-waves is rigorously demonstrated. In fact, in one space dimension, all solutions with the same values at spatial infinity, (at $t = 0$), will time asymptotically decay to the same solution. The conclusion we draw from the mathematical theory of shock-waves is that the pure conservation law $divT = 0$ contains *all of the information* about the zero viscosity limit, that in this limit irreversibility and dissipation are encoded in the structure of the shock-waves, and that shock-waves will always form in solutions except in very special cases when the fluid is everywhere expansive. We believe that this conclusion carries over to the Einstein equations (1) for a perfect fluid (3). Indeed, in locally Lorentzian coordinates at a point, (coordinates in which $g_{ij} = \eta_{ij} = diag(-1,1,1,1)$, and $g_{ij,k} = 0$ for all $i,j,k = 0,...,3$), the covariant divergence equals the classical divergence; and thus, to within higher order errors, the perfect fluid in general relativity satisfies the relativistic Euler equations at each point. Thus if shock-waves were to never form in T, then in essence, the gravitational field would have the effect of removing all of the compressive action of the Euler equations, and thereby would have the effect of eliminating the dissipation in fluids! This we find physically impossible, and thus we maintain that shock-waves are as fundamental to solutions of (1), (3) as they are to the classical Euler equations. This suggests that when viscosity and heat conduction are neglected, (as they are in the case of a perfect fluid), it is the shock-waves in the sources T that fundamentally introduce irreversibility into the dynamics of the gravitational field. This point of view seems to have essentially been overlooked, [12, 5]. In [15], we organize the general theory of matching two metric solutions of (1) across an interface from the point of view of the theory of shock-waves, and we apply this theory to explicitly construct spherically symmetric shock-wave solutions of the Einstein equations. This construction solves the problem first posed by Oppenheimer and Snyder in 1939 of generalizing their model to the case of nonzero pressure.

If the fluid variables shock, the entries in the stress tensor T become discontinuous, and thus by the Einstein equation, the metric loses continuity in the second derivative. In Section 3 of our paper [15], we present a self-contained treatment of the general theory of shock-waves for the Einstein equations by considering the general problem of allowing the metric to be only Lipschitz continuous across a hypersurface Σ in spacetime. This treatment re-organizes and extends previous results of Israel. (See also [9] and references therein where this topic is referred to by the heading *Junction Conditions*). The result is the the following theorem:

Theorem 1 *Let Σ denote a smooth, 3-dimensional shock surface in spacetime with spacelike normal vector* $\mathbf{n}$. *Assume that the components g_{ij} of the gravitational metric g are smooth on either side of Σ, (continuous up to the boundary on either side separately), and Lipschitz continuous across Σ in some fixed coordinate system. Then the following statements are all equivalent:*
(i) $[K] = 0$ at each point of Σ.
(ii) The curvature tensors R^i_{jkl} and G_{ij}, viewed as second order operators on the metric components g_{ij}, produce no delta function sources on Σ.
(iii) There exist locally Lorentzian coordinate frames at each point $P \in \Sigma$.
(iv) For each point $P \in \Sigma$ there exists a $C^{1,1}$ coordinate transformation defined in a neighborhood of P, such that, in the new coordinates, (which can be taken to be the Gaussian normal coordinates for the surface), the metric components are $C^{1,1}$ functions of these coordinates.

Moreover, if any one of these equivalencies hold, then the Rankine-Hugoniot jump conditions

$$[G]_i^\sigma n_\sigma = 0, \tag{6}$$

(which express the weak form of conservation of energy and momentum across Σ when $G = \kappa T$), hold at each point on Σ, a covariant statement.

Here $[K]$ denotes the jump in the second fundamental form (extrinsic curvature) K across Σ, (this being determined by the metric separately on each side of Σ because g_{ij} is only Lipschitz continuous across Σ), and by $C^{1,1}$ we mean that the first derivatives are Lipschitz continuous. Theorem 1 should be credited mostly to Israel, who obtained these results in Gaussian normal coordinates. Our contribution was to identify the covariance class of $C^{1,1}$ transformations, and to thereby obtain precise coordinate independent statements for (ii)-(iv). As a consequence of this, we obtain the result that the Ricci scalar curvature R *never* has delta function sources at a Lipschitz continuous matching of the metrics, as well as the following theorem that partially validates the statement that shock-wave singularities in the source free Einstein equations $R_{ij} = 0$ or $G_{ij} = 0$ can only appear as coordinate anomalies, and can be transformed away by coordinate transformation:

Theorem 2 *If a smooth shock surface Σ forms in weak solutions of $R_{\alpha\beta} = 0$ or $G_{\alpha\beta} = 0$ posed in some given coordinate system y, such that the y-components $g_{\alpha\beta}$ of the metric tensor g are Lipschitz continuous across Σ, and C^k functions of y on either side of Σ (continuous up to the boundary on either side separately), then there exists a regular $C^{1,1}$ coordinate transformation taking $y \to x$, such that, the components g_{ij} of g in x-coordinates are actually C^k functions of x in a neighborhood of each point on the surface Σ.*

Note that when there are delta function sources in G on a surface Σ, the surface should be interpreted as a surface layer (because $G = \kappa T$), and not a shock-wave, [4, 9]. In [15] we show that for spherically symmetric solutions, the weak form of conservation of energy and momentum (6) implies the absence of surface layers, so long as the areas of the spheres of symmetry match smoothly at Σ. We use this result in our construction of the shock-waves that extend the Oppenheimer-Snyder model to the case of non-zero pressure.

3 An Exact Shock-Wave Solution

We now discuss how we explicitly construct spherically symmetric shock-wave solutions of the Einstein equations (1) for a perfect fluid by matching the Robertson-Walker (R-W) metric,

$$(R-W) \qquad ds^2 = -dt^2 + R^2(t)\left\{\frac{1}{1-kr^2}dr^2 + r^2 d\Omega^2\right\}, \tag{1}$$

and Oppenheimer-Tolman (O-T) metric,

$$(O-T) \qquad d\bar{s}^2 = -B(\bar{r})d\bar{t}^2 + \left(1 - \frac{2M(\bar{r})}{\bar{r}}\right) d\bar{r}^2 + \bar{r}^2 d\Omega^2, \tag{2}$$

at shock-wave interfaces across which the metrics are Lipschitz continuous, and the conditions (i)-(iv) of Theorem 1 above hold. Here, the quantity $d\Omega^2 = d\theta^2 + sin^2\theta d\varphi^2$ denotes the standard metric on the 2-sphere. The R-W metric is homogeneous at each fixed t, and expands or contracts in time according to the Cosmological Scale Factor $R(t)$. The fluid is assumed to be co-moving with the metric, c.f. [18], and the Einstein equations (1) reduce to a system of two ODE's in two unknowns when an equation of state $p = p(\rho)$ is specified. The R-W metric has been accepted as a model for the universe as a whole. In this interpretation, $R(t)$ determines the red-shift factors for far away objects, and the existence of a singularity in the R-W metric in backwards time has been interpreted as the original *Big Bang*. The O-T metric is a time independent solution of (1) that is used to model the interior of a star. In this case the stress tensor is again taken to be that of a perfect fluid that is co-moving with the metric, and the functions $B(\bar{r})$ and $M(\bar{r})$, (the total mass inside radius $\bar{r}$), are also determined by an autonomous system of two equations in two unknowns when an equation of state $\bar{p} = \bar{p}(\bar{\rho})$ is chosen, and the Einstein equations (1) are imposed.

The idea in the matching is to define a coordinate transformation mapping $(t, r) \to (\bar{t}, \bar{r})$ such that, under this identification of coordinates, the metrics R-W and O-T agree Lipschitz continuously on a 3-dimensional shock surface. We set $\bar{r} = Rr$ in order that the spheres of symmetry match smoothly at the surface, and with this it remains to define $\bar{t} = \bar{t}(t, r)$. The following general theorem states that a Lipschitz continuous matching of metrics can be achieved for arbitrary R-W and O-T metrics,[15]

Theorem 3 *Let $B(\bar{r}), M(\bar{r}), \bar{\rho}(\bar{r}), \bar{p}(\bar{r})$ denote any O-T solution of the Einstein equations (1), and let $R(t), \rho(t), p(t)$ denote any R-W solution of (1). Then, (under non-degeneracy assumptions, [15]), there exists a coordinate mapping $(t, r) \to (\bar{t}, \bar{r})$ of the form $\bar{r} = Rr$ and $\bar{t} = \bar{t}(t, r)$ such that, under this identification, the metrics agree and are Lipschitz continuous at the shock surface*

$$M(\bar{r}) = \frac{4\pi}{3}\rho(t)\bar{r}^3. \tag{3}$$

The equation (3) defines the shock surface $\bar{r} = \bar{r}(t)$ implicitly so long as $\bar{\rho}'(\bar{r}) < 0$.

Note that for this theorem, arbitrary equations of state $p = p(\rho)$ and $\bar{p} = \bar{p}(\bar{\rho})$ can be assigned. The identity (3) can be interpreted as a global conservation of mass principle for such matchings: in words, (3) says that the total mass that the O-T solution would see inside the shock interface $\bar{r} = \bar{r}(t)$ if it were continued on into the origin $\bar{r} = 0$, equals the total mass inside a sphere of radius $\bar{r} = \bar{r}(t)$ and constant density $\rho(t)$.

A further constraint on the metrics must be imposed to ensure that the Rankine-Hugoniot jump conditions (6) for conservation hold across a shock. By the general theory, it suffices to match the second fundamental forms $[K] = 0$ across the surface, and this alone will ensure that the surface is not a surface layer, and the conditions of Theorem 1 will then all apply at the shock surface. In [15] we show that in the presence of spherical symmetry, $[K] = 0$ is equivalent to the single condition

$$[T^i_j]n_i n^j = 0. \tag{4}$$

Using this, we derive a system of two autonomous ordinary differential equations in the shock position $r(t)$ and the cosmological scale factor $R(t)$ that determine the R-W metrics that will match any given O-T metric at the shock surface (3) such that energy and momentum are conserved across the shock. In order to impose the extra constraint (4), we must allow the R-W pressure $p(t)$ to be determined by the equations, rather than by an equation of state. This is summarized in the following theorem,[15]:

Theorem 4 *Let $B(\bar{r}), M(\bar{r}), \bar{\rho}(\bar{r}), \bar{p}(\bar{r})$ denote any fixed O-T solution of the Einstein equations. Then, (under non-degeneracy assumptions), the R-W metric $R(t), \rho(t), p(t)$ will satisfy conservation across the shock surface (3), (under the identification $\bar{r} = Rr, \bar{t} = \bar{t}(t,r)$), if and only if $(r(t), R(t))$ solve the system of ODE's*

$$\dot{R}^2 = \frac{8\pi G}{3}\rho R^2 - k, \tag{5}$$

$$\alpha\dot{r}^2 + \beta\dot{r} + \gamma = 0, \tag{6}$$

where all of the functions $\rho, \alpha, \beta, \gamma$ appearing in (5) and (6) can be expressed in terms of the unknowns $(r(t), R(t))$ through the shock surface equation (3) and the identity $\bar{r} = Rr$. The R-W pressure is then given by

$$p = -\frac{\frac{d}{dt}(\rho R^3)}{3R^2\dot{R}}. \tag{7}$$

Thus, for any given O-T solution, system (5), (6) gives the conservation constraint in terms of a first order *autonomous* ODE in the unknowns $(r(t), R(t)$, and the solutions of this ODE determine the R-W metrics that will match a given O-T metric across a shock surface, such that the additional condition of conservation is maintained across the shock. The solutions that solve this ODE reduce to the Oppenheimer-Snyder solutions when M is constant and the O-T metric reduces to the empty space Schwarzschild metric. In the O-S case, the solution of (5), (6) reproduces $p \equiv 0$. Note that for a given O-T solution, we can, in principle, construct a shock-wave at radius $\bar{r} = Rr$ with arbitrary R-W pressure p assigned at that radius by specifying the appropriate initial conditions for r and R.

We obtain an explicit solution of the ODE's (5), (6) when we take $\bar{p} = \bar{\sigma}\bar{\rho}$, for $\bar{\sigma} \equiv const. < c$. Substituting this into the O-T metric (2) yields the following exact solution which represents a relativistic version of a *static, singular isothermal sphere*. Let γ be defined by

$$\gamma = \frac{1}{2\pi\mathcal{G}}\left(\frac{\bar{\sigma}}{1 + 6\bar{\sigma} + \bar{\sigma}^2}\right). \tag{8}$$

Then the following define an exact solution of O-T type:

$$\bar{p} = \bar{\sigma}\bar{\rho}, \tag{9}$$

$$\bar{\rho}(\bar{r}) = \frac{\gamma}{\bar{r}^2}, \tag{10}$$

$$M(\bar{r}) = 4\pi\gamma\bar{r} \tag{11}$$

$$A(\bar{r}) = 1 - 8\pi\mathcal{G}\gamma \equiv const, \tag{12}$$

$$B(\bar{r}) = \bar{r}^{\frac{4\bar{\sigma}}{1+\bar{\sigma}}}. \tag{13}$$

Note that when $\bar{\sigma} \to 0$, we must have $A \to 1$ and $B \to 1$, which is the Newtonian limit. Moreover, $\bar{\rho} \to \infty$ and $\bar{p} \to \infty$ as $\bar{r} \to 0$, and thus this "isothermal sphere" is "singular" in the sense that it takes an infinite pressure at $\bar{r} = 0$ to hold the configuration up. When we match this solution to the R-W metric with $k = 0$, and impose conservation, we obtain that $p = \sigma\rho$ gives the equation of state in the R-W metric where

$$\bar{\sigma} = H(\sigma) \equiv \frac{1}{2}\sqrt{9\sigma^2 + 54\sigma + 49} - \frac{3}{2}\ sigma - \frac{7}{2}, \tag{14}$$

and the shock positions $\bar{r}(t)$ and R-W density $\rho(t)$ and scale factor $R(t)$ are given by the formulas:

$$\bar{r}(t) = \pm\sqrt{18\pi\mathcal{G}\gamma(1+\sigma)(t-t_0)+\bar{r}_0}, \tag{15}$$

$$\rho(t) = \frac{3\gamma}{\bar{r}^2}, \tag{16}$$

$$R(t) = R_0\left(\frac{\bar{r}(t)}{\bar{r}_0}\right)^{\frac{2}{3+3\sigma}}. \tag{17}$$

Here the $+$ in (15) corresponds to the out-going shock-wave, which is physical because $\rho(t) = 3\bar{\rho}(\bar{r}) > \bar{\rho}(\bar{r})$ on the shock surface $\bar{r} = \bar{r}(t)$. Note also that all functions in (15)-(17) are functions of $\bar{r}(t)$, and thus we see immediately that there is a singularity in this solution in backward time $t = t^*$,

$$t^* = t_0 - \frac{\bar{r}_0}{\sqrt{18\pi\mathcal{G}\gamma(1+\sigma)}}, \tag{18}$$

at which point $\bar{r}(t^*) = 0$, $R(t^*) = 0$, $\rho(t^*) = \infty$ and $p(t^*) = \infty$.

Thus this explicit solution provides a scenario by which the *Big Bang* begins with a shock-wave explosion. We observe that the shock-wave solution (15)-(17) removes the singularity in the O-T solution for $\bar{p} = \bar{\sigma}\bar{\rho}$ after time $t = t^*$, and we interpret this as saying that *the shock-wave provides the pressure required to hold the O-T configuration up.* Also note that the Newtonian limit is acheived as $\sigma, \bar{\sigma} \to 0$.

It is not difficult to show that $H'(\sigma) > 0$ for $0 < \sigma < 1$, and that

$$\bar{\sigma} = \frac{3}{7}\sigma + O(\sigma^2)$$

as $\sigma \to 0$. Moreover, $\bar{\sigma} = \sqrt{17} - 4 \approx .1231$ when $\sigma = 1/3$; and $\bar{\sigma} = \frac{\sqrt{112}}{2} - 5 \approx .2915$ when $\sigma = 1$.

A further calculation shows that the shock speed relative to the R-W fluid particles is a constant, depending only on σ and independent of $\bar{r}(t)$. The shock speed tends to the speed of light $c = 1$ as $\sigma \to \sigma_2 = \sqrt{5}/3 \approx .745$, and the Lax shock conditions hold for $0 < \sigma < \sigma_1 \approx .458$. When $\sigma_1 < \sigma < \sigma_2$, the shock-wave is supersonic relative to the fluid on both sides of the shock. The solution (15)-(17) represents one very special solution of (5)-(6) in which $\bar{r}$ is always larger than the Schwarzschild raduis for the mass $M(\bar{r})$.

4 Conclusion

Our shock-wave solution opens up intriguing possibilities in cosmology, as well as in the classical theory of shock-waves. For cosmology, our solution opens up the possibility (with an *explicit* model) that there is a shock-wave at the edge of the universe. Consequently, the 2.7 degree Kelvin background radiation could well be due to the blackbody radiation from the surface of this shock-wave. This suggests the possibility of using our explicit formulas, (together with radiation conditions), to obtain interesting new relations between the speed of the shock, its position, and the average energy density in the universe. Moreover, our dynamical matched solution need not have the same type of singularity in the past as does the R-W metric because, in backward time, the shock-wave can reach $\bar{r} = 0$ at the same time that the singularity in the R-W metric forms. Thus, in our model, the nature of the *Big Bang* appears to depend on the equation of state and initial conditions taken in the outer O-T solution. It also suggests interesting implications for the Oppenheimer-Volkoff and Chandrasekhar limits for stable stars. Indeed, a well known theorem states that whenever the total mass inside the radius $\bar{r}$ in an O-T solution falls within 9/8 of the Schwarzschild radius for that mass, the solution is unstable because it takes an infinite pressure at $\bar{r} = 0$ to "hold the star up", [16]. This implies a maximum red-shift factor of two for light emitted from the surface of stable stars. Thus it has been argued that exotic stars like quasars, point sources of light which emit frequencies with red-shift factors much greater that two, must be very far away to account for the large red-shift, which leads to the conclusion that they are extremely large in order to explain their brightness. Our general shock-wave model applies to highly compressed O-T solutions allowing for arbitrary red-shift factors because the shock-wave in the core can generate the enormous pressures required to hold the star up. (This possibilty was realized in the explicit solution (15)-(17) which, however, contains no Schwarzschild singularity.) All of these scenarios can be studied within the context of our models which allow $p \neq 0$.

Our construction also has interesting implications for the classical theory of shock-waves. Our analysis demonstrates that there are remarkable simplifying features of the Einstein equations over the classical Euler equations that allow us to construct multi-dimensional shock-waves in general relativity. Indeed, the Einstein equations are obtained from the Euler equations by introducing the metric potentials g_{ij}, and it is easier to match the metric across a shock-wave than it is to solve the Rankine-Hugoniot jump relations for conservation directly. Once we match the metrics, the number of conservation constraints is reduced by one, and this enables us to assign a non-zero pressure behind the shock for *any* choice of O-T solution on the outside. The covariance of the Einstein equations enables us to obtain an explicit formula for the shock surface, as well as a global conservation of mass principle, and we do this without ever having to tackle the more difficult problem of expressing the metric components in a single coordinate system at the shock. It is also surprising that, by Theorem 1, for every shock-wave solution of $G = \kappa T$, there exists special coordinate systems in which the metric is $C^{1,1}$ across the shock. Thus in computing the shock-wave solution in these coordinates, one never has to difference a discontinuous function when one computes with the Einstein equations (1) instead of the Euler equations $divT = 0$. This suggests new possibilities for the numerical simulation of shock dynamics.

References

[1] D. Christodoulou, (private communication).

[2] R. Courant and K.O. Friedrichs, *Supersonic Flow and Shock-Waves*, John Wiley & Sons, New York, 1948.

[3] J. Glimm, Comm. Pure Appl. Math., **18**(1965), pp. 697-715.

[4] W. Israel, IL Nuovo Cimento, Vol. XLIV B, N. 1, 1966.

[5] W. Israel, edited by S. W. Hawking and W. Israel, Cambridge University Press, 1987, pp. 199–276.

[6] P.D. Lax, Comm. Pure Appl. Math., **10**(1957), PP. 537–566.

[7] G.C. McVittie, Astro. Phys. Jour., **140**(1964), pp. 401–416.

[8] C. Misner and D. Sharp, Phys. Rev., **26**(1964), p. 571–576.

[9] C. Misner, K. Thorne and J. Wheeler, W.H. Freeman and Co., *Gravitation*, New York, 1973.

[10] J.R. Oppenheimer and J.R. Snyder, Phys. Rev., **56**(1939), p. 455–459.

[11] J.R. Oppenheimer and G.M. Volkoff, Phys. Rev., **55**(1939), pp. 374–381.

[12] R. Penrose, *The Emperor's New Mind*, Oxford University Press, 1989.

[13] J. Smoller, *Shock-Waves and Reaction-Diffusion Equations*, Springer Verlag, 1983.

[14] J. Smoller and B. Temple, Commun. Math. Phys. 156,67-97 (1993).

[15] J. Smoller and B. Temple, *Shock-wave solutions of the Einstein equations: the Oppenheimer-Snyder model of gravitational collapse extended to the case of non-zero pressure*, (to appear), Arch. Rat. Mech. Anal.

[16] H. Stephani, *General Relativity, An Introduction to the Theory of the Gravitational Field*, Cambridge University Press, 2nd ed., 1990.

[17] R. Tolman, *Static Solutions of Einstein's Field Equations for Spheres of Fluid*, Physical Review, , **55**(1939), p. 364-374.

[18] S. Weinberg, *Gravitation and Cosmology: Principles and Applications of the General Theory of Relativity*, John Wiley & Sons, New York, 1972.

An Accuracy Test of a Cartesian Grid Method for Steady Flow in Complex Geometries

Marsha Berger* John Melton†

Abstract

We describe a method to solve the inviscid compressible Euler equations using non-body-fitted Cartesian grids. We briefly survey the geometric procedures needed for the "grid generation" in this approach and show some Cartesian grids for three dimensional configurations. The major source of error in these methods is at the irregular boundary cells. This has motivated us to study various improvements in the numerical scheme (a Godunov method) on a model problem with smooth flow in two space dimensions.

1 Introduction

We describe a Cartesian mesh method for solving the steady inviscid Euler equations for complicated geometries. This approach uses a regular Cartesian grid, with the solid boundaries carved out of the underlying grid in an essentially arbitrary way (see Figure 2.) This technique is sometimes called the "cookie cutter" approach. It leaves a border of irregular cells along the edge of a solid boundary, where special difference schemes are needed. However the rest of the grid is completely regular, allowing us to use highly efficient and accurate finite difference schemes over most of the domain. All the overhead for the geometry complexity is at the boundary. This is not the case for body-fitted or unstructured grids - the other popular ways to deal with complex geometry.

Until recently, Cartesian grid methods have received relatively less attention than the alternative approaches. Previous work includes [13], [7], [11] and [10], and of course TRANAIR [22] (using Cartesian grids for potential flow), a production workhorse at Boeing. In the time dependent case methods have been developed in [5], [6], [19], [20]. Lockheed-Ft.Worth [15] now uses a Cartesian grid method as one of their CFD tools, and a commercial package for "under the hood" automotive calculations is now available as well. Clearly, interest in these methods is growing.

The purpose of this talk is two-fold. We first give an overview of the types of geometric calculations used in generating a Cartesian grid, and present some results from these algorithms for 3D configurations. Secondly, we address the important question of accuracy. Just because we can generate a grid doesn't mean we can compute a good flow solution on it. We start with the simplest method on a 2D model problem, and measure the error in the computation. We then modify various pieces of the algorithm to try to isolate the sources of error, and improve the overall quality of the solution.

Since Cartesian grids are completely irregular at the boundaries of the domain, (where the grid intersects the body), it is not even clear what accuracy and convergence rate to expect. (There is beginning to be some literature on the question of accuracy on non-smooth grids [16], [21]). Cartesian mesh methods are usually applied to such complex problems that the sources of error may never be known in a quantitative way. We feel it is important to study *model problems* to gain an understanding of the behavior of these methods.

Our test problem was first used by [2] in a comparative study of the accuracy of quadrilateral vs. triangular body-fitted grids. We will compare our results with the corresponding body-fitted results in that work. Since for aerodynamic flows we are especially interested in the solution at the boundary, we will measure the accuracy

*Courant Institute, 251 Mercer Street, New York, NY 10012
†NASA Ames Research Center, Moffett Field, CA 94035

along the boundary as well as over the entire flow field. A previous study of accuracy was also done in [8] for a slightly different method. We obtain similar results to theirs in this work.

In the next section we give an overview of the Cartesian mesh method. We list the quantities needed for the flow solver, and the geometric calculations needed to provide those quantities. In section 3 we describe our numerical algorithm. Section 4 presents the computational results for steady supersonic flow in a circular channel. We try several modifications of the basic algorithm to improve its convergence rate at the boundary. This work is still in progress, so instead of definitive conclusions we have a brief discussion of the current state of affairs in section 5.

2 A Cartesian Grid Mesh Generator

In this section we describe the geometric operations necessary to define the flowfield geometry to the flow solver. We present this specifically in the context of solving the compressible Euler equations in integral form,

$$\frac{d}{dt}\iiint_{\Omega} U\,dx\,dy\,dz = -\oint_{\partial\Omega} \mathbf{f}\cdot\mathbf{n}\,dS \tag{1}$$

where $U = (\rho, \rho u, \rho v, \rho w, \rho E)^T$ and

$$\mathbf{f}\cdot\mathbf{n} = \begin{pmatrix} \rho\mathbf{q}\cdot\mathbf{n} \\ \rho u\mathbf{q}\cdot\mathbf{n} + pn_x \\ \rho v\mathbf{q}\cdot\mathbf{n} + pn_y \\ \rho w\mathbf{q}\cdot\mathbf{n} + pn_z \\ (\rho E + p)\mathbf{q}\cdot\mathbf{n} \end{pmatrix},$$

where $\mathbf{q} = (u, v, w)$. These equations place no restrictions on the shape of an individual control volume. Over most of the domain the volumes will be the usual Cartesian cubical cell (although since mesh refinement will be used in computing the solution, the volumes of these cells will depend on the level of refinement). Only the body-intersecting cells will be irregular. For these cells special information will be required to accurately compute the flow and impose the solid wall boundary conditions $\mathbf{q}\cdot\mathbf{n} = \mathbf{0}$, where $\mathbf{n}$ is the unit normal to the surface.

This normal vector can be approximated in a cell using the divergence theorem: the sum of the surface areas multiplied by the normal vector over all sides of the volume sum to zero. Since the other sides of the cell are portions of a Cartesian cell, this computation is rather simple, and the area of the solid surface intersecting the cell does not need to be directly computed. A similar application of the divergence theorem simplifies the calculation of the volume centroid as well. In two dimensions, this amounts to approximating the boundary of the solid body by a piecewise linear segment in each cell. Note that in three dimensions, even if the body intersects each face in a straight line, it may still be impossible to find a plane that gives those straight line intersections.

For our numerical scheme (described completely in the next section in two space dimensions), the information needed by the flow solver is:

- area of each cell face (exterior to the solid body),
- face centroid for each face,
- volume centroid for each irregular cell,
- surface normal for the body surface in each cell at the boundary.

This information is found, for the most part, by intersecting Cartesian cells with the description of the body. For the computations here, the body is described by a list of triangles defining the surface. Other representations are possible however, and we have also investigated generating the Cartesian mesh directly from a CAD/CAM system using a NURBS description of the geometry [18]. The operations themselves are conceptually simple though unfortunately not simple to program. We illustrate them pictorially here. More details are found in [17].

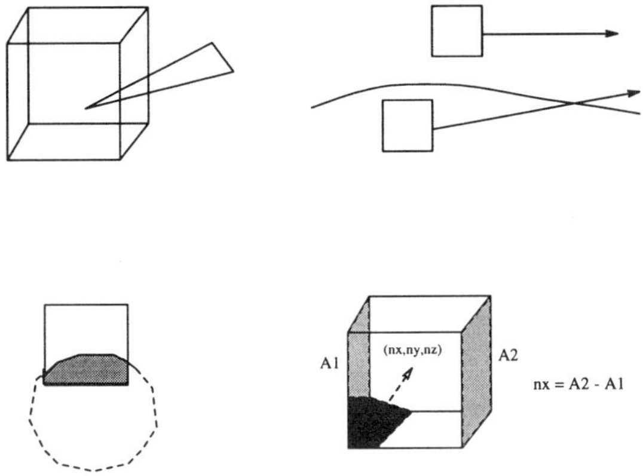

Figure 1: Illustration of some of the steps in Cartesian grid generation.

We first examine each cell to determine whether it is body-intersecting. This is illustrated in Figure 1. This step amounts to computing the intersections of lines and planes. If a cell is not body-intersecting, we determine whether it is inside or outside the flow field domain by casting a ray from the cell center out to the far-field and counting the number of intersections it makes with the surface. This is shown in Figure 1b. For cells that intersect the body, the face polygon is 'clipped' (see e.g. [12]) with the co-planar surface cross-section, using the Sutherland-Hodgman polygon clipping algorithm (illustrated in Figure 1c). The area and centroid of the resulting planar polygon is easily calculated. Applying the divergence theorem with the calculated areas of the 6 planar faces of each cell gives the integrated surface normal and area for the body surface in that cell. This is shown in Figure 1d. At present, the cell intersection information is computed for all cells at a pre-determined finest level of refinement. This is saved in what we call the "database grid". However, not all these cells need to be present in the flow solution. Using curvature criteria, also described in [17], the appropriate level of refinement to describe the body geometry is determined, before the flow solver is applied. (This is as opposed to the dynamic refinement based on the flow solution itself that occurs at various times during the flowfield solution process). It is likely that in future versions of the code, this will be done in a more efficient way.

In Figure 2 we show several views of an F16XL Cartesian grid. It has 277657 cells in all. Of these cells, 20%, or 56205 cells intersect the body. It took approximately 6 minutes on a C90 to generate the grid. We demonstrate that an actual flow solution for a complex configuration in three dimensions can be computed on such grids in Figure 3, which shows the Mach distribution for a supersonic transport test case. The flow solver in Figure 3 uses a Jameson-Schmidt-Turkel type scheme [14]. In the rest of this paper however we solve a two dimensional model problem with a Godunov type scheme to study the accuracy of flow computations on non-body-fitted grids.

3 A Cartesian Grid Flow Solver

In this section we describe a variation of a second order Godunov scheme for use on Cartesian non-body-fitted grids. The scheme is based on piecewise polynomial reconstruction in each cell. Such schemes typically give second order convergence for smooth flows on regular grids. For irregular grids however the usual truncation error analysis, based on cancellation of the leading error terms when differencing the fluxes, breaks down. We will study the convergence rate of our algorithm numerically in section 4. We first describe the basic method (in two space dimensions) applied over the regular portions of the domain. We then describe the modifications necessary for the irregular cells at the boundary of the domain. As much as possible the implementation

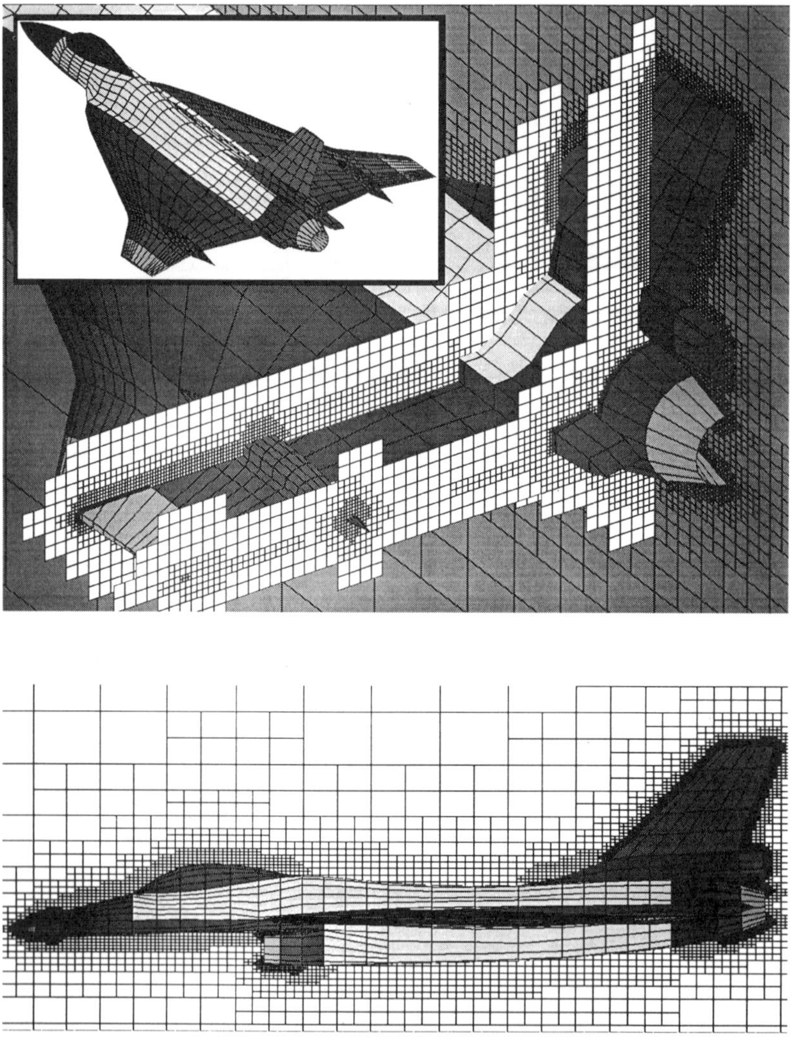

Figure 2: Cartesian Grid for an F16XL.

Figure 3: Mach distribution for supersonic transport.

proceeds in the same way: regular (vectorizable) calculations over the entire Cartesian grid (including cells in the "interior" of the body) followed by a "fix-up" pass over the irregular cells at the solid surface.

The basic scheme we use is a high resolution Godunov's method: for each cell we

- Reconstruct the solution in each cell using a piecewise linear function. We use a second order gradient based on central differencing in each direction.
- Predict a left and right state at the midpoint of each cell edge, using the gradient information.
- Solve a Riemann problem to determine a unique state at the midpoint of each edge. The Riemann solver we use is based on [9].
- Evaluate the integral of the fluxes in (1) using the midpoint rule.
- Advance the solution to steady state using a three stage Runge Kutta method with local time stepping.

These steps need to be modified for use at the irregular cells. For example, the midpoint of the cell edge for an irregular cell (centroid in 3D) needs to be calculated during the mesh generation phase described in Section 2. The actual length (in 2D) or face area (in 3D) needs to be calculated then as well. For steady state calculations, the volume of the cut cell is not needed. For the local time step for the irregular cells we take

$$\Delta t = \frac{cfl * A_{full}}{\max(u+c, v+c) * Length_{max}},$$

where $A_{full} = \Delta x \times \Delta y$ is the area of a regular (full) cell, and $Length_{max}$ is the length of the maximum side of a cut cell. For the border of cells *next* to the irregular cells, we use a one-sided second order derivative approximation instead of a central difference.

The most difficult modification concerns the reconstruction of the solution at the irregular cells themselves. Since the cell centroids generally do not line up in neighboring cut cells, we can no longer determine the differences separately in each direction. We use a least squares procedure to determine the gradient in the following way. For a given cell u_0 we try to fit a linear polynomial so that

$$\bar{u}_0 + (x_j - x_0)u_{0x} + (y_j - y_0)u_{0y} = \bar{u}_j, \tag{2}$$

where the centroid of cell 0 is (x_0, y_0) with cell average $\bar{u}_0$, and the subscript j refers to the same quantities of a neighboring cell. For linear reconstruction, we use neighboring cells that share an edge with the given cell, (unless there are not enough, for example, in the triangular cell case, where we also include diagonal neighbors that share a node). This formulation is automatically conservative, since a linear function through $\bar{u}_0$ at the centroid automatically preserves the cell average. However, for quadratic reconstruction this is not the case, and considerably more work must be done. In our tests with quadratic reconstruction, (the results of which are described in Section 4), we set up a linear system for a polynomial p in cell 0

$$p(x,y) = c_0 + c_1(x-x_0) + c_2(y-y_0) + c_3[(x-x_0)^2 - s_{xx}] + c_4[(x-x_0)(y-y_0) - s_{xy}] + c_5[(y-y_0)^2 - s_{yy}]. \tag{3}$$

In (3), the unknowns are the coefficients c_1 through c_5, and the quantities $s_{xx} = \int_{cell_0}(x-x_0)^2$, s_{xy} and s_{yy} are pre-computed to maintain conservation,

$$\int_{cell_0} p(x,y) = \bar{u}_0 A_0. \tag{4}$$

This is necessary because, unlike the linear case, $p(x_0, y_0) \neq \bar{u}_0$. Given this definition of p, the reconstruction proceeds again in a least squares sense using

$$\int_{cell_j} p(x_j, y_j) = \bar{u}_j,. \tag{5}$$

where the neighboring cells j now include two rings of cells that share edges, starting from the given cell. To compute the integral of the quadratic polynomial, both for (4) and (5), we use a third order accurate approximation. Specifically, the integral over a triangle is approximately the average of the function values at the midpoints of the sides multiplied by the area of the triangle. In two dimensions this is somewhat less work than using the divergence theorem to reduce the computation to an integral around the boundary of the cell. More in depth discussions of high order polynomial reconstruction can be found in [4, 3, 1].

Despite all the care to accurately compute quadratic reconstructions with the correct cell average properties, we found virtually no difference in the results between using (3) and simply treating the cell averages as pointwise values. Since the former is much more expensive computationally, we would probably not do this again for steady state computations. However, in all the numerical examples, we compare the computed solution $\bar{u_j}$, which is an approximation to the integral average over cell j, with the "exact" integral average over the cell (computed to third order using the approximation mentioned above).

One final point concerns the use of limiters. Since our test problem is smooth, and our goal is to test the accuracy of our reconstruction procedures, we did not use a limiter. When it is necessary, we limit the reconstructed solution over the entire cell so that it does not exceed any neighboring solution value at the cell edge. We note that [2] reports better results using a face based limiter, where the reconstructed solution has a different limiter applied at each face (in 2D, each edge is limited separately).

4 Numerical Results

The two dimensional test problem for our numerical convergence study is found in [2]. We solve the inviscid compressible Euler equations for a supersonic vortex in a channel formed by concentric circular arcs. The boundaries of the channel form one quarter of a circle, with inner radius r_i and outer radius r_o. A smooth analytic solution exists for this problem, so the errors in the computation can be evaluated. The density is:

$$\rho(r) = \rho_i[1 + \frac{\gamma-1}{2}M_i^2\{1-(\frac{r_i}{r})^2\}]^{\frac{1}{\gamma-1}}, \tag{6}$$

and the velocity varies inversely with the radius. We use the same geometry and test parameters as [2], $\rho_i = 1$, $r_i = 1$, $r_o = 1.384$, $M_i = 2.25$, and $p_i = 1/\gamma$. For all test cases, the program is run until the residuals have

Mesh Size	Second Order		First Order	
	Reg. Quad	Distorted	Reg. Quad	Distorted
31 × 6	.40	.97	16.04	15.77
61 × 11	.10	.43	7.39	7. 31
121 × 21	.03	.19	3.65	3.79
Rate	2.08	1.21	1.10	1.07

Table 1: 1st and 2nd order results of Aftosmis et al.

h	Linear				Constant			
	All cells	Ratio	Bndry. Cells	Ratio	All cells	Ratio	Bndry. Cells	Ratio
$.053\bar{3}$	9.68(-2)		2.61 (-1)		5.28 (0)		7.65(0)	
$.026\bar{6}$	2.54(-2)	3.81	9.51 (-2)	2.74	2.75 (0)	1.92	4.34(0)	1.76
$.013\bar{3}$	6.23(-3)	4.07	3.71 (-2)	2.56	1.45 (0)	1.89	2.38(0)	1.82

Table 2: 1st and 2nd order Cartesian results for entire flowfield and boundary cells only.

converged four or five orders of magnitude. We measure the actual global error in the following tables. The relevant results of [2] for structured quadrilateral grids are given here as well for comparison purposes. In all cases the error in the density field is measured. Starting from the basic algorithm outlined in the previous section, we will experiment with the reconstruction algorithm, the flux evaluations, and the representation of the boundary itself to try to improve the accuracy of the results at the solid walls.

The first numerical study examines the errors using piecewise linear reconstruction of the solution. Table 1 summarizes the percent relative errors from [2], and includes their linear reconstruction (second order) and piecewise constant (first order) results. Also included are their results using randomly distorted quadrilateral meshes as an interesting aside. In their results, the l_1 and l_2 norms gave the same convergence rates and approximate error, so we only repeat the former. The quantity measured in [2] is

$$Error = \sum_k |\frac{q_{exact} - q_k}{q_{exact}}| A_k \tag{7}$$

where the sum is over all cells k. We will use the same definition in Table 2 to present the Cartesian results. Later we will also use the following measure of relative error in the l_1 norm,

$$\frac{||q_{exact} - q_{computed}||_1}{||q_{exact}||_1} = \frac{\sum_k |q_{exact} - q_k| A_k}{\sum_k |q_{exact}| A_k}, \tag{8}$$

where A_k is the area of the kth cell. This is a direct discretization of the continuous L_1 norm. We also measure the error at the boundary only, defined as

$$\frac{||q_{exact} - q_{computed}||_1}{||q_{exact}||_1} = \frac{\sum_{k\in\partial} |q_{exact} - q_k| l_k}{\sum_{k\in\partial} |q_{exact}| l_k}. \tag{9}$$

In (9) the sum is taken only over the cut cells, and the cell error is multiplied by a length rather than the cell area. Here we take l_k to be the length of the boundary segment in that cell. Note however that a large cell could have a small boundary segment. We have also tried using $\sqrt{A_k}$ in the boundary norm with similar though less clean results. Although the use of a piecewise linear boundary representation in each cell is second order the convergence is not smooth, since the intersection of the boundary with the Cartesian mesh is not regular. We expect the error to be bounded by $const \times h^2$, but we do not expect an asymptotic expansion. Since the norm itself is not smooth, this increases the difficulty in measuring the accuracy of the approximation at the boundary cells.

In our Cartesian non-body fitted grid we use a regular grid spacing for our convergence study starting with $h = .053\bar{3}$ in each direction. There are 294 cells in the flow field on this coarsest grid, 206 full cells and 88

h	2nd Order Linear Recon	Ratio	Match Errors	Ratio	Quad at cut cells	Ratio	Quad at 2 rings	Ratio
$.053\bar{3}$	2.41 (-1)		2.31 (-1)		2.25 (-1)		1.83 (-1)	
$.026\bar{6}$	9.37 (-2)	2.57	9.17 (-2)	2.52	8.88 (-2)	2.53	6.45 (-2)	2.83
$.013\bar{3}$	4.01 (-2)	2.34	4.31 (-2)	2.13	3.98 (-2)	2.23	2.80 (-2)	2.30

Table 3: Various types of quadratic reconstruction at boundary cells.

cut cells. In the body-fitted case a grid with 31×6 nodes has 186 degrees of freedom. Thus the first entry in Table 2 corresponds to resolution slightly better than the first row of Table 1.

There are several things to note from Tables 1 and 2. The solution using regular quadrilaterals is slightly worse than the corresponding Cartesian solution, even with the flow alignment of the body- fitted grid. In fact, the extra accuracy of using regular Cartesian grids is apparent even in the first order results. The quality of the results for distorted quadrilaterals (where the nodes of the body-fitted grid are randomly perturbed a small amount) deteriorate quite a bit. In the Cartesian results, where one can view the perturbation as restricted to the boundary of the mesh, the boundary cells have similar accuracy, but the overall flowfield results do not suffer.

Table 2 also shows that the errors at the boundary are significantly worse than in the interior. Although the overall accuracy remains second order, often the errors at the boundaries are the most important in terms of measurable quantities such as lift and drag. The following experiments focus strictly on improving the errors at the boundary.

Next we try to improve the results from linear reconstruction by using quadratic reconstruction in four ways, shown in Table 3. First, we use a second order linear reconstruction instead of first order in the cut cells. To do this, we compute a quadratic in the cut cells, but only use the linear part of it to predict the edge states. This makes it more compatible with the interior scheme, which also uses second order gradient information. Secondly, we do quadratic reconstruction in the cut cells, but at edges of regular cells, we *match* the form of the error in the neighboring reconstructed edge state. This means that the cells bordering the cut cells will not suffer the loss of accuracy from the changing stencil, nor lose the local truncation error cancellation property. To use a 1D example, a Taylor series approximation to the edge state is

$$u_{edge} = u_{centroid} + \frac{h}{2}u_x + \frac{h^2}{8}u_{xx} + O(h^3)$$

If u_x is computed to second order, the leading term in the error for the edge state using linear reconstruction is

$$\frac{h^2}{8}u_{xx}. \tag{10}$$

To match the error at edges between cut cells and full cells, the cut cell edge state is approximated using the full quadratic polynomial, and then an error term of the form (10) is explicitly added in (either u_{xx} or u_{yy} depending on the edge orientation. Note however that the coefficients in 10 do not vary with cut cell size). This approach explicitly forces the first full cell to retain locally third order truncation error, since when differencing the edge states the second order errors cancel. This comes however with the penalty of reducing the locally third order accuracy of the cut cells.

Thus, in the third column we use the full quadratic reconstruction at all edges of the cut cells. In the fourth column we use quadratic reconstruction in both the cut cells and their neighbors. In all cases we only use the midpoint rule for a single flux evaluation at the cell edges. We measure only the percent relative error along the boundary using (9). The ratio between the errors in successive approximations is in the adjacent column. In all cases the convergence rate of the entire flowfield continues to be second order, as in Table 2.

Although in all cases the magnitude of the error decreases a bit over that of Table 2, the rate of convergence still falls below second order. Thus, for the next test we improve on the results in Table 3 by using 2 point Gaussian quadrature to do the flux evaluations, along with quadratic reconstruction at all edges of the cut cells. This result is summarized in the first two columns of Table 4, showing first the error over the entire flow

h	Quad+2pt.at cut cells				Quad+2pt.for whole FF			
	FF Error	Ratio	Bndry.Er.	Ratio	FF Error	Ratio	Bndry.Er.	Ratio
$.053\bar{3}$	9.45(-2)		1.44(-1)		6.62(-2)		1.19(-1)	
$.026\bar{6}$	2.59(-2)	3.65	4.31(-2)	3.34	1.09(-2)	6.07	2.73(-2)	4.35
$.013\bar{3}$	6.53(-3)	3.96	1.39(-2)	3.1	1.59(-3)	6.86	6.77(-3)	4.03

Table 4: Quadratic reconstruction with 2 pt. Gauss quadrature at cut cell edges (first 2 columns) and over the whole flow field (last 2 columns).

h	FF Error	Ratio	Bndry. Er.	Ratio	FF Error	Ratio	Bndry. Er.	Ratio
$.053\bar{3}$	4.94(-2)		1.24(-1)		6.59(-2)		1.19(-1)	
$.026\bar{6}$	1.34(-2)	3.68	3.84(-2)	3.23	1.08(-2)	6.10	2.73(-2)	4.35
$.013\bar{3}$	3.98(-3)	3.37	1.57(-2)	2.95	1.56(-3)	6.9	6.70(-3)	4.07

Table 5: Flowfield and boundary errors using multiple line segments per cell to represent the boundary. The first half of the Table uses quadratic reconstruction at the boundary; the second half uses it over the entire flowfield along with 2 pt. Gauss quadrature.

field, then the boundary errors. For comparison, we also use quadratic reconstruction with two point Gaussian quadrature over the entire flow field; this is summarized in the last two columns. There is improvement both in the actual level of the error as well as the convergence rate. (We would have hoped however that this rate would be closer to third order).

The final experiment in this series looks at the error coming from the boundary representation itself. A piecewise linear segment in each cell incurs an error of $O(h^3)$ in the area of the cell. A logical next step is to try a higher order representation of the boundary, although for 3D calculations this might be impractical unless working directly with a NURBS representation of the surface. We test the hypothesis that a large portion of the error comes from the boundary representation by using a subcell resolution of the boundary. We use multiple line segments in each cell taken from a grid that is four times finer in each direction. Everything else is as above: quadratic reconstruction with two point Gaussian quadrature at all edges of the cut cells. Other cells use linear reconstruction with a midpoint rule for the flux evaluations. The results are in Table 5. (For comparison, the last columns use quadratic reconstruction for the whole flowfield, along with the subcell boundary resolution).

Again, the actual error magnitude decreases, but it is not showing a second order rate. Hypotheses for this include the mismatch between the boundary scheme and the interior scheme, perhaps the boundary layer of error is a growing function of h, perhaps this mesh spacing is not in the asymptotic regime, or possibly a bug.

5 Discussion

This study is not complete, since we have not yet definitely isolated the major sources of error. However we have reduced the magnitude of the error by a factor of 5. Mesh refinement would have required the mesh spacing to be reduced by more than a factor of 2 in each direction, leading to more than 4 times the work. More investigation is needed to understand the accuracy tradeoffs in finding the most effective method for efficiently computing flows for complex configurations. This will be especially important in three dimensions, when mesh refinement alone will not be an efficient way to improve accuracy (since a single cell division leads to 8 times the number of cells).

6 Acknowledgements

We thank Michael Aftosmis for many helpful conversations, and for suggesting the comparison with [2]. M. Berger was supported in part by AFOSR Grant 94-1-0132, and DOE Grant DE-FG02-88ER25053. Some of this work was performed at RIACS, whose support is gratefully acknowledged.

References

[1] Remi Abgrall. On essentially non-oscillatory schemes on unstructured meshes: analysis and implementation. *J. Comp. Phys.*, 114:45–58, September 1994.

[2] M. Aftosmis, D. Gaitonde, and T. Sean Tavares. On the accuracy, stability and monotonicity of various reconstruction algorithms for unstructured meshes. *AIAA-94-0415*, January 1994.

[3] Tim Barth. Recent developments in high order k-exact reconstruction on unstructured meshes. *AIAA-93-0668*, 1993.

[4] Tim Barth and Paul Frederickson. Higher order solution of the Euler equations on unstructured grids using quadratic reconstruction. *AIAA-90-0013*, 1990.

[5] M.J. Berger and R.J. LeVeque. An adaptive Cartesian mesh algorithm for the Euler equations in arbitrary geometries. *AIAA-89-1930*, June 1989. 9th Computational Fluid Dynamics Conf., Buffalo, NY.

[6] M.J. Berger and R.J. LeVeque. Cartesian meshes and adaptive mesh refinement for hyperbolic partial differential equations. In *Proc. Third Intl. Conf. Hyperbolic Problems*, 1990.

[7] D. K. Clarke, M. D. Salas, and H. Hassan. Euler calculations for multielement airfoils using Cartesian grids. *AIAA Journal*, 24, 1986.

[8] William J. Coirier and Kenneth G. Powell. An accuracy assessment of Cartesian-mesh approaches for the Euler equations. *AIAA-93-3335*, July 1993.

[9] Phil Colella and Harland Glaz. Efficient solution algorithms for the Riemann problem for real gases. *J. Comp. Phys.*, 59:264–289, June 1985.

[10] D. DeZeeuw and K. Powell. An adaptively refined Cartesian mesh solver for the Euler equations. *AIAA-91-1542*, June 1991.

[11] B. Epstein, A. Luntz, and A. Nachschon. Cartesian Euler method for arbitrary aircraft configurations. *AIAA Journal*, 30(3):679–687, March 1992.

[12] Foley, van Dam, Feiner, and Hughes. *Computer graphics, principles and practice.* Addison Wesley, 1990.

[13] R. Gaffney, H. Hassan, and M. Salas. Euler calculations for wings using Cartesian grids. *AIAA-87-0356*, January 1987.

[14] A. Jameson, W. Schmidt, and E. Turkel. Numerical solutions of the Euler equations by finite volume methods using Runge-Kutta time-stepping schemes. *AIAA-81-1259.*

[15] S. Karman Jr. Splitflow: a 3d unstructured Cartesian prismatic grid CFD code for complex geometries. Technical report, Lockheed Fort Worth Co., 1994. Preprint.

[16] H. O. Kreiss, T. A. Manteuffel, B. Swartz, B. Wendroff, and Jr. A. B. White. Supraconvergent schemes on irregular grids. *Math. Comp.*, 47, 1986.

[17] J. Melton, M. Berger, Michael Aftosmis, and Michael Wong. 3D applications of a Cartesian grid Euler method. *AIAA-95-0853*, January 1995.

[18] J. Melton, F. Enomoto, and M. Berger. 3D automatic Cartesian grid generation for Euler flows. *AIAA-93-3386*, July 1993.

[19] R. Pember, J. Bell, P. Colella, W. Crutchfield, and M. Welcome. An adaptive Cartesian grid method for unsteady compressible flow in complex geometries. Preprint UCRL-JC-115650, Lawrence Livermore National Laboratory, 1993.

[20] J.J. Quirk. An alternative to unstructured grids for computing gas dynamics flows around arbitrarily complex two dimensional bodies. Technical Report 92-7, ICASE, 1992.

[21] B. Wendroff and A.B. White. A supraconvergent scheme for nonlinear hyperbolic systems. *Comput. Math. Appl.*, 18, 1989.

[22] D. Young, R. Melvin, M. Bieterman, F. Johnson, S. Samant, and J. Bussoletti. A locally refined rectangular grid finite element method: application to computational fluid dynamics and computational physics. *J. Comp. Phys.*, January 1991.

Shock Capturing and Global Solutions To the Compressible Euler Equations with Geometrical Structure

Gui-Qiang Chen[†] James Glimm[‡]

Abstract We are concerned with global solutions and corresponding shock capturing approximations to the Euler equations of compressible gas dynamics with geometrical structure. Recent developments in this direction are reviewed and the role of the shock capturing approach in solving the Euler equations is discussed. Some efficient shock capturing methods as well as their convergence and consistency are analyzed to compute the corresponding compressible flows and to construct correct approximate solutions.

1. Introduction

The compressible Euler equations for isentropic gases consist of conservation laws of mass and momentum, which are of the following form:

$$(1) \qquad \begin{cases} \rho_t + \nabla \cdot (\rho \vec{u}) = 0, \\ (\rho \vec{u})_t + \nabla \cdot (\rho \vec{u} \otimes \vec{u}) + \nabla p = 0, \quad \vec{x} \in \boldsymbol{R}^n, \end{cases}$$

where $\rho, \vec{u}$, and p are the density, the velocity, and the pressure of the gas, respectively. For polytropic gas, $p(\rho) = \rho^\gamma/\gamma$, where the adiabatic exponent γ is restricted to the interval $(1, 5/3]$, as is usual for gases. Many physical geometric flows reduce (1) to the systems of compressible Euler equations with geometrical source terms. Such geometric flows include transonic nozzle flow, spherically symmetric flows, and cylindrically symmetric rotating flow.

One of main features of these systems of Euler equations is the singularity of solutions which develops in a finite time even starting from smooth initial data. This feature reflects the physical phenomenon of breaking of waves and the development of shock waves. For this reason, attention focuses on solutions in the space of

† Department of Mathematics, Northwestern University, 2033 Sheridan Road, Evanston, IL 60208, USA. *Email: gqchen@math.nwu.edu*

‡ Department of Applied Mathematics and Statistics, SUNY at Stony Brook, NY 11794, USA. *Email: glimm@ams.sunysb.edu*

1980 *Mathematics Subject Classification* (1985 *Revision*). Primary 35L65, 76N10; Secondary 76H05, 76U05, 65M12

Key words and phrases. Euler equations, global solutions, shock capturing scheme, plane flow, transonic nozzle flow, symmetric flow, rotating flow discontinuous functions, where one cannot directly use many powerful classical analytic techniques that predominate in the general theory of partial differential equations. Another main feature is the resonance phenomenon among different characteristic modes or between characteristic modes and the geometrical source mode for geometrical flows, which causes extra analytical difficulty.

To overcome these difficulties, one idea is to construct appropriate shock capturing approximate solutions to the Euler equations and then to study convergence and consistency of the approximate solutions to the compressible flows. The motivation of this idea arises naturally from continuum physics and numerical analysis. The system of gas dynamics generally involves viscosity terms, although the viscosity coefficient is very small and can be ignored. Numerical computations to the Euler equations are limited to calculations of difference equations and discrete Cauchy mesh data. Examples of this approach are the Lax-Friedrichs scheme [La], the Glimm scheme [Gl], the Godunov scheme [Go], the artificial viscosity method (cf. [Ho,Ge,Ne], and the fractional step techniques (e.g. [DC2]) for various problems (e.g. [Di, L4, Te, C1, CW]). In particular, shock capturing difference methods automatically provide corresponding approximate solutions that preserve the most important feature of hyperbolic systems—finiteness of propagation speeds, guaranteed by the Courant-Friedrichs-Lewy condition [CL], which allows one to avoid global assumptions on the initial data, especially at infinity, as the artificial viscosity usually does (cf. [Di]). Such an advantage is well recognized in using the Glimm scheme [Gl] as well as the Lax-Friedrichs scheme [La] and the Godunov scheme [Go]. For example, see [Ni, L4, Te, C1-2, CW].

In Sections 2-4 we discuss some recent developments in studying the compressible Euler equations and analyze how to design appropriate shock capturing difference methods in obtaining the global solutions to the compressible Euler equations with geometrical structure.

2. Plane Flow

The compressible plane flow is governed by the one-dimensional compressible Euler equations:

$$(2)\qquad \begin{cases} \rho_t + (\rho u)_x = 0, \\ (\rho u)_t + (\rho u^2 + p(\rho))_x = 0, \end{cases}$$

where the velocity u is now scalar.

The system (2) is equivalent to the p-system in Lagrange coordinates

$$(3)\qquad \begin{cases} v_t - u_x = 0, \\ u_t + p(v)_x = 0, \end{cases}$$

with $p(v) = k^2 v^{-\gamma}$ in the sense of weak solutions (see [Wa, C3]).

The Riemann problem is the simplest Cauchy problem. Its solutions serve as building blocks for general solutions (cf. [Gl]). It was first solved in Riemann's pioneering work [Ri] in 1860 with the aid of characteristic analysis. The Cauchy problem for the general case is much more difficult. For the exponent $\gamma = 1$ and the general Cauchy data with locally finite total variation, Nishida [Ni] established the first large-data existence theorem using the Glimm method. Some large-data theorems have also been obtained for general $\gamma > 1$ in the case where the initial data is restricted to prevent the development of cavities (e.g. [ZG,NS,DZ,Li]). The difficult point in bounding the total variation norm at low densities is that the coupling between characteristic fields increases as the density decreases. This difficulty is a reflection of the resonance between different characteristic modes at the vacuum, at which the strict hyperbolicity fails, i.e., the eigenvalues and eigenvectors coalesce on the line $\rho = 0$ in the phase space.

This analytical difficulty in using the Helly framework motivated the use of weak compactness methods in modern functional analysis, especially compensated compactness. With the aid of the compensated compactness method of Tartar and Murat [Ta,Mu], DiPerna [Di] first succeeded in solving the Cauchy problem for the case $\gamma = 1 + \frac{2}{2m+1}$, $m \geq 2$ integers, using the artificial viscosity method.

A compactness framework for the approximate solutions is established in [DC1] for $\gamma = 3/2$ and in [C1-2] for the usual case of gases $1 < \gamma \leq 5/3$. This compactness framework states that uniform boundedness of the approximate density and velocity and H^{-1} compactness of the corresponding weak entropy dissipation measures implies the compactness of the approximate solutions. The analysis is based on the Young measures uniquely determined by the approximate solutions and restricted by a Tartar-Murat commutativity relation for weak entropy pairs derived from the H^{-1} compactness condition for the weak entropy pairs and the continuity theorem for the 2×2 determinant in the weak topology. Specifically, each Young measure commutes with the 2×2 determinant mapping acting on weak entropy pairs. The basic motivation is, as first observed in [Ta,Di], that the commutativity relation represents an imbalance of regularity. The operator on the left is more regular than one on the right as a result of cancelation, which forces the Young measure to reduce to a point mass. Observe that the vanishing of the derivative of a Radon measure except at one point in the Lebesgue sense implies the point mass of the measure. The key strategy is to choose the weak entropy pairs whose leading term is coercive with respect to the 2×2 determinant and show that the coercive behavior guarantees that the derivative of the Young measure vanishes except at one point. The essential difficulty is that only a subspace of entropy pairs, weak entropy pairs, can be used in the Tartar-Murat relation.

To apply this framework to the existence proof of global solutions, one has to find appropriate shock capturing methods to construct corresponding approximate solutions satisfying the conditions of the framework. For the plane flow, the Lax-Friedrichs scheme and the Godunov scheme fit into this category and the convergence and consistency of these two difference schemes to the global weak solutions to the Euler equations of plane flow are proved in [C1-2,DC1]. As a corollary, an existence theorem of global solutions of the plane flow is established for arbitrarily L^∞ large initial data with nonnegative density.

For some recent efforts on the kinetic formulation to the plane flow, we also refer the reader to Lions-Perthame-Tadmor [LP].

3. Transonic Nozzle Flow

The problem of transonic nozzle flow through a variable-area duct has been studied extensively (cf. [EM, GL, GM, L1-L3, CF, Wh]). Such a flow is governed by the following nonlinear Euler equations with geometrical source terms:

$$\begin{cases} \rho_t + (\rho u)_x = -\frac{A'(x)}{A(x)}\rho u, \\ (\rho u)_t + (\rho u^2 + p(\rho))_x = -\frac{A'(x)}{A(x)}\rho u^2, \end{cases} \tag{4}$$

where $A(x) \geq c_0 > 0$ represents the cross-sectional area at x in a variable-area duct for transonic nozzle flow.

Liu [L1-L3] systematically studied the asymptotic behavior of global solutions for such a flow and established a global existence theorem by incorporating the steady-state building blocks with the random choice method [Gl], provided that the initial data have a sufficiently small total variation and are bounded away from both sonic and vacuum states. A generalized random choice method is introduced to compute transient gas flows in a Laval nozzle [GL, GM]. Motivated by the previous work, we introduce an appropriate shock capturing scheme to obtain L^∞ estimates and weak compactness estimates of the corresponding approximate solutions. Our method is to incorporate natural building blocks from the Riemann solutions and the steady-state solutions with the Godunov scheme. Such estimates lead to the convergence of the approximate solutions using a compensated compactness framework (cf. [C1-2]) and to an existence theory of global entropy weak solutions for any measurable initial data function of large amplitude. The central difficulty to realize this strategy for such a flow lies in the associated steady state equations, which changes type from elliptic to hyperbolic at the sonic point; such steady state solutions are fundamental building blocks in this strategy and in early work on nozzle flow.

In [CG] we develop first order Godunov shock capturing schemes, with piecewise constant building blocks replaced by piecewise steady ones. The main point is to use the steady-state solutions, which incorporate geometrical source terms, to modify the wave strengths in the Riemann solutions. This construction yields better approximate solutions, and permits us to trace some important bounds, especially the uniform L^∞ bound, in contrast to the traditional fractional step building blocks for this case (cf. [C4, DC2]). There are two technical difficulties which we overcome to achieve this goal, both due to transonic phenomena. One is that no smooth steady-state solution exists in each cell in general. This problem is easily solved by introducing a standing shock at the center of the cell. The other is that the constructed steady-state solution in each cell must satisfy the following requirements:

(a) The oscillation of the steady-state solution around the Godunov value must be of the same order as the cell length to obtain the L^∞ estimate for the convergence arguments;

(b) The difference between the average of the steady-state solution over each cell and the Godunov value must be higher than first order in the cell length to ensure the consistency of the corresponding approximate solutions with the Euler equations.

These requirements are naturally satisfied by smooth steady-state solutions that are bounded away from the sonic state in the cell. The general case must include the transonic steady-state solutions. The sonic difficulty is overcome, as in experimental physics, by introducing an additional standing shock with continuous mass and by adjusting its left state and right state in the density and its location to control the growth of the density. These requirements also enable us to make H^{-1} compactness estimates for corresponding entropy dissipation measures to deduce the strong convergence of the approximate solutions with the aid of the compactness framework, discussed in Section 2 (also see [C1-2]).

4. Spherically Symmetric Flows

The spherically symmetric flows are governed by the Euler equations of the same form as (4) with surface area $A(x)$ of the sphere of radius $x = |\vec{x}|$. For example, $A(x) = 2\pi x$ for cylindrically symmetric flow and $A(x) = 4\pi x^2$ for three dimensional spherically symmetric flow. The essential difference of the spherically symmetric flows from the transonic nozzle flow is the unboundedness of the function

$A(x)$ as the waves move outward and the nonpositive lower boundedness of $A(x)$ as the waves move inward.

Such flows arise in many important physical problems, such as flow in a jet engine inlet manifold, stellar dynamics including supernovae formation, inertial confinement fusion, and underwater explosions. The similarity solutions of such problems have been discussed and reviewed in many books (cf. [CF, Wh]), which can be described by systems of ordinary differential equations. A global entropy weak solution with spherical symmetry has been constructed in [MU] for the isothermal case $\gamma = 1$ and the local existence of such a weak solution for general case $1 < \gamma \leq 5/3$ has also been discussed in [MT] in the region away from the origin.

Some evidence indicates that such global solutions blow up near the origin at some time under certain circumstance. In [C4], a framework is developed to ensure the existence of L^∞ spherically symmetric weak solutions in the large of the compressible Euler equations for a class of L^∞ Cauchy data of arbitrarily large amplitude, which model outgoing blast waves and large-time asymptotic solutions. The shock capturing method developed in [C4] is the first order fractional Godunov scheme, as used in [DC2], to capture outgoing blast waves and to yield L^∞ bound in a natural manner from the class of L^∞ initial data. A detailed analysis near the origin leads to H^{-1} compactness of the corresponding entropy dissipation measures to derive the convergence of the approximate solutions with the aid of the compactness framework discussed in Section 2. Large-time asymptotic decay of the global weak solutions is also discussed.

The central difficulty for general L^∞ Cauchy data in the unbounded domain is the reflection of waves from infinity and their strengthening as they move radially inward. In [CG] we consider the initial-boundary problem for the spherically symmetric flows outside a solid ball to understand the L^∞ stability of such a flow in the unbounded domain. We use the same strategy, as the one for the transonic nozzle flows, to develop an appropriate shock capturing scheme to construct approximate solutions for the initial-boundary problem. Then a similar careful analysis leads to the L^∞ boundedness of approximate solutions and the H^{-1} compactness of the corresponding entropy dissipation measure.

With the aid of the compactness framework discussed in Section 2, it is proved that the approximate solutions strongly converge to a global entropy weak solution of the spherically symmetric flows for arbitrary large measurable initial data.

5. Cylindrically Symmetric Rotating Flow

Another important physical geometric flow is cylindrically symmetric rotating flow outside a solid ball, for example, $|\vec{x}| \geq 1$:

$$
\begin{cases} \rho(\vec{x},t) = \rho(|\vec{x}|,t), \\ \vec{m}(\vec{x},t) = m(|\vec{x}|,t)\frac{\vec{x}}{|\vec{x}|} + \tilde{m}(|\vec{x}|,t)\frac{(-x_2,x_1)}{|\vec{x}|}. \end{cases} \tag{5}
$$

Then $(\rho(x,t), m(x,t), \tilde{m}(x,t)), x = |\vec{x}|$, are governed by the following system:

$$
\begin{cases} \rho_t + m_x = -\frac{1}{x}m, \\ m_t + (\frac{m^2}{\rho} + p(\rho))_x = -\frac{1}{x}\frac{m^2-\tilde{m}^2}{\rho}, \\ \tilde{m}_t + (\frac{m\tilde{m}}{\rho})_x = -\frac{2}{x}\frac{m\tilde{m}}{\rho}, \end{cases} \tag{6}
$$

subject to the initial-boundary condition:

$$
\begin{cases} (\rho, m, \tilde{m})|_{t=0} = (\rho_0(x), m_0(x), \tilde{m}_0(x)), \\ m|_{x=1} = 0. \end{cases} \tag{7}
$$

One can check that two characteristic fields of system (6) are genuinely nonlinear, and the other characteristic field is linearly degenerate.

The central difficulty for general L^∞ Cauchy data for the initial-boundary problem (6)-(7) in the unbounded domain is the reflection of waves from infinity, their strengthening as they move radially inward, and the interaction between linear and nonlinear waves. In [CG] we also consider such a flow to understand the L^∞ stability in the unbounded domain. We combine the approach for the spherically symmetric flows with the fractional step techniques as in [DC2] to develop a convenient shock capturing scheme to construct approximate solutions $(\rho^h, m^h, \tilde{m}^h)$ for this problem.

The method for the convergence and consistency analysis here is to combine our present approach for the spherically symmetric flows with the method of quasidecoupling (cf. [C3]) due to the linear degeneracy of the extra field generated from the angular momentum and the fractional step techniques as in [C4, DC2]. With the aid of the compactness framework discussed in Section 2 and the approach we use for the spherically symmetric flows, we first obtain the convergence of the approximate sequence of $(\rho^h(x,t), m^h(x,t))$ to $(\rho(x,t), m(x,t))$ in strong topology. To show the convergence of $\tilde{m}^h(x,t)$, we apply the method of quasidecoupling (cf. [C3]) to deduce that the family of Young measures $\mu_{x,t}$, determined by the approximate solutions $(\rho^h(x,t), m^h(x,t), \tilde{m}^h(x,t))$, is a family of Dirac masses for $(x,t) \in (1,\infty) \times (0,\infty) - \{\rho(x,t) = 0\}$ as long as the corresponding family of initial Young measures $\mu_{x,0}$ has such a property. The consistency is proved by using similar arguments in [C4, DC2] for the fractional step schemes and the techniques developed in [CG].

Similarly, one can obtain global entropy solutions to the systems that describes a spherically symmetric rotating flow surrounding a solid star with mass M.

Finally, we remark that the approaches and ideas developed here can be applied to more general systems, especially the compressible Euler equations with a more general equation of state.

Acknowledgments: Gui-Qiang Chen was supported in part by the Office of Naval Research grant N00014-91-J-1384, by the National Science Foundation grants DMS-9207080, and by an Alfred P. Sloan Foundation Fellowship. James Glimm was supported in part by the Applied Mathematics Subprogram of the U.S. Department of Energy DE-FG02-90ER25084, by the Army Research Office grant DAAL03-92-G-0185 through the Mathematical Sciences Institute of Cornell University under subcontract to the University at Stony Brook, ARO contract number DAAL03-91-C-0027, and by the National Science Foundation grant DMS-9201581.

References

[C1] Chen, G.-Q., *Convergence of the Lax-Friedrichs scheme for isentropic gas dynamics (III)*, Acta Math. Sci. **6**, 75-120(1986); **8**, 243-276 (1988) (in Chinese).

[C2] Chen, G.-Q., *The compensated compactness method and the system of isentropic gas dynamics*, MSRI Preprint 00527-91 (1990).

[C3] Chen, G.-Q., *The method of quasidecoupling to conservation laws*, Arch. Rat. Mech. Anal. **121** (1992), 131-185.

[C4] Chen, G.-Q., *Remarks on spherically symmetric solutions to the compressible Euler equations*, Preprint (1993).

[CG] Chen, G.-Q. & Glimm, J., *Global solutions to the compressible Euler equations with geometrical structure*, Preprint (1994).

[CW] Chen, G.-Q. & Wagner, D., *Large time, weak solutions to reacting Euler equations,*, In: *Nonlinear Hyperbolic Problems: Theoretical, Applied, and Computational Aspects*, edited by A. Donato and F. Oliveri, *Notes on Numerical Fluid Mechanics*, Vol. 43, 144-149, Vieweg, Braunschweig 1993.

[CF] Courant, R. & Friedrichs, K. O., *Supersonic flow and shock waves*, Springer-Verlag, New York, 1948.

[CL] Courant, R., Friedrichs, K. O., & Lewy, H., *Uber die partiellen Differenzengleichungen der mathematisches Physik,*, Math. Ann. **100** (1928), 32-74.

[Da] Dafermos, C., *Polygonal approximations of solutions of the initial-value problem for a conservation law*, J. Math. Anal. Appl. **38** (1972), 33-41.

[DC1] Ding, X., Chen, G.-Q., & Luo, P., *Convergence of the Lax-Friedrichs scheme for the isentropic gas dynamics*, Acta Math. Sci. **5**, 483-500,501-540; **7**, 467-480; **8**, 61-94 (in Chinese).

[DC2] Ding, X., Chen, G.-Q., & Luo, P., *Convergence of the fractional step Lax-Friedrichs scheme and Godunov scheme for the isentropic system of gas dynamics*, Commun. Math. Phys. **121** (1989), 63-84.

[DZ] Ding Shia-Shi, Zhang-Tung, Wang Ching-Hua, Hsiao-Ling, & Li Tsai-Chung, *A study of the global solutions for quasilinear hyperbolic systems of conservation laws*, Scientica Sinica **16** (1973), 317-335.

[Di] DiPerna, R., *Convergence of the viscosity method for isentropic gas dynamics*, Commun. Math. Phys. **91** (1983), 1-30.

[EM] Embid, P., Goodman, J., & Majda, A., *Multiple steady state for 1-D transonic flow*, SIAM J. Sci. Stat. Comp. **5** (1984), 21-41.

[Ge] GEL'FAND, L., *Some problems in the theory of quasilinear equations*, Usp. Mat. Nauk. **14** (1959), 87-158.

[GL] GLAZ, H. & LIU, T.-P., *The asymptotic analysis of wave interactions and numerical calculation of transonic nozzle flow*, Adv. Appl. Math. **5** (1984), 111-146.

[Gl] GLIMM, J., *Solutions in the large for nonlinear hyperbolic systems of equations*, Commun. Math. Phys. **18** (1965), 697-715.

[GP] GLIMM, J., MARSHALL, G., & PLOHR, B., *A generalized Riemann problem for quasi-one-dimensional gas flow*, Adv. Appl. Math. **5** (1984), 1–30.

[GM] GLIMM, J. & MAJDA, A., *Multidimensional Hyperbolic Problems and Computations*, IMA Volume **29**, Springer-Verlag, New York, 1991.

[Go] GODUNOV, S., *A difference method for numerical calculation of discontinuous solutions of the equations of hydrodynamics*, Mat. Sb. **47(89)** (1959), 271-360.

[Hf] HOFF, D., *Spherically symmetric solutions of the Navier-Stokes equations for compressible, isothermal flow with large, discontinuous initial data*, Indiana U. Math. J. **41** (1992), 1225-1302.

[Ho] HOPF, E., *The partial differential equations $u_t + uu_x = \mu u_{xx}$*, Comm. Pure Appl. Math. **3** (1950), 201-230.

[La1] LAX, P. D., *Weak solutions of nonlinear hyperbolic equations and their numerical computation*, Comm. Pure Appl. Math. **7** (1954), 159-193.

[La2] LAX, P. D., *Hyperbolic Systems of Conservation Laws and the Mathematical Theory of Shock Waves*, CBMS. 11, SIAM, 1973.

[Li] LIN, L.-W., *Vacuum states and equidistribution of the random sequence for Glimm's scheme*, J. Math. Anal. Appl. **124** (1987), 117-126.

[LP] LIONS, P. L., PERTHAME, B., & TADMOR, E., *Kinetic formulation of the isentropic gas dynamics and p-systems*, J. Amer. Math. Soc. **7** (1994), 169-192.

[L1] LIU, T.-P., *Quasilinear hyperbolic system*, Commun. Math. Phys. **68** (1979), 141–172.

[L2] LIU, T.-P., *Nonlinear stability and instability of transonic gas flow through a nozzle*, Commun. Math. Phys. **83** (1983), 243-260.

[L3] LIU, T.-P., *Nonlinear resonance for quasilinear hyperbolic equation*, J. Math. Phys. **28** (1987), 2593–2602.

[L4] LIU, T.-P., *Solutions in the large for the equations of non-isentropic gas dynamics*, Ind. Univ. Math. J. **26** (1977), 147-177.

[MU] MAKINO, T., MIZOHATA, K., & UKAI, S., *Global weak solutions of the compressible Euler equations with spherical symmetry I, II*, Japan J. Industrial Appl. Math. **9** (1992), 431-449.

[MT] MAKINO, T. & TAKENO, S., *Initial-Boundary Value Problem for the spherically symmetric motion of isentropic gas*, Preprint (1993).

[M] MORAWETZ, C. S., *On a weak solution for a transonic flow problem*, Comm. Pure Appl. Math. **38** (1985), 797-818.

[Mu] MURAT, F., *L'injection du cone positif de H^{-1} dans $W^{-1,q}$ est compacte pour tout $q < 2$*, J. Math. Pures Appl. **60** (1981), 309-322.

[Ne] VON NEUMANN, J., *Theory of Shock Waves*, Collected Works, Vol. 5-6, Pergamon Press: Oxford, 1963.

[Ni] NISHIDA, T., *Global solution for an initial-boundary-value problem of a quasilinear hyperbolic system*, Proc. Jap. Acad. **44** (1968), 642-646.

[NS] NISHIDA, T.& SMOLLER, J., *Solutions in the large for some nonlinear hyperbolic conservation laws*, Comm. Pure Appl. Math. **26** (1973), 183-200.

[Sl] SLEMROD, M., *Resolution of the spherical piston problem for compressible isentropic gas dynamics via a self-similar viscous limit*, Preprint, Wisconsin-Madison, 1994.

[Ta] TARTAR, L., *Compensated compactness and applications to partial differential equations*, Research Notes in Mathematics, Nonlinear Analysis and Mechanics, ed. R.J. Knops, vol. 4, Pitman Press, New York, 1979.

[Wa] WAGNER, D. H., *Equivalence of Euler and Lagrangian equations of gas dynamics for weak solutions*, J. Diff. Eqs. **68** (1987), 118-136.

[Wh] WHITHAM, G. B., *Linear and Nonlinear Waves*, John Wiley & Sons, New York, 1974.

[ZG] ZHANG TONG & GUO YU-FA, *A class of initial-value problem for system of aerodynamic equations*, Acta Math. Sinica **15** (1965), 386-396.

Theory of Conservation Laws in China

Xiaxi Ding[1]

Since 1953, just 4 years after the foundation of P. R. China, we began research work on partial differential equations. The main characteristic of P.D.E. investigation in China is its close relationship with practical problems. People searched for problems from industrial construction, new technologies, meanwhile keeping an eye open for international development trends.

We began with the theory of nonlinear hyperbolic conservation laws at the end of the 1950's. We met the system of Saint Vennat equations in the investigation of bores in the Yangtze River, and the system of gas dynamics equations in the explosion of atomic bombs. We needed to treat solutions with discontinuities. At that time many Chinese universities selected talented youths to participate in the groups working on these topics. We organized a group of scholars to make some observations on the bores of the Yangtze River. From then on, teachers in many mathematical departments started to teach hyperbolic conservation laws. At the same time many Chinese students, e.g., C. Gu, and Y. Chow and others come back from Russia, and introduced the newest results on conservation laws. Hence we touched the early work of E. Hopf, O. Oleinik, P. D. Lax, etc. In particular we took Hopf's paper (1950) as the most fundamental reference and discussed it again and again in seminars for several months. After that, we gave lectures to students of the 1963 class of mathematics at USTC (Univ. of Science and Technology of China), on the most fundamental ideas and knowledge of hyperbolic conservation laws. That class included the students L. Hsiao, C. Wang, C. Li and others. Afterwards, we also read "Supersonic flow and Shock Waves", the famous book of Courant and Friedrichs.

In the following we shall talk about some work of Chinese scholars on conservation laws theory.

Firstly I need to declare that in this lecture I would like to focus on the global theory of discontinuous solutions of conservation laws. Much important work of C. Gu and T. Li on local existence theorems of discontinuous solutions, and that of L. Lin and T. Li on global smooth solutions, will not be discussed in detail here.

---

[1]Inst. of Math., Shantou Univ., Shantou, Guangdong 515063, P. R. C. Telephone: 0086-754-8510000 ext 33040 (Shantou), 01-2561754 (Beijing). Fax: 0086-754-8510505. Email: `dxx%stumis@hkucnt.hku.hk`

I. Scalar Conservation Laws

It is well-known that E. Hopf [1], O. A. Oleinik [2] and P. Lax [3] considered global discontinuous solutions of the Cauchy problem for scalar conservation laws in 1950's,

$$\begin{cases} u_t + f(u)_x = 0 & (x,t) \in R \times R^+, \\ u(x,0) = u_0(x) & x \in R, \end{cases} \tag{1.1}$$

where the flux function $f(u)$ is assumed to be convex. Both existence and uniqueness have been solved completely. In 1963, Z. Wu [5] proved the existence of global solution of (1.1) for nonconvex $f(u)$ under the assumption that $u_0(x)$ is monotone. He used the Douglis framework and successfully extended the Douglis ordering principle from the convex case to the nonconvex one.

In 1965, L. Lin [7] extended Wu's results to nonmonotone initial data and obtained the first complete existence theorem for a Cauchy problem of nonconvex conservation laws. Lin's paper was submitted to the journal Acta Mathematica Sinica. Because of the so-called Cultural Revolution in China, it was not published until 1979. An analogous result was obtained by D. Ballou [8] in 1970.

As suggested by R. Wang, Z. Wu [9](1963) introduced the concept of generalized characteristics and then proved the uniqueness of solutions to the Cauchy problem in the class of piecewise smooth functions for nonconvex scalar conservation laws.

II. Systems of Conservation Laws (A)
The work before 1979

In 1860, B. Riemann [10] considered the Cauchy problem with the simplest discontinuous data for isentropic flow as follows

$$\begin{cases} u_t + p(v)_x = 0 \\ v_t - u_x = 0 \end{cases} \quad (x,t) \in R \times R^+ \tag{2.1}$$

$$(u,v)(x,0) = \begin{cases} (u^-, v^-) & x < 0 \\ (u^+, v^+) & x > 0 \end{cases} \tag{2.2}$$

where the constitutive function satisfies $p'(v) < 0$ (hyperbolicity) and $p''(v) > 0$ (convexity).

Riemann's result was extended in several aspects in China, as follows:

1. A group of students, directed by S. Ou considered initial data possessing two jumps in their bachelor thesis [12]. This problem had not been solved until that time.

2. Based on the work of Ou and others, the Cauchy problem (2.1), (2.2) with a class of initial data which only produced forward rarefaction and backward compressive waves was solved by T. Zhang and Y. Guo [13] in 1963. This is perhaps the first global result involving discontinuities of systems other than that in the Riemann problem. J. Smoller and J. Johnson [14] extended this result to some general 2×2 systems. V. Liapijefskee [15] proved the uniqueness.

3. In 1973, using Glimm's scheme, X. Ding and others proved a global existence theorem for (2.1) and (2.2) with more general initial data, overtaking of shocks and rarefaction waves of the same type, and includes the initial values used by Zhang and Guo as special cases. The overtaking of shock and rarefaction waves of the same kind was previously studied by C. Moler and J. Smoller [17] and J. Greenberg [18], but we did not know their work at that time.

4. Directed by T. Zhang, C. Li and others (1963) extended Riemann's result to the nonconvex case in their Bachelor's thesis [25]. Analogous results to [25] were obtained by B. Wendroff [27] and L. Leboriech (1974) independently. The entropy condition for a general nonconvex system was suggested by T. P. Liu(1974) [30]. L. Hsiao and T. Zhang (1978)[31] extended [25] to a more general nonconvex 2×2 system under suitable hyperbolicity assumptions independently.

Zhang, Li and Hsiao (1975)[32] studied the interaction of waves for the nonconvex case and successfully dropped the convexity assumption needed for Zhang and Guo's result.

5. The Riemann problem for one-dimensional adiabatic flow

$$\begin{cases} u_t + p_x = 0 \\ v_t - u_x = 0 \\ E_t + (pu)_x = 0 \end{cases} \tag{2.5}$$

where $E = e + \frac{1}{2}u^2$, $e = \frac{pv}{\gamma - 1}$, $\gamma > 1$, was solved by Courant and Friedrichs [36]. Moreover, they obtained fundamental results on interaction of elementary waves, to which some supplements were contributed by Zhang and Hsiao [37] in 1979. Independent of B. Wendroff [29] (1972) and T. P. Liu [39] (1975), Zhang and Hsiao [38] (1979) investigated the Riemann problem for the nonconvex adiabatic case (2.5).

III. Systems of Conservation Laws (B) Some work of chinese scholars visiting in U.S.A.

Since 1979, the situation in China has changed a lot. Some Chinese mathematicians have had a chance to visit the west and to do cooperative work with westerners. Many new results have been obtained. We will mention some of them.

1. For guaranteeing uniqueness, the discontinuous solution has to satisfy some entropy criterion. There are several entropy criteria and one of them is the so-called entropy rate admissibility criterion suggested by C. Dafermos [42] (1979). L. Hsiao [43](1980) proved that this criterion is equivalent to others when $\gamma \geq \frac{5}{3}$(monotonic gas), but not equivalent when $\gamma < \frac{5}{3}$ and the shocks are very strong.

2. L. Ying and J. Wang [44](1988), proved the existence of a global solution to the Cauchy problem for system (2.1) with $p(v) = \frac{c}{v}$ and with an inhomogeneous term on which only a rather simple restriction is assumed. A generalization of Glimm's method is proposed in the proof. This is one of the earliest results for inhomogeneous systems of conservation laws. Other early works are due to T. P. Liu [46](1979), and C. Dafermos and L. Hsiao [47] (1982).

3. J. Wang and C. Li [48](1982), proved the Riemann solution of polytropic gas is globally stable under a small perturbation of the initial data.

4. T. P. Liu and J. Wang [49](1985), proved the existence of global solutions for

$$\begin{cases} u_t + (\phi u)_x = 0 \\ v_t + (\phi v)_x = 0 \end{cases} \tag{3.1}$$

$$(u, v)(x, 0) = (u_0(x), v_0(x)) \tag{3.2}$$

where $\phi = \phi(u, v)$. More interesting is that they clarified some of the behavior of solutions. This is early work on nonstrictly hyperbolic systems.

5. C. Li and T. P. Liu [51] (1983), studied the asymptotic states for hyperbolic conservation laws with a moving source. The scalar model and transonic gas flow through a duct with varying cross section is studied. For the latter case, it is shown that stability of a wave pattern depends on the geometry of the duct but not on the equation of the state.

6. Chinese scholars have done a lot of work on global smooth solutions. One of the most interesting is L. Lin's contribution to the vacuum problem. Lin(1987) [23] showed that vacuum states cannot appear in rarefaction solutions of (3.1) unless the vacuum is present at time $t = 0+$.

IV. Systems of Conservation Laws (C) Work of the past ten years.

1. Compensated Compactness and Hyperbolic Conservation Laws.

It is well known that Sobolev spaces are successfully used in various topics of PDE excluding the theory of discontinuous solutions of hyperbolic conservation laws, because discontinuous solutions do not have generalized derivatives in the sense of Sobolev. This situation has changed since the introduction of compensated compactness theory in the early 1980's [52][53] by L. Tartar and R. Diperna. In this theory, a Sobolev space of negative order $H^{-1}(\Omega)$ is used. Using compensated compactness theory, X. Ding, G. Chen and P. Luo [54-6] (1987-89), proved the convergence of the Lax–Friedrichs and Gudunov difference schemes, and then obtained a global existence theorem of the Cauchy problem for the equations of isentropic flow of a polytropic gas with adiabatic exponent $\gamma \in (1, \frac{5}{3}]$. The special case $\gamma = 1 + \frac{2}{2m-1}$, $m(> 2)$ being an integer, was obtained by R. Diperna [53] (1983), using the viscosity method.

G. Chen and Y. Lu [57] (1988), proved the existence theorem for a scalar conservation law by compensated compactness methods without using measure theory.

Y. Lu also studied the global existence theorem and the interactions of elementary waves for the system

$$\begin{cases} u_t + \frac{1}{2}(3u^2 + v^2)_x = 0 \\ v_t + (uv)_x = 0 \end{cases}$$

2. Combustion Theory

L. Ying and Z. Teng systematically studied the Riemann problem, Cauchy problem and the asymptotic behavior of solutions for a simple model equation of combustion theory.

T. Zhang, C. Wang and Y. Lu have made some generalizations and supplements to the theory. In addition, T. Zhang studied the Riemann problem of combustion theory for gas dynamics equations.

3. 2-dimensional Riemann Problem

Recently T. Zhang and his students have developed a systematic theory for 2-dimensional Riemann problems. For 2-dimensional isentropic gas dynamics, they made a serious of conjectures, but these are difficult to prove. Instead, they study the model system

$$\begin{cases} u_t + (u^2)_x + (uv)_y = 0 \\ v_t + (uv)_x + (v^2)_y = 0. \end{cases}$$

The solutions of this system reveal a variety of geometric structures: 57 exact solutions and the corresponding criteria have been obtained. T. Zhang will explain them in detail.

4. δ–wave Solution

T. Zhang and his students [58] find so-called δ–wave solutions when they study the 2-dimensional Riemann problem. We explain it for the simplest system

$$u_t + uu_x = 0 \tag{4.1}$$

$$v_t + (uv)_x = 0. \tag{4.2}$$

We introduce the potential $\omega(x,t)$ of (4.2):

$$\omega(x,t) = \oint_{(0,0)}^{(x,t)} vdx - uvdt.$$

Then we have

$$\omega_x = v, \quad \omega_t = -uv,$$

and

$$\omega_t + u\omega_x = 0. \tag{4.3}$$

Now

$$\omega(x,0) = \int_0^x v(x,0)dx.$$

In the case of a Riemann problem for (4.1), (4.2), it's easy to construct the solution (u,ω) of (4.1), (4.3) by the characteristic method. In general $\omega(x,t)$ is discontinuous along the discontinuity of $u(x,t)$, so the solution $v = \omega_x$ of (4.2) is not a function but a δ–measure.

We say (u,ω) is a generalized solution of the system (4.1), (4.3), equivalently (u,ω_x) is a generalized solution of (4.1), (4.2) if

$$\iint (u\varphi_t + \frac{u^2}{2}\varphi_x)dxdt = 0, \quad \text{and}$$

$$\iint \omega\psi_t dx dt - \iint u\psi d\omega(x,t)dt = 0.$$

for all $\varphi, \psi \in C_0^\infty(R_+^2)$, where the last integral is in the sense of Lebesgue-Stieltjes.

With this definition, we solved the Cauchy problem (4.1), (4.2) completely [60].

References

[1] Hopf, E. (1950), The partial differential equation $u_t + uu_x = \mu u_{xx}$. Comm. Pure Appl. Math. 3:201-230.

[2] Oleinik, O. A. (1957), Discontinuous solution of nonlinear differential equations.Usp. Mat. Nauk 12:3-73. English Trans. in Amer. Soc. Transl. Ser. 2, 26:95-172.

[3] Lax, P. (1954), Weak solutions of nonlinear hyperbolic equations and their numerical computation. Comm. Pure Appl. Math.7:159-193.

[4] Oleinik, O. A. and Kalashnikov, A. C. (1960), A class of discontinuous solutions for first order quasilinear equations. Proc. Conf. Diff.Equat .133-137, Jerevon (in Russian).

[5] Wu, Zhuoqun (1963), Existence and uniqueness of generalized solution of Cauchy problem for first order quasilinear differential equation without convexity condition, Acta Math. Sinica, 13:515-530(in Chinese).

[6] Douglis, A. (1959), An ordering principle and generalized solutions of certain quasi-linear partial differential equations, Comm. Pure Appl .Math., 12:87-112.

[7] Lin, Longwei (1979), The global solutions of Cauchy problem for quasi -linear hyperbolic equation without convexity condition, Acta Jilin Univ. .2:17-26. (in Chinese with English abstract).

[8] Ballou, D. (1970), Solution to nonlinear hyperbolic Cauchy problems without convexity conditions, Tran. AMS. 152:441-460.

[9] Wu, Zhouqun (1963), The ordinary differential equations with discontinuous rigth-hand members and the discontinuous solutions of the quasilinear partial differential equations. Acta Math. Sinica 13:515-530. English trans. in Sciential Sinica 13:1901-1907.

[10] Riemann, B. (1896), Gesammelte Werke.

[11] Lax, P. (1957), Hyperbolic system of conservation laws II. Comm. Pure Appl. Math. 10:537-566.

[12] Zhang, Chenjun; Sun, Geng; Xiong, Jiwu; Liao, Yingxin; Jia, Shufeng and Yang, Tongzu (1963), Initial value problem with three constant states as initial data for an aerodynamic system. Bachelor thesis of the Chinese Science and Technology Univ.

[13] Chang, Tong and Guo, Yufa (1965), A class of initial value problems for system of aerodynamic equations. Acta Math. Sinica 15:386 -396; English trans. in Chinese Math. 7(1965):90-101.

[14] Smoller, J. and Johnson, J. (1969), Global solutions for an extended class of hyperbolic system of conservation laws. Arch. Rat. Mech. Anal. 32:160-189.

[15] Liapijefskee, V. (1974), On uniqueness of generalized solution for aerodynamic equations. Dokl. Akad. Nauk. SSSR 215:535-538(in Russian).

[16] Ding, Xiaxi; Chang, Tong; Wang, Chinghua; Hsiao, Ling and Li, Caizhong (1973), A study of the global solutions for a quasilinear hyperbolic system of conservation laws. Scientia Sinica 16:317-335.

[17] Moler, C. and Smoller, J. (1970), Elementary interaction in quasilinear hyperbolic systems. Arch. Rat. Mech. Anal. 37:309-322.

[18] Greenberg, J. (1970), On the interaction of shocks and simple waves of the same family. Arch. Rat. Mech. Anal. 37:136-160.

[19] Gu, Chaohao et. al. (1962), A global solution without shocks for quasilinear hyperbolic system of equations. Math. Anthology of Fudan Univ. 36-39(in chinese).

[20] Lin, Longwei (1963), On the global existence of the continuous of the reducible quasilinear hyperbolic systems. Acta Jilim Univ. 4:83-96.

[21] Lax, P. D. (1964), Development of singularities of solutions of linear hyperbolic partial differential equations. J. Math. Phys. 5:611 -613.

[22] Yamaguti, M. and Nishida, T. (1968), On some global solutions for quasilinear hyperbolic equations. Funkcialaj Ekvacioj 11:51-57.

[23] Lin, Longwei (1987), On the vacuum state for the equations of isentropic gas dynamics. J. Math. Anal. Appl. 121:406-425.

[24] Gelfand, I. (1959), some problems in the theory of quasilinear equations, Usp. Mat. Nauk. 14:87-158(in Russian).

[25] Li, Caizhong; Hsiao, Ling; Yang, Shaoqi and Yuan, Zuwen (1963), Riemann problem for typical quasilinear hyperbolic system without convexity. Bachelor thesis of the Chinese Science and Technology Univ.

[26] Chang, Tong and Hsiao, Ling (1977), Riemann problem and discontinuous initial-value problem for typical quasilinear hyperbolic system without convexity. Acta Math. Sinica 20:229-231 (in Chinese).

[27] Gu, Chaohao; Li, Daqian; Yu, Wenci and Hou, Zongyi (1961–1962), Discontinuous initial-value problem for hyperbolic system of quasilinear equations (I), (II), (III). Acta Math. Sinica 11. 314- 323 ;324-337;12:132-143 (in Chinese).

[28] Hsiao, Ling and Zhang, Tong (1981), Perturbation of the Riemann problem in gas dynamics. J. Math Anal. Appl. 79:436-460.

[29] Wendroff, B. (1972), The Riemann problem for materials with nonconvex equations of state I.Isentropic Flow. II. General flow; J. Math. Anal. Appl. 38:454-466; 640-658.

[30] Liu, T. P. (1974), The Riemann problem for general 2×2 nonservation laws.Trans. AMS. 199:89-112.

[31] Hsiao, Ling and Zhang, Tong (1978), Riemann problem for 2×2 quasilinear hyperbolic system without convexity. Kexue Tongbao 8: 465–469 (in Chinese).

[32] Chang, Tong; Li, Caizhong and Hsiao, Ling (1975), Global solution for a class of initial-value problem of a typical quasilinear hyperbolic system without convexity. Kexue Tongbao, 20: 506-510 (in Chinese).

[33] Wu, Zhouqun (1963), Uniqueness for discontinuous solution with centered rarefaction waves for aerodynamic equations. Acta Jilin Univ. 4 :35-49 (in Chinese with an English abstract).

[34] Oleinik, O. A. (1957), Uniqueness of generalized solutions of Cauchy problem for a quasilinear system of equations appearing in mechanics. Usp. Mat. Nauk. 11:169-176. (in Russian).

[35] Wang, Rouhwai and Wu, Zhuoqun (1963), On mixed initial boundary value problem for quasilinear hyerbolic equations in two independent variables. Acta Jilin Univ. 2:459- 502 (in Chinese with an English abstract).

[36] Courant, R. and Friedrichs, K. O. (1948), Supersonic Flow and Shock Waves. Wiley-Interscience, New York.

[37] Hsiao, Ling and Zhang, Tong (1979), Interaction of elementary waves in one-dimensional adiabatic flow. Acta Math. Sinica 22:596-619 (in Chinese with an English abstract).

[38] Chang, Tung and Hsiao, Ling (1979), Riemann problem for one -dimensional adiabatic flow without convexity. Acta Math. Sinica 22:229 -231 (in Chinese with an English abstract).

[39] Liu, T. P. (1975), Existence and uniqueness theorems for Riemann problem. Trans. AMS. 213:3755-3782.

[40] Chang, Tong and Hiao, Ling (1989), The Riemann problem and interaction of waves in gas dynamics Pitman Monographs and Surveys in Pure and Appl. Math. 41, Longman Scien. Tech.

[41] Ding Xiaxi and Wu Yonghui (1993), Acta Math.Sci.13:2.

[42] Dafermos, C. (1979), The entropy rate admissibility criterion of solutions for solutions of hyperbolic conservation laws. J. Diff. Equat . 14:202-212.

[43] Hsiao, Ling (1980), The entropy rate a admissibility criterion in gas dynamics, J. Diff. Equat. 38:226-238.

[44] Ying, Lungan and Wang, Chinghua (1980), Global solutions of the Cauchy problem for a nonhomogenous quasilinear hyperbolic system. Comm . Pure Appl. math. 33:579-597.

[45] Glimm, J. (1965), Solutions in the large for nonlinear hyperbolic systems of equations. Comm. Pure Appl. Math. 18:697-715.

[46] Liu, T. P. (1979), Quasilinear hyperbolic system, Comm. Math. Phys ., 68:141-172.

[47] Dafermos, C. and Hsiao, Ling (1982), Hyperbolic system of balance laws with inhomogeneity and dissipation. Indiana Univ. Math. J. 331:471 -491.

[48] Wang, Chinghua and Li, Caizhong (1982), Study of global solutions for nonlinear conservation laws. J. Math. Anal. Appl. 85:236-256.

[49] Liu, T. P. and Wang, Chinghua (1985), On a nonstrictly hyperbolic system of conservation laws. J. Diff. Equat. 57:1-14.

[50] Keyfitz, B. and Kranzer, H. (1980), A system of nonstrictly hyperbolic conservation laws.

[51] Li, Caizhong and Liu, T. P. (1983), Asymptotic states for hyperbolic conservation laws with a moving source. Advances Appl. Math. 4:353-379.

[52] Tartar, L. (1979), Compensated compactness and applications to partial differential equations. Research Notes in Math., nonlinear analysis and mechanics: Heriot Watt Symp., Vol.4 Knops, R. J. (ed.), New York, Pitman Press.

[53] Diperna, R. (1983), Convergence of the viscosity method for isentropic gas dynamics. Comm. Math. Phys. 91:1-30.

[54] Ding, Xiaxi; Chen, Guiqiang and Luo, Peizhu (1987-8), Convergence of the Lax-Friedrichs scheme for isentropic flow, Acta Math. Sinica (I) 7:483-540; (II) 8:61-94. (in Chinese).

[55] Chen, Guiqiang(1988), ibid (III) 8:101-134 (in Chinese).

[56] Ding, Xiaqi; Chen, Guiqiang and Luo, Peizhu (1989), Convergence of the fractional step Lax–Friedrichs and Godunov scheme for the isentropic system of gas dynamics. Comm. Math. Phys. 121:63-84.

[57] Chen, Guiqiang and Lu, Yungguang (1988), Study of application route of compensated compactness theory, Kexue Tongbao, 33:641-644 (in Chinese).

[58] D. Tan, T. Zhang and Y. Zheng, Delta-shock waves as limits of vanishing viscosity for hyperbolic systems of Conservation Laws. Math. Science Research Institute, Berkeley, 013-92.

[59] Ding Xiaxi (1993), On a Non-Strictly Hyperbolic System, Preprint, Dept. of Math. University of Jyraskyla, No. 169.

[60] Ding Xiaxi and Wang Zhen, Generalized Solutions of Hyperbolic Systems Defined by Lebesgue-Stieltjes Integral. Preprint.

Stability of non-classical shock waves

Heinrich Freistühler[1]

Abstract

This paper presents results about the short time stability of ideal shock waves. Some of these results refer to certain algebraically overdetermined ("overcompressive") shock waves, others to certain algebraically underdetermined ("undercompressive") shock waves.

1 Formal classification of ideal shock waves

Take a hyperbolic system of conservation laws

$$u_t + f(u)_x = 0, \tag{1.1}$$

determined by its flux f, a smooth function mapping an open set $U \subset \mathbf{R}^n$ into $\mathbf{R}^n$ such that $Df(u)$ is $\mathbf{R}$-diagonalizable for every $u \in U$. Consider henceforth a non-characteristic shock wave

$$u(x,t) = \begin{cases} u^-, & x < st, \\ u^+, & x > st, \end{cases} \tag{1.2}$$

associated with (1.1), determined by any triple $(u^-, u^+, s) \in U \times U \times \mathbf{R}$ which satisfies the Rankine-Hugoniot jump conditions

$$(u^+ - u^-)s = f(u^+) - f(u^-)$$

and the non-characteristicity assumption that s is neither an eigenvalue of $Df(u^-)$ nor one of $Df(u^+)$. For $(u,s) \in U \times \mathbf{R}$, let

$$R^\pm(u,s) \equiv \sum_{\pm(\lambda - s) > 0} \ker(Df(u) - \lambda I).$$

Definition 1 *The shock wave (1.2) is algebraically stable (or Laxian) if*

$$R^-(u^-,s) \oplus R^+(u^+,s) \oplus \mathbf{R}(u^+ - u^-) = \mathbf{R}^n; \tag{LS}$$

[1]Institut für Mathematik, RWTH Aachen, 52062 Aachen, Germany

it is algebraically overdetermined if

$$(LO)\qquad R^-(u^-,s)+R^+(u^+,s)+\mathbf{R}(u^+-u^-)\neq\mathbf{R}^n;$$

it is algebraically underdetermined if

$$(LU)\qquad R^-(u^-,s)+R^+(u^+,s)+\mathbf{R}(u^+-u^-)\quad \textit{is not a direct sum.}$$

Obviously, a Laxian shock wave is neither overdetermined nor underdetermined; a non-Laxian shock wave may be either purely overdetermined, i. e., overdetermined and not underdetermined, or purely underdetermined, i. e., underdetermined and not overdetermined, or both over- and underdetermined.

Note that at the present level, the above distinction is purely formal in the sense that it classifies all solutions of the form (1.2), regardless of their mathematical or physical significance. For example, a shock wave can be algebraically stable while it violates the Oleinik-Liu condition (E); a "rarefaction shock" (obtained form a Laxian shock by interchanging the roles of u^- and u^+) is identified as underdetermined. For example, for scalar equations, both these types of shock waves exist, but they do not occur in solutions which are viscosity limits. Meaningful examples of non-Laxian shock waves occur in various important physical systems. For example, standard magnetohydrodynamics allows, besides Laxian shocks of course, for purely overdetermined shock waves, combined overdetermined-underdetermined shock waves, and, in its coplanar version, also for purely underdetermined shock waves, all of them being viscosity limits.

We motivate the above general definition and conditions (LS), (LO), (LU) as follows. Interpret (1.2) as a classical solution (u,σ) of the system

$$\begin{aligned} u_t(x,t)+(f\circ u)_x(x,t) &= 0, \quad t\geq 0, x\neq\sigma(t),\\ -(u(\sigma(t)+0,t)-u(\sigma(t)-0,t))\sigma'(t) & \\ +(f(u(\sigma(t)+0,t))-f(u(\sigma(t)-0,t))) &= 0, \quad t\geq 0, \\ u(x,0)={}^0u(x),\quad x\neq 0;\qquad \sigma(0) &= 0. \end{aligned} \tag{1.3}$$

Switching to a co-moving coordinate frame by letting $\tilde{u}(x,t)=u(x+\sigma(t),t)$, and linearizing around the reference solution

$$\tilde{u}(x,t)=\begin{cases} u^-, & x<0,\\ u^+, & x>0,\end{cases}$$

corresponding to (1.2), we obtain (cf., e. g., [3])

$$
\begin{aligned}
\bar{u}_t(x,t) + (Df(u^\pm) - sI)\bar{u}_x(x,t) &= 0, \quad t \geq 0, x \neq 0, \\
-(u^+ - u^-)\bar{\sigma}'(t) + (Df(u^+) - sI)\bar{u}^+(t) & \\
-(Df(u^-) - sI)\bar{u}^-(t) &= 0, \quad t \geq 0, \qquad (1.4)\\
\bar{u}(.,0) = {}^0\bar{u}, \quad \bar{\sigma}(0) &= 0,
\end{aligned}
$$

where $\bar{u}, \bar{\sigma}$ are the linearized quantities corresponding to $\tilde{u}$ and σ, respectively, and $\bar{u}^\pm \equiv \bar{u}(0\pm,.)$. Decomposing $\bar{u}^\pm$ as

$$
\begin{aligned}
\bar{u}^- &\equiv \bar{u}^-_- + \bar{u}^-_+, \quad \bar{u}^-_\pm \in R^\pm(u^-,s) \\
\bar{u}^+ &\equiv \bar{u}^+_- + \bar{u}^+_+, \quad \bar{u}^+_\pm \in R^\pm(u^+,s)
\end{aligned}
$$

and writing the second equation of (1.4) in the form

$$
\begin{aligned}
-(u^+ - u^-)\bar{\sigma}' + (Df(u^+) - sI)\bar{u}^+_+ - (Df(u^-) - sI)\bar{u}^-_- &= \\
-(Df(u^+) - sI)\bar{u}^+_- + (Df(u^-) - sI)\bar{u}^-_+ &\quad , \qquad (1.5)
\end{aligned}
$$

we obtain the following.

Proposition 1 *The linearization (1.4) of system (1.3) around its shock wave solution (1.2) has a unique solution for all data ${}^0\bar{u} \in C^0(\mathbf{R}, \mathbf{R}^n)$ with ${}^0\bar{u}(0) = 0$ if and only if condition (LS) holds. It has no solution at least for certain such data if and only if condition (LO) holds. It has several solutions at least for certain such data if and only if condition (LU) holds.*

Note that this well-/ over-/ underdeterminacy of the linearized dynamics is relative to the assumptions that (a) initial perturbations which vanish at the shock lead to nothing else than a basically "isomorphic" one-shock configuration and that (b) the Rankine-Hugoniot jump conditions are the only intrinsic relation linking the two boundary states of a shock wave together. Of course, either of these assumptions may sometimes be inappropriate.

2 Stability notions for ideal shock waves

Fix $T > 0$ arbitrarily and and $U' \subset\subset U$ appropriately and let $\mathcal{L}_0, \mathcal{L}$ be the sets of all Lebesgue measurable U'-valued functions defined on $\mathbf{R}, \mathbf{R} \times [0,T]$, respectively, considered as metric spaces with respective metrics d_0, d defined through

$$d_0(w_0, \tilde{w}_0) = \sup_{y\in\mathrm{R}} \int_{|x-y|\le 1} |w_0(x) - \tilde{w}_0(x)| dx,$$
$$d(w, \tilde{w}) = \sup_{y\in\mathrm{R}} \int_0^T \int_{|x-y|\le 1} |w(x,t) - \tilde{w}(x,t)| dx dt.$$

Consider a fixed reference solution $u^* \in \mathcal{L}$ of (1.1) with data ${}^0u^* \in \mathcal{L}_0$.

Definition 2 *u^* is dynamically stable if a continuous map $S_0 : {}^0u \mapsto u \in \mathcal{L}$ exists at least on a neighborhood of ${}^0u^*$ in $\mathcal{L}_0$ such that each $u = S_0({}^0u)$ solves (1.1) with respective data ${}^0u$ and in particular $S_0({}^0u^*) = u^*$.*

In addition to (1.1), consider also a family of systems

$$u_t + (f(u))_x = D[u] \tag{2.1}$$

where D ranges in some class $\mathcal{D}$, a collection of appropriately defined and interpreted *"dissipative"* operators. $\overline{\mathcal{D}} := \mathcal{D} \cup \{0\}$ is supposed to be a metric space, with metric δ. Speaking of $\mathcal{D}$ reflects the fact that realistic continuum mechanical models are typically endowed with whole families of different conceivable or physically realizable dissipation mechanisms.

Definition 3 *u^* is a viscosity limit (with respect to $\mathcal{D}$) if there exists a sequence $\{u^k, D_k\}_{k\in\mathrm{N}} \subset \mathcal{L} \times \mathcal{D}$ such that u_k solves (2.1) with $D = D_k$ and data ${}^0u_k$, and $d(u_k, u^*) + d_0({}^0u_k, {}^0u^*) + \delta(D_k, 0) \to 0$ when $k \to \infty$.*

Definition 4 *u^* is a uniformly stable viscosity limit (with respect to $\mathcal{D}$) if a continuous map $S : ({}^0u, D) \mapsto u \in \mathcal{L}$ exists at least on a neighborhood $\{({}^0u, D) \in \mathcal{L} \times \overline{\mathcal{D}}; d_0({}^0u, {}^0u^*) < \varepsilon, \delta(D, 0) < \varepsilon\}, \varepsilon > 0$, of $({}^0u^*, 0)$ such that each $u = S(u_0, D)$ solves (2.1) with the respective D and data ${}^0u$ and in particular $S({}^0u^*, 0) = u^*$.*

The rather auxiliary notion given in the next definition refers to the case of self-similar entropy solutions. A solution is *self-similar* if it is a function of x/t alone. Relative to a given entropy η with associated entropy flux q, u is an *entropy solution* if the inequality $\eta(u)_t + q(u)_x \le 0$ holds. Let $\mathcal{R}_0 \subset \mathcal{L}_0, \mathcal{R} \subset \mathcal{L}$ denote the sets of Riemann data and of piecewise self-similar solutions, respectively, considered as metric subspaces of $\mathcal{L}_0, \mathcal{L}$.

Definition 5 *In case u^* is of the form (1.2), it is said to be stable as a self-similar entropy solution if a continuous map $S_{\mathcal{R}} : {}^0u \mapsto u \in \mathcal{R}$ exists at least on a neighborhood of ${}^0u^*$ in $\mathcal{R}_0$ such that each $u = S_{\mathcal{R}}({}^0u)$ solves (1.1) with the respective data ${}^0u$, in particular $S_{\mathcal{R}}({}^0u^*, 0) = u^*$, and each $u = S_{\mathcal{R}}({}^0u)$ is an entropy solution.*

3 Instability of certain algebraically overdetermined shock waves

Results obtained in [1, 2, 3, 5] can be summarized as follows, cf.also in particular [4, 6].

Theorem 1 *Assume that the flux of (1.1) is equivariant with respect to an appropriate action of the orthogonal group* $\mathbf{O}(m)$ *on* $\mathbf{R}^n$ *with some* $m \geq 2$. *Consider a symmetry invariant state* u_* *at which the flux is no more degenerate than is necessarily the case due to this symmetry. Then there exist purely overdetermined shock waves as well as simultaneously over- and underdetermined shock waves, both with values arbitrarily close to* u_*. *Assume further that the flux allows for an appropriate entropy. Then, these shock waves are self-similar entropy solutions but are not stable as such.*

Corollary 1 *In this same situation, consider also an associated family (2.1) of dissipative systems having the property that for arbitrarily fixed Riemann data, (2.1) has a unique solution* u^D *for each* $D \in \mathcal{D}$, *which converges, as* $D \searrow 0$, *towards a piecewise continuous self-similar entropy solution of (1.1). Then, the shock waves considered in Theorem 1 cannot be uniformly stable viscosity limits (while they typically are viscosity limits).*

The above results apply to elasticity theory, nonlinear electromagnetism, magnetohydrodynamics, and other models. They apply in particular to the following general situation, under which at least the magnetohydrodynamics and elasticity theory cases can be subsumed.

Example 1. (Cf. [5].) Consider the system

$$\begin{aligned} v_t - w_x &= 0 \\ w_t - (g(v,\eta))_x &= 0 \\ (e(v,\eta) + \frac{1}{2}|w|^2)_t - (w^\top g(v,\eta))_x &= 0, \end{aligned}$$

consisting of $n = 2d+1$ equations, where v ranges in a certain open set $V \subset \mathbf{R}^d$, w in $\mathbf{R}^d$, and η in some w-dependent part of $\mathbf{R}$. The system is induced by a smooth function e of v and η through $g = e_v$. In addition to the basic properties $D^2 e > 0$, $e_\eta < 0$, the following symmetry and nondegeneracy assumptions are made on e. V is of the form $V = \mathbf{R}^m \times \mathcal{Z}$ with some $\mathcal{Z} \subset \mathbf{R}^{d-m}$ and some m satisfying $2 \leq m \leq d$. Writing $v = (y,z)$ with $y \in \mathbf{R}^m, z \in \mathcal{Z}$ correspondingly, e is of the form

$$e(v,\eta) = \hat{e}(|y|, z, \eta)$$

with a smooth function $\hat{e}$ which is even in its first argument. We denote this argument by $\hat{y}$ and write $\hat{v} = (\hat{y}, z)$ with $\hat{y} \in \mathbf{R}, z \in \mathcal{Z}$. At points of the form $(0, z, \eta) \in \mathbf{R} \times \mathcal{Z} \times \mathbf{R}$, where due to the even symmetry of $\hat{e}$ with respect to $\hat{y}$,

$$\hat{e}_{\hat{v}\hat{v}} = \begin{pmatrix} \hat{e}_{\hat{y}\hat{y}} & 0 \\ 0 & \hat{e}_{zz} \end{pmatrix}$$

the number $\hat{e}_{\hat{y}\hat{y}}$ is not an eigenvalue of the matrix $\hat{e}_{zz}$. Those eigenvalues of $\hat{e}_{\hat{v}\hat{v}}$ which coincide at these points with the eigenvalues of $\hat{e}_{zz}$ are either genuinely nonlinear or linearly degenerate near these points.

It is certainly obvious to the reader that in contrast to those overdetermined shock waves considered above, certain others are very natural ingredients of the hyperbolic theory.

Example 2. Consider the system $(u_j)_t + (u_j^2)_x = 0$, $j = 1, \ldots, n$, with $n \geq 2$, and its shock wave solutions (1.2) with $u^- = (\alpha_1, \ldots, \alpha_n)$, $u^+ = (-\alpha_1, \ldots, -\alpha_n)$, and $s = 0$, where $\alpha_1, \ldots, \alpha_n > 0$. These shocks are algebraically overdetermined, but they are dynamically stable and even uniformly stable as viscosity limits as follows immediately from Hopf's theory of the Burgers equation. The algebraic overdeterminacy is, of course, related to the fact that under generic perturbation of the data, such a shock wave splits into several separate Laxian shocks. In other words, assumption (a) from the end of Section 1 is not directly appropriate in this case. Still, as in many other interesting situations, the $\mathcal{L}$-closure of (LS) [1] (more generally, that of (LS)$'$, cf. Sec. 4) provides an appropriate "admissibility criterion."

4 Stability of certain algebraically underdetermined shock waves

Consider a triple $(u_0^-, u_0^+, s_0) \in U \times U \times \mathbf{R}$ with $u_0^- \neq u_0^+$ and such that s_0 is not an eigenvalue of $Df(u_0^-)$ nor of $Df(u_0^+)$.

Definition 6 (u_0^-, u_0^+, s_0) *is said to satisfy condition* (LS)$'$ *with respect to a function* g *if* g, *being of the form*

$$g(u^-, u^+, s) = \begin{pmatrix} f(u^+) - f(u^-) - s(u^+ - u^-) \\ h(u^-, u^+, s) \end{pmatrix} \quad \textit{with some } h : \mathcal{N} \to \mathbf{R}^{k-n},$$

maps an open neighborhood $\mathcal{N} \subset U \times U \times \mathbf{R}$ *of* (u_0^-, u_0^+, s_0) *smoothly into* $\mathbf{R}^k$ *for a* $k \geq n$*, such that* $g(u_0^-, u_0^+, s_0) = 0$ *and*

$$\frac{\partial g}{\partial u^-}(u_0^-, u_0^+, s_0)R^-(u_0^-, s_0) \oplus \frac{\partial g}{\partial u^+}(u_0^-, u_0^+, s_0)R^+(u_0^+, s_0) \oplus \mathbf{R}\frac{\partial g}{\partial s}(u_0^-, u_0^+, s_0) = \mathbf{R}^k.$$

To interpret the function g in an interesting way, consider also the regularized systems

$$u_t + (f(u))_x = \epsilon(B(u)u_x)_x, \quad \epsilon > 0,$$

where $B : U \to \mathbf{R}^{n \times n}$ is an appropriate viscosity matrix. A viscous profile, with respect to B, for the triple (u^-, u^+, s) is any solution $\phi : \mathbf{R} \to U$ of the differential equation

$$B(\phi)\phi' = f(\phi) - s\phi - c$$

which satisfies the boundary conditions

$$\phi(\pm\infty) = u^\pm;$$

the existence of ϕ implies $f(u^-) - su^- = f(u^+) - su^+ = c$.

Definition 7 (u_0^-, u_0^+, s_0) *is said to satisfy condition* (LS)$'$ *with respect to* B *if it satisfies* (LS)$'$ *with respect to a function* g *as in Definition 6 which has the additional property that the equivalence*

$$(u^-, u^+, s) \text{ has a viscous profile with respect to } B \quad \Leftrightarrow \quad g(u^-, u^+, s) = 0$$

holds in an open neighborhood of (u_0^-, u_0^+, s_0) *in* $U \times U \times \mathbf{R}$.

Obviously, this definition does not depend on the choice of g. Fix $p \in \mathbf{N}, p \geq 1$. The following two theorems are proved in [7].

Theorem 2 *Assume that the eigenvalues of* $Df(.)$ *have locally constant multiplicity near* u_0^- *and* u_0^+. *Let* (u_0^-, u_0^+, s_0) *satisfy* (LS)$'$ *with respect to* g *[B]. With two* C^p *functions* $v^- : (-\infty, 0] \to U, v^+ : [0, \infty) \to U$ *satisfying* $v^\pm(0) = u_0^\pm$*, let*

$$v(x) = \begin{cases} v^-(x), & x < 0 \\ v^+(x), & x > 0. \end{cases} \tag{4.1}$$

Then there exists, on a neighborhood of $(x,t) = (0,0)$ *in* $t \geq 0$*, a piecewise* C^p *function* u *such that* u *solves (1.1) weakly with initial data (4.1),* u *is continuous except along a* C^p *curve* $x = \sigma(t)$*, and for each* t*,* $(u(\sigma(t) - 0, t), u(\sigma(t) + 0, t), \sigma'(t))$ *satisfies (LS)$'$ with respect to* g *[B]. Such* u *is unique if attention is restricted to functions assuming values in the union of small neighborhoods of* u_0^-, u_0^+.

Theorem 3 *Assume that locally near u_0^-, u_0^+, all eigenvalues of $Df(.)$ are either genuinely nonlinear or linearly degenerate. Let (u_0^-, u_0^+, s_0) satisfy* (LS)$'$ *with respect to g [B]. Consider data (4.1) with two C^p functions $v^- : (-\infty, 0] \to U, v^+ : [0, \infty) \to U$ satisfying*

$$|v^\pm(0) - u_0^\pm| < \delta.$$

If $\delta > 0$ is sufficiently small, then there exists, on a neighborhood of $(x,t) = (0,0)$ in $t \geq 0$, a piecewise C^p function u such that u solves (1.1) weakly with initial data (4.1), u is continuous except along C^p curves $\sigma, \xi_1, \ldots, \xi_N$, and for each $t > 0$, $(u(\sigma(t) - 0, t), u(\sigma(t) + 0, t), \sigma'(t))$ satisfies (LS)$'$ *with respect to g [B], while $(u(\xi_j(t) + 0, t), u(\xi_j(t) - 0, t), \xi_j'(t))$ are small classical shocks or contact discontinuities. Such u is unique if attention is restricted to functions assuming values in the union of small neighborhoods of u_0^-, u_0^+.*

The proofs result from straightforward adaptation of Li and Yu's theory of boundary value problems for quasilinear hyperbolic systems. For the proof of Theorem 3, the following easy proposition is another ingredient.

Proposition 2 *Under the assumptions of Theorem 3, the initial value problem for (1.1) with Riemann data*

$$u(x,0) = \begin{cases} u_l, & x < 0 \\ u_r, & x > 0. \end{cases}$$

has a self-similar piecewise smooth solution when $|u_l - u_0^-|, |u_r - u_0^+|$ are sufficiently small. This solution u contains a shock which satisfies (LS)$'$ with respect to g [B], and $k-1$ classical shocks, rarefaction waves, or contact discontinuities. Such u is unique if attention is restricted to functions assuming values in the union of small neighborhoods of u_0^-, u_0^+.

Examples 3. While (LS)$'$ precludes (LO), it is a frequent property of shock waves satisfying (LS) or (LU). Examples of shock waves which satisfy (LU) and satisfy (LS)$'$ with respect to appropriate and natural viscosities B are provided by certain shock waves in quadratic models, but also by certain magnetohydrodynamic shock waves (for the unsimplified original equations; cf. Theorem 3 of [8]).

Remark. Of course, the idea of introducing (LS)$'$ is directly motivated by situations for which assumption (b) from the end of Section 1 seems inadequate. The property of satisfying (LS)$'$ may be referred to as (*relative*) algebraic stability with respect to g $[B]$. Similarly to (LS)$'$, one might, by an obvious continuation of Definition 1 on the one and Definitions 6,7 on the other hand, define conditions (LO)$'$ and (LU)$'$ which express (*relative*) algebraic over-(under)determinacy respectively with respect to appropriate g $[B]$.

5 Formal classification of shock waves revisited

Continuing Section 1 directly, with $l = \dim R^-(u^-, s) + \dim R^+(u^+, s) + 1$ and $r = \dim (R^-(u^-, s) + R^+(u^+, s) + \mathrm{R}(u^+ - u^-))$ the number of unknowns and the rank, respectively, of (1.5), let $\underline{\kappa} = l - r \geq 0$ and $\overline{\kappa} = n - r \geq 0$ denote the *degrees* of under-(over)determinacy respectively of this linear algebraic system, and call the ordered pair

$$(\underline{\kappa}, \overline{\kappa})$$

the *algebraic type* of the shock wave. In terms of this quantification, a shock wave is Laxian if and only if its algebraic type is $(0, 0)$. The shock waves of Theorem 1 are of types $(0, m-1)$ and $(1, m-1)$, respectively. Finally, shock waves obeying condition (LS)$'$, have type $(\underline{\kappa}, 0)$ with some $\underline{\kappa} \geq 0$.

Remark. The following list of references is restricted to a minimum of some papers by this author, and one joint article with Peter Szmolyan, which are logically sufficient to cover details that have been omitted above. This option has been chosen *only for the sake of brevity*. Other pertinent work is referenced and properly described in the individual papers listed.

References

[1] H. Freistühler, *Rotational degeneracy of hyperbolic systems of conservation laws*, Arch. Rational Mech. Anal. 113 (1991), 39-64.

[2] —, *Instability of vanishing viscosity approximation for hyperbolic systems of conservation laws with rotational invariance*, J. Differ. Equations 87 (1990), 205-226.

[3] —, *Dynamical stability and vanishing viscosity: a case study of a non-strictly hyperbolic system*, Commun. Pure Appl. Math. 45 (1992), 561-582.

[4] —, *Non-uniformity of vanishing viscosity approximation*, Appl. Math. Lett. 6 (1993), 35-41.

[5] —, *Small amplitude intermediate MHD shock waves*, preprint Aachen 1993.

[6] —, *Contributions to the mathematical theory of magnetohydrodynamic shock waves*, Proceedings of International Conference on Nonlinear Evolutionary Partial Differential Equations Beijing 1993, to appear.

[7] —, *A short note on the persistence of ideal shock waves*, Arch. Math., to appear.

[8] — and P. Szmolyan, *Existence and bifurcation of viscous profiles for all intermediate magnetohydrodynamic shock waves*, SIAM J. Math. Anal., to appear.

THE APPLICATION OF FRONT TRACKING TO THE SIMULATION OF SHOCK REFRACTIONS AND SHOCK ACCELERATED INTERFACE MIXING

JOHN W. GROVE, RICHARD L. HOLMES, AND DAVID H. SHARP

ABSTRACT. Front tracking simulations of Richtmyer-Meshkov instability produce significantly better agreement with experimentally measured growth rates than obtained in non-tracking computations. Careful analysis of the early stages of the shock acceleration process shows that nonlinearity and compressibility play a critical role in the behavior of the shocked interface and are responsible for the deviations from the linear theories.

1. INTRODUCTION

The quantitative understanding of Richtmyer-Meshkov (shock accelerated) interface instability has challenged scientists for over 30 years. Interest in this instability is motivated by its importance in subjects such as inertial confinement fusion and supernovae dynamics.

The growth of perturbations at a shocked interface between two fluids was first studied in Richtmyer's seminal 1960 paper. Richtmyer studied the solution of the linearized (compressible) Euler equations and found that after an initial rapid increase, the growth rate in the linearized solutions oscillated about a limiting value. He proposed an analytic formula for the growth rate based on an approximation to the full linear theory. The predictions of the analytic formula, known as the impulsive model, agreed quite well with the limiting value of the linear theory. In the time since Richtmyer's work this problem has been the subject of extensive investigation. The earliest experimental studies of this instability were performed by Meshkov [8] who measured growth rates that were significantly smaller than the predictions of the impulsive model. More recent experiments by Benjamin [1, 2] also showed growth rates that were nearly half the value predicted by Richtmyer's theory. In sharp contrast however, perturbation growth rates obtained by numerical simulation of experiments agreed quite well with the linear theory [2, 4, 9]. Other simulations of the Richtmyer-Meshkov instability also gave growth rates close to the values predicted by the impulsive model. Several examples can be found in [3, 5, 7] and interesting exceptions are found in [10] and [12].

Here we report substantially improved agreement between experiment, theory and computation. Our main tool is front tracking simulations of experiments of Benjamin [2]. Our computed growth rates are in excellent agreement with those found in Benjamin's experiments.

The simulations raise a number of important theoretical issues aside from the matter of agreement with experiment. Among these is the relationship between the nonlinear theory and the small amplitude theory, derived by linearizing the equations of motion. These theories agree for small initial amplitude perturbations and sufficiently early times, as expected. For larger amplitudes or later times there is an abrupt departure of the nonlinear theory from the linear theory. We explain

---

1991 *Mathematics Subject Classification.* 76N10,76E30.

Key words and phrases. Front Tracking, Richtmyer-Meshkov.

We give our thanks to James Glimm for his encouragement and guidance. This work has been supported by the U. S. Army Research Office the Mathematical Sciences Institute of Cornell University under subcontract to SUNY Stony Brook, ARO contract number DAAL03-91-C-0027, the National Science Foundation, Grant no. DMS-9057429, the National Science Foundation, Grant no. DMS-9201581, the U. S. Army Research Office, Grant no. DAAL03-92-G-0185 and the U. S. Department of Energy.

this behavior in terms of a re-acceleration of the interface by shocks produced by the (nonlinear) self-interaction of the reflected and transmitted waves.

2. Experimental configurations

Benjamin conducted his experiments in a rectangular shock tube where two gases were initially separated by a thin membrane with an imposed single wavelength sinusoidal perturbation. The membrane was ruptured by a plane shock propagating down the shock tube. Figure 2.1 shows a schematic representation of the geometry.

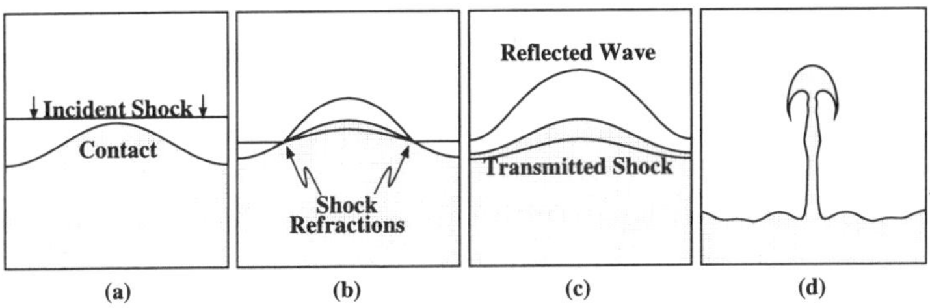

FIGURE 2.1. A schematic of the geometry in Richtmyer-Meshkov experiments. A shock wave is refracted by the material interface, producing a transmitted shock and either a reflection shock or reflected rarefaction wave. The instability consists of the growth in time of perturbations on the material interface. The last frame shows a possible late time configuration with a spike of heavy gas being injected into the lighter gas.

Benjamin's experiments used a Mach 1.2 shock in air incident on an interface with sulfur hexafluoride (SF_6). The interface perturbation had an amplitude of 0.24 cm and a wavelength of 3.75 cm. As these experiments were performed at the high altitude Los Alamos National Laboratory we used ambient pressures of 0.8 bar and densities of 0.95 g/l for air and 4.85 g/l for SF_6 in our calculations. This corresponds to an initial Atwood ratio of 0.67. We used perfect gas equations of state with $\gamma_{air} = 1.4$ and $\gamma_{SF_6} = 1.09$. The rear wall of the tube was located sufficiently far from the interface that measurements of the growth rate could be made before the arrival of the shock reflected off this rear wall.

The quantities of interest in this study are the amplitude of the perturbations on the interface, $a(t)$, defined to be one-half the total inter-penetration width, and the amplitude growth rate, $\dot{a}(t)$. Other diagnostics include pressure and density fields of the flow and the interface position. We also measured the flow velocities at the tips of the heavy gas spike and the light gas bubble. Note that $\dot{a}(t) = (v_{spike} - v_{bubble})/2$, where v denotes the velocity at the interface.

3. Comparison of the linear and nonlinear theories

Richtmyer [11] proposed a model for the growth rate of interface perturbations based on an analogy of the shock acceleration process to an impulsive body force in otherwise incompressible Rayleigh-Taylor instability. If the perturbation is sinusoidal with wave number k, Richtmyer's formula for the perturbation amplitude growth rate is

$$\dot{a}(t) = kA\Delta u a(0+). \tag{3.1}$$

Here $A = (\rho_2^* - \rho_1^*)/(\rho_2^* + \rho_1^*)$ is the Atwood ratio using post-shock densities ρ_i^*, and Δu is the change in translational velocity of the interface due to the action of the shock and $a(0+)$ is the shock-compressed amplitude. Formula 3.1 is assumed hold as long as $ka(t) \ll 1$. Note that this formula explicitly assumes that compressibility is negligible after the initial shock refraction. Meyer and Blewett [9] proposed a modification of formula 3.1 for reflected rarefactions in which $a(0+)$ is replaced by the average of the pre- and post-shock amplitudes.

For $ka \ll 1$ the Euler equations can be linearized about the solution of the Riemann problem for a planar shock refracting through a planar material interface. This approach is also originally due to Richtmyer. More recently this method was extended to the case of a reflected rarefaction [12]. Note that this linearization makes no incompressibility assumption.

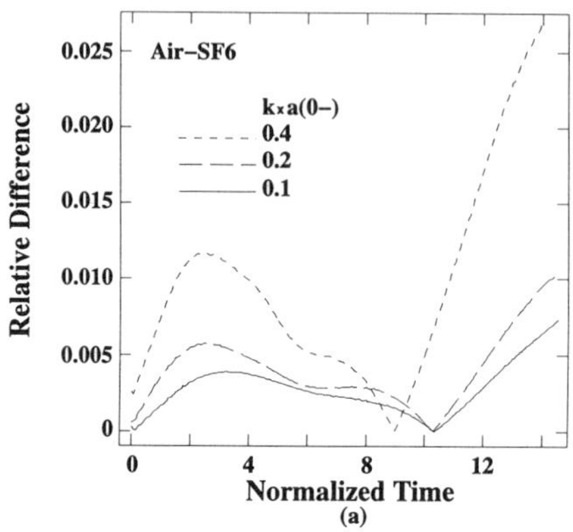

FIGURE 3.1. Convergence of the nonlinear simulations to the linearized solution for small amplitudes. Convergence is measured in terms of the relative difference between the nonlinear and linear solutions, $|(a(t) - a_{lin}(t))|/|(a_{lin}(t)| + |a(0-))|$ where $a(0-)$ is the initial amplitude of the perturbation. The horizontal axis is in dimensionless time units kc_0M_0t, where k is the wave number of the perturbation, c_0 is the sound speed in the air ahead of the incident shock and M_0 is the incident shock Mach number. One dimensionless time unit is about $14\mu s$.

We investigated the convergence of the nonlinear front tracking simulations to the linearized system by computing solutions for small initial amplitudes. Figure 3.1 shows the results of these computations. We obtain good convergence at early times providing an important validation of our calculations.

Figure 3.2 shows amplitude and amplitude growth rates from the impulsive model, numerical solution of the linear theory, nonlinear simulations and experiment. We focus on two regions of the growth rate curves where divergence between the linear and nonlinear theories occurs. The first region of divergence begins at $t \approx 70\mu s$ with a reversal in the growth rate $\dot{a}(t)$ followed by a second period of rapid increase. The second region of divergence occurs shortly after the maximum of $\dot{a}(t)$ is reached, at $t \approx 200\mu s$. Here the solution to the linearized equations saturates near this maximum value. In contrast, the nonlinear solution experiences a marked deceleration at this point.

The rapid deceleration of the shocked interface at time $200\mu s$ is an inherently nonlinear and compressible phenomenon. It can be understood in terms of a series of re-accelerations of the material interface by secondary shocks generated by self-interactions at the reflected and transmitted wave edges. As seen in the grey scale representations of the pressure field in figure 3.3, curvature in the reflected and transmitted waves generates additional shocks via nonlinear self-interaction. At $33\mu s$ these waves are clearly seen near the edges of the reflected shock (figure 3.3a). As the interaction proceeds the compression fronts steepen into shocks. These shocks collide near the tip of the spike

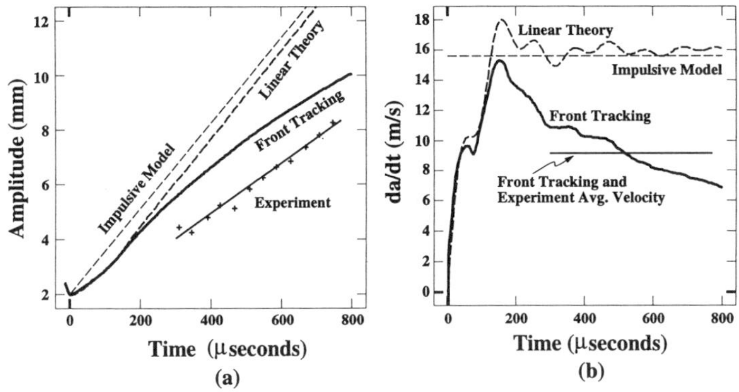

FIGURE 3.2. Perturbation amplitude, $a(t)$, and amplitude growth rate, $\dot{a}(t)$, of a shocked air-SF_6 interface. This graph compares experimental averages, front tracking simulation, linear theory and Richtmyer's impulsive model. Also shown is a least squares fit to the front tracking amplitude data over the period of experimental observation. Note that the front tracking average growth rate is indistinguishable from the experimental value in figure (b). The plus marks (+) show the results of one particular experiment.

at about $68\mu s$ (figure 3.3(b)), producing a high pressure pulse near the interface that causes the growth rate to decrease. This deceleration corresponds to the first blip in the growth rate graph figure 3.2(b). At later times the waves emanating from the reflected and transmitted edges produce a series of criss-cross shock reflections. At approximately $207\mu s$ (figure 3.3c) a strong pulse crosses the interface, leading to the severe (and permanent) deceleration of the bubble. That this decrease is due primarily to action at the bubble side of the interface is clear from figure 3.4(a). By time $t = 344\mu s$ we see that the strong wave action has moved away from the interface region and the growth rate curve has smoothed considerably. Note the pair of Mach triple points at the far right of this figure.

Figures 3.3e-f show pressure plots of the solution to the linearized equations and of the solution to the nonlinear Euler equations at time $t = 195\mu s$, near the time of the most extreme divergence of the two solutions. We call attention to two important features of linear solutions: they do not allow the focusing of characteristics and the linearization constrains the geometry of the wave fronts to be sinusoidal. These two restrictions prevent the production of the additional shocks that are present in the nonlinear solution, *i.e.*, there is no cascade of reflected shocks from the self-interaction of the transmitted and reflected waves. The linear solution displays a series of acoustic waves, but they do not sharpen into shocks and are thus much weaker than the corresponding waves of the nonlinear solution. Note that the strong waves in the nonlinear solution come in pairs and that there are two pressure maxima across the tube in the x-direction. This is in contrast to the linear theory in which all quantities are assumed to have sinusoidal perturbations and, hence, have a single maximum and minimum value.

The analysis of secondary waves can be followed further in time to account for the plateau in $\dot{a}(t)$ during the observation period. During the observational window this ringing of waves near the interface subsides.

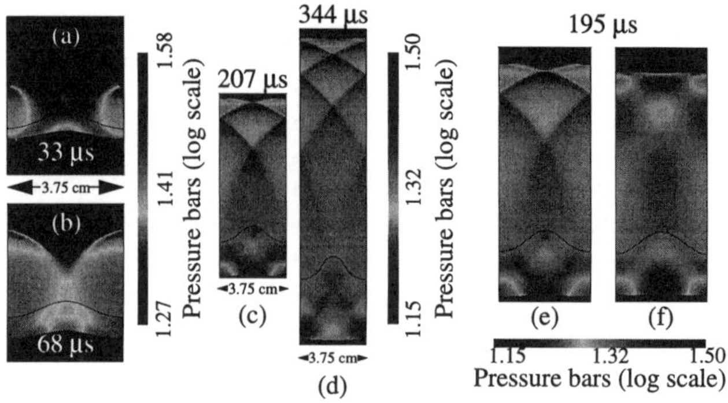

FIGURE 3.3. (a-d) A cascade of shock waves generated by the self-interaction of the transmitted and reflected waves propagate back toward the interface and affect the perturbation growth rate at early and intermediate times. For reference we have labeled in the first frame the transmitted shock (T), the interface (I) and the reflected shock (R). Note that the grey scale map has been adjusted so that the full range of variation occurs within the region of interest (between the transmitted and reflected shocks). The regions at the left and right have pressures of 0.8 bars and 1.1 bars, respectively. Linear and nonlinear air-SF_6 solutions at $t = 195$ are shown on the right. We see that the linear solution does not show the strong wave action that affects the early and intermediate time growth rate in the nonlinear solution.

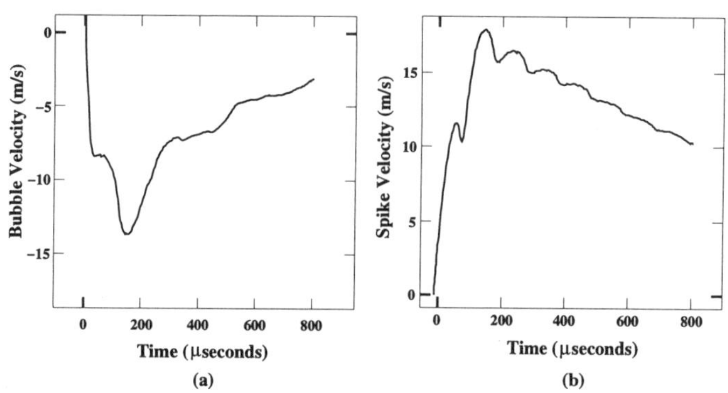

FIGURE 3.4. (a) Bubble and (b) spike velocities at a shocked air-SF_6 interface. The full amplitude growth rate shown in Figure 3.2 is $\dot{a}(t) = (v_{spike} - v_{bubble})/2$. Note how certain events in the amplitude growth rate occur at either the spike or the bubble, but not both.

4. COMPARISON TO EXPERIMENTS

Comparison of simulations to experiments results in excellent agreement for the growth rates (see figure 3.2 and [6]). However, there is a significant offset in the value of the absolute amplitude of the perturbations.

The improved agreement between the front tracking simulations and experiment is due to the sharp resolution of the interface. Tests using untracked shock waves show that the growth rates are very similar to those found in the fully tracked simulations except at early times. This is consistent with the analysis given in section 3 of the wave action generated by shock self-interaction. Differences in the transmitted and reflected shocks will be felt most strongly when the waves are closest to the interface and thus the effects of an untracked shock will be most significant at early times.

We know of no model that provides quantitative estimates of the effect of the membrane in Meshkov's and Benjamin's experiments. We have shown that it is not necessary to invoke membrane effects to achieve agreement with Benjamin's measured growth rate. This does not mean membrane effects have been ruled out. They could well have an effect on the early time behavior (not covered in Benjamin's data) and possibly account for the residual discrepancy in the amplitude. The best way to resolve the issue of membrane effects is to perform Richtmyer-Meshkov instability experiments in which membranes are not used.

A detailed discussion of membrane effects is beyond the scope of this paper. We note, however, that the computational approach to estimating membrane effects is not promising. Analyzing the deformation and fragmentation of a thin elasto-plastic membrane would add severe demands to an already demanding computation. There is no established and bench-marked methodology for modeling the fragmentation of such a membrane. Even if one did come up with a model of membrane dynamics, it would be very difficult to know that it is correct. Since such a model is likely to require phenomenological parameters, simply fitting these parameters to obtain agreement with experiment would accomplish very little.

5. Conclusion

The analysis of the simulations shows that compressibility and nonlinear wave interactions play an important role in the early time development of the Richtmyer-Meshkov instability, up to and including the time measured in experiments. These wave interactions are not captured in the linearized theories commonly used to estimate perturbation growth rates.

Front tracking and the proper timing of growth rate measurements lead to improved agreement between experiment and simulation perturbation growth rates. It appears worthwhile to study the effect of mass diffusion on the early time growth of the perturbations to aid in understanding the still unresolved difference in amplitudes.

References

1. R. Benjamin, *Experimental observations of shock stability and shock induced turbulence*, Advances in Compressible Turbulent Mixing (W.P. Dannevik, A.C. Buckingham, and C.E. Leith, eds.), National Technical Information Service, U.S. Department of Commerce, 5285 Port Royal Rd. Springfield VA 22161, 1992, pp. 341–348.
2. R. Benjamin, D. Besnard, and J. Haas, *Shock and reshock of an unstable interface*, LANL report LA-UR 92-1185, Los Alamos National Laboratory, 1993.
3. D. Besnard, C. Cavailler, L. Dagens, P. Figeac, M. de Gliniasty, J.F. Haas, P.A. Holstein, J. Montigny, C. Parisot, V. Rupert, B. Sitt, and N. Wilke (eds.), *Proceedings of third international workshop on the physics of compressible turbulent mixing at Royaumont, France*, CEA DAM, 1991.
4. L. D. Cloutman and M. F. Wehner, *Numerical simulation of Richtmyer-Meshkov instabilities*, Phys. Fluids A **4** (1992), 1821–1830.
5. W.P. Dannevik, A.C. Buckingham, and C.E. Leith (eds.), *Advances in compressible turbulent mixing*, National Technical Information Service, U.S. Department of Commerce, 5285 Port Royal Rd. Springfield VA 22161, 1992.
6. J. Grove, R. Holmes, D. H. Sharp, Y. Yang, and Q. Zhang, *Quantitative theory of Richtmyer-Meshkov instability*, Phys. Rev. Lett. **71** (1993), no. 21, 3473–3476.
7. P. F. Linden, D. L. Youngs, and S. B. Dalziel (eds.), *Proceedings of the 4th international workshop on the physics of compressible turbulent mixing*, Cambridge University Press, Cambridge, England, 1993.
8. E. E. Meshkov, *Instability of a shock wave accelerated interface between two gases*, NASA Tech. Trans. **F-13** (1970), 074.

9. K. A. Meyer and P. J. Blewett, *Numerical investigation of the stability of a shock-accelerated interface between two fluids*, Phys. Fluids **15** (1972), 753–759.
10. K. O. Mikaelian, *Growth rate of the Richtmyer-Meshkov instability at shocked interfaces*, Phys. Rev. Lett. **71** (1993), no. 18, 2903–2906.
11. R. D. Richtmyer, *Taylor instability in shock acceleration of compressible fluids*, Comm. Pure Appl. Math. **13** (1960), 297–319.
12. Y. Yang, Q. Zhang, and D. H. Sharp, *Small amplitude theory of Richtmyer-Meshkov instability*, Phys. Fluids **6** (1994), no. 5, 1856–1873.

Department of Applied Mathematics and Statistics, University at Stony Brook, Stony Brook, NY 11794-3600

E-mail address: `grove@ams.sunysb.edu`

Courant Institute of Mathematical Sciences, New York University, 251 Mercer St., New York, NY 10012

E-mail address: `rich@cims.nyu.edu`

Complex Systems Group, Theoretical Division, Los Alamos National Laboratory, Los Alamos, NM 87545

E-mail address: `dhs.lanl.gov`

Images and movies scale-spaces

F. Guichard, J-M. Morel[1]

Introduction. Define an image as a bounded real function on $\mathbb{R}^N$. Most practical image analysis devices assume a preliminary step in the processing of the pictures, which consists in passing from the original picture to smoothed versions, which still contain the significant information. At this stage of the processing, the image is transformed into new images, and the main parameter of this preliminary transform is the "scale", a parameter which measures the degree of smoothing, or more trivially, the size of the neighbourhoods which are taken into account for giving an estimate of the brightness of the picture at a given point. A "multiscale analysis" or "scale-space", as it is generally called in image processing theory, tends to give a less local and therefore more reliable information on the grey level than the original fluctuating "pixel".

We define a "multiscale analysis" as a family of transforms $(T_t)_{t\geq 0}$, which, applied to the original picture $u(\mathbf{x})$, yield a sequence of pictures $u(t,\mathbf{x}) = (T_t f)(\mathbf{x})$. The scale space image processing transforms must satisfy a list of formal requirements. We classify them into two categories : "architectural axioms" such as locality, causality in the scale space, the comparison principle; and "geometrical axioms" which correspond to shape preserving properties (rotation invariance, scale invariance, etc.). A classification of all image multiscale transforms satisfying these requirements has been done in [1]. This classification yields a characterization of all classical models and includes new ones, which all are Partial Differential Equations. We examine more precisely the case of the 2D images and the case of the movies.

1 Axioms for scale-space.

1.1 Architectural axioms.

We first consider an axiom which is called in vision theory the "pyramidal architecture" property. According to this assumption, T_t can be computed from T_s for any $s \leq t$, and T_0 is, of course, the identity. This is natural, since a coarser analysis of the original picture is likely to be deduced from a finer one without any dependence upon the original picture. The finest picture analysis is the identity.

We include $T_t = T_{t,0}$ into a family of transition operators $T_{s,t}$ indexed by $0 \leq s,t < \infty$ and satisfying

[Pyramidal architecture] $T_{t+s} = T_{t+s,s} o T_s$ for all $0 \leq s,t < \infty$.

From the theory of perception viewpoint, [Pyramidal architecture] is a sound hypothesis, if the image perceptual analysis is a sequence of filters which are applied sequentially. Since new images are constantly arriving to the retina, the image analysis process is thought of as a flow of the picture through different filters, each associated with a scale t.

We also introduce an axiom stating a natural assumption of continuity of T_t.

[1]CEREMADE, Univ. PARIS Dauphine, Place du Marechal de Lattre de Tassigny, 75775 Paris Cedex, Fax: (33) 44.05.45.99, E-mail: fguichar or morel@ceremade.dauphine.fr

[Regularity] Let $u(x,y,\theta) = u(\mathbf{x}) = \frac{1}{2}\langle A\mathbf{x}, \mathbf{x}\rangle + \langle p, \mathbf{x}\rangle + c$ be a quadratic form of $\mathbb{R}^3$. We assume that the expression

$$(T_{t+\Delta t,t}u - u)(\mathbf{x})/(\Delta t)$$

has a limit $\mathcal{A}(u)(\mathbf{x},t)$, when Δt tends to zero. Then, there exists a function $F(A,p,\mathbf{x},c,t)$ such that $\mathcal{A}(u)(\mathbf{x},t) = F(A,p,\mathbf{x},c,t)$. We shall assume, in addition, that F is continuous in all its arguments. (This assumption can be weakened.)

Comparison principle. The comparison principle is an obvious order-preserving property (the "maximum principle"). It means that no enhancement is made but just a smoothing of the original image. Thus if an image g is everywhere brighter than another f, this ordering will be preserved. In the following we shall assume the locality. We directly state a variant of the comparison principle by adding the locality aspect. We denote by $\mathrm{B}(\mathbf{x},r)$, the ball with center $\mathbf{x}$ and radius r.
[local maximum principle] $\forall f,g,t,\mathbf{x}$ if $\forall \mathbf{y} \in \mathrm{B}(\mathbf{x},r)/\{\mathbf{x}\}$ $f(\mathbf{y}) > g(\mathbf{y})$ then

$$T_{t+h,t}f(\mathbf{x}) \geq T_{t+h,t}g(\mathbf{x}) \text{ for } h \text{ small enough.}$$

This just means that if an image f is, around a point $\mathbf{x}$, brighter than g, then this property remains true at the point $\mathbf{x}$ for an infinitesimally larger scale.

1.2 Geometrical and Morphological axioms.

The next axioms state that image analysis must be invariant under fluctuations of light, change of position, orientation and scale of the planar shapes. A first essential fact in the understanding of images and shapes is to take into account how arbitrary the grey scale of perceptual or digital pictures is. This is due to the fluctuations of illumination of the perceptum : indeed, sun, clouds, artifical lights, reflections. The state of the percipiens can also be altered : Aperture, focus, trace of anterior perception, etc. of the perception device also drastically change the effect of the light on the photosensitive surface. In conclusion, the image analysis must be independent of the (arbitrary) grey-level scale. We shall always assume the following weak form of this axiom :
[Grey level shift inv.] $T_t(0) = 0, T_t(f+C) = T_t(f)+C$ for any f and any constant C.

The contrast invariance axiom can be written as,
[Morphological Inv.] $T_t(h(f)) = h(T_t(f))$ for all f and all $t \geq 0$, where h is any nondecreasing real function.

The function h simply is an order-preserving rearrangement of the grey level. A stronger form of the axiom assumes the same relation when h is nondecreasing or nonincreasing.

We shall now list a series of axioms which traduce the invariance of image analysis under the respective positions of the percipiens and perceptum. We first impose that the multiscale analysis is translation invariant, that is
[Transl. Inv.] $T_t(\tau_h.f) = \tau_h.(T_t f)$ for all h in $\mathbb{R}^N$, $t \geq 0$, where $(\tau_h.f)(\mathbf{x}) = f(\mathbf{x}+h)$.

This only means that all points of the space are *a priori* equivalent. There is no *a priori* knowledge about the location of any feature of the picture.
[Isom. invariance] $T_t(R.f) = R.T_t(f)$ for all $f, t \geq 0$ and for all transforms R defined by $(R.f)(\mathbf{x}) = f(R\mathbf{x})$ where R is an orthogonal transform of $\mathbb{R}^N$.

The next geometrical axiom is the scale invariance of image analysis. Set $D_\lambda f(\mathbf{x}) = f(\lambda\mathbf{x})$. The scale invariance can be stated as

[Scale Inv.] For any λ and t, there exists t' such that $D_\lambda T_{t'} = T_t D_\lambda$. We assume also that $t'(t, \lambda)$ is differentiable with respect to λ at $\lambda = 1$, that the function $g(t) = \partial t'/\partial\lambda(t, 1)$ is continuous and positive for $t > 0$, and that the map $t \to T_t$ is one to one.

The scale invariance axiom means that the result of the multiscale analysis T_t is independent of the size of the analyzed features : this is very important in the world where we live, since the same object can be seen at very different distances and therefore at very different scales.

Finally, we state the affine invariance axiom (which implies [Isom. invariance], [Scale Inv.].) Set for any such transform $Af(\mathbf{x}) = f(A\mathbf{x})$. With the same formalism as in [Scale Inv.], we get
[Affine Inv.] For any A, $det(A) \neq 0$, and t, there exists $t'(t, A)$ such that $AT_{t'} = T_t A$.

2 Main classification theorem.

Under the architectural axioms, we characterize each multiscale filtering as the solution of a generic partial differential equation. We obtain in Theorem 1 a result explaining why the main multiscale image processing models are Parabolic Partial Differential Equations of order 2.

Theorem 1 *Under the axioms* [Pyramidal architecture], [local maximum principle], *and* [Regularity], *there exists a function F such that*

$$((T_{t+h,t}f - f)/h)(\mathbf{x}) \to F(D^2 f(\mathbf{x}), Df(\mathbf{x}), f(\mathbf{x}), \mathbf{x}, t) \tag{1}$$

as h tends to 0^+ for all f and $\mathbf{x}$ where f is C^2. In addition, F is a continuous and non-decreasing function with respect to its first argument, that is

$$\textit{If } A \geq B, \textit{ for the ordering of symmetric matrices,} \tag{2}$$

$$\textit{then,} \quad F(A, p, c, \mathbf{x}, t) \geq F(B, p, c, \mathbf{x}, t)).$$

Moreover, if [Grey level shift inv.] *holds, then the function F does not depend on c.*

We define $BUC(I\!\!R^N)$ or $\mathcal{X}$ the space of the functions which are uniformly continuous and bounded. Moreover, we shall say that $u(t, \mathbf{x})$ is in $BUC_x([0, +\infty[*I\!\!R^N)$ if $\forall t$, $u(t, .)$ is in $BUC(I\!\!R^N)$. Recall the concept of viscosity solutions for parabolic equations (see [3]). $u \in BUC_x([0, +\infty[*I\!\!R^N)$ is a viscosity subsolution (resp. a supersolution) at the point $\mathbf{x}_0$, and scale t_0 if for all ϕ in $C_b^\infty(I\!\!R^N * [0, T])$ such that $(t_0, \mathbf{x}_0)$ is a global maximum point of $u - \phi$ (resp. of $\phi - u$), we have

$$\frac{\partial\phi}{\partial t}(t_0, \mathbf{x}_0) - F(D^2\phi(t_0, \mathbf{x}_0), D\phi(t_0, \mathbf{x}_0), t) \leq 0. \qquad \text{resp. } \geq 0$$

At last, we say that u is a viscosity solution at the point $\mathbf{x}_0$, and scale t_0 if u is a viscosity subsolution and supersolution.

Theorem 2 *Assume* [Pyramidal architecture], [local maximum principle], [Regularity], [Grey level shift inv.], *and* [Transl. Inv.]. *Assume also that for any initial datum* u_0 *in* $BUC(\mathbb{R}^N)$*, the equation defined by*

$$\frac{\partial u}{\partial t} = F(D^2 f(\mathbf{x}), Df(\mathbf{x}), t) \qquad u(0,\mathbf{x}) = u_0(\mathbf{x}) \tag{3}$$

has a unique viscosity solution in $BUC_x([0,+\infty[*\mathbb{R}^N)$*. Then,* T_t *maps* $BUC(\mathbb{R}^N)$ *into itself, and* $u(t,\mathbf{x}) = T_t u_0(\mathbf{x})$ *is the viscosity solution of Equation (3).*

3 Examples of multiscale analysis.

3.1 Axiomatization of the Marr-Hildreth-Koenderink-Witkin theory.

We can easily deduce from Theorem 3 a characterization of the heat equation $\partial u/\partial t = \Delta u$ as the unique linear and isometrically invariant multiscale model.

The classical model comes from Marr-Hildreth [8] and has been formalized by Witkin [13], Koenderink [5]. The basic step of the multiscale analysis is the convolution of the original image with Gaussians of increasing variance. Koenderink [5] noticed that the convolution of the signal with gaussians at each scale was equivalent to the solution of the heat equation with the signal as initial datum. Denoting by u_0 this datum, the "scale space" analysis associated with u_0 consists of solving the heat equation. The solution of this equation for an initial datum with bounded quadratic norm is $u(t,x,y) = G_t * u_0$ where $G_t(x,y) = (4\pi)^{-1}t^{-1}exp(-(x^2+y^2)/4t)$ is the Gauss function.

Theorem 3 *Let* T_t*,* $t \geq 0$ *be a multiscale analysis satisfying* [Pyramidal architecture], [Regularity], [local maximum principle], *and* [Transl. Inv.]. *If the* T_t *are linear and satisfy* [Isom. invariance], *then (up to a rescaling* $t' = h(t)$*),* $u(t,\mathbf{x}) = (T_t u_0)(\mathbf{x})$ *is the solution of the heat equation* $\partial u/\partial t - c\Delta u = 0$ *in* $\mathbb{R}^N * [0,\infty[$ *,* $u(0,.) = u_0(.)$ *in* $\mathbb{R}^N$*, where c is some positive constant.*

3.2 Basic operators of classical Mathematical Morphology.

The basic operations of mathematical morphology, on which all other ones are built, are dilations and erosions.

Let B be a bounded subset of $\mathbb{R}^N$, which is called in mathematical morphology "structuring element". Let $t \geq 0$ be a scale parameter. The basic operations of mathematical morphology are called dilation and erosion. The dilation at scale t of function f with structuring element B is defined by

$$D_t f(\mathbf{x}) = \sup\{f(\mathbf{y}), \mathbf{y}-\mathbf{x} \text{ in } tB\}$$

Similarly, the erosion at scale t of function f with structuring element B is defined by taking the inf instead of the sup.

Assume that B is convex and contains 0, these families can be obtained by a partial differential equation. We define $||\mathbf{y}||_B = \sup_{\mathbf{z}\in B}\langle \mathbf{y}.\mathbf{z}\rangle$, where $\langle \mathbf{y}.\mathbf{z}\rangle$ denotes the euclidean scalar product of $\mathbf{y}$ and $\mathbf{z}$.

The multiscale analyses $D_t f(\mathbf{x})$ and $E_t f(\mathbf{x})$ satisfy all above mentioned axioms (except affine invariance). Set $u(t,\mathbf{x}) = D_t f(\mathbf{x})$ (resp. $E_t f(\mathbf{x})$). Then $u(t,\mathbf{x})$ is the viscosity solution of

$$\partial u/\partial t = ||Du||_B.$$

$(resp. \partial u/\partial t = -||Du||_B.)$

3.3 The anisotropic diffusion of Perona and Malik.

An important improvement of the classical linear multiscale analysis, with a more accurate edge detection, was proposed by Perona and Malik [7] who also propose a PDE model. The main idea of Perona and Malik is to introduce a part of the edge detection step in the filtering itself, allowing an interaction between scales from the beginning in the algorithm.

They propose to replace the heat equation by a nonlinear equation :

$$\frac{\partial u}{\partial t} = div(g(|Du|)Du) \tag{4}$$

$$u(0) = u_0.$$

In this equation, g is a smooth, nonincreasing function with $g(0) = 1, g(s) > 0$ and $g(s)$ tending to zero at infinity. The idea is that the smoothing process obtained by the equation is "conditional": if $Du(x,y)$ is large, then the diffusion will be small, and therefore, the exact localization of the "edges" will be kept. If $Du(x,y)$ is small, then the diffusion will tend to regularize still more around (x,y). Thus the choice of g corresponds to a sort of thresholding.

3.4 The Osher and Rudin constrained smoothing.

Osher and Rudin [9, 11] proposed a variational method for denoising images. In order to denoise an image u_0, they propose to minimize $\int |Du(\mathbf{x})|d\mathbf{x}$ under constraints corresponding to the *a priori* knowledge on the statistics of the noise (for instance, the variance). The minimization algorithm leads to the evolution equation

$$\frac{\partial u}{\partial t} = div(Du/|Du|) + \lambda(u)(u - u_0)$$

where λ is a Lagrange multiplier depending on u.

3.5 The Alvarez-Lions-Morel "pure" anisotropic diffusion : mean curvature motion.

In [2], Alvarez, Lions and Morel propose and study a class of nonlinear parabolic differential equations for image processing which originates from the Perona and Malik original idea and Osher and Sethian's "mean curvature motion" :

$$\frac{\partial u}{\partial t} = |Du|div(Du/|Du|) = |Du|curv(u), \qquad u(0,\mathbf{x}) = u_0(\mathbf{x}) \tag{5}$$

Roughly speaking, the interpretation of the terms of the equation are as follows. The term $|Du|div(Du/|Du|)$ represents the degenerate diffusion term which diffuses u in the direction orthogonal to its gradient Du and does not diffuse at all in the direction of Du. This equation has recently received a lot of attention because of its geometrical interpretation. Indeed, at least formally (see S. Osher and J. Sethian [10], L. C. Evans and J. Spruck [4]) the level sets of the solution move in the normal direction with a speed proportional to their mean curvature (curvature in two dimensions).

4 Classification of image scale-space under geometrical axioms.

We shall consider a multiscale analysis defined by an equation $\partial u/\partial t = F(D^2u, Du, t)$, for a function F nondecreasing with respect to its first argument and continuous where $Du \neq 0$.

In any points where $Du \neq 0$, η represents the direction of the gradient and ξ the orthogonal direction of the gradient. We shall consider $u_{\xi\xi} = D^2u(Du^\perp/|Du|, Du^\perp/|Du|)$, $u_{\eta\xi} = D^2u(Du/|Du|, Du^\perp/|Du|)$, and $u_{\eta\eta} = D^2u(Du/|Du|, Du/|Du|)$.

Theorem 4 *(N=2)*
Let T_t be a multiscale analysis satisfying [Affine Inv.] *(and the architectural axioms), then, with a possible rescaling $t' = h(t)$,*

$$\frac{\partial u}{\partial t} = \frac{1}{t} G(t^4|Du|^3 curv(u), t^4|Du|^2(u_{\eta\eta} - \frac{(u_{\eta\xi})^2}{u_{\xi\xi}}) \tag{6}$$

where G is a continuous function , and $curv(u)$ is given by

$$curv(u) = div(Du/|Du|) = |Du|^{-1}u_{\xi\xi} \tag{7}$$

(This term can be interpreted as the curvature of the level line $\{\mathbf{y}, u(\mathbf{y}) = u(\mathbf{x})\}$ passing by $\mathbf{x}$.)

We then continue with the morphological axiom,

Theorem 5 *Let $N = 2$ and T_t be a multiscale analysis satisfying* [Morphological Inv.]. *Then T_tu staisfies the equation*

$$\frac{\partial u}{\partial t} = |Du|G(curv(u), Du/|Du|, t), \tag{8}$$

for some function G on $I\!R * I\!R^2$, continuous on $I\!R * (I\!R^2/\{0\})$, such that $G(\lambda, p)$is nondecreasing with respect to $\lambda \in I\!R$, for all $p \in I\!R^2, p \neq 0$.

Equation 8 can be translated into an equation applied on each level curve of the image u. If we parametrize for a λ in $I\!R$ fixed, the level surface $u(\mathbf{x}, t) = \lambda$ by $\mathbf{C}(s, t)$, with s the arc length, we have

$$\frac{\partial \mathbf{C}(s,t)}{\partial t} = \mathbf{N}(s,t)G(curv(u), Du/|Du|, t)$$

where $\mathbf{N}$ denotes the normal vector of the curve. We clearly see that the function G can be interpreted as the speed of the motion of the level curve. This kind of equation contains the basic operations of mathematical morphology. Indeed, dilation and erosion can be interpreted as a (forward or backward) motion of level curves at constant speed in the direction of the normal. We also find the mean curvature motion, $(G(curv(u), Du/|Du|, t) = curv(u))$, which corresponds to the motion of the level curve with a speed equal to the curvature.

We finish by the combination of morphological axiom and affine invariance.

Theorem 6 $(N = 2)$
(i) Let T_t be a multiscale analysis defined by an equation $\partial u/\partial t = |Du|G(curv(u), t)$ (which derives from the fact that the multiscale analysis satisfies [Morphological Inv.], [Isom. invariance]. *) If the multiscale analysis satisfies* [Scale Inv.], *then, with a possible rescaling $t = h(t)$, the equation of the multiscale analysis is*

$$\frac{\partial u}{\partial t} = |Du|\beta(tcurv(u)), \tag{9}$$

where β is a continuous nondecreasing real function.
(ii) If we assume in addition [Affine Inv.], *then the only possible equation is, up to a rescaling,*

$$\frac{\partial u}{\partial t} = |Du|(tcurv(u))^{1/3}. \tag{10}$$

Returning to the formulation in terms of curve evolution, Equation 10 can be written as

$$\frac{\partial \mathbf{C}}{\partial t} = k^{\frac{1}{3}}\mathbf{N} \tag{11}$$

(where k is the curvature), By analogy with the euclidian case, Sapiro and Tannenbaum have proven (see [12]) that this equation is equivalent to

$$\frac{\partial \mathbf{C}}{\partial t} = \frac{\partial^2 \mathbf{C}}{\partial s^2}, \tag{12}$$

if s is a special parameter. The affine length s, which is defined by the relation $[\mathbf{C}_s, \mathbf{C}_{ss}] = 1$, where $[,]$ stands for the determinant of two vectors. That is why Equation 10 is often assimilated to an intrinsic heat equation.

5 MOVIES

5.1 Axioms for movie scale-spaces.

We shall consider the same axioms as these for image scale-spaces, but because of the time special rule, we must make adaptation of some of the geometrical axioms.

First, we consider euclidian and affine invariance only for the two first coordinates (that is, the spatial variables.) Indeed, a rotation involving time and space has no signification.

Secondly we state two new axioms :
[Time acceleration inv.] For any e in $I\!R$ we define by S_e the linear map $S_e(x, y, \theta) = (x, y, e\theta)$ There exists $t''(t, e)$ such that $S_e(T_{t''}u) = T_t(S_e u)$ (t'' can be different from the t' introduced in the spatial scale invariance axiom.)

We finally assume that the analysis is invariant under "traveling". Motion of a whole single picture with constant velocity v does not alter the analysis. This is an essential assumption if we want to analyse each velocity in the same way.

We denote by $\vec{i}$ and $\vec{j}$ spatial unit vectors, by $\vec{k}$ temporal unit vectors, and by B_v the galilean translation operator

$$B_v u(x,y,\theta) = u(x - \langle v,\vec{i}\rangle\theta, y - \langle v,\vec{j}\rangle\theta, \theta)$$

[Galilean Inv.] For any v and t, we have

$$B_v(T_{t''}u) = T_t(B_v u)$$

5.2 Equations for movie scale-spaces.

Let $u(x,y,\theta)$ or $u(\mathbf{x})$ be a movie. θ denotes the time variable.
We note $\nabla u = (u_x, u_y, 0)$, the spatial gradient of $u(x,y,\theta)$.
We denote (when $\nabla u \neq 0$) by $curv(u)$ the spatial curvature of the level sets.
Let $\mathcal{G}$ be the gaussian curvature of the level sets, (when $Du \neq 0$).
We define

$$accel(u) = \left(\frac{|Du|}{|\nabla u|}\right)^{\frac{1}{4}} \frac{\mathcal{G}}{curv(u)}$$

Theorem 7 *Let T_t be a multiscale analysis satisfying the axioms* [Pyramidal architecture], [Regularity], [Transl. Inv.], [Isom. invariance], [Morphological Inv.], *and* [Galilean Inv.]. *Then, the multiscale analysis of movies is governed by an equation of the type*

$$\frac{\partial u}{\partial t} = |\nabla u|\; F(curv(u), accel(u), t) \tag{13}$$

If we add [Affine Inv.] *and* [Time acceleration inv.], *the only possible scale-space equations are*

$$(AMG) \qquad \frac{\partial u}{\partial t} = |\nabla u|\; curv(u)^{\frac{1-q}{3}}\; (sgn(curv(u))accel(u)^q)^+ \tag{14}$$

for some $q \in]0,1[$, or

$$(q=0) \qquad \frac{\partial u}{\partial t} = |\nabla u| curv(u)^{\frac{1}{3}}, \quad or$$

$$(q=1) \qquad \frac{\partial u}{\partial t} = |\nabla u| sgn(curv(u))(sgn(curv(u))accel(u))^+$$

In the above formulae, we use the convention that the power preserves the sign, that is $a^q = |a|^q \mathrm{sgn}(a)$. And we set $x^+ = sup(0,x)$.

5.3 Interpretation of the equation.

We begin to focus on the new operator $accel(u)$. We can interpret $accel(u)$ as an "apparent acceleration".

Let us consider a point $\mathbf{x} = (x,y,\theta)$ of the movie. We denote the set of vectors $\mathcal{W}$ by

$$\mathcal{W} = \{(\alpha,\beta,1),\;\; \alpha,\beta \in I\!R\}$$

This set is the set of the possible speed vectors. Indeed, if $v \in \mathcal{W}$ is the speed of the point $\mathbf{x}$, this means that at time $\theta + d\theta$ the point $\mathbf{x}$ is on the point $\mathbf{x} + vd\theta = (x + \alpha d\theta, y + \beta d\theta, \theta + d\theta)$.

Lemma 1 *At any point where u is C^2, and, $\nabla u \neq 0$, we have,*

$$|\nabla u|(sgn(curv(u))\, accel(u))^+ = \tag{15}$$

$$min_{V_b,V_a \in \mathcal{W}} \frac{1}{\Delta\theta^2}(|u(\mathbf{x} - V_b\Delta\theta) - u(\mathbf{x})| + |u(\mathbf{x} + V_a\Delta\theta) - u(\mathbf{x})| + \Delta\theta|\langle\nabla u, V_b - V_a\rangle|) + o(1)$$

In the above formula, we can consider V_a and V_b as "candidate" speed vectors respectively between θ and $\theta + d\theta$, and between $\theta - d\theta$ and θ. Then,

$|u(\mathbf{x} - V_b\Delta\theta) - u(\mathbf{x})| + |u(\mathbf{x} + V_a\Delta\theta) - u(\mathbf{x})|$ corresponds to the variation of grey level value of the point with the candidate speed vectors V_a and V_b.

The second term, $|\langle\nabla u, V_b - V_a\rangle|)$, corresponds to the acceleration in the direction of the spatial gradient for going from V_b into V_a.

Now, let us make the classical assumption that the point does not change its grey level value during its motion. If we choose V_a and V_b equal to the speed of the point, the first term is then equal to zero. Only the second term remains which corresponds to the acceleration in the direction of the spacial gradient.

We now come back to the second part of the equation (14),

$$|\nabla u|\ curv(u)^{\frac{1-q}{3}}\ (\mathrm{sgn}(curv(u))accel(u)^q)^+$$

We can cut it into two terms, the first one $|\nabla u|\ curv(u)^{\frac{1-q}{3}}$ corresponding to a spatial diffusion, like the scale-spaces of images that we have already seen. The second one can be seen as the speed of the execution of this spatial smoothing. The larger is the second term, the spatial smoothing. We deduce that a point with large acceleration will be smoothed very quickly, and in the other way, a point in uniform motion will be kept at any scale.

This equation is an adaptative spatial smoothing depending on the acceleration. We used this equation for movie denoising :

Up : Original noisy movie of a square and a point in uniform motion, 24 images 64 × 64. All the 24 images of the movie are presented in a table 6*4. The first image is up-left, the last is down-right. The images are classed left to right and up to down.

Down : The same movie smoothed. No boundary conditions have been chosen for the discretization of the PDE. We simply impose no evolution of the first and the last image. This provides some consequences in the evolution, but not too much due to the local aspect of AMG equation.

Spurious trajectories due to noise are eliminated. Only the true trajectories remain : the square and the point.

References

[1] L. Alvarez, F. Guichard, P.L. Lions and J.M. Morel. *Axioms and fundamental equations of image processing. Report 9216, CEREMADE. Université Paris Dauphine, 1992.* Accepted at Arch. for Rat. Mech., July 1992.

[2] L. Alvarez, P.L. Lions and J.M. Morel. *Image selective smoothing and edge detection by nonlinear diffusion (II).* SIAM J. Num. Analysis. 1992.

[3] M.G. Crandall, H. Ishii and P-L. Lions. *User's guide to viscosity solution of second order partial differential equation.* Preprint, CEREMADE, 1990.

[4] L.C. Evans and J. Spruck. *Motion of level sets by mean curvature*, I., preprint.

[5] J.J. Koenderink. *The structure of images*, Biol Cybern. 50, p 363-370, 1984.

[6] C. Lopez, J.M. Morel. *Axiomatisation of shape analysis and application to texture hyper-discrimination.* Procedings of the Trento Conference on Motion by Mean Curvature and Related Models. Springer 1992.

[7] P. Perona and J. Malik *A scale space and edge detection using anisotropic diffusion.* Proc. IEEE Computer Soc. Workshop on Computer Vision, 1987.

[8] D.Marr and E. Hildreth, *Theory of edge detection.* Proc. Roy. Soc. Lond. B207 p 187-217, 1980.

[9] S. Osher and L. Rudin, *Feature-oriented image enhancement using shock filters.* SIAM J. on Numerical Analysis, to appear.

[10] S. Osher and J. Sethian. *Fronts propagating with curvature dependent speed : algorithms based on the Hamilton-Jacobi formulation.* J. Comp. Physics, 79,p.12-49, 1988.

[11] L. Rudin, S. Osher, E. Fatemi *Nonlinear Total Variation based noise removal algorithms*, Physica D., Proceedings of 11th Conf. on Experimental Mathematics, Sept. 1992.

[12] G. Sapiro and A. Tannenbaum. *On affine plane curve evolution.* Departement of electrical engineering. Technion Israel Institute of Technology. Haifa. Israel. EE Pub 821. February 1992, submitted.

[13] A.P. Witkin *Scale-space filtering.* Proc. of IJCAI, Karlsruhe, p 1019-1021, 1983.

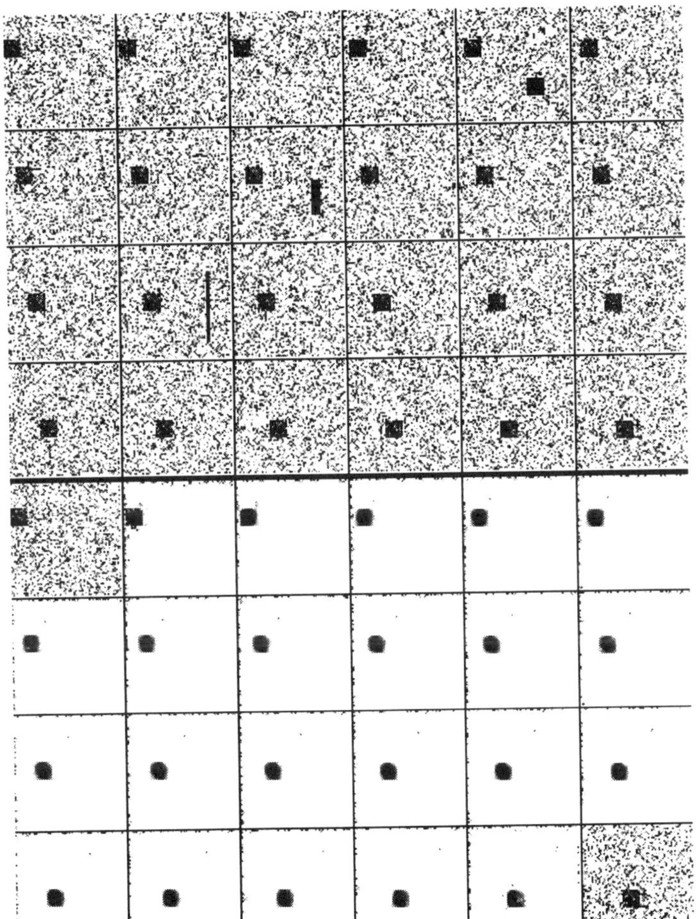

DISCONTINUOUS SOLUTIONS OF THE NAVIER-STOKES EQUATIONS FOR MULTIDIMENSIONAL, ISENTROPIC FLOW

DAVID HOFF*

We discuss some recent global existence results for weak solutions of the Navier-Stokes equations

$$\rho_t + \mathrm{div}\,(\rho u) = 0 \tag{1}$$

$$(\rho u^j)_t + \mathrm{div}\,(\rho u^j u) + P(\rho)_{x_j} = \varepsilon \Delta u^j + \lambda \mathrm{div}\, u_{x_j} \tag{2}$$

in two and three space dimensions when the Cauchy data (ρ_0, u_0) is small but possibly discontinuous. Here $t \geq 0$ is time, $x \in \mathbf{R}^n, n = 2, 3$, is the spatial coordinate, and $\rho(x,t), u(x,t) = \left(u^1(x,t), \ldots, u^n(x,t)\right)$, and $P = P(\rho)$ represent the fluid density, velocity, and pressure, $\varepsilon > 0$ and $\lambda \geq 0$ are viscosity constants, and div and Δ are the usual spatial divergence and Laplace operators. Specifically, we fix a positive constant density $\tilde{\rho}$, and we assume that $\rho_0 - \tilde{\rho}$ is small in L^∞, that $\rho_0 - \tilde{\rho}$ and u_0 are small in L^2 (when $n = 2$ the L^2 norm must be weighted slightly), and that u_0 is bounded in L^{2^n}. In particular, the initial data may be discontinuous across a hypersurface of $\mathbf{R}^n$.

The results discussed here extend our earlier work [7], in which we proved global existence for such data when the pressure is given by an isothermal equation of state $P(\rho) = K\rho$. An essential difficulty prevented our extending that result to more general equations of state, however. To be specific, we obtained solutions in the isothermal case as limits of smooth solutions (ρ^δ, u^δ) corresponding to mollified initial data $\rho_0^\delta = j_\delta * \rho_0, u_0^\delta = j_\delta * u_0$. A number of important a priori estimates, independent of δ, were obtained, and these were sufficient to yield a limiting pair (ρ, u) in the limit as $\delta \to 0$. Most of these a priori estimates concerned various (partial) regularity properties of the u^δ; the only

* This research was supported in part by the NSF under Grant No. DMS-9201597.

estimates available for the ρ^δ, however, were uniform pointwise bounds. As a result, we were able to conclude only that $\rho^\delta \rightharpoonup \rho$ weakly, and this was insufficient to pass the convergence under a nonlinear pressure term. We therefore could not conclude that the limiting pair (ρ,u) is a weak solution in the required sense, except in the isothermal case $P(\rho) = K\rho$.

Here we describe an additional argument, disjoint from the analysis of [7] (but starting from the a priori bounds of that paper, which were derived under very mild assumptions on P), which proves that, when $P(\rho) = K\rho^\gamma, \gamma > 1$, the convergence $\rho^\delta \to \rho$ is in fact strong. This result is somewhat surprising, given the absence of any uniform regularity information for the approximate densities ρ^δ. It then follows that $P(\rho^\delta) \to P(\rho)$, so that the pair (ρ,u) is indeed a weak solution in the required sense. An outline of this argument will be given below, following the statement of our main theorem. Complete technical details are given in [8].

We now give a precise formulation of our result. We fix constants $\varepsilon > 0$ and $\lambda \geq 0$, and assume that $\lambda/\varepsilon < b_n$, where

$$b_n = \begin{cases} 1+\sqrt{2}, & n = 2 \\ (1+\sqrt{3})/3, & n = 3. \end{cases} \tag{3}$$

The pressure will be given by $P(\rho) = K\rho^\gamma$, where K is a positive constant and $\gamma > 1$, and the initial data (ρ_0,u_0) will be measured in the norm given by

$$C_0 = \| \rho_0 - \tilde{\rho} \|^2_{L^\infty(\mathbf{R}^n)} + \int_{\mathbf{R}^n} \left[(\rho_0(x) - \tilde{\rho})^2 + | u_0(x) |^2 \right] (1 + | x |^2)^\beta dx. \tag{4}$$

Here β is positive but arbitrarily small when $n = 2$, and $\beta = 0$ when $n = 3$. (This weighting of the L^2 norm serves to compensate for the growth of the fundamental solution of the Laplacian at infinity in two space dimensions.) Our main existence result may then be stated as follows:

Theorem: *Let $n = 2$ or 3, fix $\varepsilon, \lambda, P = K\rho^\gamma, \tilde{\rho}$, and β as above, and let $N > 0$ be given (N may be arbitrarily large). Then there are positive constants a and C, depending on*

$K, \gamma, \tilde{\rho}, \beta, N$, and on upper bounds for ε, λ, and $(b_n\varepsilon - \lambda)^{-1}$ (see (3)), and there is a global positive constant θ, such that, given initial data (ρ_0, u_0) satisfying

$$C_0 \le a \text{ and } \|u_0\|_{L^{2n}} \le N, \tag{5}$$

the Cauchy problem for (1)–(2) with data (ρ_0, u_0) has a global weak solution (ρ, u) for which

$$\|\rho - \tilde{\rho}\|_{L^\infty(\mathbf{R}^n \times [0,\infty))} \le CC_0^\theta \tag{6}$$

and

$$\begin{aligned} &\sup_{t\ge 0} \int \left[|u(x,t)|^2 + \sigma(t)|\nabla u(x,t)|^2 \right] dx \\ &\qquad + \operatorname{ess\,sup}_{t>0} \int \left[(\rho(x,t) - \tilde{\rho})^2 + \sigma(t)^n |\dot{u}(x,t)|^2 \right] dx \\ &\qquad + \int_0^\infty \int \left[|\nabla u|^2 + \sigma|\dot{u}|^2 + \sigma^n |\nabla \dot{u}|^2 \right] dxdt \\ &\le CC_0^\theta, \end{aligned} \tag{7}$$

where $\sigma(t) = \min\{1,t\}$ and $\dot{u}$ is the convective derivative $\dot{u}^j = u_t^j + u \cdot \nabla u^j$.

Inequalities (6) and (7) reflect uniform regularity properties of the approximate solutions (ρ^δ, u^δ) which are inherited by the limiting pair (ρ, u). Additional regularity properties of the solution are given in [7]. These properties are delicately interrelated, and seem to be optimal for data in the given regularity classes, but nearly minimal for a rigorous existence theory. See the introduction to [7] for an extensive discussion at the heuristic level in which these properties are derived and interpreted in terms of the Rankine–Hugoniot conditions applied to (1)–(2).

We now give an outline of the proof of the above theorem. First, as discussed above, we let (ρ^δ, u^δ) be approximate solutions corresponding to mollified initial data $(\rho_0^\delta, u_0^\delta)$, and we attempt to extract a limiting solution (ρ, u) in the limit as the mollifying parameter δ

goes to zero. Now, the solutions (ρ^δ, u^δ) are shown in [7] to satisfy a number of a priori estimates, independently of δ, among which we record the following:

$$C^{-1} \leq \rho^\delta \leq C \tag{8}$$

$$\sup_t \left(\|u^\delta(\cdot,t)\|^2 + \|\rho^\delta(\cdot,t) - \tilde{\rho}\|^2 \right) + \int_0^\infty \|\nabla u^\delta(\cdot,t)\|^2 \, dt \leq C \tag{9}$$

$$\sup_{t \geq \tau} \|\nabla u^\delta(\cdot,t)\|^2 + \int_\tau^\infty \left(\|\dot{u}^\delta(\cdot,t)\|^2 + \||\nabla u^\delta(\cdot,t)|^2\|^2 \right) dt \leq C(\tau) \, , \tau > 0 \tag{10}$$

$$\langle u^\delta \rangle_{\{t \geq \tau\}} \leq C(\tau) \, , \tau > 0 \, . \tag{11}$$

Here $\|\cdot\|$ is the usual $L^2(\mathbf{R}^n)$ norm, and $\langle \cdot \rangle$ denotes a modulus of Hölder continuity. These estimates are derived under fairly mild conditions on P: it is assumed only that $P, P' > 0$ near $\tilde{\rho}$. The bounds (8)–(11) then provide sufficient compactness to extract a limiting sequence $\delta \to 0$ for which

$$\rho^\delta \rightharpoonup \rho \tag{12}$$

$$u^\delta \to u \tag{13}$$

$$\nabla u^\delta \rightharpoonup \nabla u \, . \tag{14}$$

Here the symbols $\to$ and $\rightharpoonup$ denote respectively strong and weak convergence in $L^2_{loc}(\mathbf{R}^n \times (0,\infty))$. It follows easily that the limiting pair (ρ, u) is a weak solution of the mass equation (1). On the other hand, we cannot conclude that the momentum equation (2) holds until we show that $P(\rho^\delta) \rightharpoonup P(\rho)$. This is obvious from (12) for the isothermal case $P(\rho) = K\rho$, which is the case treated in [7]. For the isentropic case $P(\rho) = K\rho^\gamma$, $\gamma > 1$, under consideration here, however, we can conclude only that, modulo a further subsequence, $P(\rho^\delta)$ is weakly convergent, say to p:

$$P(\rho^\delta) \rightharpoonup p \geq P(\rho)\ , \tag{15}$$

the last inequality following from the convexity of P. We therefore need to show that $p = P(\rho)$, and this will be a consequence of (and in fact equivalent to) the *strong* convergence $\rho^\delta \to \rho$. To prove this strong convergence, it suffices to exhibit a strictly convex, C^2 function $g = g(\rho)$ for which $g(\rho^\delta) \rightharpoonup g(\rho)$; for then, given any compact set $S \subseteq \{t \geq 0\}$,

$$\iint_S [g(\rho^\delta) - g(\rho)]\, dxdt \geq \iint_S g'(\rho)(\rho^\delta - \rho)\, dxdt$$

$$+ C^{-1} \iint_S (\rho^\delta - \rho)^2\, dxdt\ .$$

The integral on the left goes to zero by the weak convergence $g(\rho^\delta) \rightharpoonup g(\rho)$, and the first integral on the right goes to zero by (12). It then follows that $\rho^\delta \to \rho$ strongly in $L^2(S)$, as required. With the benefit of hindsight, we find that the appropriate function to take in the above argument is $g(\rho) = \rho^{1-\gamma}$. Appealing again to (8), as well as to the convexity of g, we may then assert that

$$(\rho^\delta)^{1-\gamma} \rightharpoonup r \geq \rho^{1-\gamma}\ . \tag{16}$$

The proof will therefore be complete if we show that $r \leq \rho^{1-\gamma}$.

At the heart of the matter is a form of compensation arising from the extra regularity of a distinguished variable $F = (\varepsilon+\lambda)\mathrm{div}\, u - P(\rho)$, which we have termed the "effective viscous flux." Heuristic considerations based upon the Rankine-Hugoniot conditions indicate that, at least formally, the jump in F across a surface of discontinuity of the solution (ρ, u) should be zero; see the introduction to [7] both for this argument as well as for a precise formulation of the regularity properties of F. For our present purposes it suffices to record the fundamental pde satisfied by F: for smooth solutions (ρ, u) of (1)–(2) we may write

$$\rho \dot{u}^j + P(\rho)_{x_j} = (\varepsilon + \lambda)\mathrm{div}\, u_{x_j} + \varepsilon(u^j_{x_k} - u^k_{x_j})_{x_k}.$$

Upon differentiating with respect to x_j and summing over j, we then obtain the elegantly simple and very important relation

$$\Delta F = \operatorname{div}(\rho\dot{u}). \tag{17}$$

Now, the weak convergence results (14) and (15) show that

$$F^\delta \equiv (\varepsilon+\lambda)\operatorname{div} u^\delta - P(\rho^\delta) \rightharpoonup f \equiv (\varepsilon+\lambda)\operatorname{div} u - p\ .$$

The distinguished role of the effective viscous flux F allows us to conclude that this convergence is in fact strong:

Lemma 1: $F^\delta \to f$.

To prove the lemma we apply (10) and (17) to obtain that, for $\tau > 0$,

$$\int_\tau^\infty \int_{\mathbf{R}^n} |\nabla F^\delta|^2\, dxdt \le C \int_\tau^\infty \int_{\mathbf{R}^n} |\dot{u}^\delta|^2\, dxdt \le C(\tau)\ ,$$

independently of δ. A similar bound holds for $\int\int (F_t^\delta)^2$. The result then follows from the Kondrachov theorem.

Returning now to (16), we shall compare r and $\rho^{1-\gamma}$ by examining their respective pde's. We first compute from the mass equation (1) that

$$\begin{aligned}\frac{\partial}{\partial t}\left[(\rho^\delta)^{1-\gamma}\right] + \operatorname{div}\left[(\rho^\delta)^{1-\gamma}u^\delta\right] &= \gamma(\rho^\delta)^{1-\gamma}\operatorname{div} u^\delta \\ &= \frac{\gamma}{\varepsilon+\lambda}\left[(\rho^\delta)^{1-\gamma}F^\delta + K\rho^\delta\right].\end{aligned} \tag{18}$$

Upon taking weak limits as $\delta \to 0$ and appealing to the results (12), (13), (16), and Lemma 1, we then conclude:

Lemma 2: *r is a weak solution of the initial-value problem*

$$\begin{cases} r_t + \operatorname{div}(ru) = \alpha\,[rf + K\rho] \\ r(\cdot,0) = \rho_0^{1-\gamma}\ , \end{cases} \tag{19}$$

where $\alpha = \dfrac{\gamma}{\varepsilon + \lambda}$.

To obtain the pde satisfied by $\rho^{1-\gamma}$, we pursue an idea of DiPerna and Lions [1], that certain *nonlinear* changes of dependent variable are admissable in the mass equation (1), even for weak solutions, owing to the fact that this equation is *linear* in ρ for given u. A fairly technical argument based on space-only mollifiers allows us to obtain:

Lemma 3: *The limiting density ρ is a weak solution of the initial-value problem*

$$(20)\qquad \begin{cases} (\rho^{1-\gamma})_t + \operatorname{div}(\rho^{1-\gamma}u) = \gamma\rho^{1-\gamma}\operatorname{div} u = \alpha\big[\rho^{1-\gamma}\big((\varepsilon+\lambda)\operatorname{div} u - P(\rho)\big) + K\rho\big] \\ (\rho^{1-\gamma})(\cdot,0) = \rho_0^{1-\gamma} . \end{cases}$$

We now define $w = r - \rho^{1-\gamma}$ and subtract (20) from (19) to obtain that, weakly,

$$\begin{aligned} w_t + \operatorname{div}(wu) &= \alpha\big[(\varepsilon+\lambda)w\operatorname{div} u - rp + \rho^{1-\gamma}P(\rho)\big] \\ &= \alpha\big[(\varepsilon+\lambda)w\operatorname{div} u - (r - \rho^{1-\gamma})p + \rho^{1-\gamma}(P - p)\big] \qquad (21) \\ &\le \alpha w f . \end{aligned}$$

because $P - p \le 0$, by (15). The differential inequality (21), together with the initial condition $w(\cdot,0) = 0$, suggest that $w \le 0$ a.e., which is precisely the inequality $r \le \rho^{1-\gamma}$ we are seeking. The rigorous argument is quite a bit more difficult technically, however. To explain this, we let $\phi = \phi(x,t)$ be an appropriate nonnegative test function. Then from (21),

$$(22)\qquad \int_{\mathbf{R}^n} w(x,\cdot)\phi(x,\cdot)\,dx\Big|_0^T - \int_0^T\int_{\mathbf{R}^n} w(\phi_t + \nabla\phi\cdot u + \alpha f\phi)\,dxdt \le 0 .$$

We may therefore attempt to deduce information about the sign of w by choosing ϕ to satisfy the adjoint problem

$$(23)\qquad \begin{cases} \phi_t + \nabla\phi\cdot u + \alpha f\phi = -G , \ t \le T \\ \phi(x,T) = 0 , \end{cases}$$

where $G = G(x,t)$ is nonnegative, say with compact support. Substituting this into (22), we then obtain

$$\iint wG \leq 0 \ .$$

If this holds for all such G, we could then conclude that $w \leq 0$ a.e. The difficulty, however, is that the regularity of the coefficients in (23) degrades at $t = 0$, and, owing to the lack of information concerning $\nabla\rho$, we have no estimate of the form $\|w(\cdot,t)\| \to 0$ as $t \to 0$, in any norm. Indeed, if G has compact support bounded below by the line $t = t_1 > 0$, then the solution ϕ of (23) satisfies

$$\phi(x(t),t) = \phi(x(t_1),t_1)\exp\left(\int_t^{t_1} \alpha f(x(s),s)\,ds\right) \ , \ t < t_1 \ ,$$

where $(x(t),t)$ is a streamline of the flow field u. Elementary considerations indicate that the term on the right here blows up like $\exp(t^{-q})$, where $q = q(n)$.

To circumvent this difficulty we construct explicitly an approximate adjoint solution ϕ from a function which is known a priori to be bounded as $t \to 0$, namely, the limiting density ρ itself. By appropriately cutting off, mollifying, and taking powers of ρ, we can construct a nonnegative test function ϕ which satisfies (23) with $G = p$, suitably cut off and mollified, and this is sufficient for our purpose. The technical details are somewhat involved, and are given in [8]. We can then conclude that $w \leq 0$ a.e., that $r = \rho^{1-\gamma}$, and therefore that $\rho^\delta \to \rho$ strongly, as required.

We remark that the (apparently) fortuitous inequalities in (15) and (16), which are crucial in the proof, are direct consequences of the hyperbolicity of the associated inviscid system (1)–(2) with $\varepsilon = \lambda = 0$, and of the specific character of the pressure in an isentropic flow. The other crucial ingredient in the proof is the role of the effective viscous flux F, whose regularity provides the compensation necessary to preclude oscillations in the sequence ρ^δ.

The questions addressed here and in [7] and [8] are motivated by the fact that discontinuous solutions are fundamental in the mathematical theory of inviscid models for compressible flow. It is natural, therefore, to seek a rigorous theory for viscous models which accomodates these solutions. Such a theory has been attained for the case of one space dimension, even for nonisentropic, nonisothermal flows (see Hoff [2] – [5], Serre [10] – [11], and Shelukhin [12]), and for multidimensional, spherically symmetric flows in the isothermal case (Hoff [6]). Existence results for multidimensional flows are obtained in Hoff [7] and Lions [9]. Lions proves the existence of solutions of (1)–(2) with *large*, discontinuous initial data for the isentropic case $P(\rho) = K\rho^{\gamma}$, when γ is sufficiently large, depending on the space dimension n.

REFERENCES

[1] R.J. DiPerna and P.L. Lions, Ordinary differential equations, transport theory, and Sobolev spaces, Inventiones Mathematicae 98 (1989), 511–547.

[2] David Hoff, Construction of solutions for compressible, isentropic Navier-Stokes equations in one space dimension with nonsmooth initial data, Proc. Royal Soc. Edinburgh, Sect. A103 (1986), 310–315.

[3] David Hoff, Global existence for 1D, compressible, isentropic Navier-Stokes equations with large initial data, Trans. AMS 303, no. 11 (1987), 169–181.

[4] David Hoff, Discontinuous solutions of the Navier-Stokes equations for compressible flow, Archive Rational. Mech. Ana. 114 (1991), 15–46.

[5] David Hoff, Global well-posedness of the Cauchy problem for nonisentropic gas dynamics with discontinuous initial data, J. of Differential Equations 95, No.1 (1992), 33–73.

[6] David Hoff, Spherically symmetric solutions of the Navier-Stokes equations for compressible, isothermal flow with large, discontinuous initial data, Indiana Univ. Math. J. 41 (1992), 1225–1302.

[7] David Hoff, Global solutions of the Navier-Stokes equations for multidimensional, compressible flow with discontinuous initial data, to appear in J. Diff. Eqns.

[8] David Hoff, Strong convergence to global solutions for compressible, viscous, multidimensional flow, with polytropic equations of state and discontinuous initial data, submitted to Archive Rational. Mech. Ana.

[9] P.L. Lions, Existence globale de solutions pour les équations de Navier-Stokes compressibles isentropiques, C.R. Acad. Sci. Paris 316 (1993), 1335–1340.

[10] Denis Serre, Solutions faibles globales des équations de Navier-Stokes pour un fluide compressible, C.R. Acad. Sc. Paris 303 (1986), 639–642.

[11] Denis Serre, Sur l'équation monodimensionnelle d'un fluide visqueux, compressible et conducteur de chaleur, C.R. Acad. Sc. Paris 303 (1986), 703–706.

[12] V.V. Shelukhin, On the structure of generalized solutions of the one dimensional equations of a polytropic viscous gas, Prikl. Matem. Mekhan. 48 (1984), 912–920.

Nonlinear diffusive phenomena of solutions for the system of compressible adiabatic flow

L. Hsiao

We study the following systems of compressible adiabatic flow with external friction acting on, such as a flow through a porous media.

$$\begin{cases} v_t - u_x = 0 \\ u_t + p(v,s)_x = -\alpha u, \alpha > 0, \alpha > 0 \\ [e(v,s) + \frac{u^2}{2}]_t + (pu)_x = -\alpha u^2 \end{cases} \tag{1}$$

where v denotes the specific volume, u is velocity, s stands for entropy, e denotes the specific internal energy for which $e_s \neq 0$ and $e_v + p = 0$ holds due to the second law of thermodynamics, and p denotes pressure with $p_v < 0$ for $v > 0$.

The system (1) is equivalent to the following system (2) for smooth solutions.

$$\begin{cases} v_t - u_x = 0 \\ u_t + p(v,s)_x = -\alpha u, \alpha > 0 \\ s_t = 0 \end{cases} \tag{2}$$

which is strictly hyperbolic with eigenvalues $\lambda_1 = -\sqrt{-p_v}, \lambda_2 \equiv 0$ and $\lambda_3 = \sqrt{-p_v}$.

We are interested in the influence, caused by the damping mechanism in (1) or (2), to the qualitative behavior of the processes in consideration.

In a special situation when $s \equiv \bar{s}$ ($\bar{s}$ is a constant) initially, the system (2) becomes into

$$\begin{cases} v_t - u_x = 0 \\ u_t + p(v,\bar{s})_x = -\alpha u, \end{cases}$$

which is the system for isentropic gas flow with damping. It has been proved by Hsiao and Liu in [1] that the solution of the Cauchy problem

$$\begin{cases} v_t - u_x = 0 \\ u_t + p(v)_x = -\alpha u, \alpha > 0, p'(v) < 0 \text{ for } v > 0 \\ v(x,0) = v_0(x), u(x,0) = u_0(x), \text{ with } \lim_{x \to \mp\infty} (v_0(x), u_0(x)) = (v^\mp, u^\mp) \end{cases} \tag{3}$$

Academia Sinica, Institute of Mathematics, Beijing 100080, China

This work is supported partially by National Natural Science fundation of China.

can be described by the solution of the following problem

$$\begin{cases} v_t = -\frac{1}{\alpha}p(v)_{xx} \\ u = -\frac{1}{\alpha}p(v)_x \\ v(x,0) = \overline{v}^*(x+\overline{x}) \end{cases} \tag{4}$$

time-asymptotically, where $\overline{v}^*(\frac{x}{\sqrt{t+1}})$ is the similarity solution of $(4)_1$, with boundary condition $\overline{v}^*(\eta) \to v^{\mp}$ as $\eta \to \mp\infty$, and $\overline{x} \in R$ is a constant, determined uniquely by

$$\int_{-\infty}^{\infty} \{v_0(x) - \overline{v}^*(x+\overline{x})\}dx = \frac{u_+ - v_-}{-\alpha}$$

The system in (4) is obtained from the one in (3) by approximating the momentum equation in (3) with Darcy's law.

Moreover, it is proved that the L_2-norm and L_∞-norm of the difference between these two solutions tend to zero with a rate $t^{-1/2}$ as time tends to infinity.

This shows that certain nonlinear diffusive phenomena occur for the solutions of the original hyperbolic problem which is caused by the damping mechanism. It was very essential to establish the decay rate of L_2-norm on the corresponding required derivatives of the similarity solution $\overline{v}^*(\frac{x}{\sqrt{t+1}})$ in obtaining of the result in [1].

For general situation when s is not a constant initially, we have to handle the system (1) or (2) for which there was no result either on the global existence or large-time behavior of smooth solutions untill recently. The first result on global existence is obtained by Hsiao and Serre in [3] which deals with the initial data $(u,v,s)(x,0) = (u_0,v_0,s_0)(x)$ such that $(u_0(x), v_0(x)-\overline{v}) \in C^1$ with a compact support and $(s_0(x)-\overline{s}) \in C^2$ with a compact support, where $\overline{v} > 0, \overline{v}$ and $\overline{s}$ are constants. It is proved in [3] that there exists a globally defined classical solution for the Cauchy problem if the C^1-norm of $(u_0(x), v_0(x) - \overline{v})$ and the C^2-norm of $(s_0(x) - \overline{s})$ are small. This means that the damping dissipation is strong enough to preserve the smoothness of the initial data when it is of smallness. This global existence result is established by combining of the L_2-estimates on the solutions with the L_∞-estimates on the first derivatives.

The second result on the Cauchy problem of (2) is concerned with both existence and large-time behavior of solutions, given by Hsiao and Serre in [4] where the solution of (2) with initial data $(u,v,s)(x,0) = (u_0(x), v_0(x), s_0(x))$ is compared to the solution of

$$\begin{cases} v_t = -\frac{1}{\alpha}p(v,s)_{xx} \\ u = -\frac{1}{\alpha}p(v,s)_x \\ s_t = 0 \end{cases} \tag{5}$$

with $v(x,0) = v_0(x)$ and $s(x,0) = s_0(x)$, such that $\lim_{x\to\mp\infty} u_0(x) = 0$, $\lim_{x\to\mp\infty} v_0(x) = \overline{v} > 0$ and $\lim_{x\to\mp\infty} s_0(s) = \overline{s}$. It is shown in [4] that for any given data $(u_0(x), v_0(x), s_0(x))$ with $u_0(x) \in H^2(\mathbb{R}), (v_0(x) - \overline{v}) \in H^5(\mathbb{R})$ and $(s_0(x) - \overline{s}) \in H^4(\mathbb{R})$, there exists a unique globally defined classical solution for (2) which tends to the solution of (5) as $t \to \infty$ in L_∞-norm and L_2-norm, provided the initial data is of smallness.

In contrast with the case of $s_0(x) \equiv \overline{s}$, there is no similarity solution for $(5)_1$ in the case of $s_0(x) \not\equiv \overline{s}$. However, the function $\hat{v}(x)$ determined by $p(\hat{v}(x), s(x)) = p(\overline{v}, \overline{s})$ is just a special solution of $(5)_1$ which is used to be compared with the solution of Cauchy problem for $(5)_1$ to obtain the corresponding L_2-estimates on this solution of $(5)_1$ which is essential in getting the result in [4].

A more important problem is to consider the initial data with

$$\lim_{x\to\mp\infty} u_0(x) = u^\mp, \lim_{x\to\mp\infty} v_0(x) = v^\mp > 0 \text{ and } \lim_{x\to\mp\infty} s_0(x) = \overline{s}, \tag{6}$$

where v^- is not equal to v^+. This causes serious difficulties since there is no similarity solution for $(5)_1$ to be used and the special solution $\hat{v}(x)$ is not useful anymore because $v^- \neq v^+$.

Hsiao and Luo solved this problem recently in [2] under certain assumptions on the variety of $s_0(x)$ and $v_0(x) \to v^\mp$ fast enough.

Compare the solution of (2) with initial data $(u_0(x), v_0(x), s_0(x))$, denoted by (u_1, v_1, s_1), to the solution of (5) with initial data $v(x,0) = \tilde{v}(x+x^*, 0)$ and $s(x,0) = s_0(x)$, denoted by (u_2, v_2, s_2), where $\tilde{v}(x,t)$ is defined by

$$p(\tilde{v}(x,t), s(x)) = p(v^*(\frac{x}{\sqrt{t+1}}), \overline{s}), \tag{7}$$

and $v^*(\frac{x}{\sqrt{t+1}})$ is the similarity solution of

$$\begin{cases} v_t = -\frac{1}{\alpha} p(v, \overline{s})_{xx} \\ v^*(\eta) \to v^\mp \text{ as } \eta \to \mp\infty \end{cases} \tag{8}$$

while $x^* \in R$ is chosen uniquely by

$$\int_{-\infty}^{\infty} [v_0(x) - \tilde{v}(x+x^*, 0)]dx = \frac{u^+ - u^-}{-\alpha} \tag{9}$$

Take any smooth function $m_0(x)$ with a compact support and $\int_{-\infty}^{\infty} m_0(x)dx = 1$, and define $m(x,t) \equiv -\frac{(u^+-u^-)}{\alpha} m_0(x) e^{-\alpha t}$. Let $\hat{u}(x,t) \equiv u^- e^{-\alpha t} + \int_{-\infty}^{x} m_t(\xi, t)d\xi$. It is proved in [2] that

$$v_1 - v_2 - m \to 0$$

and

$$u_1 - u_2 - \hat{u} \to 0$$

as $t \to \infty$ in the L_2-norm and L_∞-norm, provided the initial data is of smallness.

This shows that the solution of the original hyperbolic problem with (2) converges to nonlinear diffusion waves essentially as time tends to infinity.

The approach to get the result in [2] is to introduce

$$\begin{cases} w = v_1 - v_2 - m \\ z = u_1 - u_2 - \hat{u} \end{cases}$$

and to get a system on (w, z) first, namely,

$$\begin{cases} w_t - z_x = 0 \\ z_t + [p(w + v_2 + m, s) - p(v_2, s)]_x + \alpha z - \frac{1}{\alpha} p(v_2, s)_{xt} = 0 \end{cases}$$

where $s = s(x) = s_1(x) = s_2(x) = s_0(x)$, for $x \in R$.

Let $y(x,t) = \int_{-\infty}^{x} w(\xi, t) d\xi$ then, to get the Cauchy problem for y due to $y_x = w$ and $y_t = z$.

$$\begin{cases} y_{tt} + [p(y_x + v_2 + m, s) - p(v_2, s)]_x + \alpha y_t - \frac{1}{\alpha} p(v_2, s)_{xt} = 0 \\ y(x, 0) = y_0(x) \\ y_t(x, 0) = y_1(x) \end{cases} \tag{10}$$

where

$$y_0(x) = \int_{-\infty}^{x} \{v_0(\xi) - \tilde{v}(\xi + x^*, 0) + \frac{(u^+ - u^-)}{\alpha} m_0(\xi)\} d\xi,$$

$$y_1(x) = u_0(x) + [p(\tilde{v}(x + x^*, 0), s(x))]_x - u^- - (u^+ - u^-) \int_{-\infty}^{x} m_0(\xi) d\xi$$

For any given initial data $\{u_0(x), v_0(x), s_0(x)\}$ such that $y_0(x) \in H^3(\mathbb{R})$ and $y_1(x) \in H^2(\mathbb{R})$, it is proved that the Cauchy problem (10) has a unique smooth solution globally in time in the space X_3, defined by

$$X_3 = \begin{Bmatrix} y(t) \in L_\infty(t, H^3), y_t \in L_\infty(t, H^2), y_{tt} \in l_\infty(t, H^1), \\ 0 \le t \le T \text{ for any } T > 0 \end{Bmatrix},$$

provided the initial data is small. Furthermore, the solution y decays to zero in the L_∞-norm, and its derivatives y_t and y_x decay to zero in the L_2-norm and L_∞-norm as $t \to \infty$, which gives the result that $v_1 - v_2 - m \to 0$ and $u_1 - u_2 - \hat{u} \to 0$ in the L_2-norm and L_∞-norm as $t \to \infty$. Namely, the solution of the system of compressible adiabatic flow with damping converges to nonlinear diffusion waves essentially as $t \to \infty$. In other words, nonlinear diffusive phenomena occurs for the solution of the system of compressible adiabatic flow, caused by damping mechanism.

To prove the result mentioned above, it is essential to solve the Cauchy problem on v_2 and to establish the required L^2-estimates on v_2 particularly. Turn to gas dynamics and take $\alpha = 1$ for convenience, the Cauchy problem on v_2 becomes

$$\begin{cases} v_t = -p(v, s)_{xx} \\ v(x, 0) = e^{\frac{1}{\gamma}[s(x) - \bar{s}]} v^*(x + x^*) \end{cases} \tag{11}$$

where $p(v,s) = (\gamma-1)v^{-\gamma}e^{s}(1<\gamma<3), s = s(x)$ and $v^*(\frac{x}{\sqrt{t+1}})$ is the similarity solution of

$$\begin{cases} v_t = -p(v,\bar{s})_{xx} \\ v(\eta) \to v^{\mp} \text{ as } \eta \to \mp\infty, \end{cases}$$

and $v^* \in R$ is determined by

$$\int_{-\infty}^{\infty} \left\{ v_0(x) - e^{\frac{1}{\gamma}[s(x)-\bar{s}]} v^*(x+x^*) \right\} dx = -(u^+ - u^-).$$

In addition to the assumption (6), it is also assumed in [2] that

$$\begin{aligned} v_0(x) - v^+ &= O(x^{-k_1}), \quad \text{as } x \to +\infty \\ v_0(x) - v^- &= O((-x)^{-k_2}), \quad \text{as } x \to -\infty \end{aligned} \tag{12}$$

for certain positive constants $k_i > 1 (i = 1,2)$, while $(s_0(x) - \bar{s})$ possesses a compact support, say

$$\text{support}(s_0(x) - \bar{s}) \subset [-N, N] \tag{13}$$

for a positive constant N. Furthermore, $s_0(x), v^+$ and v^- are given so that there are constants $k_0 \geq N, a_0 \geq 0$ and $x_0 \in R$ such that if v_3 is a positive solution of the following problem

$$\begin{cases} v_t = -p(v,x)_{xx} \\ v(x,0) = e^{\frac{1}{\gamma}[s(x)-\bar{s}]} v^*(x+x_0) \end{cases} \tag{14}$$

then it holds either

$$v_3(x,t) \leq v^* \left(\frac{x+x_0+a_0}{\sqrt{t+1}} \right) \quad \text{for } |x| \geq k_0$$

or (15)

$$v_3(x,t) \geq v^* \left(\frac{x+x_0-a_0}{\sqrt{t+1}} \right) \quad \text{for } |x| \geq k_0$$

The property expressed in (15) has been verified in [2] under certain assumptions on the variety of $s_0(x)$.

Under the assumptions (6), (12), (13) and (15), it is able to obtain the global existence and uniqueness of smooth solution v_2 for which the expected estimates hold, namely,

$$\begin{aligned} &\|v_x\|_{H^2}^2 + \|v_t\|_{H^2}^2 + \|v_{tt}\|_{H^1}^2 + |v_x|_{C^1}^2 + |v_t|_{C^1}^2 + |v_{tt}|_{C^0}^2 \\ &+ \int_0^t \int_{-\infty}^{\infty} \{v_t^2 + v_{xt}^2 + v_{tt}^2 + v_{xxt}^2 + v_{ttx}^2\}(x,\tau) dx d\tau \\ \leq\ & G \left\{ \|s_0(x) - \bar{s}\|_{H^5}^2 + \sup_{x\in[-N,N]} |e^{\frac{1}{\gamma}[s(x)-\bar{s}]} - 1| + [v_\eta^*(0)]^2 + |v^+ - v^-| \right\}^{1/3}, \end{aligned} \tag{16}$$

and

$$a_1 \leq v(x,t) \leq a_2, \quad \text{for any } t \geq 0 \tag{17}$$

where G, a_1 and a_2 depend only on $v^{\mp}$, $\sup\limits_R s_0(x)$ and $\inf\limits_R s_0(x), v(x,t)$ is the unique solution of (11).

The above result on the Cauchy problem (11) is established in the section 2 of [2] which presents a careful analysis, combining of energy method with comparison method. Where the global existence and uniqueness of smooth solution v_3 for (14) is obtained first and the L^2-estimates on the corresponding derivatives of this solution is established also which is essential in getting the global existence of smooth solution v_2 for (11) and the expected decay estimates on the corresponding derivatives of v_2.

By using the result on v_2, one is able to solve the Cauchy problem (10) for y in the space X_3.

It can be proved that the classical local existence theorem gives this solution in the space X_3 locally in time. For the global existence in $t > 0$ one only needs the a priori estimates in the C^2-norm of y for which the a priori estimates in the norm of X_3 is sufficient.

With the help of the following lemma, proved in the section 3 of [2], the Theorem of y can be obtained.

Choose $\hat{r} > 0$ such that $0 < \hat{r} < a_1$

Lemma There exists a constant $\varepsilon = \varepsilon(\hat{r})$ with $0 < \varepsilon \leq \max\{1, \frac{\hat{r}}{2}\}$ such that if $y(t) \in X_3$ with $y_x \in [-\hat{r}, \hat{r}]$ to the Cauchy problem (10) is small as

$$\|y(t)\|_2 < \varepsilon \quad \text{in } 0 \leq t \leq T$$

and it holds that

$$\|s(x) - \bar{s}\|_{H^5} < \varepsilon, \quad |v^+ - v^-| < \varepsilon$$

then, one has the a prori estimate

$$\begin{aligned} E(t) \leq \quad & F_0 E(0) + F_1 \Big\{ \|s(x) - \bar{s}\|^2_{H^5} + \sup_{x \in [-N,N]} |e^{\frac{1}{\gamma}[s(x)-\bar{s}]} - 1| \\ & + [v^*_\eta(0)]^2 + |v^+ - v^-| \Big\}^{1/3} \quad \text{in } 0 \leq t \leq T \end{aligned}$$

where F_0 and F_1 depend only on $\hat{r}, v^{\mp}$, $\sup s_0(x)$, $\inf s_0(x)$ and $p.E(t) \stackrel{\text{def}}{=} \sum_{j=1}^{3} E_j(t)$ with

$$E_1(t) = \frac{1}{2}\int_{-\infty}^{\infty} \left\{\frac{1}{2}(y \cdot y_t + \frac{y^2}{2}) + y_t + [-p_v(\sigma y_x + v, s)]y_x^2\right\}(x,t)dx$$

$$E_2(t) = \frac{1}{2}\int_{-\infty}^{\infty} \{y_{tt}^2 + [1 - p_v(y_x + v, s)]y_{tx}^2 + [-p_v(y_x + v, s)]y_{xx}^2\}(x,t)dx$$

$$E_3(t) = \frac{1}{2}\int_{-\infty}^{\infty} \{y_{ttx}^2 + y_{xxt}^2\}(x,t)dx$$

and $0 < \sigma < 1$.

Theorem Under the hypotheses (6), (12), (13) and (15), the Cauchy problem (10) has a unique smooth solution in the large in time, provided that $E(0), |v^+ - v^-|$ and $\|s(x) - \bar{s}\|_{H^s}$ are suitably small. Moreover, the solution $y(t)$ decays to zero in the L_∞-norm as $t \to +\infty$, y_t and y_x decay to zero in the L_2-norm and L_∞-norm as $t \to \infty$.

References

[1] L. Hsiao and T.P. Liu, Convergence to nonlinear diffusion wave for solutions of a system of hyperbolic conservation laws with damping, Commun. Math. Phys. 143(1992) 599–605.

[2] L. Hsiao and T. Luo, Nonlinear diffusive phenomena of solutions for the system of compressible adiabatic flow through porous media. (to appear)

[3] L. Hsiao and D. Serre, Global existence of solutions for the system of compressible adiabatic flow through porous media. (to appear)

[4] L. Hsiao and D. Serre, Large-time behavior of solutions for the system of compressible adiabatic flow through porous media. (to appear).

Numerical calculation of the flux across an interface between two rock types of a porous medium for a two-phase flow

Jérôme Jaffré[1]

Abstract

When neglecting capillary effects, two-phase flow in porous media is modeled by a nonlinear hyperbolic equation. We consider the case where there is a discontinuity in the medium. Then the flux function of the conservation law is discontinuous with respect to the space variable, and we address the problem of calculating a numerical flux at the interface between the two rock types. Several solutions are discussed.

Keywords : flow in porous media, conservation laws, nonlinear hyperbolic equations, discontinuous flux function, finite differences.

1 Introduction

In the simulation of two-phase flow in porous media, the porous medium may be inhomogeneous. The domain of calculation is divided into several subdomains corresponding to different types of rock and separated by lines or surfaces along which, not only porosity and absolute permeabilities change, but relative permeabilities and capillary pressure functions differ too from one subdomain to the other.

Usually the domain of calculation is discretized so boundaries between subdomains with different rock types coincide with intercell boundaries. If one uses cell-centered finite difference or finite volume methods one has to calculate along these boundaries the numerical flux representing transfer between cells.

In the following we shall neglect capillarity effects. Then two-phase flow in porous media is modeled by a nonlinear hyperbolic equation. With different rock types this is a case where the problem is unstable and of limited physical value as shown in [8]. Yet, since very small capillary pressure might

[1]INRIA, BP 105, 78153 Le Chesnay Cédex, France

be introduced in codes, it is still important to check how numerical fluxes perform in the limiting case without any capillary pressure.

We shall also restrict ourselves to the one-dimensional case. In the multidimensional case most numerical methods still use one-dimensional flux calculations in the direction normal to the boundaries of the discretization cells.

The numerical flux that is the most widely used in petroleum engineering is the upstream mobility flux. This flux was designed only for two-phase flow from physical considerations and was justified mathematically in the case where the medium is homogeneous [9, 1]. But there is no justification in the heterogenous case.

In this paper we will show, experimentally, that this flux gives a correct answer by comparing it to another one [2], based on Godunov's technique, and for which a mathematical justification can be given in terms of a Riemann solver.

2 Two-phase flow equations

Capillary-free two-phase incompressible flow is modeled by the following scalar nonlinear hyperbolic equation

$$\phi\frac{\partial S}{\partial t}+\frac{\partial f}{\partial x}=0 \tag{1}$$

where ϕ is the porosity of the rock, and $S = S_1$ is the saturation of phase 1 which lies in a bounded interval $[S_m, S_M]$. The flux function f is the Darcy velocity φ_1 of phase 1, and it has the form

$$f=\varphi_1=\frac{\lambda_1}{\lambda_1+\lambda_2}[q+(g_1-g_2)\lambda_2]. \tag{2}$$

Here $q = \varphi_1+\varphi_2$ denotes the total Darcy velocity where $\varphi_\ell, \ell = 1, 2$, denotes the Darcy velocity of phase ℓ with, for the second phase,

$$\varphi_2=\frac{\lambda_2}{\lambda_1+\lambda_2}[q+(g_2-g_1)\lambda_1]. \tag{3}$$

Since the flow is assumed to be incompressible, the total Darcy velocity q is independent of the space variable x.

The quantities λ_ℓ, $\ell = 1,2$ may be called effective mobilities. These are products of the absolute permeability K by the mobilities k_ℓ :

$$\lambda_\ell = Kk_\ell,\ \ell = 1, 2.$$

The absolute permeability K may depend on x, and the quantities k_ℓ and λ_ℓ are functions of S which satisfy the following properties :

$$(4)\qquad \begin{array}{l} k_1 \text{ and } \lambda_1 \text{ are increasing functions of } S,\ k_1(S_m) = \lambda_1(S_m) = 0, \\ k_2 \text{ and } \lambda_2 \text{ are decreasing functions of } S,\ k_2(S_M) = \lambda_2(S_M) = 0. \end{array}$$

We also shall assume that these functions are smooth functions of the saturation S and so is the flux function f.

The gravity constants g_ℓ, $\ell = 1, 2$ of the phases are

$$g_\ell = g\rho_\ell \frac{dx}{dz},\ \ell = 1, 2,$$

with g the acceleration due to gravity, ρ_ℓ the density of phase ℓ and z is the vertical position of the point of abscissa x.

The most popular way to discretize equation (1) is to use a cell-centered finite difference (or finite volume) method

$$(5)\qquad \frac{\phi_i}{\Delta t}(S_i^{n+1} - S_i^n) + \frac{1}{h}(f_{i+1/2}^n - f_{i-1/2}^n) = 0,$$

where i denotes the cell index, $i + 1/2$ and $i - 1/2$ are indices for the end points of cell i, Δt and h are the time and space discretization intervals which satisfy a CFL condition.

The numerical flux at the interface between two cells is a function of the left-hand and right-hand values of the saturation :

$$f_{i+1/2} = F(S_i, S_{i+1}).$$

It is calculated by solving exactly or approximately the Riemann problem, denoted by (S_i, S_{i+1}, f), associated with the initial data S_i, S_{i+1} and the flux function f. For the associated finite difference scheme to be convergent, the function F must be consistent, monotone and Lipshitz continuous [7, 3].

In the following, we will consider two numerical flux functions. The Godunov flux is defined by

$$(6)\qquad F^G(a, b) = \begin{cases} \min\limits_{[a,b]} f & \text{if } a < b, \\ \max\limits_{[a,b]} f & \text{if } a \geq b. \end{cases}$$

It is obtained by solving exactly the associated Riemann problem[6]. Here, and throughout the rest of this paper, a and b will denote, respectively, the left-hand and the right-hand values of the saturation.

The other numerical flux function that we consider is the upstream mobility flux. It is an ad-hoc flux for two-phase flow in porous media, invented by petroleum engineers from simple physical considerations, and it corresponds to an approximate solution to the Riemann problem. It is given by the following formula :

$$(7)\qquad \begin{aligned} F^{UM}(a,b) &= \frac{\lambda_1^*}{\lambda_1^* + \lambda_2^*}[q + (g_1 - g_2)\lambda_2^*], \\ \lambda_\ell^* &= \begin{cases} \lambda_\ell(a) & \text{if } \varphi_\ell > 0, \\ \lambda_\ell(b) & \text{if } \varphi_\ell \leq 0. \end{cases} \end{aligned}$$

As we can see, the flux is calculated using the mobilities of the phases which are upstream with respect to the flow of the phases. This flux has been shown to have all the desired properties for convergence of the associated finite difference scheme [9, 1].

When the two phases are flowing in the same direction, the Godunov's flux and the upstream mobility flux give the same answer and coincide with standard upstream weighting, but they differ when the phases are flowing in opposite directions.

3 The case of two rock types : a solution based on the Godunov flux

In the following, we consider a one-dimensional domain divided into two regions, with rock type I on the left and rock type II on the right (see Fig. 1). We also assume that the boundary between the two rock types coincides with a discretization point. The discontinuities in the physical constants and in

Rock type I	**Rock type II**
$\phi^{\mathrm{I}}, K^{\mathrm{I}}, k_1^{\mathrm{I}}, k_2^{\mathrm{I}}$	$\phi^{\mathrm{II}}, K^{\mathrm{II}}, k_1^{\mathrm{II}}, k_2^{\mathrm{II}}$
$S_m^{\mathrm{I}} \leq S \leq S_M^{\mathrm{I}}$	$S_m^{\mathrm{II}} \leq S \leq S_M^{\mathrm{II}}$
$f \equiv f^{\mathrm{I}}$	$f \equiv f^{\mathrm{II}}$

Figure 1: Constants, mobilities and fluxes for two rock types.

the mobilities result in a flux function f which is discontinuous from one cell to the other. However since the mobilities satisfy properties (4), the two flux functions associated with the two rock types take the same values at their end points (see Fig. 2):

$$f^{\mathrm{I}}(S^{\mathrm{I}}_m) = f^{\mathrm{II}}(S^{\mathrm{II}}_m) = 0, \quad f^{\mathrm{I}}(S^{\mathrm{I}}_M) = f^{\mathrm{II}}(S^{\mathrm{II}}_M) = q. \tag{8}$$

We will use this property when defining a numerical flux below.

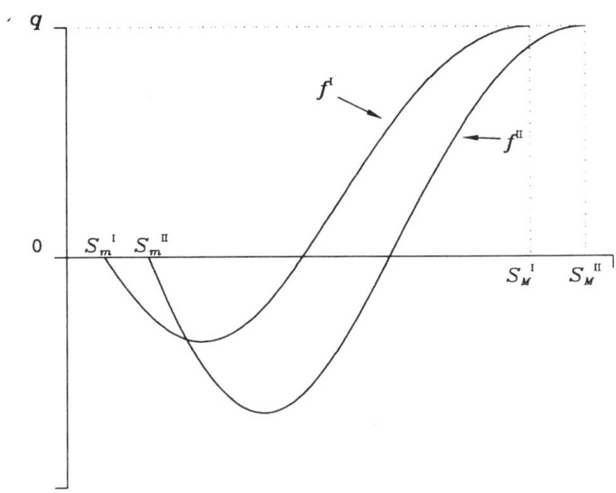

Figure 2: Two flux functions at the interface between two rock types

At the interface between two rock types, the saturation may be discontinuous but the Darcy velocities must be continuous in order to preserve the mass balance of the two phases.

When thinking in terms of Godunov fluxes, formulas (6) cannot be applied at the interface between two rock types since we do not know which flux function to use. So the problem is now to define a numerical flux there.

A solution was given in [2] from the following considerations. Introduce a saturation $\overline{S}$ in an infinitely small transition zone between the two rock types. Calculating the flux in the domain with rock type I would give a numerical flux $F^{G,\mathrm{I}}(a,\overline{S})$, and calculating the flux in the domain with rock type II would give a numerical flux $F^{G,\mathrm{II}}(\overline{S},b)$. Since the Darcy velocity must be continuous, we are looking for a number $F^G(a,b)$ such that

$$F^G(a,b) = F^{G,\mathrm{I}}(a,\overline{S}) = F^{G,\mathrm{II}}(\overline{S},b). \tag{9}$$

The existence of such a number is proven by the following theorem.

Theorem 1 *There exists a unique number $F^G(a,b)$ satisfying equation (9).*

Proof : First, since F^G is a monotone flux, $F^{G,\mathrm{I}}(a,S)$ is decreasing, and $F^{G,\mathrm{II}}(S,b)$ is increasing with respect to S. Then it remains to show that the ranges of the functions $F^{G,\mathrm{I}}(a,\cdot)$ and $F^{G,\mathrm{II}}(\cdot,b)$ have a nonempty intersection. From the definition of the Godunov flux (6), we have

$$F^{G,\mathrm{I}}(a,S) \in [\min_{[a,S_M^{\mathrm{I}}]} f^{\mathrm{I}}, \max_{[S_m^{\mathrm{I}},a]} f^{\mathrm{I}}], \quad F^{G,\mathrm{II}}(S,b) \in [\min_{[S_m^{\mathrm{II}},b]} f^{\mathrm{II}}, \max_{[b,S_M^{\mathrm{II}}]} f^{\mathrm{II}}].$$

Since the two flux functions have the same endpoints (see equalities (8)), the bounds of their ranges satisfy

$$\begin{array}{ccccccccc} \min\limits_{[a,S_M^{\mathrm{I}}]} f^{\mathrm{I}} & \le & f^{\mathrm{I}}(S_M^{\mathrm{I}}) & = & q & = & f^{\mathrm{II}}(S_M^{\mathrm{II}}) & \le & \max\limits_{[b,S_M^{\mathrm{II}}]} f^{\mathrm{II}}, \\ \min\limits_{[S_m^{\mathrm{II}},b]} f^{\mathrm{II}} & \le & f^{\mathrm{II}}(S_m^{\mathrm{II}}) & = & 0 & = & f^{\mathrm{I}}(S_m^{\mathrm{I}}) & \le & \max\limits_{[S_m^{\mathrm{I}},a]} f^{\mathrm{I}}. \end{array}$$

Therefore their ranges have a nonempty intersection.

Fig. 3 shows how to calculate the numerical flux $F^G(a,b)$ in the case where the flux functions are not monotone, as when gravity effects are large. In the simpler case where the flux functions are monotone, say increasing, as when the phases are both flowing in the same direction, it is easy to check that $F^G(a,b) = f^{\mathrm{I}}(a)$; that is, this numerical flux coincides with the upstream flux. One should notice that the number $\overline{S}$ corresponding to $F^G(a,b)$ in equation (9) may not be unique. Also one can easily check that when $f^{\mathrm{I}} \equiv f^{\mathrm{II}}$ (continuous case), the above calculation gives the standard Godunov flux.

The construction above is valid for any smooth flux function as long as equation (8) holds. Therefore, it can be applied to conservation laws which may arise from physical problems other than two-phase flow in porous media.

4 Interpretation as a Riemann problem

The calculation of the numerical flux proposed in the previous section has been obtained from heuristic considerations. However the Riemann problem with a discontinuous flux function has been independently studied in [4] and the Cauchy problem is studied in [5].

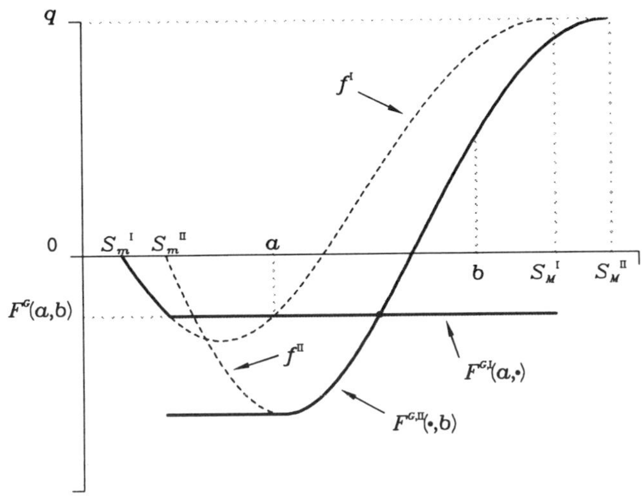

Figure 3: Calculation of the numerical flux $F^G(a,b)$ as the value at the intersection of the functions $F^{G,\mathrm{I}}(a,\cdot)$ and $F^{G,\mathrm{II}}(\cdot,b)$

In order to construct the solution of the Riemann problem, in [4] the authors consider the two Riemann problems (a, S, f^{I}) and (S, b, f^{II}), and introduce the sets

$$\begin{aligned} H^{\mathrm{I}}(a) &= \{S;\, (a, S, f^{\mathrm{I}}) \text{ is solved by waves of nonpositive speed only}\} \cup \{a\}\,, \\ H^{\mathrm{II}}(b) &= \{S;\, (S, b, f^{\mathrm{II}}) \text{ is solved by waves of nonnegative speed only}\} \cup \{b\}. \end{aligned}$$

Since $[S_m^{\mathrm{I}}, S_M^{\mathrm{I}}] \cap H^{\mathrm{I}}(a)$ and $[S_m^{\mathrm{II}}, S_M^{\mathrm{II}}] \cap H^{\mathrm{II}}(b)$ are closed sets, one may define

$$\begin{aligned} c^{\mathrm{I}}(a, S) &= \{S' \in H^{\mathrm{I}}(a); |S - S'| \,\text{minimum}\}\,, \quad h^{\mathrm{I}}(a, S) = f^{\mathrm{I}}(c^{\mathrm{I}}(S))\,, \\ c^{\mathrm{II}}(b, S) &= \{S' \in H^{\mathrm{II}}(b); |S - S'| \,\text{minimum}\}\,, \quad h^{\mathrm{II}}(b, S) = f^{\mathrm{II}}(c^{\mathrm{II}}(S)). \end{aligned}$$

The relationship of these constructions with that of the previous section is given by the following equalities which can be easily checked :

$$F^{G,\mathrm{I}}(a,\cdot) \equiv h^{\mathrm{I}}(a,\cdot)\,, F^{G,\mathrm{II}}(\cdot,b) \equiv h^{\mathrm{II}}(b,\cdot)\,.$$

Therefore the construction described in section 3 gives a flux which is that given by the exact solution to the Riemann problem with a discontinuous flux function. This solution is obtained by piecing together waves traveling to the left produced by the left-hand side Riemann problem (a, S, f^{I}) and waves traveling to the right produced by the right-hand side Riemann problem (S, b, f^{II}). In other words we remove the waves crossing the interface between the two rock types, replacing them by waves with zero speed.

5 The case of two rock types : the upstream mobility flux

The same method as described above for the Godunov flux can be applied to the upstream mobility flux (7). Again introducing an infinitely small transition zone where the saturation is $\overline{S}$, we look for a number $F^{UM}(a,b)$ such that

$$F^{UM}(a,b) = F^{UM,\mathrm{I}}(a,\overline{S}) = F^{UM,\mathrm{II}}(\overline{S},b). \tag{10}$$

Existence and uniqueness of the number $F^{UM}(a,b)$ can be shown in a manner similar to Theorem 1 since the upstream mobility flux is monotone.

Calculation of the flux given by equation (10) is fairly complicated to implement, and there is no advantage in it compared to that given by (9) using the Godunov flux. However, even though there is no mathematical support for it in the case of two rock types, there is a more convenient way to adapt the upstream mobility flux defined by equations (7) to this case :

$$\begin{aligned} F^{UM}(a,b) &= \frac{\lambda_1^*}{\lambda_1^* + \lambda_2^*}(q + (g_1 - g_2)\lambda_2^*), \\ \lambda_\ell^* &= \begin{cases} \lambda_\ell^{\mathrm{I}}(a) & \text{if } \varphi_\ell > 0, \\ \lambda_\ell^{\mathrm{II}}(b) & \text{if } \varphi_\ell \leq 0, \end{cases} \qquad \ell = 1,2. \end{aligned} \tag{11}$$

This is the way that petroleum engineers usually implement the calculation of the upstream mobility in their reservoir simulation codes when different rock types are present.

6 Numerical experiments

Since the upstream mobility flux (see equation (11)) and that based on the Godunov flux (equation (9)) give the same answer when the two phases flow in the same direction, we will simulate only the case with counter-currents. Therefore, we consider an idealized experiment with a vertical closed core, initially saturated with the heavy fluid in the upper half and with the light fluid in the lower half. In the upper half there is one rock type with absolute permeability K_T and in the lower half with higher permeability K_B (see Fig. 4). During the experiment, the heavy phase moves downward while the light phase moves upward. The data that are used are the following :

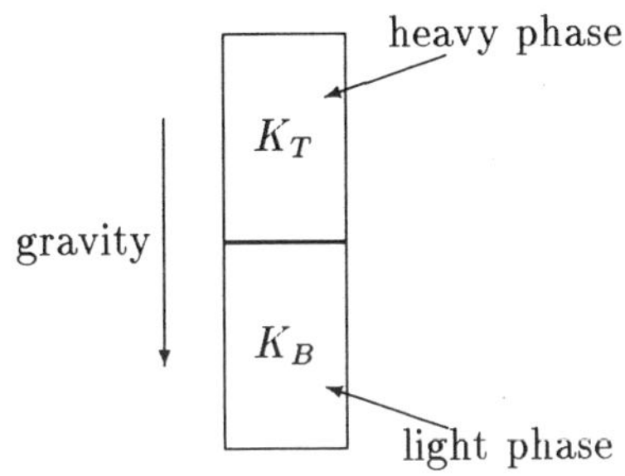

Figure 4: An heteregeneous vertical core with permeability K_T in the top half and K_B in the bottom half. At initial time, the top half is saturated with the heavy phase and the bottom part with the light phase.

$$\phi = 1,\ q = 0,\ g_1 = 1,\ g_2 = 2,\ \lambda_1 = \tfrac{1}{2}KS^2,\ \lambda_2 = \tfrac{1}{2}K(1-S)^2,$$
$$K = K_T \text{ if } x < 0 \text{ and } K = K_B \text{ if } x < 0.$$

In this section we compare experimentally three flux calculations of practical interest : that based on the exact solution of the Riemann problem and using the Godunov flux (see equation (9)), that using the upstream mobilities (see equation (11)) and a third flux that one would naturally want to try. This third flux is the regular Godunov flux for the flux function obtained by averaging the right-hand and the left-hand flux functions at the interface between the two rock types (assuming that they have the same domain).

In the following figures, the saturation of the heavy phase is shown at $t = 2$. In order to illustrate the convergence of the numerical method, in each figure there are two pictures corresponding to calculations made with two different space discretization steps h.

In Figure 5 we present first in the homogeneous case when $K_T = K_B = 1$. We observe, as expected, that the scheme using the Godunov flux gives results which converge faster than when using the upstream mobility flux. However, the latter also gives fairly good results, but the Courant-Friedrich condition requires time steps which are about two thirds of those required when using the Godunov flux.

In Figure 6, we increase the absolute permeability in the bottom part to $K_B = 1000$. One notices that whatever the flux calculations, all curves converge to the same solution. Those corresponding to the exact solution to the Riemann problem are the fastest to converge. Even though there is no mathematical justification for the two other methods, they are also converging: those using an average permeability with the Godunov flux are converging more slowly, and those using upstream mobilities are the slowest to converge.

The choice of the method depends on how much accuracy one wants. If one needs a very accurate solution, the flux corresponding to the exact Riemann solver should be used. However this is not the case, most of the time, when simulating flow in porous media, since there are so many physical uncertainties, so the upstream mobility flux, which is easier to implement and less costly, is usually sufficient.

7 Conclusion

We showed how to calculate the numerical flux at the interface between two rock types, corresponding to the exact solution of the associated Riemann problem. This flux is based on the Godunov flux.

This numerical flux has been compared, in a finite difference calculation, to the upstream mobility flux that is commonly used by petroleum engineers. The latter is less complicated to implement, and even though it is not justified mathematically, it gives results that also converge to the right solution, though more slowly.

This comparison demonstrates experimentally the validity of the upstream mobility flux in the case with two rock types. When very accurate results are desired, the numerical flux based on Godunov flux may be preferred.

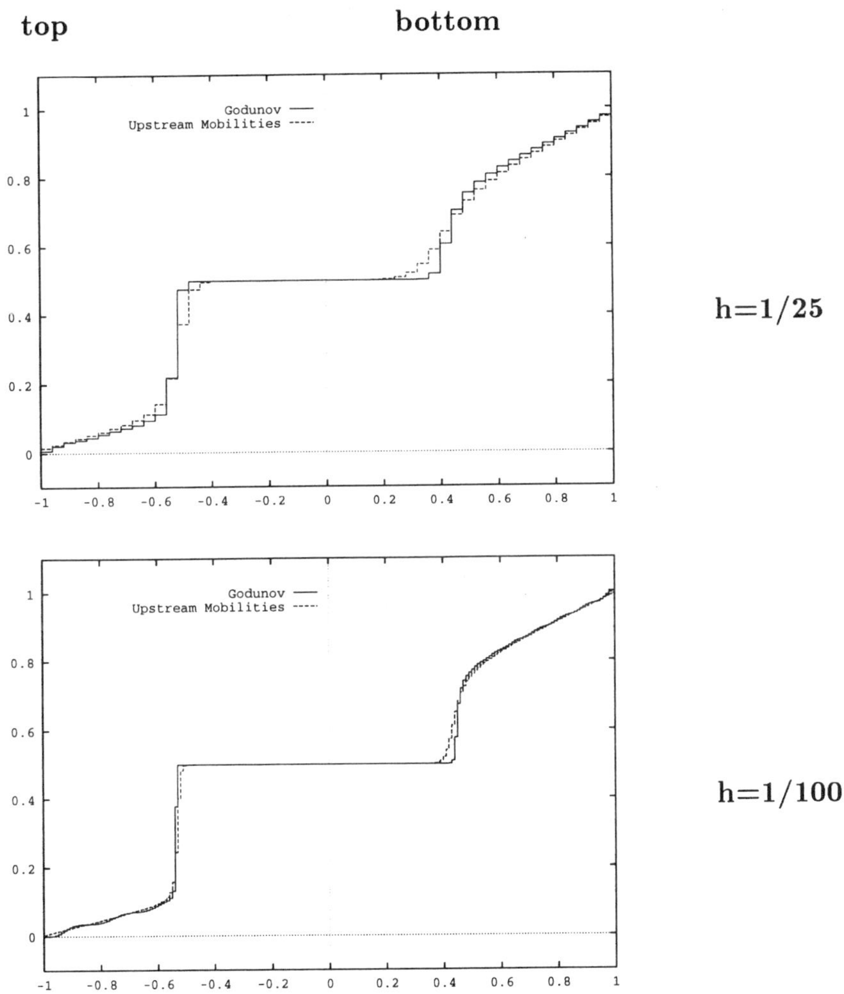

Figure 5: Comparaison of solutions obtained with the Godunov flux and with the upstream mobilities flux when $K_T = K_B = 1$ (homogeneous case).

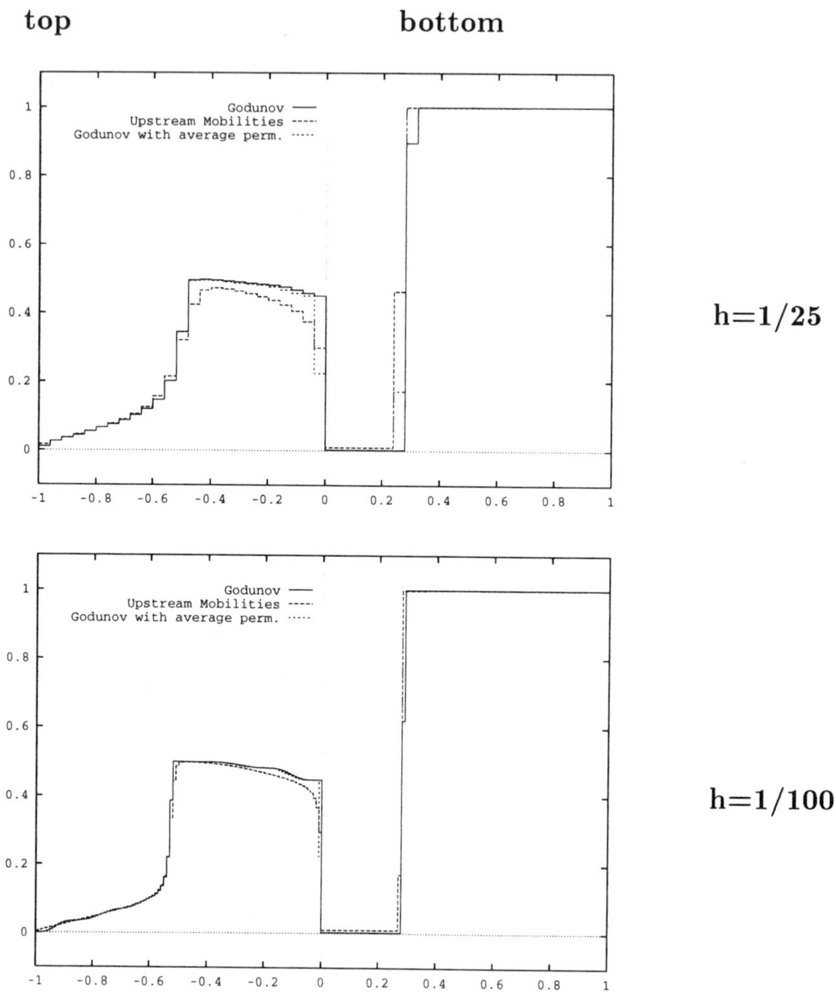

Figure 6: Comparaison of solutions obtained with the Godunov flux, using the method of Section 3 or an average permeability, and of solutions obtained with the upstream mobility flux (dotted line) when $K_T = 1, K_B = 1000$.

References

[1] Y. Brenier and J. Jaffré, *Upstream differencing for multiphase flow in reservoir simulation*, SIAM J. Num. Anal. **28** (1991), 685–696.

[2] G. Chavent, G. Cohen, and J. Jaffré, *A finite element simulator for incompressible two-phase flow*, Transport in Porous Media **2** (1987), 465–478.

[3] M.G. Crandall and A. Majda, *Monotone difference approximations for scalar conservation laws*, Math. Comp. **34** (1980), 1–21.

[4] T. Gimse and N.H. Risebro, *Riemann problems with a discontinuous flux function*, Third International Conference on Hyperbolic Problems, Uppsala, Sweden (June 1990), Vol. I (B. Engquist and B. Gustafsson, eds.), Studenlitteratur, Chartwell-Bratt, Uppsala, 1991, pp. 488–502.

[5] ———, *Solution of the Cauchy problem for a conservation law with a discontinuous flux function*, SIAM J. Math. Anal. **23** (1992), 685–696.

[6] S.K. Godunov, *Finite difference methods for numerical computation of discontinuous solutions of the equations of fluid dynamics*, Math. Sbornik **47** (1959), 271–306.

[7] N.N. Kuznecov and S.A. Volosin, *On monotone difference approximations for a first order quasi-linear equation*, Soviet Math. Dokl. **17** (1976), 1203–1206.

[8] H.P. Langtangen, A. Tveito, and R. Winther, *Instability of Buckley-Leverett flow in heterogeneous media*, Transport in Porous Media **9** (1992), 165–185.

[9] P.H. Sammon, *An analysis of upstream differencing*, SPE Reservoir Engrg. **3** (1988), 1053–1056.

A Bifurcation Diagram for Oblique Shock Interactions in the Unsteady Transonic Small Disturbance Equation

Sunčica Čanić[1], Barbara Lee Keyfitz[2], and David H. Wagner[2]

ABSTRACT. The UTSD equation is used to model the transition from regular to Mach reflection for weak shocks. Here we examine a related initial value problem: initial data consisting of two oblique shocks in the upper half plane with a symmetry boundary condition on the axis. The slopes, a and $-b$, of the two shocks parameterize the resulting Riemann problem. We identify regions in parameter space corresponding to different qualitative behavior of solutions.

1. INTRODUCTION

The unsteady transonic small disturbance (UTSD) equation

$$\begin{aligned} u_t + uu_x + v_y &= 0, \\ -v_x + u_y &= 0, \end{aligned} \tag{1}$$

has been proposed as a model to study the transition between Mach and regular reflection for weak shocks and small wedge angles. It arises as an asymptotic limit of the Euler equations describing the flow in the neighborhood of the region of interaction of waves. To shed light on the interaction of the incident and reflected shocks, Brio and Hunter, [2], analysed (1) with Riemann data consisting of three states in the upper half plane, separated by shocks. The boundary data on the x-axis are $u_y = 0 = v$. The study in [2] explains the derivation of the model and reports on some numerical simulations. Among their results, Brio and Hunter show that up to an affine change of coordinates one may take

$$\begin{cases} U_0 &= (0,0) \\ U_1 &= (1,-a) \\ U_2 &= (1+a/b, 0) \end{cases} \tag{2}$$

for the data, as in Figure 1, which is what we shall use in this paper.

We will refer to the two discontinuities as the "incident" and "reflected" shocks, although in this version of the problem they are given independently. Throughout, $a > 0$, $b > 0$. Though the values of a and b are related to the angle of the wedge and the strength of the shock, Tabak and Rosales have shown, [12], that the shock reflection problem has only a single parameter. Nonetheless, it is interesting to study this Riemann problem in its own right, and the solution to the shock reflection problem could be recovered by tracing a path through a, b-space.

Taking the initial discontinuities to be along $x = ay$ and $x = -by$ results in discontinuities which propagate as shocks. Hence, the basic problem attacked in this paper can be termed the "oblique shock interaction" problem. This contrasts to a study by Zhang and Zheng, [13], which

[1]Mathematics Department, Iowa State University, Ames, Iowa 50011 canic@iastate.edu

[2]Department of Mathematics, University of Houston, Houston, TX 77204-3476 keyfitz@uh.edu wagner@math.uh.edu

The research of the first author was supported by the Texas Advanced Research Program under Grant 003652-124-ARP and the Department of Energy, grant DE-FG02-94ER25220. Research of the second author supported by the Texas Advanced Research Program under Grant 00365-2102-ARP and the Department of Energy, grant DE-FG03-94ER25222. Research of the third author supported by the Texas Advanced Research Program under Grant 00365-2102-ARP.

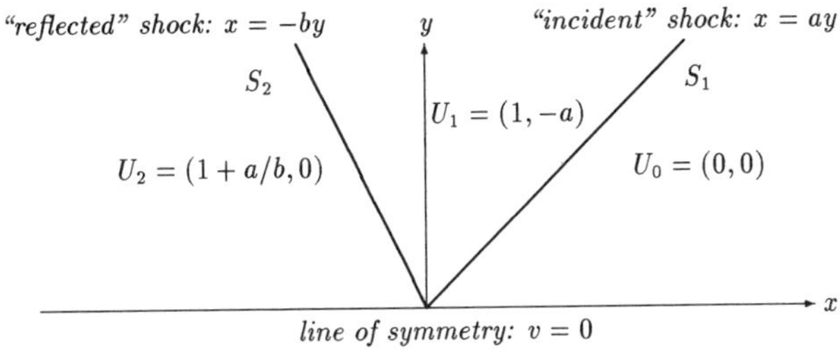

FIGURE 1. Riemann Data for Oblique Shock Reflection Problem

examines only waves parallel to the coordinate axes. Our thesis is that the two-parameter problem displays qualitative features seen in studies of shock interaction problems — both experimental studies on shock reflection by straight wedges in real gases, theoretical predictions based on shock polar analysis, and numerical studies using models for real and ideal gases, [1], [7], [8], [10], [11]. Shock polar analysis is insufficient, since it assumes constant solutions away from shocks. For weak shocks, small variations in the solutions are no longer negligible. What needs to be substituted is an analysis which takes into account the behavior of the solution in regions where it need not be a constant state or a simple wave; the solution in subsonic regions is particularly important.

The contribution of this paper lies in formulating basic principles for subsonic interactions. We have not yet constructed complete solutions. In this paper we describe how to reduce the self-similar problem to the point where subsonic problems can be identified and studied. In other work, [3], [4], we have begun an analysis of the subsonic problem.

Our results are confined to the simplified model above, for which analysis of the subsonic region seems to be tractable. However, it may apply to the nonlinear wave equation, [6], and the unsteady transonic full-potential equation, [9]. More significant extensions would be necessary to incorporate the effects of linear waves, which are present and influential in the full Euler equations of ideal or real gas dynamics.

Similarity analysis will not give a unique solution for every configuration of data. We believe that all the solutions we discuss are stable to small perturbations of the Riemann data, to perturbation by smoothing of the data, or to viscous perturbation of the equations. This conjecture must also be tested, but this lies beyond the scope of the present paper.

In self-similar coordinates, $\xi = x/t$, $\eta = y/t$, equation (1) becomes

$$\begin{aligned} (u-\xi)u_\xi - \eta u_\eta + v_\eta &= 0\,, \\ -v_\xi + u_\eta &= 0\,. \end{aligned} \tag{3}$$

This can be written as a quasilinear system

$$A(U,\Xi)U_\xi + B(U,\Xi)U_\eta = 0 \tag{4}$$

or in conservation form

$$\partial_\xi F(U,\Xi) + \partial_\eta G(U,\Xi) = S\,, \tag{5}$$

with $U = (u, v)$ and $\Xi = (\xi, \eta)$. Because the coordinate functions appear explicitly in the coefficients of (4) and (5), this system is not standard for hyperbolic conservation law theory. Some basic analysis of shock and rarefaction waves, their construction and admissibility, is carried out in [5]. We summarize the results we need here.

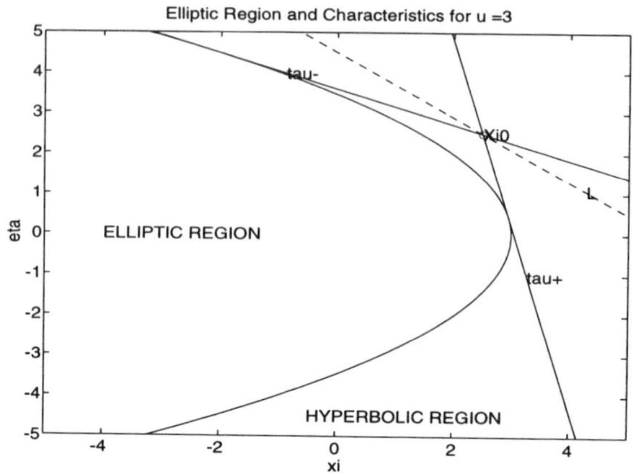

FIGURE 2. The Hyperbolic and Elliptic Regions in (ξ, η)

Equation (3) is of mixed type; it is hyperbolic outside and elliptic inside the parabola

$$P_u : \qquad \xi + \frac{\eta^2}{4} = u.$$

When the equation is linearized about a constant state, the characteristics in the hyperbolic region are straight lines tangent to P_u as sketched in Figure 2.

Shocks between constant states have the form $\xi = \kappa\eta + \omega$ where

$$\kappa \equiv -\frac{[v]}{[u]} = -\frac{v_L - v_R}{u_L - u_R}; \qquad \omega \equiv \kappa^2 + \bar{u} = \kappa^2 + \frac{u_L + u_R}{2}.$$

In the next section we identify the geometry of the elementary figures corresponding to the states in the problem (1), (2).

2. THE GEOMETRY OF PRIMARY INTERSECTIONS

We define the following objects.

Shocks: $\mathbf{S}_1$ (between U_0 and U_1): $\xi = a\eta + (1/2 + a^2)$
$\mathbf{S}_2$ (between U_1 and U_2): $\xi = -b\eta + (1 + a/2b + b^2)$
Parabolas: $\mathbf{P}_0$ (corresponding to U_0): $\xi = -\eta^2/4$
$\mathbf{P}_1$ (corresponding to U_1): $\xi = -\eta^2/4 + 1$
$\mathbf{P}_2$ (corresponding to U_2): $\xi = -\eta^2/4 + 1 + a/b$
Axis of Symmetry: W (wall): $\eta = 0$

These six objects define the *primary curves* of the problem. We study the reduced problem (3) with quasi-one-dimensional data given on the parabolic arc

$$\xi + \eta^2/4 = C \tag{6}$$

for large C. This is consistent with the forward timelike direction in (3) inherited from the original Riemann data for the two-dimensional problem. The data consist of the constant states in (2) with discontinuities at S_1 and S_2.

The *domain of influence* of a point is the union of the compact forward wave cone through the point and the interior of the parabola. For a hyperbolic point (ξ, η) one can also define the *domain*

Intersection	Name	Point
$S_1 \cap S_2$	Ξ_I	$(1/2 + a/2b + ab, 1/2b + b - a)$
$S_2 \cap P_2$	Ξ_0	$(1 + a/2b - b^2 - \sqrt{2ab}, 2b + \sqrt{2a/b})$
	Ξ_*	$(1 + a/2b - b^2 + \sqrt{2ab}, 2b - \sqrt{2a/b})$
$S_1 \cap P_1$	Ξ_1	$(-a^2 + a\sqrt{2} + 1/2, -2a + \sqrt{2})$
$S_1 \cap W$	Ξ_A	$(\omega_1, 0) = (1/2 + a^2, 0)$
$S_2 \cap W$	Ξ_B	$(\omega_2, 0) = (1 + a/2b + b^2, 0)$

TABLE 1. The Primary Intersections

of dependence by following the characteristics of (3) back to (6). For piecewise constant data on (6) there is a region of (ξ, η) space outside of any intersections of the primary curves. We have

Definition 1. *The* primary intersections *are all the intersections of primary curves which could generate new waves. These are $S_1 \cap S_2$, $S_i \cap W$ and $S_i \cap P_i$ for $i = 1$ and 2.*

We name these points in Table 1, for reference.

Remark: It is clear that intersections of one shock with another and of shocks with the wall generate new waves. But the point where a shock intersects the parabola corresponding to one state of the shock may mark a transition from uniform to nonuniform flow. Extending the catalog to include this possibility is the important first step in our approach.

Not all the primary intersections occur in the solution of the Riemann problem; some never occur, and some occur or not depending on the values of a and b. We define those which we expects to occur, and then give a partial classification. Because the characteristics in the hyperbolic region are easily computed, it is straightforward to find the forward domain of influence of the data. However, the solution may contain states of larger magnitude than any of the data at infinity, and hence a computation of the backward domain of dependence of any point is not possible if we do not know the state at that point. We therefore introduce a term for those intersections which are determined *a priori* by the data at infinity, while acknowledging that they may not actually occur.

Definition 2. *An* admissible intersection *is one which appears to be completely determined by the data at infinity.*

We can determine the admissible intersections for each a and b.

Proposition 1. *If Ξ_I is upstream from Ξ_0 then it is admissible.*

Proof: In this case, Ξ_I is in the quasi-one-dimensional domain of dependence of the data at infinity. ∎

Now, Ξ_I lies on P_2 if

$$a^2 + (2b - \frac{3}{b})a + \left(\frac{1}{2b} - b\right)^2 = 0 \, .$$

This equation has two solutions $I(a) < K(a)$ for each $a > 0$; $\Xi_I = \Xi_0$ on I, $\Xi_I = \Xi_*$ on K.

Proposition 2. *The shocks meet at the wall if $a = b + 1/2b$.*

Proof: This occurs if $\Xi_I = \Xi_A = \Xi_B$, or $\eta_I = 0$. This gives the equation

$$a = b + \frac{1}{2b} \equiv J(b).$$

∎

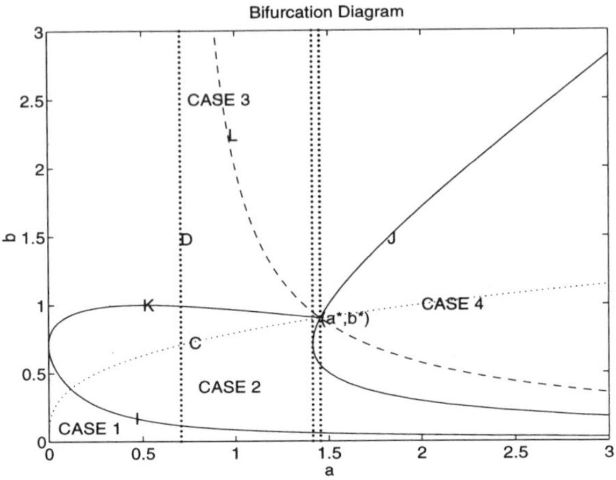

FIGURE 3. Classification of Parameter Space by Position of Ξ_I

In this case the solution of the Riemann problem is a uniform regular reflection. The curve J does not intersect I, but intersects K at a point (a^*, b^*). Since points outside the upper half plane have no meaning in the problem, we consider the curve K to terminate at this point. Classification by position of Ξ_I thus leads to four open regions in the a, b plane:

Case 1: The region below I: Ξ_I above P_2.
Case 2: The region between I and K and left of J: Ξ_I inside P_2, above W.
Case 3: The region above K, left of J: Ξ_I beyond P_2, above W.
Case 4: The region right of J: Ξ_I below W.

See Figure 3, where the curves I, J and K are solid lines.

A classification based on the position of Ξ_I does not determine the qualitative behavior of the flow (for example, Mach or regular reflection), since Ξ_I may not be an admissible point. We note the following elementary results.

Proposition 3. *In Case 1, Ξ_I is the only admissible intersection.*

Proposition 4. *In Cases 2, 3, or 4, Ξ_0 is an admissible intersection.*

Proof: The entire interior of P_2 is downstream from Ξ_0. In Case 2, all the other primary intersections are inside P_2. ■

In Cases 3 and 4 there may be one or two admissible points, depending on whether Ξ_A is inside or outside P_2. We see that Ξ_A is outside P_2 if

$$a > \frac{1}{2b} + \sqrt{\frac{1}{4b^2} + \frac{1}{2}} \equiv L(b).$$

The curve L contains the point (a^*, b^*). We have

Proposition 5. *In either Case 3 or Case 4, if (a, b) lies to the right of $a = L(b)$, there are two admissible primary intersections, Ξ_0 and Ξ_A. If (a, b) lies to the left of this curve, there is only one admissible primary intersection, Ξ_0.*

In Case 1, one or both shocks are nearly vertical, and one can construct a Mach reflection solution by means of quasi-one-dimensional Riemann problems.

3. SOLVABLE QUASI-ONE-DIMENSIONAL RIEMANN PROBLEMS

A quasi-one-dimensional Riemann problem has *data* consisting of a *center* Ξ_0 and two states, U_1, U_2 on a spacelike line, L, through Ξ_0. The *solution* is defined *downstream* (towards the parabola) from L. It is a function of $t \equiv (\xi - \xi_0)/(\eta - \eta_0)$ defined in a forward half-space below L.

A shock on the line $\xi - \xi_0 = t(\eta - \eta_0)$ satisfies, from (5), the *Rankine-Hugoniot relation*

$$F(U, \Xi_0) - F(U_0, \Xi_0) = t(G(U, \Xi_0) - G(U_0, \Xi_0)) \tag{7}$$

where U and U_0 are the constant states on either side of the shock. If (U_0, Ξ_0) is hyperbolic, then eliminating t from (7) defines the *shock polar*; it is a loop, defined for $u \leq u_M(\Xi_0)$; there are two values of v for each $u \neq u_0$, u_M.

Centered simple waves are the other type of local solution $U(t)$. They satisfy

$$(A - tB)\frac{dU}{dt} = 0. \tag{8}$$

Solving (8), we obtain $R(U_0, \Xi_0)$, the *rarefaction polar* connecting to U_0, centered at Ξ_0. Both U and U_0 must be hyperbolic at Ξ_0. For a given point Ξ_0, the curves are defined for $u \leq u_s \equiv \eta_0^2/4 + \xi_0$. Unlike the shock polar, the rarefaction polar does not form a loop, but has two finite and two semi-infinite branches.

Quasi-one-dimensional Riemann problems do not have solutions for all choices of U_1, U_2, and Ξ_0. Nor will the solution always be uniquely defined. The precise result is as follows.

Define the *downstream wave locus*, $D(U_1)$, of a state U_1, with respect to a fixed Ξ_0, as the union of all states which can be joined to U_1 by a shock or rarefaction in which U_1 is the upstream state. The downstream wave locus divides the half-plane $u \leq u_s$ into four regions, labelled $1 - 4$ in Figure 4; we label the region above U_1 as '1' and continue counterclockwise. We state

Theorem 1. *Given a Riemann data triple (U_1, U_2, Ξ_0) with hyperbolic states U_1 on the lower half and U_2 on the upper half of a spacelike line L through Ξ_0, let $D(U_1)$ be the downstream wave locus of U_1. Then if U_2 lies in any of the regions 1, 2, or 4 in the complement of D, the Riemann problem has a unique admissible solution. If U_2 lies in region 3, there may be none, one or two admissible solutions.*

Proofs of this and the other results in this section can be found in [5].

Theorem 1 establishes the form of the solution of the hyperbolic Riemann problem which occurs when (a, b) lies below the curve I (Case 1 data).

Theorem 2. *If oblique shock data (2) given for equation (1) are such that $a \leq I(b)$, then a hyperbolic quasi-one-dimensional Riemann problem occurs at Ξ_I. Furthermore, the data for this problem lead to a region 4 configuration, and there is a unique admissible solution, which consists of a rarefaction wave and a shock.*

This result gives a solution to the shock interaction problem for Case 1 data at all supersonic points in the (ξ, η) plane. In [3] and [4] we pursue the subsonic part of the solution. In the first paper, we have solved the degenerate elliptic equation for a fixed boundary, taking account of the fact that a square-root type singularity occurs in the solution at the degenerate boundary. The second paper concerns the free boundary problem which determines the nonuniform part of the Mach stem.

The problem in Case 1 is simplified by the fact that Ξ_I is the only admissible primary intersection, and it organizes the entire problem.

For other cases, the behavior at Ξ_I does not determine the solution; in fact, Ξ_I may not even occur. There are other types of shock interaction where we can formulate quasi-one-dimensional Riemann problems. We have not yet solved the corresponding elliptic (subsonic) problems for these cases.

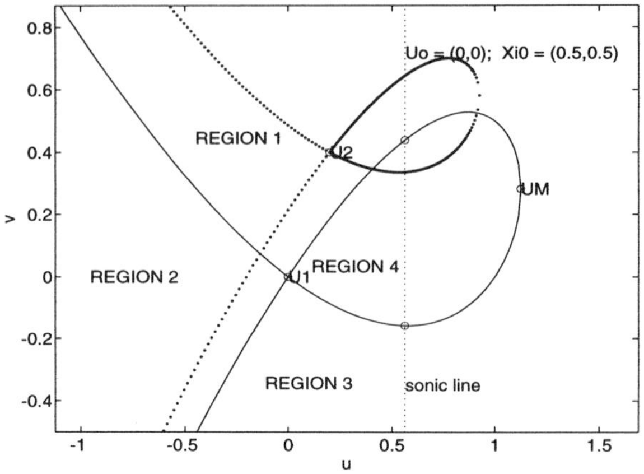

FIGURE 4. Solution of Quasi-One-Dimensional Problems

3.1. Symmetric Shock Reflection at the Wall. Proposition 5 shows that Ξ_A is a primary intersection if (a,b) lies to the right of the curve $L(b)$ in Figure 3. The interaction of the incident shock, S_1, with the wall may be reduced to a standard Riemann problem by considering as Riemann data at Ξ_A a spacelike line L (vertical through Ξ_A) and data U_1 on the upper half of the line and the reflected state $U_1^* \equiv (1,a)$ on the lower half.

Proposition 6. *This is a hyperbolic problem if* $a \geq 1/\sqrt{2}$.

For data of this form, if we consider the downstream locus, $D(U_1^*)$, then we are in Region 3 of Theorem 1, where we are not guaranteed existence or uniqueness of a solution. In fact, we have

Proposition 7. *The quasi-one-dimensional Riemann problem with the data triple* (U_1^*, U_1, Ξ_A) *has a solution if and only if* $a \geq \sqrt{2}$. *If* $a > \sqrt{2}$, *there are two solutions,* $U_R = (1 + a^2 - a\sqrt{a^2-2}, 0)$, *and* $U_F = (1 + a^2 + a\sqrt{a^2-2}, 0)$.

We can formulate the perturbed regular reflection equations for the region of parameter space $\mathcal{R}$ where $a \geq \sqrt{2}$ and $a > I(b)$. Only if $a > L(b)$ is Ξ_A an admissible primary intersection, but it is always *locally admissible*: that is, it lies outside of and upstream from P_1.

For points (a,b) near $J(b)$ there may be solutions close to the respective uniform solutions defined in Proposition 2. We can define two perturbation problems at each point in $\mathcal{R}$, one based on each of the two Riemann solutions of Proposition 7. In both cases, the solution will be piecewise constant outside an elliptic region, in which the solution will satisfy a degenerate elliptic free boundary problem. We conjecture that only for some of these problems will solutions exist; thus, there may be two, one or even no regular reflection type solutions for (a,b) in $\mathcal{R}$.

4. Shock Reflection Problems for Other Parameter Values

4.1. The UTSD Prototype for von Neumann Reflection. There is another region of parameter space where we can reduce the problem to solving an elliptic equation. This is the case where $\Xi_1 \equiv S_1 \cap P_1$ is admissible. This requires the two conditions $\eta_1 > 0$ (so that Ξ_1 exists) and $a > I(b)$, so that Ξ_I is not admissible. The first condition, from Table 1, is $a < 1/\sqrt{2}$. We shall call this region $\mathcal{N}$.

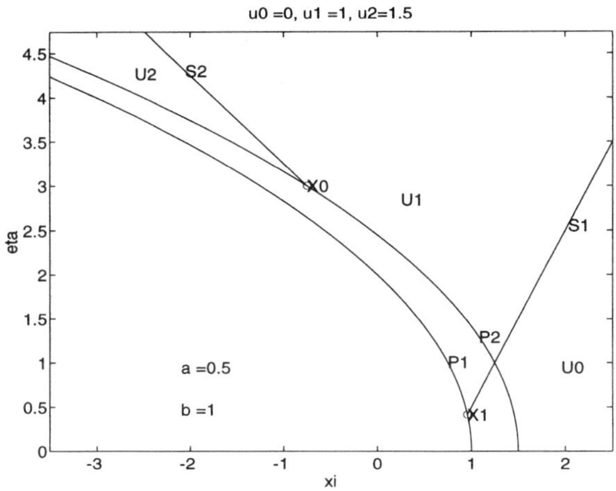

FIGURE 5. The Hyperbolic Solution in von Neumann Reflection

For $(a,b) \in \mathcal{N}$, we again have a reduction of the problem to a degenerate elliptic system. Figure 5 pictures the important variables.

The solution has the constant value U_2 left of the shock S_2 up to the parabola P_2; the shock then becomes nonuniform, decaying and turning until it attains zero strength as it becomes tangent to P_1. The shock S_1 is uniform until it reaches P_1, then it forms a free boundary between a nonuniform, subsonic solution and the upstream state U_0. The elliptic problem in this case has two degenerate parts of the boundary – at P_2 and at P_1 – and two free boundary segments. Details remain to be worked out.

4.2. A Mechanism for Mach Reflection with a Kink. Suppose that Ξ_0 is admissible (that is, we are outside Case 1) and that Ξ_A is inside P_2 (that is, $a < L(b)$). In this case, we conjecture that another type of solution may appear: there may be nonuniform *hyperbolic* solutions $(u(\xi,\eta), v(\xi,\eta))$ inside P_2 which approach U_2 at P_2 but with a singularity

$$u \sim -w\sqrt{\rho_2 - \rho}$$

near P_2. Here, $\rho \equiv \eta^2/4 + \xi$ is a parabolic coordinate, and $w > 0$. Singular solutions of the degenerate hyperbolic problem of this form exist, but it is not clear whether the degenerate hyperbolic free boundary problem is well-posed. The free boundary is now the perturbed continuation of S_2, with upstream data the constant state U_1. The Rankine-Hugoniot conditions give sufficient data to determine the nonuniform flow completely. We conjecture that the singular hyperbolic IV-BVP has a unique solution and determines the continuation of S_2 for some distance inside P_2. Then either the continuation of S_2 meets S_1 and a hyperbolic Mach reflection, as in Case 1, occurs or S_2 meets P_1 and disappears, as in von Neumann reflection.

The solution suggested here will look like the prototype Mach reflection but with an extra kink: a bend in the S_2 shock above the intersection point, and a nonuniformity (and singularity) in the solution along P_2. We can describe this prototype as *kinky Mach reflection* (KMR). It shares some features with the phenomenon named 'Transitional Mach Reflection' (TMR) by Ben-Dor, [1].

4.3. Double Mach Reflection in the UTSD Equation. If we compare the proposed wave interactions, by region, with theoretical and experimental predictions (see Ben-Dor, [1], for example), there is a rough correspondence between the 'bifurcation diagram' in the (M_s, θ_w^C) plane and the one that we are building in the (a, b) plane. When the (M_s, θ_w^C) diagram is rotated 90 degrees counterclockwise and then reflected in a vertical axis, the correspondence becomes apparent. As a qualitative picture of what this model may be capable of showing, namely the relation between different types of two-dimensional wave interactions as parameters vary, the picture so far shows some success: von Neumann reflection and regular reflection are far apart; in fact, they span the parameter space and are separated by all the other modes of interaction. The major discrepancy is that there is a large region devoted to 'double Mach reflection' in Ben-Dor's experimental and theoretical descriptions, and a big hole in our (a, b) plane, for $1/\sqrt{2} < a < \sqrt{2}$ and b larger than the values which will allow construction of the kinky Mach reflection solution of the preceding section.

Double Mach reflection is difficult to describe in our simplified model. Even the 'simple Mach reflection' of Section 3 (Theorem 2 and the comments following) does not contain a genuine triple point. Mach reflection in this model is, in fact, a hyperbolic quasi-one-dimensional Riemann problem in which there are four states rather than three and a small rarefaction wave as well as three shocks. There should be a relationship between this simplified Mach stem and that in the full gas dynamics equation model. We have certainly neglected some important features, such as the interaction of linear and nonlinear waves, but we expect to be able to show that the simplified model mimics qualitative features of transitions in real flows.

In our model, a 'double Mach reflection' would have to contain two hyperbolic quasi-one-dimensional problems. We conjecture that one such problem will occur near each of Ξ_0 and Ξ_A in the case when each is admissible but neither gives rise to a hyperbolic Riemann solution $(1/\sqrt{2} < a < \sqrt{2}$ and $a > L(b))$.

In fact, because of the possibility that states U of larger magnitude may appear in the subsonic region, one cannot state *a priori* that any primary interaction is determined by the data at infinity. In particular, the symmetric reflection problem at the wall has no solution, if $a < \sqrt{2}$, irrespective of what other conditions are imposed in the rest of the physical plane. We can find a candidate for a solution by splitting the shock S_1 at a point upstream from Ξ_A and supposing that the shock S_2 splits slightly upstream from Ξ_0. These two triple points, connected by a shock, give a double Mach reflection. As in the other cases we have considered, there is a degenerate elliptic problem which will determine the solution in the subsonic region and will describe what happens to the downstream waves at the second Mach point. The connecting shock is hyperbolic at both end points and transonic for a range in the middle. In this, it resembles the uniform shock which occurs when a lies on the upper branch of $J(b)$.

5. Conclusions

In this paper, we have analysed a simplified model for two-dimensional interactions of weak shocks. We studied a two parameter family of initial data which gives rise to different shock interaction patterns. We showed that the initial value problem for the model equation reduces to solving a quasi-one-dimensional Riemann problem on a curve near infinity. To carry out the analysis, we developed a theory of solutions of quasi-one-dimensional Riemann problems which allows us to construct a solution in the supersonic (hyperbolic) regime. The understanding of a subsonic (elliptic) regime requires solving a series of free boundary problems for a degenerate elliptic equation. When this program is completed, it will provide a full classification of wave interactions, and a precise description of transition criteria between classes of solutions with different qualitative behavior. To the extent of our present knowledge, we present the general bifurcation diagram in Figure 6.

References

1. G. Ben-Dor, *Shock wave reflection phenomena*, Springer-Verlag, New York, 1992.

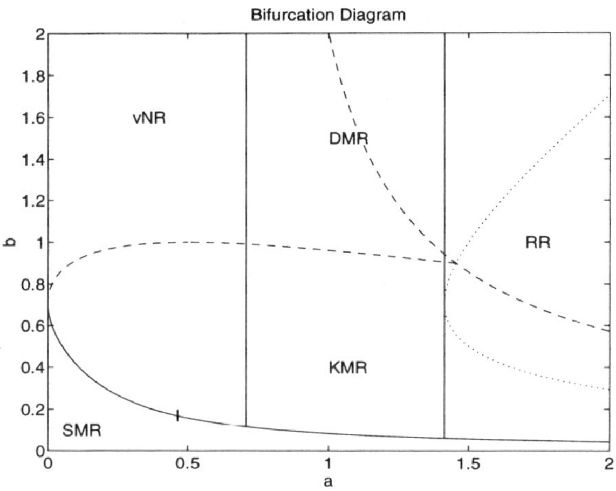

FIGURE 6. Classification of Parameter Space

2. M. Brio and J. K. Hunter, *Mach reflection for the two dimensional Burgers equation*, Physica D **60** (1992), 194–207.
3. S. Čanić and B. L. Keyfitz, *An elliptic problem arising from the unsteady transonic small disturbance equation*, Journal of Differential Equations, submitted.
4. ———, *A free-boundary problem for the two-dimensional unsteady transonic small disturbance equation*, in preparation.
5. ———, *Riemann problems for the two-dimensional unsteady transonic small disturbance equation*, in preparation.
6. ———, *The two-dimensional nonlinear wave equation: Notebook*, in preparation.
7. P. Colella and L.F. Henderson, *The von Neumann paradox for the diffraction of weak shock waves*, Journal of Fluid Mechanics **213** (1990), 71–94.
8. J. Glimm, *Nonlinear waves: Overview and problems*, Multidimensional Hyperbolic Problems and Computations (New York) (J. Glimm and A. Majda, eds.), vol. IMA, 29, Springer, New York, 1991, pp. 89–106.
9. Cathleen Synge Morawetz, *Potential theory for regular and Mach reflection of a shock at a wedge*, Communications on Pure and Applied Mathematics, to appear.
10. A. Sakurai, L.F. Henderson, K. Takayama, Z. Walenta, and P. Colella, *On the von Neumann paradox of weak Mach reflection*, Fluid Dynamics Research **4** (1989), 333–346.
11. A. Sasoh, K. Takayama, and T. Saito, *A weak shock wave reflection over wedges*, Shock Waves **2** (1992), 277–281.
12. E. G. Tabak and R. R. Rosales, *Weak shock focusing and the von Neumann paradox*, Physics of Fluids A, to appear.
13. T. Zhang and Y.-X. Zheng, *Conjecture on the structure of solutions of the Riemann problem for two-dimensional gas dynamics systems*, SIAM Journal of Mathematical Analysis **21** (1990), 593–630.

CLAWPACK — A software package for solving multi-dimensional conservation laws

Randall J. LeVeque[1]

ABSTRACT. **clawpack** is a collection of Fortran subroutines for solving hyperbolic systems of conservation laws in one and two space dimensions, both with and without source terms. High resolution flux-limiter finite volume methods are used on uniform rectangular grids. In two dimensions an unsplit multidimensional algorithm using transverse wave propagation has been implemented. A variety of example problems are included in the package, which is intended both for research and instructional purposes.

1. Introduction

This paper contains a brief description of `clawpack` (Conservation **LAW**s **PACK**age), a collection of Fortran subroutines for solving hyperbolic systems of conservation laws in one and two space dimensions. This package can handle both homogeneous systems, e.g.,

$$u_t + f(u)_x + g(u)_y = 0$$

in two dimensions, and also systems with source terms,

$$u_t + f(u)_x + g(u)_y = \psi(u).$$

Problems with source terms are solved using a Strang splitting in which the solution of the homogeneous equation is alternated with solutions of the ODE $u_t = \psi(u)$.

The methods used are finite volume methods on uniform rectangular grids, e.g., in two dimensions,

$$U_{ij}^{n+1} = U_{ij}^n - \Delta t \left[\frac{1}{\Delta x}(F_{i+1/2,j} - F_{i-1/2,j}) + \frac{1}{\Delta y}(G_{i,j+1/2} - G_{i,j-1/2}) \right].$$

The value U_{ij}^n is an approximation to the cell average of $u(x, y, t_n)$ over the (i, j) cell and F and G are numerical flux functions.

High resolution flux-limiter methods are used that are based on solving one-dimensional Riemann problems at each interface between adjacent grid cells. A "Riemann problem" is simply the one-dimensional problem $u_t + f(u)_x = 0$ or $u_t + g(u)_y = 0$ normal to the interface with piecewise constant data. For a hyperbolic system of `meqn` equations satisfying standard constraints, the solution to the Riemann problem is a similarity solution (constant along rays x/t = constant if the initial discontinuity is located at $x = t = 0$) consisting of `meqn` waves (shock waves, rarefaction waves, etc.). The solution to the Riemann problem is fundamental to the study of conservation laws and to high resolution numerical methods of the type employed here. See, e.g., [8, 11, 15].

Use of the `clawpack` routines requires that the user specify a Riemann solver for the particular equations being solved, as described further below. Riemann solvers for several commonly occurring systems of equations are included in the package. A library of applications is also being developed that contains additional examples.

The fact that the package only applies to uniform rectangular grids in more than one dimension is a major limitation in terms of its applicability to practical problems. However, it is possible to solve some problems on quadrilateral grids in logically rectangular domains by incorporating

[1]Department of Applied Mathematics and Department of Mathematics, University of Washington, Seattle, WA 98195. rjl@amath.washington.edu

an appropriate coordinate transformation in the equations. A future goal is to develop an expanded package for arbitrary geometries (in two dimensions, at least) using the same finite volume methods on Cartesian grids but with additional routines to handle arbitrary boundaries cutting through the grid cells, based on the techniques developed by Berger and LeVeque[1, 2].

The main goal of the present package, however, is not to provide software capable of handling general engineering applications, but rather to provide an educational and research tool for the study of hyperbolic systems of conservation laws and the future development of high resolution numerical methods for their solution. Some of the particular goals and features of this package are listed below.

- The routines are, I hope, not only easy to use but also to read and understand, so that they can be used by students in connection with a course on numerical methods for conservation laws. I am currently writing an expanded version of [11] that includes extensive discussion of the multi-dimensional methods used in this package, and that will include numerous exercises and examples based on the package.
- The routines are written in a modular form that should allow students and researchers to modify or replace sections of the code in order to test alternative methods. By using the same basic framework and driver programs it will be possible to efficiently compare different methods on the same problem.
- The `clawpack` package and the associated applications library should provide a test bed of standard problems that is widely available to aid in the evaluation of new algorithms. Running the `clawpack` routines on these problems gives a sample solution for comparison purposes. (Performing better than `clawpack` should be viewed as a necessary, but far from sufficient, condition for considering a new algorithm to be a significant advance in the field. Many other algorithms are already available which may perform at least as well or better on particular applications.)
- The package should also provide a tool for experimenting with the behavior of solutions to systems of conservation laws. This could be useful for students in developing better intuition for physical systems such as the Euler equations, and also for researchers investigating a newly developed or poorly understood system of equations in some applications area.

In one space dimension the method used is a fairly standard flux-limiter method. In two dimensions the multi-dimensional method developed in [12] (see also [9]) is implemented. This method is similar in some respects to various other methods, e.g., [4, 13], but is constructed in a "wave-propagation" framework that allows it to be developed as a very natural generalization of the one-dimensional algorithm. This makes it relatively easy to understand and explain to students, and indeed pedagogical needs were a primary motivation in the development of these methods. At the same time, the method performs quite well in practice on many problems, and appears to be competitive with more complicated methods, at least in the context of a general software package that is widely applicable. Naturally for any particular application it is possible to develop specialized methods with various bells and whistles that will perform much better. A useful list of other software packages available for CFD work can be found on the Web at `http://icemcfd.com/cfd/CFD_codes.html`. This includes, for example, some PPM codes and LCPFCT[3].

Obtaining `clawpack`. The package is available by from `netlib`. Send a mail message to `netlib@research.att.com` with the body

```
send pdes/claw/index
```

to get the index of available files. These can be requested by e-mail or obtained by anonymous ftp from `netlib.att.com`. In particular, the file `netlib/pdes/claw.tar` is a tar file of everything. Alternatively, you can obtain files by ftp through Mosaic using the URL

`ftp://netlib.att.com/netlib/pdes/claw/index.html`.

It is also possible to browse through the library using Mosaic from

`ftp://amath.washington.edu/pub/leveque/programs/claw/index.html`.

2. The structure of `clawpack`

This description is valid for Version 1.0 of `clawpack`, available at the time of writing. The structure may change in future versions.

The netlib directory `pdes/claw` contains both the basic `clawpack` directory and also directories for auxiliary examples (see Section 5). The `clawpack` directory contains the following subdirectories:

1d: One-dimensional subroutines and examples.
2d: Two-dimensional subroutines and examples.
limiters: Limiter functions used in both 1d and 2d routines.
matlab: m-files for creating graphics using matlab for various examples.

The directories `1d` and `2d` each contain the following subdirectories:

claw1/claw2: The basic package in 1d/2d
rp: Sample Riemann solvers
bc: Sample boundary value routines
ode: Sample ode solvers for source terms
examples: Example driver routines
misc: Additional subroutines used in some of the examples

The examples provided are intended to illustrate the use of the package on a variety of standard test problems. From these examples it should be quite easy to modify the driver program and select the appropriate Riemann solver and boundary condition routines (or write new ones if necessary) to solve a wide variety of problems.

As an illustration of how the example directories are set up, the directory `clawpack/1d/examples/ex1eu1` contains a program to solve a shock tube problem for the Euler equations of gas dynamics. The file `driver.f` contains the main driver program, which calls a subroutine `ic` to set up the initial conditions, and then calls `claw1` repeatedly to advance the solution by a specified time increment 0.05. The routine `out1eu` is called to output the solution after each call to `claw1` in a form suitable for the matlab graphics routines. Within each call to `claw1`, variable time steps are automatically chosen to reach the next output time with the Courant number roughly equal to 0.9, as specified in the driver. (Alternatively, one can specify a given fixed time step rather than the desired Courant number.) Each intermediate solution is output into a different file. The initial data at $t = 0$ goes into `fort.100`, the solution at $t = 0.05$ goes into `fort.101`, etc. These can then be viewed graphically in matlab by invoking the m-file `ex1eu1`, or by reading the output files into another graphics package preferred by the user.

The directory `clawpack/1d/examples/ex1eu1` also contains an executable file `compile`. Typing `compile` at the Unix prompt causes all of the appropriate subroutines to be compiled together, creating the executable `a.out`. If `a.out` is then executed, the driver program will run and prompt the user for various parameters such as `dx`, the grid spacing, and the data values such as `rhol, ul, pl`, and `rhor, ur, pr` in the shock tube problem (the values the density, velocity, and pressure to the left and right of the membrane in the shock tube).

The file `data` contains the appropriate values for Sod's shock tube problem[16], a standard test case. Simply typing `a.out < data` will cause this to run. Several examples are also available. Some of the other examples now available in `clawpack/1d/examples` are:

ex1bu1: Burgers' equation with sinusoidal initial data and periodic boundary conditions.
ex1eu1: Euler equations with piecewise constant initial data (Riemann problem) and solid wall boundary conditions.
ex1it1: Isothermal equations with piecewise constant initial data.
ex1sw1: Shallow water equations with solid walls.

Several examples in two space dimensions are available in `clawpack/2d/examples`, including:

ex2bu1: 2D Burgers' equation with periodic boundary conditions.
ex2eu1: Euler equations with piecewise constant initial data with one value inside a circle and another value outside. (radially symmetric).
ex2eu2: Euler equations with piecewise constant initial data in four quadrants ("2D Riemann problem").
ex2eu3: The Euler equations with radially symmetric data is solved on a sequence of grids and the solution in each case is compared with the "exact" solution computed by solving the radial problem with claw1so on a fine 1D grid. The order of accuracy and error constants are then approximated based on the reduction in error each time the grid is refined by a factor of 2.

3. The `claw1` routines

In one space dimension the main subroutines called by the user are `claw1` (for homogeneous problems) and `claw1so` (for problems with source terms). Both of these routines are in `clawpack/1d/claw1` and contain complete documentation on their use and all parameters in their calling sequences. Here, for example, is the calling sequence of `claw1so`:

```
      subroutine claw1so(maxmx,meqn,mwaves,mbc,mx,
     &           q,dx,tstart,tend,dtv,cflv,nv,method,mthlim,verbose,
     &           qwork,work,mwork,info,bc1,rp1,ode1)
```

In the code, q is used for the vector of conserved quantities rather than u, in order to avoid confusion with the standard notation for the x-component of velocity. A brief description of each parameter follows, based on a shortened form of the documentation in the code. See the code itself for more details.

```
c     maxmx is the maximum number of interior grid points in x.
c
c     meqn is the number of equations in the system.
c
c     mwaves is the number of waves that result from the
c            solution of each Riemann problem.
c
c     mbc is the number of "ghost cells" that must be added on to each
c        side of the domain to handle boundary conditions.  (usually 2)
c
c     mx is the number of grid cells in the x-direction.
c
c     q(1-mbc:maxmx+mbc, meqn)
```

```
c         On input:  initial data at time tstart.
c         On output: final solution at time tend.
c
c     dx = grid spacing in x.
c
c     tstart = initial time.
c
c     tend = Desired final time (on input).
c          = Actual time reached (on output).
c
c     dtv(1:5) = array of values related to the time step
c
c     cflv(1:4) = array of values related to Courant number
c
c     nv(1:2) = array of values related to the number of time steps
c
c     method(1:2) = array of values specifying the numerical method to use
c          method(1) = 0 if fixed size time steps are to be taken.
c                    = 1 if variable time steps are to be used.
c
c          method(2) = 1 if Godunov's method is to be used, with no 2nd order
c                        corrections.
c                    = 2 if second order correction terms are to be added, with
c                        a flux limiter as specified by mthlim.
c                    = 3 if ``third order'' corrections are to be added,
c                        an experimental option.
c
c     mthlim(1:mwaves) = array of values specifying the flux limiter to be used
c                        in each wave family.
c
c     verbose = .true.   to print dt and Courant number every time step
c
c     qwork(1-mbc:maxmx+mbc, meqn) =
c         Array of same dimensions as q needed for work space.
c
c     work(mwork) = double precision work array of length at least mwork
c
c     mwork = length of work array. Must be at least
c               (maxmx + 2*mbc) * (meqn + mwaves + meqn*mwaves)
c
c     info = output value yielding error information
c
c     bc1 = user-supplied subroutine that specifies the boundary conditions.
c          This subroutine should extend the values of q from cells
c          1:mx to the mbc ghost cells along each edge of the domain.
c
c     rp1 = user-supplied subroutine that implements the Riemann solver
c
c     ode1 = user-supplied subroutine that solves the ODE q_t = psi(q) in
c              each grid cell for the source terms.
c
```

The algorithm. There is only room here for a very brief description of the algorithms. See the papers [9, 10, 12] for more complete descriptions. The slides from a talk on the subject, which are in `claw/doc/slides.ps.Z`, also provides a brief overview with many figures.

A finite volume method is used, of the form

$$U_i^{n+1} = U_i^n - \frac{\Delta t}{\Delta x}(F_{i+1/2}^n - F_{i-1/2}^n).$$

The numerical flux $F_{i+1/2}$ is computed based on the solution to Riemann problems at each cell interface. The Riemann solver is the fundamental user-supplied routine that specifies the conservation law being solved and provides the mechanics of solving each Riemann problem. The subroutine `rp1` need not compute the exact solution to the Riemann problem, and in fact typically will not. An approximate Riemann solution such as the one proposed by Roe[14] is suitable. The calling sequence of the routine `rp1` is

```
      subroutine rp1(maxmx,meqn,mwaves,mbc,mx,ql,qr,wave,s,f0)
```

where

```
c         On input, ql and qr contain the data for the Riemann problems.
c             The values ql(i,:) and qr(i,:) are values at the left and
c                right edge of the i'th cell.
c             The values qr(i-1,:) and ql(i,:) are used to solve the
c                i'th Riemann problem
c         On output,
c             wave(i,m,p) is the m'th component of the jump across
c                             wave number p in the i'th Riemann problem.
c             s(i,p) is the wave speed of wave number p in the
c                             i'th Riemann problem.
c             f0(i,m) is m'th component of the Godunov flux in the i'th
c                     Riemann problem, which is defined as the value of
c                     f(x_0) where f is the flux function and x_0 is the
c                     value of the Riemann problem solution along x/t = 0.
```

For the methods currently implemented, `rp1` is always called with `ql = qr`, since these methods first solve the Riemann problem based on the cell averages and then apply second-order slope corrections based on information about the wave structure coming from the solution to the Riemann problems. The additional flexibility of allowing `ql` $\neq$ `qr` is included so that one can easily experiment with other methods using the same Riemann solvers. Methods such as MUSCL[20], PPM [5] or Colella's multi-dimensional method[4] start by computing slopes in each cell and then use the resulting edge values to solve Riemann problems for the fluxes.

If the exact Riemann solution is implemented, then `f0` returns the value $f(u_0)$ where u_0 is the value of u along $x/t = 0$ in the exact similarity solution to the Riemann problem. The classical Godunov method would then consist of setting

$$F_{i-1/2} = f0(i, \cdot). \tag{1}$$

The arrays `wave` and `s` are used only to implement second order corrections. To implement these corrections, it is assumed that solving the Riemann problem also yields `mwaves` waves denoted by $W_{i,p}$ for $p = 1, 2, \ldots,$ `mwaves`, moving at speeds $s_{i,p}$. ($W_{i,p}$ represents the vector with mth component given by `wave(i,m,p)`). Each wave is assumed to be a discontinuity in u, with

$$\sum_{p=1}^{\texttt{mwaves}} W_{i,p} = U_i - U_{i-1}.$$

For a system of meqn equations, it will often be the case that mwaves = meqn, but this need not be so. For example, the approximate Riemann solver proposed by Harten, Lax and van Leer[7] always approximates the solution to any Riemann problem by only two waves, one moving in each direction, and so mwaves = 2 independent of the number of equations.

With the Roe solver, mwaves = meqn and the waves are determined by solving the Riemann problem for a linearized system of equations

$$u_t + A_i u_x = 0$$

where the matrix A_i is determined locally at each interface by the data U_{i-1} and U_i. The vector $U_i - U_{i-1}$ is split into eigenvectors $r_{i,p}$ of A_i as

$$U_i - U_{i-1} = \sum_{p=1}^{\texttt{meqn}} \alpha_{i,p} r_{i,p}$$

and $W_{i,p} = \alpha_{i,p} r_{i,p}$. The speed $s_{i,p}$ is the corresponding eigenvalue of A_i. The value of f0 might then be set to $f(u_0)$, where

$$u_0 = U_{i-1} + \sum_{\text{leftgoing}} W_{i,p},$$

with the sum over all p for which $s_{i,p} < 0$. Alternatively, the formula

$$\texttt{f0} = f(U_{i-1}) + \sum_{\text{leftgoing}} s_{i,p}\alpha_{i,p} r_{i,p}$$

could be used. The two are equivalent for a linear system but not, in general, for a nonlinear problem.

In practice it is sometimes necessary to modify f0 to include an "entropy fix", so that transonic rarefactions are handled more properly, as discussed in [11]. An example can be seen in clawpack/1d/rp/rp1eu.f, where the Roe solver for the one-dimensional Euler equations is implemented with the entropy fix of [6]. (For comparison purposes, it is also possible to turn off this entropy fix.)

If method(2) = 1 in the call to claw1, then the first order Godunov method is used and $F_{i-1/2} = \texttt{f0}(i, \cdot)$ is used. In this case the values returned in wave and s are not used. If method(2) = 2, then second order corrections are applied as described in [12]. First a limiter is applied to the waves $W_{i,p}$ by comparing each wave to the neighboring wave $W_{i-1,p}$ (if $s_{i,p} > 0$) or $W_{i+1,p}$ (if $s_{i,p} < 0$). This is accomplished by a call to the subroutine limiter, which loops over all i and p to limit all of the $W_{i,p}$. For each set of i and p, the vector $W_{i\pm1,p}$ is projected onto the vector $W_{i,p}$ and the magnitude of this projection is compared to the 2-norm of $W_{i,p}$ itself. The function philim is applied to these two values to produce a limited value, and $W_{i,p}$ is replaced by the vector

$$W_{i,p} \;:=\; \texttt{philim}(\|W_{i,p}\|,\; W_{i\pm1,p}^T W_{i,p}/\|W_{i,p}\|) \cdot \frac{W_{i,p}}{\|W_{i,p}\|}.$$

This vector points in the same direction in phase space as the original $W_{i,p}$, but has its magnitude limited by philim. Several standard limiters are implemented in philim, including minmod, superbee, and the limiters of van Leer (see, e.g., [11, 17, 18, 19]). The value of mthlim(p) passed into claw1 determines which limiter is used in the pth wave family. Different limiters can be used in each family, e.g., one could use superbee for contact discontinuities and minmod for the nonlinear families in the Euler equations.

After applying the limiter, the flux (1) is then corrected for high resolution by performing

$$F_{i+1/2} \;:=\; F_{i+1/2} + \frac{1}{2}\sum_{p=1}^{\texttt{mwaves}} |s_{i,p}| \left(1 - \frac{\Delta t}{\Delta x}|s_{i,p}|\right) W_{i,p}.$$

If `mthlim(p) = 0`, then no limiter is applied, *i.e.*, `philim`$(a, b) = a$, and the method reduces to a Lax-Wendroff style second order accurate method. With limiters, this becomes a form of the flux-limiter method discussed, for example, in [11, 17].

4. The claw2 routines

The two-dimensional routines have a similar style. For example, `claw2so` has the calling sequence

```
      subroutine claw2so(maxmx,maxmy,meqn,mwaves,mbc,mx,my,
     &                  q,dx,dy,tstart,tend,dtv,
     &                  cflv,nv,method,mthlim,verbose,
     &                  qwork1,qwork2,work,mwork,info,bc2,rp2A,rp2B,ode2)
```

where most of the parameters are exactly analogous to the one-dimensional case. Complete details can be found in the documentation of the code.

The major difference is in the Riemann solver. There are now two routines required, `rp2A` and `rp2B`. The first of these is directly analogous to the one-dimensional routine `rp1` and will be discussed first. The latter is needed for performing wave splitting in the transverse direction (orthogonal to the direction of the Riemann problem), as required for the multi-dimensional algorithm. The calling sequence for `rp2A` is

```
      subroutine rp2A(ixy,maxm,meqn,mwaves,mbc,mx,q1d,wave,s,f0,dfl,dfr)
```

This routine works on a one-dimensional slice of the data, stored in `q1d`. If `ixy = 1` then this is assumed to be a slice in the x-direction along $y =$ const, so the Riemann problem $u_t + f(u)_x = 0$ must be solved at each interface. If `ixy = 2` then `q1d` contains a slice in the y-direction and $u_t + g(u)_y = 0$ must be solved at each interface. Otherwise the routine is essentially identical to `rp1`, except that it must also return two arrays `dfl` and `dfr` that give the flux-difference splitting at each interface into leftward and rightward moving pieces. These should be set by

$$\begin{aligned}\texttt{dfl}(i,m) &= \texttt{f0}(i,m) - f(i-1,m)\\ \texttt{dfr}(i,m) &= f(i,m) - \texttt{f0}(i,m)\end{aligned}$$

where $f(j, m)$ represents the mth component of the flux vector $f(U_j)$. Note that the notation assumes `ixy = 1`. If `ixy = 2` then the flux is g rather than f.

The flux differences are used only in doing transverse corrections, and are the main data passed into `rp2B` to be decomposed into eigenvectors of a linearized Roe matrix in the orthogonal direction. For example, if `ixy = 1` then `rp2A` splits each $U_i - U_{i-1}$ into eigenvectors of a matrix A_i as described in Section 3, where $u_t + A_i u_x = 0$ is a linearization of $u_t + f(u)_x = 0$ based on U_{i-1} and U_i. (These values are really $U_{i-1,j}$ and U_{ij} for some fixed j determining the particular slice.) This same data U_{i-1} and U_i can also be used to define a matrix B_i so that $u_t + B_i u_y = 0$ is a linearization of $u_t + g(u)_y = 0$ valid near the interface. The routine `rp2B` should then split any vectors `df` into eigenvectors of B_i. For example, with the Euler equations the Roe-averaged velocity, enthalpy, etc. that are used to define A_i can also be used to define B_i. This makes the transverse splitting very efficient since the bulk of the work is already done in `rp2A` and then `rp2B` is essentially just a matrix-vector multiply. Hence the transverse propagation can be added with very little increase in computational effort beyond what is required to solve the Riemann problems in the normal direction. Besides giving full second order accuracy, the transverse derivatives also increase the stability limit so that a Courant number of 1 can be used, rather than 1/2 as would be required with normal propagation alone. The calling sequence for `rp2B` is

```
      subroutine rp2B(ixy,maxm,meqn,mwaves,mbc,mx,q1d,df,waveb,sb)
```

where on input `df` contains the array to be split and the output consists of the arrays `waveb` and `sb`, where

$$\texttt{df}(i,m) = \sum_{p=1}^{\texttt{mwaves}} \texttt{waveb}(i,m,p)$$

is the splitting of `df` into eigenvectors of B_i and `sb(i,p)` contains the corresponding eigenvalues.

Denote the waves obtained by splitting `dfl` by $W^L_{i,p}$ with speeds $s^L_{i,p}$ for $p = 1, 2, \ldots, mwaves$. Then the transverse propagation is accomplished by updating the fluxes $G_{i-1,j+1}$ and $G_{i-1,j}$ to include the upward and downward moving waves, respectively. (Again the notation assumes `ixy = 1.`). These updates are:

$$\begin{aligned} G_{i-1,j} &:= G_{i-1,j} + \frac{1}{2}\frac{\Delta t}{\Delta x} \sum_{s^L_{i,p}<0} s^L_{i,p} W^L_{i,p} \\ G_{i-1,j+1} &:= G_{i-1,j+1} + \frac{1}{2}\frac{\Delta t}{\Delta x} \sum_{s^L_{i,p}>0} s^L_{i,p} W^L_{i,p} \end{aligned}$$

Similarly, the rightgoing waves $W^R_{i,p}$ are used to modify the fluxes above and below the cell to the right of the interface:

$$\begin{aligned} G_{ij} &:= G_{ij} + \frac{1}{2}\frac{\Delta t}{\Delta x} \sum_{s^R_{i,p}<0} s^R_{i,p} W^R_{i,p} \\ G_{i,j+1} &:= G_{i,j+1} + \frac{1}{2}\frac{\Delta t}{\Delta x} \sum_{s^R_{i,p}>0} s^R_{i,p} W^R_{i,p} \end{aligned}$$

This transverse propagation algorithm is described in more detail in [12]. See also the slides from my talk (available in `claw/doc`) for several figures illustrating the ideas.

It has recently been found that applying transverse propagation to the correction waves is also possible and often yields better resolution. This is described in the scalar case in [9] and implemented in the algorithm in the case when `method(3) = 2` in the `claw2` code.

The multidimensional algorithm carries over to three space dimensions quite easily. A three-dimensional code `claw3` is currently being developed with Jan-Olav Langseth of the University of Oslo.

5. Applications

The netlib directory `pdes/claw` also contains some applications beyond the examples included in `clawpack`. Currently only a few are available, but more will be added in the future. Contributions of other applications by users would be gratefully accepted. The current contents of `pdes/claw` is:

clawpack: The basic package.
doc: Documentation and papers on the package and applications (including this manuscript).
advection: Two-dimensional advection algorithms from [9], implemented using `clawpack`, with most of the test problems from that paper and some new ones.
nozzle: A quasi-one-dimensional nozzle computed using `claw1so`.
vBurgers: viscous Burgers' equation, with diffusion equation as source.
L-domain: L-shaped domains, including the forward facing step problem.

Acknowledgments. Numerous colleagues and students have influenced my development of this software, and provided considerable encouragement. In particular Marsha Berger has had an enormous impact through our countless hours of writing and debugging code together, and her careful critiques of earlier versions of this software. This work was supported in part by NSF Grants DMS-8657319, DMS-9204329, and DMS-9303404, and DOE grant DE-FG06-93ER25181.

References

1. M. Berger and R. J. LeVeque, *Cartesian meshes and adaptive mesh refinement for hyperbolic partial differential equations*, Proc. Third Int'l Conf. Hyperbolic Problems, Uppsala, 1990.
2. ———, *A rotated difference scheme for Cartesian grids in complex geometries*, AIAA Conference on Computational Fluid Dynamics, CP-91-1602, 1991.
3. J. P. Boris, A. M. Landsberg, E. S. Oran, and J. H. Gardner, *LCPFCT — A flux-corrected transport algorithm for solving generalized continuity equations*, Naval Research Laboratory NRL/MR/6410–93-7192, Washington, DC, 1993.
4. P. Colella, *Multidimensional upwind methods for hyperbolic conservation laws*, J. Comput. Phys. **87** (1990), 171–200.
5. P. Colella and P. Woodward, *The piecewise-parabolic method (PPM) for gas-dynamical simulations*, J. Comput. Phys. **54** (1984), 174–201.
6. A. Harten and J. M. Hyman, *Self-adjusting grid methods for one-dimensional hyperbolic conservation laws*, J. Comput. Phys. **50** (1983), 235–269.
7. A. Harten, P. D. Lax, and B. van Leer, *On upstream differencing and Godunov-type schemes for hyperbolic conservation laws*, SIAM Review **25** (1983), 35–61.
8. P. D. Lax, *Hyperbolic systems of conservation laws and the mathematical theory of shock waves*, SIAM Regional Conference Series in Applied Mathematics, #11, 1972.
9. R. J. LeVeque, *High-resolution conservative algorithms for advection in incompressible flow*, SIAM J. Numer. Anal. **to appear**, (`ftp://amath.washington.edu/pub/leveque/papers/advection.ps.Z`).
10. ———, *High resolution finite volume methods on arbitrary grids via wave propagation*, J. Comput. Phys. **78** (1988), 36–63.
11. ———, *Numerical methods for conservation laws*, Birkhäuser-Verlag, 1990.
12. ———, *Simplified multi-dimensional flux limiter methods*, Numerical Methods for Fluid Dynamics 4 (M. J. Baines and K. W. Morton, eds.), Oxford University Press, 1993, pp. 175–190.
13. Yu. B. Radvogin, *Quasi-monotonous multidimensional differential schemes with second order accuracy*, Soviet Academy of Sciences Preprint (in Russian), 1991.
14. P. L. Roe, *Approximate Riemann solvers, parameter vectors, and difference schemes*, J. Comput. Phys. **43** (1981), 357–372.
15. J. Smoller, *Shock waves and reaction-diffusion equations*, Springer, 1983.
16. G. Sod, *A survey of several finite diffeence emthods for systems of nonlinear hyperbolic conservation laws*, J. Comput. Phys. **27** (1978), 1–31.
17. P. K. Sweby, *High resolution schemes using flux limiters for hyperbolic conservation laws*, SIAM J. Num. Anal. **21** (1984), 995–1011.
18. B. van Leer, *Towards the ultimate conservative difference scheme II. Monotonicity and conservation combined in a second order scheme*, J. Comput. Phys. **14** (1974), 361–370.
19. ———, *Towards the ultimate conservative difference scheme IV. A new approach to numerical convection*, J. Comput. Phys. **23** (1977), 276–299.
20. ———, *Towards the ultimate conservative difference scheme V. A second order sequel to Godunov's method*, J. Comput. Phys. **32** (1979), 101–136.

On the Hydrodynamic Model for Semiconductors and the Relaxation to the Drift–Diffusion Equation

P.A. MARCATI† R. NATALINI[1]‡

Abstract. We investigate an hyperbolic hydrodynamic model for semiconductors and its relaxation to the classical drift–diffusion equations as the relaxation time tends to zero.

1. Introduction

In recent years the theory of semiconductor device modeling has become an area of increasing interest in applied mathematics. For the moment, to deal with the basic kinetic transport equations remains too expensive for real life applications. Neverthless, from the transport equations, it is possible to derive simpler fluid dynamical equations for macroscopic (integral) quantities like particle, current or energy densities. They represent a good compromise between physical accuracy and reduction of computational cost.

Fluid dynamical models are obtained from the infinite hierarchy of the moment equations, derived from the kinetic transport equation, by a suitable truncation procedure. This requires finding appropriate expressions for the nth order moment in terms of the previous ones and modeling the production terms on the right-hand side of the moment equations which arise from the collision term in the kinetic transport equation. According to the different closures, we recover different limit models and, in particular, the drift–diffusion equations and the hydrodynamic Euler–Poisson system, see [Va], [Bl], [BW], the reference book of Markovich, Ringhofer and Schmeiser [MRS] and the literature quoted therein.

The bipolar drift–diffusion model was first derived in [Va] and is the most popular model for simulations in semiconductor devices. In [Po] this model was derived by the Hilbert expansion of the solutions in powers of the mean free path. This model works very well under the assumptions of low carrier densities and small electric fields. By contrast, the hydrodynamic models are usually considered to describe high field phenomena or submicronic devices. More recently Anile and Pennisi [AP1,2] and Anile and Muscato [AM] derived some more general hydrodynamic models by using the extended thermodynamics.

† Dipartimento di Matematica Pura e Applicata, Università degli Studi dell'Aquila, I–67000 L'Aquila, Italia

‡ Istituto per le Applicazioni del Calcolo "M. Picone", C.N.R., Viale del Policlinico 137, I–00161 Roma, Italia

Here, we first consider a (scaled) unipolar hydrodynamic isentropic Euler–Poisson system for semiconductors. We consider the unidimensional Euler system

$$\begin{cases} \partial_t n + \partial_x j = 0 , \\ \partial_t j + \partial_x \left(\frac{j^2}{n} + p(n)\right) = nE + C , \end{cases} \tag{E}$$

supplemented by the classical Poisson equation:

$$\partial_x E = n - b(x) , \tag{P}$$

for all $(x,t) \in \mathbb{R} \times [0,\infty)$, $\tau > 0$; here n is the electron density, j is the electron current density, E stands for the (negative) electric field and $b \in L^1$ for the density of the fixed (positively charged) background ions. The pressure–density relation is given for the electron pressure by $p(n) = \frac{1}{\gamma} n^\gamma$ with $1 < \gamma \leq 5/3$. The notation $u := \frac{j}{n}$ for the electron velocity is also used. The equations (E) are an hyperbolic system of balance laws describing the motion of a gas of electrons in a semiconductor under the action of the electric field E. Then, by the general theory of the quasilinear hyperbolic equations [La2], [Sm], the solutions can develop discontinuous states (shock waves) and need to be considered just in a weak sense.

The electron current relaxation term C models the effects of the collisions of the electrons with the semiconductor crystal lattice. Here we assume the following relaxation type form [Bl], [BW]

$$C = -\frac{j}{\tau}, \tag{1.1}$$

where the constant $\tau > 0$ stands for the current relaxation time.

We are also concerned with the alternative model given by the following version of the (unipolar) drift–diffusion equations [Ma], [MRS], [Jü]

$$\mathcal{N}_S + (\mathcal{N}\mathcal{E} - p(\mathcal{N})_x)_x = 0 \tag{DD}$$

where the (negative) electric field $\mathcal{E}$ is still given by the Poisson equation (P), i.e.:

$$\partial_x \mathcal{E} = \mathcal{N} - b .$$

Again $\mathcal{N}$ stands for the concentration of electrons. The density of the electron current is then recovered by the usual relation

$$\mathcal{J} = \mathcal{N}\mathcal{E} - p(\mathcal{N})_x .$$

Observe that the equation (DD) is a degenerate parabolic system, and the solutions could be non-regular in the region $\{\mathcal{N} = 0\}$, see [Gi], [DiB], [Jü] and the references therein.

Here we establish a connection between these models. Let (n, j, E) be a weak solution to the Euler–Poisson system. We introduce the scaled variables $\mathcal{N}^\tau(x,s) = n\left(x, \frac{s}{\tau}\right)$, $\mathcal{J}^\tau(x,s) = \frac{1}{\tau} j\left(x, \frac{s}{\tau}\right)$ and $\mathcal{E}^\tau(x,s) = E\left(x, \frac{s}{\tau}\right)$. Our main result here is the convergence of this scaled sequence to the solutions of the drift–diffusion model. In fact, we have (Section 3) that, as τ tends to zero, $\mathcal{N}^\tau$ converges to a density profile $\mathcal{N}$, strongly in L^p_{loc}, $p < +\infty$, $\mathcal{J}^\tau$ converges weakly in L^2_{loc} to a current profile $\mathcal{J}$

and $\mathcal{E}^\tau$ converges uniformly on compact sets to a limit electric field $\mathcal{E}$. The limit profiles $(\mathcal{N}, \mathcal{E})$ are a solution for the drift–diffusion model (DD)–(P).

We recall that the weak solutions to the hydrodynamic model (E)–(P) have been investigated by many authors, in particular [AMPS], [Ga] and [MP], for the stationary case, and [MN1] in the general (isentropic) case. Notice that the result concerning the Cauchy problem for the system (E)–(P) obtained in [MN1] needs to be strongly improved here in order to investigate the relaxation limit. Indeed, these results were obtained by using the classical fractional step Lax–Friedrichs and Godounov schemes, in the spirit of [DCL], and the *a priori* L^∞ and energy estimates were not uniform with respect to the relaxation time τ.

Therefore, we introduce here a different version of the fractional step method, which provides the required bounds. Next, in order to deal with weak solutions, we perform our limiting process by using the methods of compensated compactness following [MM] and [MMS], where a similar analysis was provided for the porous media flow and the nonlinear heat conduction. More details and proofs can be found in [MN2]. Finally, let us mention that these results have been also extended in [Na] to the bipolar system [Ma], [MRS]

$$\begin{cases} \partial_t n + \partial_x j_n = 0 \ , \\ \partial_t p + \partial_x j_p = 0 \ , \\ \partial_t j_n + \partial_x \left(\frac{j_n^2}{n} + r_n \right) = nE + C_n \ , \\ \partial_t j_p + \partial_x \left(\frac{j_p^2}{p} + r_p \right) = -pE + C_p \ , \end{cases} \tag{BE}$$

supplemented by the classical Poisson equation:

$$\partial_x E = n - p - b(x) \ , \tag{P'}$$

where n is again the electron density, p the positively charged hole density and j_n and j_p are respectively the electron and the hole current densities.

2. The uniform bounds

Let us define the entropy solutions to the Cauchy problem for the system (E)–(P) with the initial conditions

$$\begin{aligned} n(x,0) &= n_0(x) \ , \quad j(x,0) = j_0(x) \\ E_0(x) &= \int_{-\infty}^{x} (n_0(y) - b(y))\, \mathrm{d}y + E^- \ , \end{aligned} \tag{2.1}$$

where E^- is a given constant.

Set

$$U = (n, j), \ F(U) = (j, \frac{j^2}{n} + p(n)), \ G(U, E) = (0, nE - \frac{j}{\tau}) \ .$$

Then the system (E) reads now

$$\partial_t U + \partial_x F(U) = G(U, E) \ . \tag{2.2}$$

We recall that an entropy–entropy flux pair for a system of balance laws is a couple (η, q) defined by the relation

$$\nabla q = \nabla \eta \nabla F$$

where ∇ denotes gradient with respect to the state variable U.

We say that the vector–valued function (U, E) is an entropy solution of the system (E)–(P) if the Poisson equation is taken in the weak (distributional) sense and for any convex entropy function $\eta = \eta(U)$, one has

$$\partial_t \eta + \partial_x q + S \leq 0 \qquad \text{in} \quad \mathcal{D}' , \tag{2.3}$$

for any locally bounded measurable function $S(x,t)$, such that

$$S + (\nabla n)^T \cdot G(E, U) \leq 0.$$

We next define, to give uniform estimates with respect to τ, our finite differencc scheme which combines the classical Lax–Friedrichs [La1] scheme with a suitable modification of the fractional step method. Consider a partition of $\mathbb{R} \times [0, +\infty)$ into horizontal layers

$$S_k = \{(x,t) | kh \leq t < (k+1)h\} ,$$

for the time mesh–length $\Delta t = h > 0$ and for any integer $k \geq 0$. If $I_k = \{i \,| k + i$ is even$\}$ let us set

$$Q_{i,k} = \{(x,t) | (i-1)l < x < (i+1)l\} \cap S_k ,$$

for any $i \in I_k$ and for the space mesh–length $\Delta x = l$, such that the ratio $\frac{h}{l}$ satisfies the CFL condition. We show how to get the approximate solution $U^l := (n^l, j^l)$ in S_k by means of the solution at $t = kh$, for any $k \geq 0$.

For any $i \in I_k$ define

$$U^{i,k} = \frac{1}{2l} \int_{(i-1)l}^{(i+1)l} U^l(x, kh - 0) \, \mathrm{d}x . \tag{2.4}$$

We denote by $\tilde{U}^l$ the solution of the Riemann problem for the homogeneous gas–dynamics in $Q_{i,k}$ for $i \in I_k$, by taking as initial datum at $t = kh$

$$U_k^l(x) = \begin{cases} U^{i-1,k} , & (i-1)l < x < il , \\ U^{i+1,k} , & il < x < (i+1)l . \end{cases}$$

We also set

$$\tilde{E}^l(x,t) = \int_{-\infty}^{x} [\tilde{n}^l(y,t) - b(y)] \, \mathrm{d}y + E^- ,$$

where E^- is given by (2.1).

Then, we set in S_k

$$\begin{cases} n^l(x,t) = \tilde{n}^l(x,t) , \\ j^l(x,t) = \tilde{n}^l(x,t) \int_{kh}^{t} e^{-\frac{(t-s)}{\tau}} \tilde{E}^l(x,s) \, \mathrm{d}s + e^{-\frac{(t-kh)}{\tau}} \tilde{j}^l(x,t) . \end{cases} \tag{2.5}$$

The approximate electric field E^l is given by

$$E^l(x,t) = \int_{-\infty}^{x} [n^l(y,t) - b(y)]\,\mathrm{d}y + E^- \ . \tag{2.6}$$

We require also the classical CFL condition, namely

$$\sup|\lambda_i(U^l)| \leq \frac{l}{2h} \ , \qquad 1 = 1,2,$$

where λ_1, λ_2 are the characteristic velocities for the system (E). Notice that in the same way it is possible to deal with a modified version of the fractional step Godunov scheme, see [MN1,2].

Our main existence result, with the uniform bounds, is the following one.

Theorem 2.1. *Assume that n^0, j^0 are some initial data with compact support which satisfy, for some positive constants C_1, C_2:*

$$0 \leq n^0(x) \leq C_1 \ , \quad |u^0(x)| = \left|\frac{j^0(x)}{n^0(x)}\right| \leq C_2 \ , \tag{2.7}$$

a.e. $x \in \mathbb{R}$ and assume that $b \in L^1$. Let E^- be a given constant and set

$$E^0(x) = \int_{-\infty}^{x} (n^0(y) - b(y))\,\mathrm{d}y + E^- \ . \tag{2.8}$$

Then, there exists a globally bounded entropy solution to the Cauchy problem (E)-(P)-(2.1) and the L^∞-bounds do not depend on τ.

First, we establish some sharp uniform bounds on the approximate solutions (n^l, j^l, E^l) given by the scheme (2.5)–(2.6).

Proposition 2.2. *Under the assumption of the Theorem 2.1, there exist some constants C_0, $\overline{E} > 0$ and $l^* > 0$ such that*

$$\begin{aligned} 0 \leq n^l(x,t) &\leq [\theta(C_0 + \tau\overline{E})]^{1/\theta} \ , \\ |u^l(x,t)| = \left|\frac{j^l(x,t)}{n^l(x,t)}\right| &\leq C_0 + \tau\overline{E} \ , \end{aligned} \tag{2.9}$$

$$|E^l(x,t)| \leq \overline{E} \ , \tag{2.10}$$

for a.e. $(x,t) \in \mathbb{R} \times [0,\infty)$ and any $l \in (0,l^*)$.

We notice that, by using the classical fractional step method, as in [MN1] and [DCL], it was possible to establish only the following kind of estimates

$$\begin{aligned} 0 \leq n^l(x,t) &\leq (\theta(C_0 + tC_1))^{1/\theta} \ , \\ |u^l(x,t)| = \left|\frac{j^l(x,t)}{n^l(x,t)}\right| &\leq C_0 + tC_1 \ , \end{aligned}$$

where the constant C_0, C_1 depends only on the data of the problem. These estimates are not stable under the relaxation scaling $t = \frac{s}{\tau}$.

Then, by arguing as in [MN1,2], we state the result of convergence of the sequence of approximate solutions. The proof follows by using the compensated compactness arguments [Ta], [DiP]. In the case of the nonhomogeneous isentropic gas-dynamics, this study was accomplished in [DCL]. As is well-known, the main step is in proving that the sequence

$$\Lambda^l := \partial_t \eta(U^l) + \partial_x q(U^l)$$

is relatively compact in H^{-1}_{loc} for any entropy-entropy flux pair (η, q). This follows by using the energy estimates on the approximate state variables given in [MN1].

Proposition 2.3. *Under the assumptions of the Theorem 2.1, there is a convergent subsequence $(n^{l\nu}, j^{l\nu}, E^{l\nu})$ and some bounded measurable functions (n, j, E) such that*

$$(n^{l\nu}, j^{l\nu}) \to (n, j)$$

for a.e. $(x,t) \in \mathbb{R} \times [0, \infty)$, and

$$E^{l\nu} \to E^l$$

uniformly on the compact subsets of $\mathbb{R} \times [0, \infty)$.

To complete the proof of the Theorem 2.1, we show the consistency of the approximate sequence $\{U^l, E^l\}$.

Proposition 2.4. *The limit functions (n, j, E), given by the Proposition 2.3, are a bounded entropy solution of the problem (E)-(P)-(2.1).*

3. Relaxation to the Drift-Diffusion equations

Consider the entropy solution (n, j, E) to the system (E)-(P), for fixed τ. We introduce the scaled variables

$$\mathcal{N}^\tau(x,s) := n\left(x \frac{s}{\tau}\right), \quad \mathcal{J}^\tau(x,s) := \frac{1}{\tau} j\left(x, \frac{s}{\tau}\right), \quad \mathcal{E}^\tau(x,s) := E\left(x, \frac{s}{\tau}\right) .$$

These new variables satisfy the following system:

$$\begin{cases} \mathcal{N}^\tau_s + \mathcal{J}^\tau_x = 0 \ , \\ (\tau^2 \mathcal{J}^\tau)_s + \left(\tau^2 \frac{(\mathcal{J}^\tau)^2}{\mathcal{N}^\tau} + p(\mathcal{N}^\tau)\right)_x = \mathcal{N}^\tau \mathcal{E}^\tau - \mathcal{J}^\tau \ , \\ \mathcal{E}^\tau_x = \mathcal{N}^\tau - b \ . \end{cases} \tag{EP$_\tau$}$$

In this section we are concerned with the limiting behaviour, as $\tau \to 0^+$, of the solutions to $(\mathrm{RS})_\tau$. An important tool is the Tartar and Murat's celebrated "div-curl" lemma, from the theory of compensated compactness (see [Ta]).

To apply this result, we have to use the L^∞ estimates given in Section 2 and the following result.

Proposition 3.1. *All uniformly bounded solutions to* $(\mathrm{RS})_\tau$, *with a compact support, verify the bounds*

$$\int_0^T \int \frac{(\mathcal{J}^\tau)^2(x,t)}{\mathcal{N}^\tau(x,t)}\,\mathrm{d}x \le C_T \ , \tag{3.1}$$

for all $T > 0$, *where the constant* C_T *depends only on* T.

Next, we have

Proposition 3.2. *The sequences* $\{\mathcal{N}^\tau\}$, $\{\tau\mathcal{J}^\tau\}$ *and* $\{(\mathcal{J}^\tau)^2/\mathcal{N}^\tau\}$ *are uniformly bounded in* $L^\infty(\mathbb{R}\times\mathbb{R}_+)$, $L^\infty(\mathbb{R}\times\mathbb{R}_+)$ *and* $L^1(\mathbb{R}\times[0,T])$ *respectively, for all* $T > 0$. *Therefore, if* $\mathcal{N} := w^* - \lim\mathcal{N}^\tau$, $\tilde{p} = w^* - \lim p(\mathcal{N}^\tau)$ *as* $\tau \to 0^+$,

$$\tilde{p} = p(\mathcal{N}) \ , \tag{3.2}$$

$$\mathcal{N}^\tau p(\mathcal{N}^\tau) \rightharpoonup N p(\mathcal{N}) \quad in \quad \mathcal{D}' \ . \tag{3.3}$$

Here we use the well known Minty's argument and the monotonicity of the function p, in the spirit of [MM], to show that $\tilde{p} = p(\mathcal{N})$.

The next step is devoted to showing the strong convergence of $\{\mathcal{N}^\tau\}$ in L^p_{loc}, for all $p \in (1,+\infty)$. To this end we follow the ideas of Tartar [Ta] and Di Perna [DiP] and characterize the weak convergence by means of the L.C. Young's probability measures.

Given a sequence $\{f^\varepsilon\}$ converging to a limit f in L^∞ weak-$*$, we have a family $\{\nu_{(x,t)}\}$ of probability measures such that for any continuous function $F(\cdot)$, the limit of $\{F(f^\varepsilon)\}$ is given by the expected value of F with respect to these measures; namely if $\overline{F} = w^* - \lim F(f^\varepsilon)$, then

$$\overline{F}(x,t) = \int F(\lambda)\nu_{(x,t)}(\mathrm{d}\lambda) \quad \text{a.e.}$$

The strong convergence is then achieved if $\nu_{(x,t)} = \delta_{f(x,t)}$. In this case one has the pointwise convergence that, combined with L^∞ bounds, provides strong convergence in L^p_{loc} for all $p \in (1,+\infty)$.

In our case one has

Lemma 3.3. *Let* $p \in C^2(\mathbb{R})$ *be such that* $p''(\lambda) > 0$ *for all* $\lambda > 0$ *and* $\{\mathcal{N}^\tau\}$ *any sequence of* L^∞, *such that*

$$p(\mathcal{N}^\tau) \rightharpoonup p(\mathcal{N}) \quad in \ L^\infty \ weak\text{-}* \tag{3.4}$$

where $\mathcal{N} = w^* - \lim\mathcal{N}^\tau$, *as* $\tau \downarrow 0$. *Then* $\mathcal{N}^\tau \to \mathcal{N}$ *in* L^p_{loc} *strongly, for all* $p \in (1,+\infty)$.

Now we state our main result on the relaxation of the hydrodynamic model.

Theorem 3.4. *Let* $\{\mathcal{N}^\tau, \mathcal{J}^\tau\}$ *be the sequence of the solutions to* $(EP)_\tau$. *Assume* $\{\mathcal{N}^\tau\}$ *and* $\{\tau\mathcal{J}^\tau\}$ *are uniformly bounded in* L^∞, *and* $\{\mathcal{J}^\tau/\sqrt{\mathcal{N}^\tau}\}$ *is uniformly bounded in* $L^2(\mathbb{R}\times[0,T])$, *for any* $T > 0$. *Then there exists* $\mathcal{N} \in L^\infty$ *such that* $\{\mathcal{N}^\tau, \tau\mathcal{J}^\tau\}$ *converges in* L^p_{loc} *strongly to* $\{\mathcal{N}, 0\}$, *for all* $p \in (1,+\infty]$ *and there exists a Lipschitz continuous function* $\mathcal{E}$ *such that* $\{\mathcal{E}^\tau\}$ *converges uniformly to* $\mathcal{E}$ *as* $\tau \to 0^+$.

Moreover let $\mathcal{J} = w - \lim\mathcal{J}^\tau$ *in* L^2, *as* $\tau \downarrow 0$, *it follows that*

$$\mathcal{J} = \mathcal{N}\mathcal{E} - \partial_x p(\mathcal{N}) \ . \tag{3.5}$$

The limit densities $\mathcal{N}$, $\mathcal{E}$ *are the weak solutions of the Cauchy problem (DD)–(P).*

REFERENCES

[AMPS] U.M. ASCHER, P.A. MARKOWICH, P. PIETRA, C. SCHMEISER, "A phase plane analysis of transonic solutions for the hydrodynamic semiconductor model", *Math. Mod. Math. Appl. Sci.*, **1** (3) (1991), 347–376.

[AP1] A.M. ANILE and S. PENNISI, "Extended thermodynamics of the Blotekjaer hydrodynamical model for semiconductors", *Continuum Mech. Therm.*, **4** (1992), 187–193.

[AP2] A.M. ANILE and S. PENNISI, "Thermodynamic derivation of the hydrodynamical model for charge transport in semiconductors", *Physical rev. B: condensed matter*, **46** (1992), 13186–13193.

[AM] A.M. ANILE and O. MUSCATO, "An improved hydrodynamical model for carrier transport c in semiconductors", *submitted to Physical rev.*, (1994).

[Bl] K. BLØTEKJÆR , "Transport equations for electrons in two–valley semiconductors", *IEEE Trans. Electron. Devices*, **ED–17** (1970), 38–47.

[BW] G. BACCARANI and M.R. WORDERMAN, "An investigation of steady–state velocity overshoot effects in Si and GaAs devices", *Solid State Electr.*, **28** (1985), 407–416.

[DiB] E. DI BENEDETTO, "Degenerate parabolic equations", Springer–Verlag, New York (1993).

[DCL] X. DING, G.–Q. CHEN, P. LUO, "Convergence of the fractional step Lax–Friedrichs scheme and Godounov scheme for isentropic gas dynamics", *Comm. Math. Phys.*, **121** (1989), 63–84.

[DiP] R. DI PERNA, "Convergence of approximate solutions of conservation laws", *Arch. Rat. Mech. Anal.*, **82** (1983), 27–70.

[Ga] I.M. GAMBA, "Stationary transonic solutions of a one–dimensional hydrodynamic model for semiconductors", *Comm. P.D.E.*, **17 (3&4)** (1992), 553–577.

[Gi] B.H. GILDING, "Improved Theory for a Nonlinear Degenerate Parabolic Equation", *Ann. Sc. Norm. Sup. Pisa Cl. Sci.* **16** (1989), 164–224.

[Ju] A. JÜNGEL, "A degenerate drift–diffusion system with electric convection arising in semiconductor and electrophoretic modeling", *preprint*, (1994).

[La1] P.D. LAX, "Weak solutions of nonlinear hyperbolic equations and their numerical computation", *Comm. Pure Appl. Math.*, **7** (1954), 159–193.

[La2] P.D. LAX, "Hyperbolic systems of conservation laws and the mathematical theory of shock waves", *Conf. Board Math. Sci.*, **11**, SIAM (1973).

[MM] P.A. MARCATI and A. MILANI, "The one–dimensional Darcy's law as the limit of a compressible euler flow", *J. Diff. Eq.*, **84** n.1 (1990), 129–147.

[MMS] P.A. MARCATI, A. J. MILANI and P. SECCHI, "Singular convergence of weak solutions for a quasilinear nonhomogeneous hyperbolic system", *Manuscripta Math.*, **60** (1988), 49–69.

[MN1] P.A. MARCATI and R. NATALINI, "Weak solutions to a hydrodynamical model for semiconductors: the Cauchy problem", *Quaderno IAC*, **2** (1993); to appear in Proc. Roy. Soc. Edinburgh.

[MN2] P.A. MARCATI and R. NATALINI, "Weak solutions to a hydrodynamical model for semiconductors and the relaxation to the drift–diffusion equation", to appear in Arch. Rational Mech. Anal..

[Ma] P.A. MARKOWICH, "On steady-state Euler-Poisson model for semiconductors", to appear in ZAMP, 1991.

[MP] P.A. MARKOWICH and P. PIETRA, "A non-isentropic Euler-Poisson model for a collisionless plasma", *Math. methods appl. sci.*, **16** (1993), 409–442.

[MRS] P.A. MARKOWICH, C. RINGHOFER, C. SCHMEISER, *Semiconductor Equations*, Springer–Verlag, Wien–New York (1990).

[Na] R. NATALINI, "The bipolar hydrodynamical model for semiconductors and the drift–diffusion equations", *preprint*, (1994).

[Po] F. POUPAUD, "Derivation of a hydrodynamic systems hierarchy from the Boltzmann equation", *Appl. Math. Letters*, **4** (1992), 75–79.

[Sm] J. SMOLLER, "Shock waves and reaction–diffusion equations", Springer–Verlag, New York (1983).

[Ta] L. TARTAR, "Compensated compactness and applications to partial differential equations", Research notes in mathematics, nonlinear analysis and mechanics: Heriot–Watt Symposium, Vol. **4**, ed. R. J. Knops, New York, Pitman Press (1979), 136–212.

[Va] W.V. VAN ROOSBROECK, "Theory of flow of electrons and holes in germanium and other semiconductors", *Bell Syst. Techn. J.*, **29** (1950), 560–607.

Kinetic discretization of gas dynamics using Fluctuation-Splitting schemes

B. Perthame[1], Y. Qiu[1], B. Stoufflet[2]

Abstract

The "fluctuation-splitting" schemes have been introduced by Roe and Sildikover to solve advection equations on rectangular grids, and they are extended to triangular grids by Roe, Deconinck, Struij, et al. We show how the kinetic approach allows to reduce the discretization of gas dynamics equations to a parametrized advection equation. Then, we use the fluctuation-splitting schemes to solve two dimensional gas dynamics equations on a triangular grid. Several numerical results show that this method is promising if second order schemes can be used.

1 Introduction

The kinetic approach to gas dynamics is derived from Boltzmann's theory. It consists in approximating the system of Euler equations for inviscid compressible flows by an advection equation in which the kinetic parameter represents all possible waves speeds of the flow. This wave decomposition has been used, at the numerical level, by several authors combined with a finite volume discretization of the physical space: Deshpande [1], De Vuyst [3], Khobalatte and Perthame [5], Perthame [7], Pullin [11], Perthame and Qiu [9], Xu and Prendergast [14].

Since this approach also greatly simplifies the stability analysis of the schemes, it is easy to use it together with other space discretizations. A new idea for discretizing advection equations on two dimensional triangular grids has been proposed recently by Roe, Sildikover, Deconinck, et al. [12] under the name of "fluctuation-splitting" schemes (FSS in short). These schemes are built on a multidimensional approach and have the advantage, compared to finite volumes, to be compact (the algorithm only uses the triangles as data structure). They are also less diffusive, at least for steady states computations, and they use a finite element data structure. Extensions to higher order accuracy and nonlinear systems are also proposed in the above references. Also, from a theoretical point of view, the convergence of these schemes, in the linear case, is proved by one of the authors [7] using the classical approach for finite volumes on triangles.

In this paper, we will show that when using appropriate kinetic equilibrium functions, it is possible to derive kinetic FSS schemes for gas dynamics which enjoy some weak stability properties: nonnegativity of the density and pressure, a conservative entropy inequality and even the maximum principle on the entropy. We present also numerical results that show the efficiency of this method. For shock tubes they are comparable to first order finite volumes schemes, but for stationary flows, they are, in general better, than the finite volume schemes based on a kinetic approach.

The paper is organized as follows. We first recall the idea of the kinetic wave decomposition. In Section 2, we describe the FSS and in Section 3, we present the way to combine FSS with a kinetic approach. Numerical results are presented in Section 4.

[1]Université Paris 6 et CNRS, Lab. d'Analyse Numérique, T55/65, 5ème étage, 4 pl. Jussieu, F75752 Paris Cedex 05, and INRIA-Rocquencourt, MENUSIN, BP. 105, F78153 Le Chesnay Cedex.

[2]Dassault Aviation, DGT/DEA, 78 quai Marcel Dassault, F92214 St Cloud

2 Kinetic approximation of gas dynamics

In this paper, we will consider the two dimensional gas dynamics equations

$$(2.1) \qquad \begin{cases} \partial_t \rho + div\rho u = 0, \quad t \geq 0, x \in \mathbb{R}^2, \\ \partial_t \rho u + div\rho u \otimes u + \nabla p = 0, \\ \partial_t E + div[(E+p)u] = 0, \end{cases}$$

u is the two dimensional velocity field, and we use a perfect gas law for the presssure

$$(2.2) \qquad E = \frac{1}{2}\rho|u|^2 + \frac{1}{\gamma-1}\rho T, \quad p = \rho T.$$

The idea of the kinetic approximation is to replace (2.1) on a timestep $[n\Delta t, (n+1)\Delta t]$ by two transport equations

$$(2.3) \qquad \begin{cases} \partial_t f + \xi.\nabla_x f = 0, \quad x, \xi \in \mathbb{R}^2, \\ \partial_t f + \xi.\nabla_x g = 0, \end{cases}$$

where the "initial values" of f and g at time $n\Delta t$ are given through equilibria related to the macroscopic quantities in (2.1)

$$(2.4) \qquad f^n(x,\xi) = \frac{\rho^n(x)}{T^n(x)}\chi\left(\frac{\xi - u^n(x)}{\sqrt{T^n(x)}}\right), \; g^n(x,\xi) = \rho^n(x)\zeta\left(\frac{\xi - u^n(x)}{\sqrt{T^n(x)}}\right).$$

Then, we recover, from the kinetic functions f and g, the macroscopic approximations to (2.1) using

$$(2.5) \qquad \begin{cases} \rho^{n+1}(x) &= \int_{\mathbb{R}^2} f^{n+1}(x,\xi)d\xi, \\ \rho^{n+1}u^{n+1} &= \int_{\mathbb{R}^2} \xi f^{n+1}(x,\xi)d\xi, \\ E^{n+1}(x) &= \int_{\mathbb{R}^2} (\frac{|\xi|^2}{2} f^{n+1} + g^{n+1})d\xi. \end{cases}$$

Notice that f^{n+1}, g^{n+1} should, more rigorously, be replaced by $f^{n+1,-}, g^{n+1,-}$ in (2.5) because, at $n\Delta t$, they have time discontinuities since we have defined f^n, g^n by (2.4). Nevertheless, ρ, u, E will be continuous, because (2.5) holds for $f^n, g^n, \rho^n, u^n, E^n$ in (2.4), as soon as we impose that χ, ζ are even functions and

$$(2.6) \qquad \int \chi(w)(1, w_i w_j)dw = (1, \delta_{ij}), \; \int \zeta(w)dx = \frac{2-\gamma}{\gamma-1}.$$

Here δ_{ij} denotes Kronecker's symbol. The most classical choice is the Maxwellian distributions $\chi = \frac{1}{2\pi}e^{-|w|^2/2}, \zeta = \frac{2-\gamma}{\gamma-1}\chi$. For numerical purposes, it is better, however, to avoid the appearance of large velocities, and we choose

$$(2.7) \qquad \chi(w) = \frac{1}{4\pi} I_{\{|w|\leq 2\}}, \quad \zeta = \frac{2-\gamma}{\gamma-1}\chi,$$

where I denotes the indicator function. This choice is particularly simple. A more complicated choice is given in [5] with an additional theoretical property: it also satisfies the maximum principle on the entropy.

This approximation is consistent with the compressible Euler equations (2.1) because, in addition to the relations (2.5), the fluxes are consistant at time t^n. Indeed, thanks to the conditions (2.6)

$$
(2.8) \qquad \begin{cases} \rho u &= \int_{\mathbb{R}^2} \xi \frac{\rho}{T} \chi(\frac{\xi - u}{\sqrt{T}}) d\xi, \\ \rho u \otimes u + pId &= \int_{\mathbb{R}^2} \xi \otimes \xi \frac{\rho}{T} \chi(\frac{\xi - u}{\sqrt{T}}) d\xi, \\ (E+p)u &= \int_{\mathbb{R}^2} \left(\frac{|\xi|^2}{2} \frac{\rho}{T} \chi(\frac{\xi - u}{\sqrt{T}}) + \rho \zeta(\frac{\xi - u}{\sqrt{T}}) \right) d\xi. \end{cases}
$$

3 Fluctuation Splitting Schemes (FSS)

In order to solve numerically an advection equation such as (2.3), we can use the FSS schemes. To do so, we introduce a triangulation $\mathcal{T}_h$ of $\mathbb{R}^2$, with nodes $(x_i)_{i \in \mathbb{N}}$. For each triangle T of $\mathcal{T}_h$, we call generically (x_i, x_j, x_h) its vertices, and we let n_i, or n_i^T, denote the interior normal to T, opposite to the vertex x_i, with length $|n_i| = |x_j - x_k|$ (see Figure 1). Finally, we denote $\underline{C}_i$ the median-based dual cell centered at the node x_i and C_i its surface. f_i^n will be the averaged value of $f^n(x)$ (here we skip the dependancy upon ξ) on $\underline{C}_i$. Then, in Ref. [13] the following terminology is introduced.

Definition 3.1 *(i) the triangle T is a "one-target of type i" if*

$$
(3.1) \qquad \xi \cdot n_i > 0, \quad \xi \cdot n_j \le 0, \quad \xi \cdot n_k \le 0,
$$

and we define

$$
(3.2) \qquad \xi(i,T) = \xi, \quad \xi(j,T) = \xi(k,T) = 0.
$$

(ii) the triangle T is a "two-target of type i, j" if

$$
(3.3) \qquad \xi \cdot n_i > 0, \quad \xi \cdot n_j > 0, \quad \xi \cdot n_k \le 0
$$

and we define uniquely three vectors by

$$
(3.4) \qquad \xi(k,T) = 0, \quad \xi(i,T) + \xi(j,T) = \xi, \quad \xi(i,T) \cdot {}_j = \xi(j,T) \cdot n_i = 0,
$$

(iii) $\mathcal{A}(i)$ denotes the set of all triangles around x_i. □

The identity $n_i + n_j + n_k = 0$ shows that every triangle is either a one- or two-target (see Figure 1). Also, for functions f that are linear on the triangle T, one has

$$
(3.5) \qquad \phi_T := \int_T \xi \cdot \nabla x f = \frac{1}{2} \xi . (n_i f_i + n_j f_j + n_k f_k).
$$

The FSS, also called N-scheme in Ref. [13], is defined by

$$
(3.6) \qquad C_i f_i^{n+1} = C_i f_i^n - \frac{\Delta t}{2} \sum_{T \in \mathcal{A}(i)} \xi(i,T)(n_i^T f_i^n + n_j^T f_j^n + n_k^T f_k^n).
$$

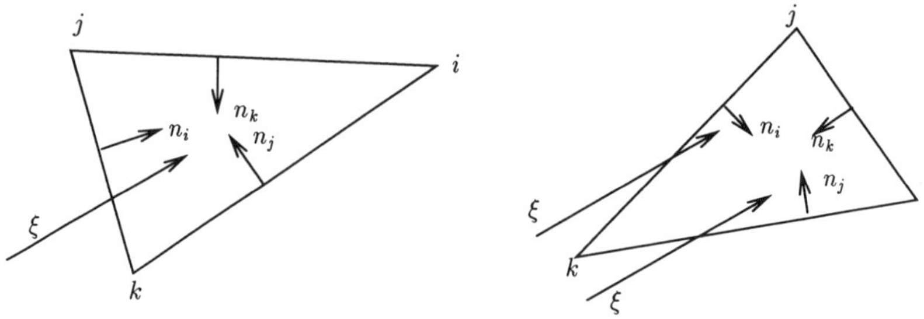

(a) a one-target triangle of type i (b). a two-target triangle of type (i,j)

Figure 1: Classification of triangles.

It can be interpreted as follows. The total flux ϕ_T is sent on the vertex x_i if T is a one-target of type i. In the two-target case, we decompose $\xi \cdot \nabla_x f = \xi(i,T) \cdot \nabla_x f + \xi(j,T) \cdot \nabla_x f$. One easily checks that the triangle T is one-target for both velocities $\xi(i,T)$ and $\xi(j,T)$. Then, the flux splitting above reduces the problem to two one-target cases and gives rise to Eq. (3.6).

The main motivation for introducing this splitting is

Lemma 3.1 *Under the CFL condition*

$$\Delta t \sum_{T \in \mathcal{A}(i)} |\xi . n_i| \leq 2C_i, \quad \forall i \in \mathbb{N}, \tag{3.7}$$

f_i^{n+1} is a convex combination of the $(f_j^n)_{j \in \mathbb{N}}$. Therefore it does not increase the l^p norms and preserve positivity. It is a conservative scheme. □

This can easily be seen because in Eq. (3.6), and from Definition 1, we have that $-\xi(i,T) \cdot n_j \geq 0$ for $j \neq i$. We refer to Refs. [13,6] for details and for the practical implementation which turns out to be extremely simple.

4 Kinetic Flux Splitting Schemes (KFSS)

It is natural to try to combine the kinetic approach with a FSS discretization of the space. A first step in that direction is performed by Eppard and Grossman [4]. They use Maxwellian distributions and a special type of grid which allows practical computations in ξ.

Let us describe here a general approach, on general triangulations, using the equilibrium functions (2.7). To simplify notations, we just take $\gamma = 2$ here, which allows one to solve the equation for f in Eq. (2.3) because $g^n = 0$. The resolution of g could be performed in the same way to treat other values of γ. We introduce, with the notation used in Section 2,

$$U_i^n = (\rho_i^n \rho u_i^n, E_i^n) = \int_{\mathbb{R}^2} (1, \xi, \frac{|\xi|^2}{2}) f_i^n(\xi) d\xi. \tag{4.1}$$

Then, the (FSS) scheme (3.6) allows to compute the evolution of (U_i^n) using

$$(4.2)\qquad \begin{aligned} C_i U_i^{n+1} &= C_i U_i^n - \tfrac{\Delta t}{2} \sum_{T\in\mathcal{A}(i)} \int_{\mathbb{R}^2} \xi(i,T)(n_i^T f_i^n(\xi) \\ &\quad + n_j^T f_j^n(\xi) + n_k^T f_k^n(\xi)) \begin{pmatrix} 1 \\ \xi \\ \frac{|\xi|^2}{2} \end{pmatrix} d\xi. \end{aligned}$$

We first show some theoretical properties of this scheme, and, in a second subsection, we indicate how to compute explicitly the integral in Eq. (4.2).

4.1 Theoretical properties of KFSS

We obtain for the KFSS the same weak stability properties as in the finite volume method. We state them in general

Theorem 4.1 *Under the CFL condition*

$$(4.3)\qquad \Delta t(|u_i| + 2\sqrt{T_i}) \sum_{T\in\mathcal{A}(i)} |n_i^T| \leq 2C_i, \forall i \in \mathbb{N},$$

the KFSS given by formula (4.2) satisfies

(i) $\rho_i^n \geq 0, T_i^n \geq 0$ *for all* n, i

(ii) the entropy inequality

$$\begin{aligned} C_i S(U_i^{n+1}) &\leq C_i S(U_i^n) - \Delta t \sum_{T\in\mathcal{A}(i)} \int_{\mathbb{R}^2} \xi(i,T)[n_i H(f_i^n(\xi)) \\ &\quad + n_j H(f_j^n(\xi)) + n_k H(f_k^n(\xi))] d\xi \end{aligned}$$

whenever the equilibrium χ *in (2.4), (2.6) minimizes the kinetic entropy associated with* $H(.)$,

(iii) the maximum principle on the specific entropy $s_i^n = \rho_i^n / T_i^n$ *whenever the equilibrium* χ *is chosen to satisfy this principle as in [5].* □

Remark 4.1 *This theorem fully applies in particular for the case* $\gamma = 2$, χ *being given by (2.7). Then, we have, in fact, infinitely many choices of singular convex entropies* S *in (ii).*

We will not write the proof of Theorem 3, which is a variant of proofs in the three papers [5,8,9] for finite volumes. The main reason that the proof extends to KFSS is that $f_i^{n+1}(\xi)$ is a convex combination of the $f_j^n(\xi)$, as stated in Section 2.

4.2 Practical implementation

The implementation of the scheme is performed triangle by triangle. For each T, the velocity space is decomposed in six subdomains depending on the signs of $\xi.n_i, \xi.n_j, \xi.n_k$ (remind that $n_i + n_j + n_k = 0$). These domains are, choosing the indices i, j, k appropriately, of the form

$$(4.4)\qquad E_1 = \{\xi; \xi \cdot n_i > 0, xi \cdot n_j \leq 0, \xi \cdot n_k \leq 0\},$$

or

$$E_2 = \{\xi; \xi \cdot n_i > 0, \xi \cdot n_j > 0, \xi \cdot n_k \geq 0\}, \tag{4.5}$$

The triangle T is a one-target of type i for the velocities ξ in E_1; it is a two-target of type (i,j) in E_2. In the first case, $\xi(i,T) = \xi$ and the corresponding part of the integral is sent on U_i^{n+1}. In the second case, two integrals occur, with $\xi(i,T)$ and $\xi(j,T)$ which are sent on U_i^{n+1} or U_j^{n+1}. But we have, from (3.4)

$$\xi(i,T) \cdot n_j = 0, \xi(i,T) \cdot n_i = \xi \cdot n_i, \xi(i,T) \cdot n_k = -\xi \cdot n_j.$$

Therefore the implementation of scheme (4.2) is reduced to calculating integrals such as

$$\int_{E_\alpha \cap \{|\xi - u| \leq 2\sqrt{T}\}} \xi_1^p \xi_2^q d\xi \tag{4.6}$$

where ξ_1, ξ_2 are now the coordinates of ξ in a given basis, α stands for 1 or 2 and p, q are integers between 0 and 3. This integration can be performed, for instance, using Green's formula to reduce the integrals to integrations along lines or arcs of circles.

5 Numerical tests

We have performed several tests in order to compare the KFSS with finite volume schemes in different situations. We present these results below.

The computer time necessary to evaluate the integrals (4.6) is rather expensive, but we did not optimize them, which seems possible but not obvious. As it is, the KFSS is ten times more expensive than a finite volume scheme based on a kinetic scheme and still more expensive than a Roe or Osher solver. Of course, such conclusions could be completely changed at second order.

In all the above results we used the same grid and the same data structures (values at the node of the triangulation).

Test 1. Reflection problem. We present in Figure 2 a comparison of the results for the reflection problem. We compared the KFSS with a finite volume scheme based on a kinetic solver or Osher solver. We can see that the KFSS and Osher schemes are comparable while the kinetic finite volume solver is slightly more diffusive.

Test 2. Lax shock tube. We present in Figure 3 a comparison for an evolution problem: Lax shock tube. In this case the KFSS is comparable to any other first order finite volume scheme (again this test is performed on a general two dimensional triangulation).

Test 3. Shear flow (Mach 5). In Figure 4, we compare the KFSS with a kinetic based finite volume calculation. A great improvment can be seen for the KFSS which is still far from a second order kinetic based finite volume method. Notice also that in such a case, a finite volume method based on Roe or Osher solvers would be better than the kinetic finite volume scheme.

Test 4. Double ellipse (Mach 25). In Figure 5, we show that the KFSS can be used indeed for strongly hypersonic flows. The results in that case are comparable, but better in some zones, than the kinetic finite volume schemes. They assert that the same stability and robustness can be achieved using finite volumes or N-schemes.

For the tests 1,2 and 4, we used the same grid as in [9] and $\gamma = 1.4$. We did not try to optimize the code, and it seems possible to win some time in the evaluation of the integrals (4.6). Also, we imposed strongly the reflection condition on the rigid boundaries. We replaced on the boundary nodes the normal velocity $u \cdot n$ by $-u \cdot n$ or we just set $u \cdot n$ to zero and increased the internal energy so as to conserve total energy. These two methods gave the same results.

We consider these results as a very encouraging first step. Indeed, for stationary computations, the KFSS gives more accurate results than the finite volume method based on a kinetic solver, while keeping the same theoretical properties. Further applications to real computations depends upon the possibility to extend the kinetic scheme, following Coquel and Liu [15] for instance, in order to get exact stationary contact discontinuities and also to extend KFSS to second order accuracy.

References

[1] S. Deshpande. A second order accurate, kinetic theory based method for inviscid compressible flows. NASA Technical paper n° 2613 (1986).

[2] H. Deconinck, H. Paillère, R. Struijs and P.L. Roe. Multidimensional upwind schemes based on fluctuation-splitting for systems of conservation laws. Computational mechanics n° 11 (1993) p. 323-340.

[3] F. De Vuyst. Thesis, Université Paris VI, in preparation.

[4] W.M. Eppard and B. Grossman. A multidimensional kinetic based upwind solver for the Euler equations. AIAA-93-3303-CP.

[5] B. Khobalatte and B. Perthame. Maximum principle on the entropy and second order kinetic schemes. Math. Comp., vol. 62 (205) pp. 119-135 (1994).

[6] H. Paillère, H. Deconinck, R. Struijs, P.L. Roe, L.M. Mesaros et J.D. Müller. Computations of inviscid compressible flows using fluctuation-splitting on triangular meshes. AIAA 93-3301.

[7] B. Perthame. Convergence of Fluctuation Splitting schemes for advection Equations. To appear in Proceedings of STAMM conference, Lisbon, 1994.

[8] B. Perthame, Second-order Boltzmann Schemes for compressible Euler equations in one and two dimensions. SIAM J. Num. Anal., vol. 29, n° 1, p. 1-19 (1992).

[9] B. Perthame and Y. Qiu. A variant of Van Leer's method for hyperbolic systems. J. Comp. Phys., vol. 1/2(2) p. 370-381 (1994).

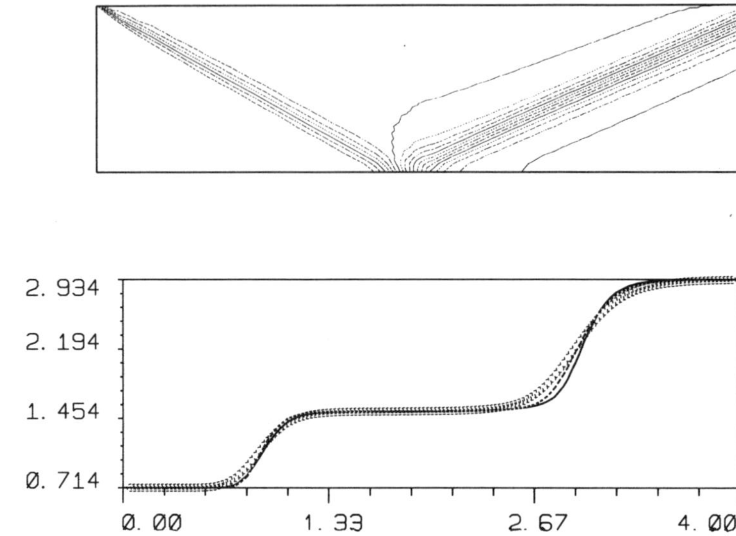

Figure 2: The reflection problem. Pressure isolines for KFSS and comparison between three schemes: KFSS(dashed line), finite volumes with Osher(continous line) and kinetic solvers(circles).

[10] B. Perthame, Y. Qiu et B. Stoufflet. Note C.R.A.S. Acad. Sci. Paris (1994). To appear.

[11] D.I. Pullin. Direct simulation method for compressible inviscid ideal-gas-flow. J. Comput. Phys., 34 (1980), p. 231-244.

[12] P.L. Roe and D. Sidilkover. Optimum positive linear schemes for advection in two and three dimensions. SIAM J. Numer. Anal. 29 (6) (1992), p. 1542-1568.

[13] R. Struijs, H. Deconinck, P. De Palma, P. Roe, A.G. Powell. Progress on multidimensional upwind Euler Solvers for unstructured grids. AIAA 91-1550.

[14] K. Xu and K.H. Prendergast. Multidimensional hydrocode from kinetic theory. Preprint.

[15] F. Coquel and M.S. Liu, Field by field hybrid upwind splitting, to appear in NASA T.M. (1994).

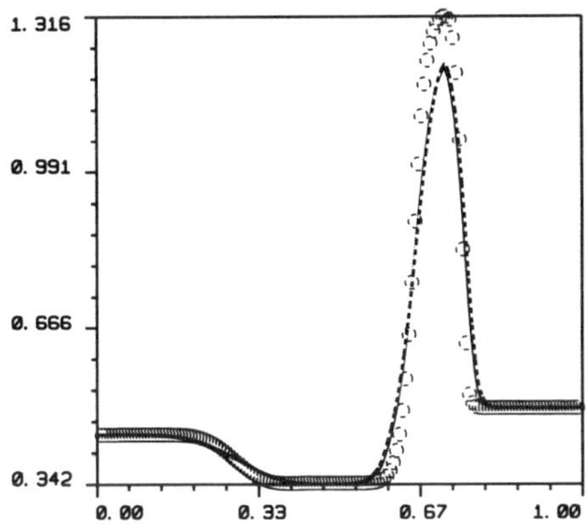

(a)

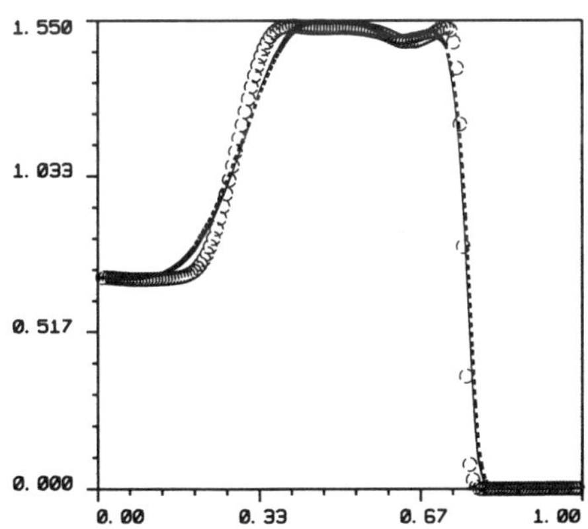

(b)

Figure 3: The Lax shock tube on a 2D random triangular grid. Comparison between KFSS(continous line), first(dashed line), and second(circles) order kinetic finite volumes: (a) density, (b) velocity.

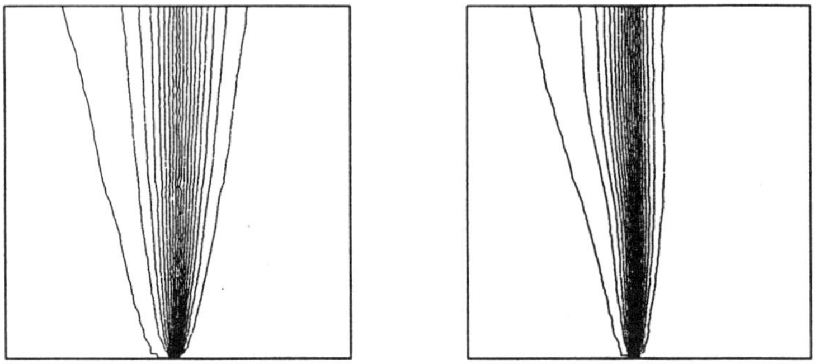

Figure 4: ***Shear flow. Left : first order kinetic finite volumes, right : KFSS.***

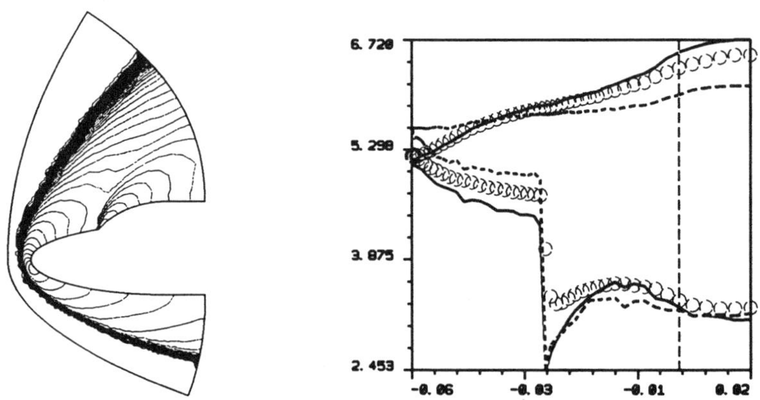

Figure 5: ***Double ellipse. Left : mach isolines, Right : entropy along the body, a comparision between KFSS(circles), first order kinetic finite volumes(continous line) and second order finite volumes(dashed line)***

A Comparison of G. I. Taylor's Inviscid Analysis and an Artificial Viscosity Approach to the Spherical Piston Problem in $N = 1,2,3$ Space Dimensions

M. Slemrod[1]

0. Introduction

In a classic paper [1] G. I. Taylor gave a method of solving the inviscid spherically symmetric expanding piston problem. The ideas have been further amplified in Chapter VI, §162 of Courant and Friedrichs [2]. Further discussion is found in the monograph of Sedov [3]. In all these works the role of viscosity is omitted. In this note I retrace Taylor's argument and show where the viscous stress contribution may be important. This has possible significance in two areas: (i) The natural physical role of considering even small viscosity when it is relevant, and (ii) the usefulness in understanding limiting small viscosity as a method of computing weak solutions to systems of hyperbolic conservation laws in $N \geq 1$ space dimensions. In line with (ii) an artificial viscosity method and result for resolution of the inviscid piston problem for $N = 1,2,3$ is presented. Complete results are given in [4].

1. G. I. Taylor's Analysis

The equations of inviscid, compressible, spherically symmetric, isentropic, gas dynamics in three space dimensions are the equations of balance of mass and momentum given by

$$\frac{\partial \rho}{\partial t} + u\frac{\partial \rho}{\partial r} + \rho\Big(\frac{\partial u}{\partial r} + \frac{2u}{r}\Big) = 0 \tag{1.1}$$

and

$$\rho\Big(\frac{\partial u}{\partial t} + u\frac{\partial u}{\partial r}\Big) + \frac{\partial p}{\partial r} = 0. \tag{1.2}$$

Taylor considers the case of a polytropic gas with $p(\rho) = A\rho^\gamma$, $\gamma \geq 1$, A a positive constant, and searches for solutions dependent on the similarity variable $\xi = \frac{r}{t}$. A simple application of the chain rule yields for smooth solutions the equivalent forms for (1.1), (1.2):

$$(u-\xi)\frac{du}{d\xi} = -\frac{1}{(\gamma-1)}\frac{d}{d\xi}c^2 \tag{1.3}$$

$$\frac{(u-\xi)}{\rho}\frac{d\rho}{d\xi} + \frac{du}{d\xi} + \frac{2u}{\xi} = 0 \tag{1.4}$$

where $c^2(\rho) = A\gamma\rho^{\gamma-1}$ and hence

$$\frac{1}{\rho}\frac{d\rho}{d\xi} = \frac{1}{(\gamma-1)c^2}\frac{dc^2}{d\xi}. \tag{1.5}$$

Taylor combines (1.4), (1.5) to find

$$\Big[1 - \frac{(u-\xi)^2}{c^2}\Big]\frac{du}{d\xi} + \frac{2u}{\xi} = 0. \tag{1.6}$$

---

[1]Center for the Mathematical Sciences, University of Wisconsin-Madison, Madison, WI 53715-1149.

This research was supported in part by Grant No. N00014-93-1-0015 from the Office of Naval Research, and Grant No. DMS-9006945 from the National Science Foundation.

Equations (1.3), (1.6) provide a non-autonomous system of two ordinary differential equations in the two unknowns u, c^2 for the self-similar dynamics.

Taylor then makes a new choice of dependent variables $U = u/\xi$, $C = c/\xi$, so that (1.3), (1.6) become

$$(U-1)\xi\Big[\xi\frac{dU}{d\xi}+U\Big] = -\frac{1}{\gamma-1}\Big[\xi^2\frac{dC^2}{d\xi}+2\xi C^2\Big], \tag{1.7}$$

$$\Big[1-\frac{(U-1)^2}{C^2}\Big]\Big[\xi\frac{dU}{d\xi}+U\Big]+2U = 0, \tag{1.8}$$

and hence

$$[C^2-(U-1)^2]\xi\frac{dU}{d\xi} = -3UC^2+U(U-1)^2. \tag{1.9}$$

Substitution of (1.8) into (1.7) then yields

$$(U-1)\Big[\frac{-2UC^2}{C^2-(U-1)^2}\Big]+\frac{2C^2}{\gamma-1} = -\frac{\xi}{\gamma-1}\frac{dC^2}{d\xi}$$

or

$$(\gamma-1)(U-1)(-2UC^2)+2C^2(C^2-(U-1)^2) = -[C^2-(U-1)^2]\xi\frac{dC^2}{d\xi}. \tag{1.10}$$

Equations (1.9), (1.10) then yield the autonomous equation

$$\frac{dC^2}{dU} = \frac{(\gamma-1)(U-1)(2UC^2)-2C^2(C^2-(U-1)^2)}{-3UC^2+U(U-1)^2}$$

or

$$\frac{dC}{dU} = \frac{(\gamma-1)(U-1)U-(C^2-(U-1)^2)}{-3C^2+(U-1)^2}\cdot\frac{C}{U} \tag{1.11}$$

which is the expression given in (162.04) of Courant and Friedrichs [2].

To resolve the piston problem, Taylor assigns constant data for a gas at rest exterior to an out-going shock, i.e., $\rho = \rho_2 > 0$, $u = 0$, for $r \geq st$, $s > 0$ the shock speed which is a free parameter. Across the shock the Rankine-Hugoniot jump conditions

$$-s[\rho]+[\rho u] = 0,\ -s[\rho u]+[\rho u^2+p] = 0$$

must be satisfied. This determines a one parameter family of data $\rho_i(s)$, $u_i(s)$ which is the limit from the interior at the shock. Taylor then solves for the self-similar wave behind the out-going shock; i.e., he solves (1.11) with initial condition

$$C_s^2 = \frac{A\gamma\rho_i(s)^{\gamma-1}}{s^2},$$

$$U_s = \frac{u_i(s)}{s},$$

on the interval $[U_s, 1]$. When $U = 1$ we have $u(\xi) = \xi$, i.e., the fluid velocity equal to the piston velocity which is the boundary condition at the piston's surface. Denote this one parameter family of solutions of (1.11) by $C(U; s)$.

Next Taylor substitutes the function $C(U;s), U_s \leq U \leq 1$ into (1.9). This determines an ordinary differential equation

$$\frac{1}{\xi}\frac{d\xi}{dU} = \frac{C^2(U;s) - (U-1)^2}{-3UC^2(U;s) + U(U-1)^2} \tag{1.12}$$

for $\xi = \xi(U;s)$. As $U \to 1$, $\xi \to \xi(1;s)$ a one parameter family of piston speeds. Thus, Taylor has produced a map $s \mapsto \xi(1;s)$ from the shock speed s to the piston speed $\xi(1;s)$. If the map is globally invertible, the piston problem is resolved for all piston speeds. The map is monotone according to Courant and Friedrichs. This is because solutions $C(U;s)$ of (1.11) satisfy $C^2(U;s)-(U-1)^2 > 0$ on $(U_s, 1)$ and so (1.12) implies $\frac{d\xi}{dU} < 0$. Since orbits of (1.12) cannot cross by uniqueness of solutions, the map is monotone decreasing and hence globally invertible for all $\gamma \geq 1$.

Taylor's analysis provides a solution of the piston problem and remains one of the few solved problems for non-planar flows.

One notices, of course, that this is a completely inviscid theory and hence issues of shock structure and boundary layer analysis at the piston surface are ignored. This may be particularly dangerous if cavitation may take place at the piston surface. For then both C^2 and $(U-1)^2 \to 0$ and the singular term $C^2 - (U-1)^2$ which so conveniently disappeared in going from (1.9), (1.10) to (1.11) remains when a viscous stress is introduced.

2. Viscosity Method

The equations of viscous, compressible, spherically symmetric, isentropic gas dynamics in N space dimensions are the equations of balance of mass and momentum given by

$$\frac{\partial \rho}{\partial t} + u\frac{\partial \rho}{\partial t} + \rho\Big(\frac{\partial u}{\partial r} + \frac{2u}{r}\Big) = 0 \tag{2.1}$$

and

$$\rho\Big(\frac{\partial u}{\partial t} + u\frac{\partial u}{\partial r}\Big) + \frac{\partial p}{\partial r} = \varepsilon\frac{\partial}{\partial r}\Big(\frac{1}{r^{N-1}}\frac{\partial}{\partial r}(r^{N-1}u)\Big) \tag{2.2}$$

where $\varepsilon > 0$.

For fixed $\varepsilon > 0$ there are interesting new results [5], [6], [7] regarding existence of solutions to (2.1), (2.2) but as the goal here is to resolve the inviscid piston problem, this topic will not be considered here.

The idea pursued here and discussed at length in [4] is to replace (2.2) with the artificially viscous equation

$$\rho\Big(\frac{\partial u}{\partial t} + u\frac{\partial u}{\partial r}\Big) + \frac{\partial p}{\partial r} = \varepsilon t\frac{\partial}{\partial r}\Big(\frac{1}{r^{N-1}}\frac{\partial}{\partial r}(r^{N-1}u)\Big). \tag{2.3}$$

Since this will allow a reduction of the viscous piston problem to a boundary value problem for ordinary differential equations, it provides the natural viscous generalization of Taylor's method given in Section 1. Of course, the idea of replacing ε by εt has been used before, originating with the independent work of Dafermos [8], Kalasnikov [9], Tupciev [10], and continued in later work by several others [9-22], though the most pertinent to the analysis given here is the recent paper of Tzavaras [22]. For it is Tzavaras' paper [22] that showed how to deal with the equations of conservation of mass and momentum where only the momentum conservation law is regularized with the artificial viscosity. This is important for two reasons. The first is obvious, the imposition of an artificial viscosity term in the momentum equation alone is a closer attempt at describing the true viscous Navier-Stokes equations than a complete regularization of both the mass and momentum equations. The second reason is more subtle, i.e., various estimates used in passage to the inviscid limit appear to be obtainable only using the true unregularized mass conservation equation. New results for the inviscid spherically symmetric Euler equations appear in [23-26]. None of the results applies to the piston problem however.

We now attempt to do a viscous self-similar analysis in the spirit of Section 1. We set $\xi = \frac{r}{t}$ and look for solutions of (2.1), (2.3) depending only on the similarity variable ξ. This yields the system

$$(u-\xi)\rho' + \rho u' + \frac{(N-1)\rho v}{\xi} = 0, \tag{2.4}$$

$$\rho(u-\xi)u' + c^2(\rho)\rho' = \varepsilon\left(\frac{1}{\xi^{N-1}}(\xi^{N-1}v)'\right)' \tag{2.5}$$

where again $c^2(\rho) = A\gamma\rho^{\gamma-1}$, $\gamma \geq 1$. If we substitute (2.4) into (2.5) to eliminate ρ' we find

$$\rho\left[\frac{(u-\xi)^2 - c^2(\rho)}{(u-\xi)}\right]u' - \frac{(N-1)c^2(\rho)\rho}{\xi(u-\xi)} = \varepsilon\Big(\frac{1}{\xi^{N-1}}(\xi^{N-1}u)'\Big)'. \tag{2.6}$$

Notice equation (2.6) is the viscous generalization of Taylor's equation (1.6). Finally, multiply (2.6) by ξ^{N-1} and use the identity

$$\left(\frac{(\xi^{N-1}u)'}{\xi^{N-1}}\right)' = \frac{(\xi^{N-1}u')'}{\xi^{N-1}} - \frac{(N-1)u}{\xi^2}$$

to write (2.6) as

$$\rho\left[\frac{(u-\xi)^2 - c^2(\rho)}{(u-\xi)}\right](\xi^{N-1}u') - \frac{(N-1)\xi^{N-2}c^2(\rho)\rho}{(u-\xi)} = \varepsilon(\xi^{N-1}u')' - \varepsilon(N-1)\xi^{N-3}u. \tag{2.7}$$

Our goal is to then solve (2.4), (2.7) for fixed $\varepsilon > 0$ subject to piston problem boundary conditions

$$\lim_{\xi\to\xi_1+} u(\xi) = \xi_1, \ \lim_{\xi\to\infty} u(\xi) = 0, \ \lim_{\xi\to\infty} \rho(\xi) = \rho_2 \tag{2.8}$$

where ξ_1, ρ_2 are prescribed positive constants, and then pass to limit as $\varepsilon \to 0+$ to produce a weak solution of the inviscid conservation laws still satisfying (2.8).

Notice that unlike Taylor's inviscid equation (1.6), the factor ρ remains on the left-hand-side of (2.6). For small density ρ and small viscosity ε, the left and right sides of (2.6) will balance indicating the difference between the approach used here and the classical inviscid theory of Taylor.

3. Viscous Limits

In order to study system (2.4), (2.7), (2.8) it is convenient to solve (2.9) explicitly and rewrite (2.7) via the variation of constants formula. This yields

$$\rho(\xi) = \frac{\rho(\xi_2)(\xi_2 - u(\xi_2))\xi_2^{N-1}}{(\xi - u(\xi))\xi^{N-1}} \exp\Big(N\int_{\xi_2}^{\xi} \frac{d\tau}{\tau - u(\tau)}\Big), \tag{3.1}$$

$$(\xi^{N-1}u'(\xi)\chi(u,\xi))' = B(u,\xi)\chi(u,\xi) \tag{3.2}$$

where $\xi_1 < \xi_2 < \infty$,

$$A(u,\xi) = \frac{(u-\xi)^2 - c^2(\rho)}{\varepsilon(\xi - u)},$$

$$B(u,\xi) = \frac{(N-1)\xi^{N-2}\rho c^2(\rho)u}{\varepsilon(\xi - u)} + (N-1)\xi^{N-2}u,$$

$$\chi(u,\xi) = \exp\left(\int_{\xi_2}^{\xi} A(u,\tau)d\tau\right),$$

where again boundary conditions (2.8) are imposed. To prove existence of solutions of (2.4), (2.7), (2.8) two steps were used. First a topological degree argument established a solution of the nonsingular finite domain problem (3.1), (3.2) with boundary data

$$u(\xi_1) = u_1 > \xi_1,\ u(\xi_2) = u_2 > 0,\ u_1 > u_2,\ \rho(\xi_2) = \rho_2 \tag{3.3}$$

on the finite interval $[\xi_1, \xi_2]$. Then equicontinuity in the parameters ξ_2, u_2 allowed passage to the limit $\xi_2 \to \infty$, $u_2 \to 0$.

More delicate, of course, was passage to the $u_1 \to \xi_1$, $\varepsilon \to 0+$ limits. The difficulty here as noted at the end of Section 2 is the strong role of viscosity in the possible presence of cavitation (a phenomenon omitted by fiat on the inviscid discussion of G. I. Taylor). It was precisely this interplay which made the recovery of the piston boundary condition $\lim_{\xi \to \xi_1 +} u(\xi) = \xi_1$ difficult in the $\varepsilon \to 0+$ limit. To recover the limit, the added condition

$$1 \le \gamma < 1 + \frac{1}{N-1} \tag{3.4}$$

was needed in the analysis of [4]. The full existence theorem given for the piston problem in [4] is as follows:

Theorem. For $1 \le \gamma < 1 + \frac{1}{N-1}$, $p(\rho) = A\rho^\gamma$, $0 < \xi_1^2 < \frac{1}{2}A\gamma\rho_2^{\gamma-1}$, (2.4), (2.7), (2.8) possesses a sequence of solutions $\{\rho^\varepsilon(\xi), u^\varepsilon(\xi)\}$, $\varepsilon > 0$. This sequence possesses a convergent subsequence also denoted by $\{\rho^\varepsilon(\xi), u^\varepsilon(\xi)\}$ and there exists a pair of bounded functions $(\rho(\xi), u(\xi))$ on $[\xi_1, \infty)$ so that $(\rho^\varepsilon(\xi), u^\varepsilon(\xi)) \to (\rho(\xi), u(\xi))$ pointwise on $[\xi_1, \infty)$.

The limit function $u(\xi)$ is monotone decreasing, $0 \le u(\xi) \le \xi_1$, and is differentiable almost everywhere possessing at most countable many points of jump discontinuity. In addition, it is locally Hölder continuous with Hölder exponent $\frac{1}{q}$, $q = \frac{1}{1-(N-1)(\gamma-1)} \ge 1$ near $\xi = \xi_1$.

The limit function $\rho(\xi)$ is also differentiable almost everywhere possessing at worst countable many jump discontinuities which are located at the same points as those of $u(\xi)$. In addition, ρ satisfies the formula

$$\rho(\xi) = \frac{\rho(\xi)(\xi_2 - u(\xi_2))}{(\xi - u(\xi))} \exp\left(N \int_{\xi_2}^{\xi} \frac{d\tau}{\tau - u(\tau)}\right),$$

for all ξ_2 in (ξ_1, ∞), and inequalities

$$\rho_2 \left(\frac{\xi - u(\xi)}{\xi}\right)^{N-1} \le \rho(\xi) \le \rho_2 \left(\frac{\xi}{\xi - u(\xi)}\right),\ 0 \le \rho(\xi) \le M < \infty,$$

for all $\xi \in (\xi_1, \infty)$.

The limit pair $(\rho(\frac{r}{t}), u(\frac{r}{t}))$ is a weak solution of the piston problem, i.e., (1.1), (1.2) are satisfied in the sense of distributions and

$$\lim_{\substack{r\to\infty \\ t>0}} \rho(\frac{r}{t}) = \rho_2, \ \lim_{\substack{r\to\infty \\ t>0}} u(\frac{r}{t}) = 0, \ \lim_{r\to\xi_1 t+} u(\frac{r}{t}) = \xi_1.$$

Furthermore, at any point ξ of jump discontinuity of (ρ, u) the Rankine-Hugoniot jump conditions

$$-\xi[\rho] + [\rho u] = 0, \ -\xi[\rho u] + [\rho u^2 + p] = 0$$

are satisfied.

(1) Taylor, G. I., The air wave surrounding an expanding sphere, Proceedings Royal Society of London, Series A, 186 (1946), 273–292.

(2) Courant, R. and K. O. Friedrichs, Supersonic Flow and Shock Waves, Applied Mathematical Sciences 21, Springer-Verlag (1976).

(3) Sedov, L. I., Similarity and dimensional methods in mechanics, 4th ed. GITTL Moscow (1957), Eng. translation, Academic Press (1959).

(4) Slemrod, M., Resolution of the spherical piston problem for compressible isentropic gas dynamics via a self-similar viscous limit, preprint.

(5) Hoff, D., Spherically symmetric solutions of the Navier-Stokes equations for compressible, isothermal flow with large, discontinuous initial data, Indiana Univ. Mathematics Journal 41 (1992), 1225–1302.

(6) Hoff, D., Global solutions of the Navier-Stokes equations for multidimensional compressible flow with discontinuous initial data, preprint, Inst. for Appl. Math. and Sci. Comp., Indiana Univ., Bloomington, No. 9306 (1993).

(7) Lions, P.-L., Existence globale de solutions pour les equations de Navier-Stokes compressible isentropiques, preprint, Ceremade, Univ. of Paris IX, Dauphine, No. 9317 (1993).

(8) Dafermos, C. M., Solution of the Riemann problem for a class of hyperbolic conservation laws by the viscosity method, Arch. Rational Mechanics and Analysis 52 (1973), 1–9.

(9) Kalasnikov, A. S., Construction of generalized solutions of quasi-linear equations of first order without convexity conditions as limits of solutions of parabolic equations with a small parameter, Dokl. Akad. Nauk. SSSR 127 (1957), 27–30 (in Russian).

(10) Tupciev, V. A., Collapse of an arbitrary discontinuity for a system of two quasilinear first order equations, Z. Vycisl. Mat. i. Fiz. 4 (1964), 817–825. Eng. trans.: USSR Comput. Math. and Math. Physics (5) 4 (1964), 36–48.

(11) Tupciev, V. A., On the method of introducing viscosity in the study of problems involving decay of a discontinuity, Dokl. Acad. Nauk SSSR 211 (1973), 55–58. English transl.: Soviet Math. Dokl. (1973), 978–982.

(12) Dafermos, C. M., Structure of solutions of the Riemann problem for hyperbolic systems of conservation laws, Arch. Rational Mechanics and Analysis 53 (1974), 203-217.

(13) Dafermos, C. M. and R. J. DiPerna, The Riemann problem for certain classes of hyperbolic systems of conservation laws, J. Diff. Eqns. 20 (1976), 90–114.

(14) Dafermos, C. M., Admissible wave fans in nonlinear hyperbolic systems, Arch. Rational Mechanics and Analysis 106 (1989), 243–260.

(15) Fan, H. T., A limiting "viscosity approach" to the Riemann problem for materials exhibiting change of phase (II), Arch. Rational Mechanics and Analysis 116 (1992), 317–338.

(16) Fan, H. T., Self-similar solutions for a modified Broadwell model and its fluid dynamic limits, to appear, S.I.A.M. J. Math. Analysis.
(17) Slemrod, M., A limiting "viscosity" approach to the Riemann problem for materials exhibiting change of phase, Arch. Rational Mechanics and Analysis 105 (1989), 327–365.
(18) Slemrod, M., A comparison of two viscous regularizations of the Riemann problem for Burgers' equation, to appear S.I.A.M. J. Math. Analysis.
(19) Slemrod, M. and A. E. Tzavaras, A limiting viscosity approach for the Riemann problem for isentropic gas dynamics, Indiana Univ. Math.. Journal 38 (1989), 1047–1074.
(20) Slemrod, M. and A. E. Tzavaras, Self-similar fluid dynamic limits for the Broadwell system, Arch. for Rational Mechanics and Analysis 122 (1993), 353–392.
(21) Tzavaras, A. E., Wave structure induced by the fluid dynamic limits in the Broadwell model, submitted to Arch. Rational Mechanics and Analysis.
(22) Tzavaras, A. E., Elastic as a limit of viscoelastic response, in a context of self-similar viscous limits, submitted J. of Diff. Eqns.
(23) Makino, T., K. Mizohata, and S. Ukai, The global weak solutions of compressible Euler equations with spherical symmetry, Japan. J. Industrial and Appl. Math. 9 (1992), 431–449.
(24) Makino, T., K. Mizohata, and S. Ukai, Global weak solutions of the compressible Euler equations with spherical symmetry (II), Res. Rep. Inf. Sci., Tokyo Inst. Tech., A–111 (1993).
(25) Chen, G.-Q. and J. Glimm, Global solutions to the compressible Euler equations with geometrical structure, preprint.
(26) Chen, G.-Q., Remarks on global solutions to the compressible Euler equations with spherical symmetry, preprint.

On the stability of rarefaction waves in viscous media

Anders Szepessy * Kevin Zumbrun †

Abstract

In this paper we review a recent result on the large time behavior of viscous rarefaction waves.

1 Introduction

In this paper we review a recent result on the large time behavior of viscous rarefaction waves. The stability of rarefaction waves is related to the question of uniqueness for systems of strictly hyperbolic conservation laws. For instance, for the $p-$system the most satisfying uniqueness result, including vanishing viscosity solutions, assumes that the solution does not have rarefaction waves that accumulate in time; see Ref. [6]. However, compression waves can produce rarefactions that accumalate. Therefore a better understanding of rarefaction waves interacting with other waves would be useful. Here we study interactions in the special case of localized initial perturbations. The behavior of numerical schemes, approximating conservation laws, is also a motivation for studying the stability of viscous rarefaction waves. All shock capturing schemes are based on adding artificial viscosity in order to avoid unphysical rarefaction-shocks. In numerical computations, the behavior for rarefaction data is critical. This is seen, e.g., in the Lax-Wendroff scheme, which can generate unphysical rarefaction-shocks for rarefaction data and give correct, but somewhat oscillatory, shock waves for shock data.

Our result, Theorem 1 below, describes the time asymptotic behavior and the pointwise convergence, towards an approximate "Burgers" rarefaction wave, of u solving the parabolic system

$$u_t + f(u)_x = u_{xx}, \quad \text{with } x \in \mathbb{R},\ t > 0,\ u(x,t) \in \mathbb{R}^m, \tag{1.1a}$$

$$u(\cdot, 0) = u_0, \tag{1.1b}$$

where f is a given smooth strictly hyperbolic flux function and the initial data u_0 is a general, localized, suitably small perturbation of rarefaction data. Here rarefaction data refers to Riemann data

$$\Re(u_-, u_+, x) \equiv \begin{cases} u_-, & x < 0, \\ u_+, & x > 0, \end{cases}$$

where the constant states $u_\pm$ are chosen such that there exists a continuous p-rarefaction wave

$$u(x,t) = \phi\left(\frac{x}{t}\right), \quad \text{with } \phi(\pm\infty) = u_\pm,$$

*Kungl Tekniska Högskolan, Stockholm.

†Indiana University, Bloomington.

satisfying the strictly hyperbolic system

$$u_t + f(u)_x = 0.$$

We define the Burgers-rarefaction wave $\tilde{\phi} \in \mathcal{C}^\infty(R \times R_+)$ by

$$\tilde{\phi}(x,t) = \phi(\lambda_p(x,t)), \tag{1.2a}$$

where $\lambda_p(x,t)$ satisfies the Burgers equation

$$\begin{aligned} \lambda_{p,t} + \left(\frac{\lambda_p^2}{2}\right)_x - \lambda_{p,xx} &= 0, \\ \lambda_p(\cdot,0) &= \Re(\lambda_p(u_-), \lambda_p(u_-) + \varepsilon, \cdot), \end{aligned} \tag{1.2b}$$

instead of the inviscid equation, $\lambda_t + \left(\frac{\lambda^2}{2}\right)_x = 0$, satisfied by $(\frac{x}{t})$.

The L_2 stability of rarefaction waves in 2×2 systems of viscous conservation laws has been analysed in Refs. [5, 14, 15, 21] using energy methods. Here, we use pointwise arguments based on Duhamel's principle, as have previously been used to study perturbations of constant states [8, 11] and to prove stability of undercompressive shocks [12, 13]. The extra control on the solution afforded by this method turns out to be critical, due to the relatively weak stability properties of rarefactions and the correspondingly large perturbations that can occur.

For example, a new feature of perturbed viscous rarefaction waves, as compared to viscous shock waves or inviscid shock and rarefaction waves (cf. Refs. [7, 9, 16]) is that the perturbation contains error waves of unbounded mass $\mathcal{O}(\log t)$ away from the corresponding inviscid rarefaction wave. A careful description of these error waves is critical in the proof. In particular, we must distinguish between the amplitude $\mathcal{O}(t^{-1/2}\log t)$ in the linear degenerate transversal fields and the smaller amplitude $\mathcal{O}(t^{-1/2})$ in genuinely nonlinear tranversal fields.

The nonlinear damping in the transverse modes is both surprising and central to our analysis. In the next two sections, we illustrate this phenomenon in simplified cases, which still give the flavor of our analysis in the transverse fields. Then we state a stability theorem for the general case.

2 An example with linear degeneracy

The viscous rarefaction wave $u = (V, U)$ for the system

$$V_t + (\tfrac{V^2}{2})_x - V_{xx} = 0, \tag{2.1a}$$

$$U_t + (V-U)_x - U_{xx} = 0, \tag{2.1b}$$

$$\begin{pmatrix} V \\ U \end{pmatrix}(x,0) = \Re\big((0,0),(\varepsilon,\log(1+\varepsilon)),x\big),$$

satisfies for $t > 1$

$$\begin{aligned} |u(x,t) - \tilde{\phi}(x,t)| &\le C\epsilon \begin{cases} \frac{1}{|x+t|} + \frac{\log(|x+t|)+1}{\sqrt{t}} e^{-(x+t)^2/5t}, & |x+t| > \sqrt{t}, \\ \frac{\log t}{\sqrt{t}} e^{-(x+t)^2/5t}, & |x+t| < \sqrt{t}, \end{cases} \\ |u(x,t) - \tilde{\phi}(x,t)| &\ge c\epsilon \begin{cases} \frac{1}{|x+t|} + \frac{\log(|x+t|)+1}{\sqrt{t}} e^{-(x+t)^2/3t}, & |x+t| > \sqrt{t}, \\ \frac{\log t}{\sqrt{t}} e^{-(x+t)^2/3t}, & |x+t| < \sqrt{t}, \end{cases} \end{aligned} \tag{2.2}$$

for constants $C, c > 0$.

Proof. The system has a linear degenerate 1-field and genuinely nonlinear 2-field with eigenvalues and eigenvectors

$$\lambda_1 = -1, \quad \lambda_2 = V,$$
$$R = \begin{pmatrix} 0 & 1 \\ 1 & \frac{1}{V+1} \end{pmatrix}.$$

In this case the Burgers 2-rarefaction wave is $\tilde{\phi} = (\lambda_2, \log(\lambda_2 + 1))$, where $\lambda_2(x,t)$ satisfies the Burgers equation (2.1a) and (1.2b). Therefore, by defining $U \equiv \log(1+\lambda_2) + n_1$ and using $V = \lambda_2$, the function n_1 solves

$$\begin{aligned} n_{1,t} - n_{1,x} - n_{1,xx} &= \tfrac{-\lambda_{2,x}^2}{(1+\lambda_2)^2}, \\ n_1(x,0) &= 0, \end{aligned} \tag{2.3}$$

which implies

$$n_1(x,t) = \int_0^t \int_R \frac{-\lambda_{2,x}^2(y,s)}{(1+\lambda_2(y,s)^2} \frac{e^{-(x-y+t-s)^2/(4(t-s))}}{\sqrt{4\pi(t-s)}} dyds.$$

By evaluating the integral we see that there is a constant $c > 0$ such that

$$-n_1(-t,t) > c\int_4^{t/4} \frac{\varepsilon e^{-s^2/(3t)}}{s\sqrt{t}} ds > c\varepsilon \frac{\log t}{\sqrt{t}},$$

and more generally, the estimate in (2.2) follows.

3 An example with genuinely nonlinearity

The viscous rarefaction wave $u = (V, U)$ for the system

$$V_t + (\tfrac{V^2}{2})_x - V_{xx} = 0, \tag{3.1a}$$
$$U_t + (V - U + U^2/2)_x - U_{xx} = 0, \tag{3.1b}$$
$$\begin{pmatrix} V \\ U \end{pmatrix}(x,0) = \Re((0,0),(\varepsilon, g(\varepsilon)), x),$$

where g is choosen such that the correponding inviscid 2-rarefaction wave exists, satisfies

$$|u(x,t) - \tilde{\phi}(x,t)| \le C \min[1, (c^0 + c^1 + \epsilon)\log t] \begin{cases} \frac{c^0+c^1+\epsilon}{|x-\lambda_i t|}, & |x - \lambda_i t| \ge C\sqrt{t}\log t, \\ \frac{1}{\sqrt{t}}, & |x - \lambda_i t| < C\sqrt{t}\log t, \end{cases} \tag{3.2}$$

$$|u(x,t) - \tilde{\phi}(x,t)| \ge c\epsilon \begin{cases} \frac{1}{|x+t|} + \frac{\min(\log(|x+t|)+1, \epsilon^{-1})}{\sqrt{t}} e^{-(x+t)^2/3t}, & |x+t| > \sqrt{t}, \\ \frac{\min(\log t, \epsilon^{-1})}{\sqrt{t}} e^{-(x+t)^2/3t}, & |x+t| < \sqrt{t}, \end{cases}$$

for a constant $c > 0$.

Proof sketch. The system has genuinely nonlinear 1 and 2 fields with eigenvalues and eigenvectors

$$\lambda_1 = -1 + U, \quad \lambda_2 = V,$$
$$R = \begin{pmatrix} 0 & 1 \\ 1 & \frac{1}{V+1-U} \end{pmatrix}.$$

The Burgers 2-rarefaction wave is $\tilde{\phi} = (\lambda_2, g(\lambda_2))$, where $\lambda_2(x,t)$ satisfies the Burgers equation (3.1a)(1.2b), and g is the solution of the differential equation $g'(\lambda_2) = 1/(1 + \lambda_2 - g(\lambda_2))$. Then, with $V = \lambda_2$ and $U \equiv g(\lambda_2) + n_1$, the function n_1 solves

$$n_{1,t} + [\lambda_1 n_1 + n_1^2/2]_x - n_{1,xx} = g''\lambda_{2,x}^2, \tag{3.3}$$
$$n_1(x,0) = 0,$$

where $\lambda_1 \equiv g(\lambda_2) - 1$. This equation can be solved using the Hopf-Cole transformation. Let

$$m_1(x,t) \equiv \int_\infty^x n_1(y,t)dy,$$

and by integration (3.3) becomes

$$m_{1,t} + \lambda_1 m_{1,x} + \frac{1}{2} m_{1,x}^2 - m_{1,xx} = a, \tag{3.4}$$

where $a \equiv \int_\infty^x g''(\lambda_2(y,t))\lambda_{2,x}^2(y,t)dy$. The Hopf-Cole transformation [4],

$$m_1 = -2\log H,$$

reduces (3.4) to

$$H_t + \lambda_1 H_x - H_{xx} + \tfrac{1}{2} aH = 0, \tag{3.5}$$
$$H(x,0) = \exp(-m_1(x,0)/2) = 1.$$

We shall solve (3.5) using the duals μ and $\tilde{\mu}$, where

$$-\mu_t - (\lambda_1 \mu)_x - \mu_{xx} - \tfrac{a\mu}{2} = 0 \qquad \text{for } t < T, \tag{3.6a}$$
$$-\tilde{\mu}_t - (\lambda_1 \tilde{\mu})_x - \tilde{\mu}_{xx} - a\tilde{\mu}/2 = 0 \qquad \text{for } t < T, \tag{3.6b}$$
$$\mu(x,T) = \delta(x - \bar{x}),$$
$$\tilde{\mu}(x,T) = \delta'(x - \bar{x}).$$

Combining these expressions, we have

$$H(\bar{x},T) = \int_R H(x,0)\mu(x,0)dx, \tag{3.7a}$$

and

$$n_1(\bar{x},T) = -2\frac{H_x}{H} = -2\frac{\int_{-\infty}^{\infty} \tilde{\mu}(x,0)H(x,0)dx}{\int_{-\infty}^{\infty} \mu(x,0)H(x,0)dx}. \tag{3.7b}$$

The estimate (3.2) of n_1 is now obtained by evaluating the integrals using the following estimates for $\mu, \tilde{\mu}$, which are proved in Ref. [17]

$$C_2S(x,t) \le \mu(x,t) \le C_1S(x,t), \tag{3.8a}$$

where for a sufficiently large constant c_2

$$S(\bar{y},s) \equiv \exp(\int_s^T \bar{a}(\tau;\bar{x},T,\bar{y},s)\,d\tau)\frac{e^{-(x-\sigma)^2/(4c_2(T-s))}}{\sqrt{T-s}} \quad \text{for } \bar{x} > -T,$$

and

$$e^{\int_s^T \bar{a}(t)dt} = \mathcal{O}(1)(\tilde{y}(s)/T)^{\bar{c}}.$$

Here $\bar{a}$ is the value of a along the characteristic path $\alpha(\cdot)$ between $(\bar{y},s)$ and $(\bar{x},T)$, defined in (3.9-10) below, and $\tilde{y}$ is the time the path α cross the rarefaction wave, i.e.,

$$\tilde{y}(s) \equiv \begin{cases} y, & \text{if } \exists y \ge s : \alpha(y) = \epsilon y/2, \\ s, & \text{otherwise,} \end{cases}$$

and

$$\bar{c} \equiv t\int_{-\infty}^{\infty} g''\lambda_{2,x}^2 dx = \mathcal{O}(\varepsilon).$$

Similarly the dual $\tilde{\mu}$, which initially is a δ' function, satifies

$$|\tilde{\mu}(x,t)| \le CS(x,t)/\sqrt{T-t}. \tag{3.8b}$$

The proof of (3.8) is based on a refined version of the parametrix method which takes into acount the convection and damping from λ_1 and a, by following the forward τ and the special backward σ characteristic curves

$$\tfrac{d\tau}{dt} = \lambda_1(\tau,t), \quad \tau(s) = \bar{y},$$
$$\tfrac{d\sigma}{dt} = \lambda_1(\tau + (\sigma-\tau)\tfrac{(t-s)}{c_2(T-t)+(t-s)}, t), \quad \sigma(T) = \bar{x}.$$

The function α, used above, is define by

$$\alpha(t) = \tau + (\sigma - \tau)\frac{(t-s)}{c_2(T-t)+(t-s)}, \tag{3.9}$$

Furthermore, let

$$\bar{a}(t;\bar{x},T,\bar{y},s) \equiv a(\tau + (\sigma-\tau)\tfrac{(t-s)}{c_2(T-t)+(t-s)}, t), \tag{3.10}$$
$$\tilde{\omega}(x,t) = \exp\left(-\int_s^t \bar{a}(s')ds'\right) S_1(x,t),$$

where $S_1(x,t) \equiv \frac{\exp[-(x-\tau(t))^2/4t]}{\sqrt{4\pi t}}$. Then, $\tilde{\omega}$ solves

$$\tilde{\omega}_t + \bar{\lambda}_i\tilde{\omega}_x - \tilde{\omega}_{xx} + \frac{\bar{a}}{2}\tilde{\omega} = 0, \qquad \tilde{\omega}(x,s) = \delta(x-\bar{y}).$$

Hence, by multiplying (3.6a) with $\tilde{\omega}$ and integrating we obtain the integral equation

$$\mu(\bar{y},s) = \tilde{\omega}(\bar{x},T) + \int_s^T\int_R \left(\lambda(x,t) - \bar{\lambda}(t)\right)\mu(x,t)\tilde{\omega}_x(x,t)dxdt \tag{3.11}$$
$$+\tfrac{1}{2}\int_s^T\int_R (a(x,t) - \bar{a}(t))\mu(x,t)\tilde{\omega}(x,t)dxdt.$$

The integrals in (3.12) can be estimated as in the standard parametrix method [1] yielding (3.8).

4 General Stability Theorem

The previous examples were chosen so that the equation governing the primary field decoupled into an exact Burgers equation, simplifying the analysis. In the general case, we can achieve an approximate decoupling by diagonalizing following Refs. [2, 9]; However, this introduces new coupling and error terms.

In particular, nonlinear source terms will occur also in the primary equation, generating errors which must be estimated at the same time as the errors in transverse fields, in an iterative scheme. The estimates in the primary field can be made using Duhamel's principle, similarly as in Example 2, with the dual waves for Burgers equation linearized around the viscous rarefaction solution (1.2). It was shown in Ref. [3] how to obtain such duals for the linearization around *any* solution of Burgers equation, by a linearized Hopf-Cole transformation. This is used carefully in the pointwise iterative scheme based on the integral equations obtained from Duhamel's principle.

The Burgers rarefaction wave defined in (1.2) satisfies

$$\tilde{\phi}_t + f(\tilde{\phi})_x - \tilde{\phi}_{xx} = -(\lambda_p(\tilde{\phi})_x)^2 r_p'(\tilde{\phi}) r_p(\tilde{\phi}),$$
$$\tilde{\phi}(\cdot,0) = \Re(u_-, u_+, \cdot),$$

where $r_p(u)$ denotes the right eigenvector of $f'(u)$, and u_+ is on the $p-$rarefaction curve passing through u_-.

As we have seen in the two examples, the error of $u - \tilde{\phi}$ is growing in L_1 as $\mathcal{O}(\log t)$. With this in mind, we split the error into two parts $u - \tilde{\phi} = \eta + w$, where $\|\eta\|_{L_1} = \mathcal{O}(\log t)$ and $\|w\|_{L_1} = \mathcal{O}(1)$. More precisely, each component of η solves an equation of the form (2.3) or (3.3) obtained by linearizing around $\tilde{\phi}$, and in addition include diagonal coupling terms of the Hessian f'', the source term $\sim \lambda_{p,x}^2$, the diagonalization and the initial perturbation. The remaining error w includes the smaller cross-coupling terms.

With this decomposition, we obtain the following generalization of the results in Examples 1 and 2. A proof of the Theorem is given in Ref. [17].

Theorem 1. *Let u be a solution of a parabolic system of conservation laws* (1.1), *with a strictly hyperbolic flux function, for which each characteristic field is either genuinely nonlinear or linear degenerate, and for which the pth field is genuinely nonlinear. Assume that there are positive constants c^0, c^1, with $c^0 + c^1 = o(\sqrt{\epsilon}/\log^5(\epsilon^{-1}))$ as $\epsilon \to 0$, such that the initial data u_0 satisfies*

$$\int_{-\infty}^{\infty} |u_0 - \tilde{\phi}(\cdot,0)| dx \le c^0,$$
$$|u_0(x) - \tilde{\phi}(x,0)| \le \frac{c^1}{(|x|+1)^\alpha}, \quad 1 \le \alpha \le 2,$$

where $\tilde{\phi}$ is the Burgers rarefaction wave defined in (1.2), with strength $\varepsilon << 1$. Then, u can be decomposed as

$$u - \tilde{\phi} = \eta_p^I r_p + \eta^{II} + w,$$

so that, for $t \geq 1$, *and some positive constants* c, C, *we have*

$$\eta_p^I = \mathcal{O}(1)\min(\tfrac{\log t}{t}, \epsilon^2)\cdot\begin{cases} e^{-x^2/(4t)}, & x < \sqrt{t},\\ 1, & \sqrt{t} < x < \varepsilon t,\\ e^{-(\varepsilon t - x)^2/(4t)}, & x > \varepsilon t,\end{cases}$$

$$\eta^{II} = \textstyle\sum_i n_i r_i(\tilde\phi) + \mathcal{O}(|\lambda_{p,x}|),$$

$$w = \textstyle\sum_i v_i r_i(\tilde\phi) + \mathcal{O}(|\lambda_{p,x}|),$$

where

$$\lambda_{p,x} = \mathcal{O}(1)\min(\tfrac{\varepsilon}{\sqrt{t}}, \tfrac{1}{t})\cdot\begin{cases} e^{-x^2/ct}, & \text{if } x < \sqrt{t},\\ 1, & \text{if } \sqrt{t} < x < \varepsilon t - \sqrt{t},\\ e^{-(x-\varepsilon t)^2/ct}, & \text{if } x > \varepsilon t - \sqrt{t},\end{cases}$$

and, for

$$\mathcal{N} = \{i: \text{ the } i\text{-field is genuinely nonlinear and } i \neq p\},$$

$$\mathcal{N}^c = \{i: \text{ the } i\text{-field is linear degenerate and } i \neq p\},$$

$$\mathcal{N}_j = \mathcal{N} \cup \{j\},$$

$$\mathcal{N}_j^c = \mathcal{N}^c \cap (\{1, \dots, m\} - \{j\}),$$

$M_\alpha = c^1$, $1 \leq \alpha < 2$, *and* $M_2 = c^1\log(t/x)$, *the diffusion waves* n_i *satisfy*

$$n_p(x,t) = \mathcal{O}(1)\begin{cases} \min\left(\frac{c^0}{t\varepsilon}, \frac{c^0}{\sqrt{t}}\right) + \frac{M_\alpha}{t^{\alpha-1}x^{2-\alpha}} + \min(\frac{1}{t}, \epsilon^{-2}), & \sqrt{t} < x < \frac{\varepsilon t}{2},\\ e^{\frac{-x^2}{4t}}(\min\left(\frac{c^0}{t\varepsilon}, \frac{c^0}{\sqrt{t}}\right) + \min(\frac{1}{t}, \epsilon^{-2})) + \frac{c^1 + c^0}{(|x|+\sqrt{t})^\alpha}, & x < \sqrt{t},\\ \text{even around } x = \epsilon t/2, & \end{cases}$$

in the primitive field,

$$n_i(x,t) = \mathcal{O}(1)\min[1, (c^0 + c^1 + \epsilon)\log t]\begin{cases} \frac{c^0 + c^1 + \epsilon}{|x - \lambda_i t|}, & |x - \lambda_i t| \geq C\sqrt{t}\log t,\\ \frac{1}{\sqrt{t}}, & |x - \lambda_i t| < C\sqrt{t}\log t,\end{cases}$$

in the nonlinear transversal fields, $i \in \mathcal{N}$,

$$n_i(x,t) = \mathcal{O}(c^0 + c^1 + \epsilon)\begin{cases} \frac{1}{|x-\lambda_i t|} + \frac{\log(|x-\lambda_i t|)+1}{over\sqrt{t}} e^{\frac{-(x-\lambda_i t)^2}{ct}}, & |x - \lambda_i t| > \sqrt{t},\\ \frac{\log t}{\sqrt{t}} e^{-(x-\lambda_i t)^2/ct}, & |x - \lambda_i t| < \sqrt{t},\end{cases}$$

in the linearly degenerate transversal fields, $i \in \mathcal{N}^c$, *and the diffusion error v satisfies, for* $j = 1, \ldots, m$,

$$v_j(x,t) = \mathcal{O}\left(\frac{\epsilon^{1/2}}{\log^5(\epsilon^{-1})}\right) \begin{cases} \frac{\log^2 t}{\sqrt{t}\,\sqrt{|\lambda_i t - x| + \sqrt{t}}}, & \frac{\lambda_{i-1}+\lambda_i}{2} t < x < \frac{\lambda_i + \lambda_{i+1}}{2} t, i \in \mathcal{N}_j, \\ \frac{\log^2 t}{t}, & \frac{\lambda_{i-1}+\lambda_i}{2} t < x < \frac{\lambda_i + \lambda_{i+1}}{2} t, i \in \mathcal{N}_j^c, \\ \frac{\log^2 t}{\sqrt{t}\,\sqrt{|x| + \sqrt{t}}}, & \frac{\lambda_{p-1} t}{2} < x < \sqrt{t}, \\ \frac{\log(t/x)}{\sqrt{t}\,\sqrt{x}}, & \sqrt{t} < x < \varepsilon t/2, \\ \frac{\log(t/(\varepsilon t - x))}{\sqrt{t}\,\sqrt{|\varepsilon t - x|}}, & \frac{\varepsilon t}{2} < x < \varepsilon t - \sqrt{t}, \\ \frac{\log^2 t}{\sqrt{t}\,\sqrt{|\varepsilon t - x| + \sqrt{t}}}, & \varepsilon t - \sqrt{t} < x < \frac{\lambda_{p+1} t}{2}, \end{cases}$$

Acknowledgements: Thanks to Jonathan Goodman for several stimulating discussions, and to Stanford University and Kungl Tekniska Högskolan, Stockholm, for their hospitality during the visits when various parts of this work were done.

This work was supported in part by NSF grant DMS-9107990 and ONR grant N00014-94-1-0456 (Zumbrun) and by TFR grant 92691 and grant AF OSR-91-0063 (Szepessy).

References

[1] Friedman A., Partial Differential Equation of Parabolic Type, Prentice-Hall, 1964.

[2] Goodman, J., Nonlinear asymptotic stability of viscous shock profiles for conservation laws, Arch. Rat. Mech. Anal. 95 (1986) 325–344.

[3] Goodman J., Szepessy A. and Zumbrun K., A remark on the stability of viscous shocks, SIAM Math. Anal., v 11, 1994.

[4] Hopf E., The partial differential equation $u_t + uu_x = \mu_{xx}$, Comm. Pure Appl. Math., v. 3, 1950, 201-230.

[5] Kawashima S, Matsumura A. and Nishihara K., Asymptotic behavoir of the solutions for the equations of a viscous heat-conductive gas. Proc. Jpn. Acad. 62, 1986, 249-252.

[6] Lefloch P. and Xin Z., Uniqueness via the adjoint problem for systems of conservation laws, preprint (1992)

[7] Liu, Linear and nonlinear large-time behavior of solutions to genereal systems of hyperbolic conservation laws, Comm. Pure Appl. Math. v. 30, 1977, 767-797.

[8] Liu T.-P., Interaction of nonlinear hyperbolic waves, in Nonlinear Analysis, Eds. F.-C. Liu & T.-P. Liu, World Scientific, 1991, 171-184.

[9] Liu,T.-P., Nonlinear Stability of Shock Waves for Viscous Conservation Laws, Vol. 328 of AMS Memoirs, Amer. Math. Soc., 1986.

[10] Liu, T.-P. Pointwise convergence to shock waves for the system of viscous conservation laws, to appear.

[11] Liu T.-P. and Zeng Y., Large time behavior of solutions for general quasilinear hyperbolic-parabolic systems of conservation laws, preprint (1994).

[12] Liu T.-P. and Zumbrun K., Nonlinear stability of an undercompressive shock of complex Burgers equation, to appear, Comm. Math. Phys.

[13] Liu T.-P. and Zumbrun K., On stability of general undercompressive viscous shock waves, preprint (1994).

[14] Matsumura A. and Nishihara K., Asymptotics towards the rarefaction waves of the solutions of a one-dimensional model system for compressible viscous gas, Jpn J. Appl. Math. 3, 1986, 1-13.

[15] Matsumura A. and Nishihara K, Global stability of the rarefaction wave of a one-dimensional model system for compressible gas, Comm. Math. Phys., 144, 1992, 335-335.

[16] Szepessy A. and Xin Z., Nonlinear stability of viscous shock waves, Arch. Rat. Mech Anal., V. 122 no 1, 1993, 53-103.

[17] Szepessy A. and K. Zumbrun., Stability of rarefaction waves in viscous media, preprint (1994).

[18] Xin Z., Asymptotic stability of rarefaction waves for 2*2 viscous hyperbolic conservation laws, J. Diff. Eq., 73, 1988, 45-77.

Uses of Local Preconditioning

Bram van Leer[1]

Abstract

Local preconditioning of the Euler equations creates a number of effects that are highly desirable in the numerical computation of steady solutions; these are discussed in some detail.

1 Introduction

I wish to report on a subject that so far has not received coverage in the conferences series on Hyperbolic Problems, namely, local preconditioning of the Euler equations. Over the past five years a wealth of insight and information has been obtained by a small group of researchers, and it is high time to present an overview to the hyperbolic research community.

The study of local preconditioning is motivated by the need to make the time-dependent Euler equations better suited to the numerical computation of steady solutions. At the level of the partial differential equations, this type of preconditioning is accomplished by multiplying either the temporal or the spatial operator by a locally evaluated matrix; for the present analysis we prefer the latter choice. To fix our thoughts, let me first introduce some notation.

For the present purpose the Euler equations are written as

$$\frac{\partial U}{\partial t} + A(U)\frac{\partial U}{\partial x} + B(U)\frac{\partial U}{\partial y} + C(U)\frac{\partial U}{\partial z} = 0, \tag{1}$$

where the state variables are defined differentially:

$$dU = (\frac{dp}{\rho a}, du, dv, dw, dS)^T; \tag{2}$$

a denotes sound speed, S entropy. These variables have the property of symmetrizing the Euler equations:

$$A(U) = \begin{pmatrix} u & a & 0 & 0 & 0 \\ a & u & 0 & 0 & 0 \\ 0 & 0 & u & 0 & 0 \\ 0 & 0 & 0 & u & 0 \\ 0 & 0 & 0 & 0 & u \end{pmatrix}, \; B(U) = \begin{pmatrix} v & 0 & a & 0 & 0 \\ 0 & v & 0 & 0 & 0 \\ a & 0 & v & 0 & 0 \\ 0 & 0 & 0 & v & 0 \\ 0 & 0 & 0 & 0 & v \end{pmatrix}, \; C(U) = \begin{pmatrix} w & 0 & 0 & a & 0 \\ 0 & w & 0 & 0 & 0 \\ 0 & 0 & w & 0 & 0 \\ a & 0 & 0 & w & 0 \\ 0 & 0 & 0 & 0 & w \end{pmatrix}. \tag{3}$$

Much of the analysis can be done assuming that the fluid moves in the positive x-direction, so that u equals the full flow speed q, and v vanishes (but not its derivatives). The matrix coefficients then

[1]W. M. Keck Foundation Laboratory for Computational Fluid Dynamics, Department of Aerospace Engineering, University of Michigan, Ann Arbor, MI 48109-2118

become

$$A(U) = a\begin{pmatrix} M & 1 & 0 & 0 & 0 \\ 1 & M & 0 & 0 & 0 \\ 0 & 0 & M & 0 & 0 \\ 0 & 0 & 0 & M & 0 \\ 0 & 0 & 0 & 0 & M \end{pmatrix}, \; B(U) = a\begin{pmatrix} 0 & 0 & 1 & 0 & 0 \\ 0 & 0 & 0 & 0 & 0 \\ 1 & 0 & 0 & 0 & 0 \\ 0 & 0 & 0 & 0 & 0 \\ 0 & 0 & 0 & 0 & 0 \end{pmatrix}, \; C(U) = a\begin{pmatrix} 0 & 0 & 0 & 1 & 0 \\ 0 & 0 & 0 & 0 & 0 \\ 0 & 0 & 0 & 0 & 0 \\ 1 & 0 & 0 & 0 & 0 \\ 0 & 0 & 0 & 0 & 0 \end{pmatrix}, \tag{4}$$

where M is the Mach number.

The preconditioned Euler equations read

$$\frac{\partial U}{\partial t} + P(U)\left(A(U)\frac{\partial U}{\partial x} + B(U)\frac{\partial U}{\partial y} + C(U)\frac{\partial U}{\partial z}\right) = 0, \tag{5}$$

where $P(U)$ is a positive definite matrix. If P is symmetric, so is P^{-1}, and the system

$$P(U)^{-1}\frac{\partial U}{\partial t} + A(U)\frac{\partial U}{\partial x} + B(U)\frac{\partial U}{\partial y} + C(U)\frac{\partial U}{\partial z} = 0 \tag{6}$$

is symmetric, just as the original system. If P is not symmetric, it is desirable that the preconditioned system be *symmetrizable*; this puts certain constraints on the elements of P [1].

In the course of our research on local preconditioning of the Euler equations the following computational benefits have been identified.

1. Local preconditioning can be designed to remove the stiffness of the system of equations caused by the range of the characteristic speeds, thus improving the convergence rate of any discrete marching scheme [2].

2. It then makes the system of equations behave more like a scalar equation, which is advantageous in designing and applying additional convergence-acceleration techniques [3].

3. It generates spatial discretizations that may retain their accuracy at low Mach number, in contrast to standard discretizations [4, 5, 6].

4. It can be designed to decouple the acoustic equations from the convective equations, allowing genuinely multidimensional discretizations [1]

These desirable properties come at the expense of robustness. Current research focuses on making the various preconditioned Euler discretizations that have been developed more reliable near flow singularities, such as sonic and stagnation points. In recent reports [7, 8] we have elaborated on the bad news, i. e. the loss of stability or convergence in computing stagnating flow with the symmetric Van Leer-Lee-Roe preconditioning [2], and how to overcome it by modifying the matrix for low Mach numbers. In the present report I wish to emphasize the good news, and devote a small section to each of the benefits of local preconditioning listed above.

2 Benefits of local preconditioning

2.1 Removing stiffness

It has been known since 1990 [2] that local preconditioning of the Euler equations can make the condition number of the characteristic speeds, i. e., the eigenvalues of the matrix $PAn_x + PBn_y + PCn_z$, where n is a unit vector, as low as $1/\sqrt{1-\min(M^2, M^{-2})}$, down from the standard value $(M+1)/\min(M, |M-1|)$. The first matrix realizing this condition number was the symmetric matrix of Van Leer et al. [2]; for the Euler equations in the flow-aligned coordinate system (cf. (4)) it reads

$$\mathbf{P}_{3D} = \begin{pmatrix} \frac{\tau}{\beta^2}M^2 & -\frac{\tau}{\beta^2}M & 0 & 0 & 0 \\ -\frac{\tau}{\beta^2}M & \frac{\tau}{\beta^2}+1 & 0 & 0 & 0 \\ 0 & 0 & \tau & 0 & 0 \\ 0 & 0 & 0 & \tau & 0 \\ 0 & 0 & 0 & 0 & 1 \end{pmatrix}, \tag{7}$$

with β and τ defined by:

$$\beta = \begin{cases} \sqrt{1-M^2}, & M < 1, \\ \sqrt{M^2-1}, & M \geq 1; \end{cases} \tag{8}$$

$$\tau = \begin{cases} \sqrt{1-M^2}, & M < 1, \\ \sqrt{1-M^{-2}}, & M \geq 1. \end{cases} \tag{9}$$

The Euler equations thus preconditioned describe a world with wave propagation that differs considerably from what we are familiar with. In particular, in subsonic flow an acoustic point disturbance in the origin spreads in time as an expanding oblate spheroid centered at the origin (not at a point moving with the flow); the equation of this acoustic front is

$$\left(\frac{x}{\beta qt}\right)^2 + \left(\frac{y}{qt}\right)^2 + \left(\frac{z}{qt}\right)^2 = 1. \tag{10}$$

In supersonic flow the acoustic front is an expanding ring moving downstream; its equation is

$$x = \beta qt, \tag{11}$$

$$\left(\frac{y}{at}\right)^2 + \left(\frac{z}{at}\right)^2 = 1. \tag{12}$$

Table 1 summarizes the effect of preconditioning on the characteristic condition number for one, two- and three-dimensional flow at arbitrary Mach number.

Numerical simulations including preconditioning have indicated significant levels of convergence acceleration, in agreement with theoretical predictions based on the improved characteristic condition number [2, 4, 9, 5, 10, 11]. One example of convergence acceleration is presented in Figures 1 and 2, taken from [2]. Shown here are the Mach-number contours of transonic flow ($M = 0.85$, angle of attack 1.0°) over a NACA 0012 airfoil, computed with the preconditioned scheme, followed by the residual histories for both this computation and a similar computation without preconditioning. The free-stream condition number is 1.85/0.15=12.3 before and $1/\sqrt{(1-0.85^2)}$=1.90 after

	original equations	preconditioned 1-D	2-D	3-D
supersonic	$\frac{M+1}{M-1}$	1	1	$\frac{1}{\sqrt{1-M^{-2}}}$
subsonic	$\frac{M+1}{\min(1-M,M)}$	1	$\frac{1}{\sqrt{1-M^2}}$	$\frac{1}{\sqrt{1-M^2}}$

Table 1: Characteristic condition number for the Euler equations before and after preconditioning.

preconditioning, suggesting a possible convergence acceleration of a factor 6.5; note, though, that condition numbers higher or lower than the free-stream value do occur. The actual convergence rate, after initial transients, shows an improvement of a factor 5.5.

2.2 Making systems behave scalar-like

The equalization of the characteristic speeds makes the system of equations behave like a single scalar equation in some respect. Particularly noticeable is the change in the eigenvalue spectrum of discretizations of the system. Figures 3 and 4 show the locus of eigenvalues ("Fourier footprint") of a first-order upwind discretization of the spatial operator at a Mach number of 0.1, before and after preconditioning. It is seen that preconditioning strongly clusters the eigenvalues of the discrete operator, making the footprint resemble one obtained from upwind discretization of a scalar advection equation, depicted in Figure 5. This clustering has proved beneficial in the development of multi-stage schemes with optimal high-frequency damping, for use in multigrid relaxation [3, 11]. Figure 6 shows the high-frequency footprint of a 2-D preconditioned Euler discretization superimposed on the level lines of the amplification factor of a four-stage scheme optimized for high-frequency damping. This scheme reduces any high frequency by a factor no greater than 0.2362. It has been shown [11] that the use of such schemes combined with preconditioning significantly improves the performance and reliability of multi-grid relaxation.

2.3 Preserving accuracy in the incompressible limit

It was discovered by Lee [4] that certain discretizations of the preconditioned Euler equations produce steady solutions that are independent of the reference Mach number as it tends to zero; likewise, the convergence history is independent of the reference Mach number. This contrasts with the observation by Volpe [12] that discretizations of the regular Euler equations loose their accuracy on a grid of fixed resolution, for vanishing reference Mach number. The preservation of accuracy for low-speed flow through preconditioning is illustrated by Figures 7, 8 and 9, taken from [4]; these show Mach-number contours for Mach 0.0001 flow over a circular bump, computed with a first-order upwind Euler scheme, its preconditioned version, and (for comparison) a potential-flow solver, respectively, on the same grid. The standard scheme converges to a completely incorrect solution, while the preconditioned scheme produces a solution with accuracy typical of a first-order scheme used on a coarse grid.

Turkel et al. [6] analyzed preconditioned Euler discretizations in the limit of $M_{\text{ref}} \to 0$, and found necessary and sufficient conditions on the elements of the preconditioning matrix so as to yield uniform accuracy in the incompressible limit. Use of preconditioning thus makes it possible to obtain, with one and the same code, accurate steady flow solutions containing both incompressible and compressible flow regions.

2.4 Decoupling acoustic from convection equations

When multiplying the preconditioning (7) with the spatial Euler operator, the equations simplify greatly. This is particularly clear in two dimensions, where the preconditioned system under supersonic conditions reduces to a nonlinear set of advection equations:

$$\begin{aligned}
(\partial_t + u\partial_x)\,S &= 0, \quad dS = dp - a^2 d\rho, &(13)\\
(\partial_t + u\partial_x)\,H &= 0, \quad dH = \frac{dp}{\rho} + \frac{1}{2}d(u^2), &(14)\\
\left(\partial_t + \frac{\beta u}{M}\partial_x + \frac{u}{M}\partial_y\right) R^+ &= 0, \quad dR^+ = \beta dp + \rho u dv, &(15)\\
\left(\partial_t + \frac{\beta u}{M}\partial_x - \frac{u}{M}\partial_y\right) R^- &= 0, \quad dR^- = \beta dp - \rho u dv. &(16)
\end{aligned}$$

Under subsonic conditions the preconditioned system reads

$$\begin{aligned}
(\partial_t + u\partial_x)\,S &= 0, &(17)\\
(\partial_t + u\partial_x)\,H &= 0, &(18)
\end{aligned}$$

$$\left\{ \begin{array}{c} \frac{\beta \partial p}{\rho u} \\ \partial v \end{array} \right\}_t + \left[\begin{array}{cc} -u\beta & 0 \\ 0 & u\beta \end{array} \right] \left\{ \begin{array}{c} \frac{\beta \partial p}{\rho u} \\ \partial v \end{array} \right\}_x + \left[\begin{array}{cc} 0 & u \\ u & 0 \end{array} \right] \left\{ \begin{array}{c} \frac{\beta \partial p}{\rho u} \\ \partial v \end{array} \right\}_y = 0. \qquad (19)$$

This system consists of two advection equations - one for entropy, one for total enthalpy - and a hyperbolic system of two time-dependent Cauchy-Riemann equations, describing the omnidirectional propagation of acoustic signals. The 3-D case is more complicated [1]: differential rather than purely algebraic preconditioning appears to be necessary for a complete decoupling of acoustic and advection equations.

The above 2-D preconditioned equations currently serve as the basis of fluctuation-split methods [8],which may be considered the genuinely multi-dimensional extension of 1-D upwind-differencing schemes. They have the promise of achieving high accuracy and fast convergence using extremely local data, making them pre-eminently suited for massively parallel computer architectures.

References

[1] E. Turkel and P. L. Roe, "Decomposition of the Euler equations." ICASE Report, in preparation, 1994.

[2] B. van Leer, W. T. Lee, and P. L. Roe, "Characteristic time-stepping or local preconditioning of the Euler equations," in *AIAA 10th Computational Fluid Dynamics Conference*, 1991.

[3] J. F. Lynn and B. van Leer, "Multi-stage schemes for the Euler and Navier-Stokes equations with optimal smoothing," AIAA Paper 93-3355-CP, 1993.

[4] W.-T. Lee, *Local Preconditioning of the Euler Equations.* PhD thesis, University of Michigan, 1991.

[5] A. G. Godfrey, *Topics on Spatially Accurate Methods and Preconditioning for the Navier-Stokes Equations with Finite-Rate Chemistry.* PhD thesis, VPI & SU, 1992.

[6] E. Turkel, A. Fiterman, and B. van Leer, "Preconditioning and the limit to the incompressible flow equations." ICASE Report 93-42, 1993.

[7] B. van Leer, L. Mesaros, C.-H. Tai, and E. Turkel, "Local preconditioning in a stagnation point." ICASE Report, in preparation, 1994.

[8] L. M. Mesaros and P. L. Roe, "Multidimensional fluctuation-splitting schemes based on decomposition methods." Submitted to AIAA 12th Computational Fluid Dynamics Conference, 1994.

[9] A. G. Godfrey, R. W. Walters, and B. van Leer, "Preconditioning for the Navier-Stokes equations with finite-rate chemistry," AIAA Paper 93-0535, 1993.

[10] A. G. Godfrey, "Steps towards a robust preconditioning," AIAA Paper 94-0520, 1994.

[11] J. F. Lynn and B. van Leer, "Multigrid Euler solutions with semi-coarsening and local preconditioning," in *Proceedings of the 14th International Conference on Numerical Fluid Dynamics*, Springer, 1995.

[12] G. Volpe, "On the use and accuracy of compressible flow codes at low Mach numbers," AIAA Paper 91-1662, 1991.

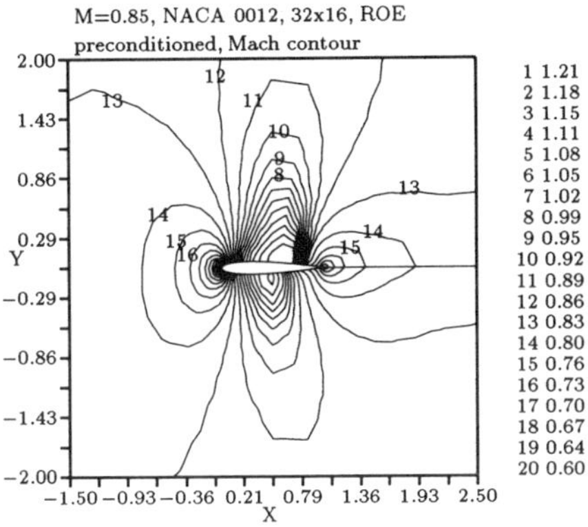

Figure 1: Steady solution of the Euler equations for flow over a NACA 0012 airfoil at $M_\infty = 0.85$, $\alpha = 1.0°$; shown are Mach-number contours. Matrix preconditioning was applied in combination with local time-stepping. Numerical flux function used: modified Roe.

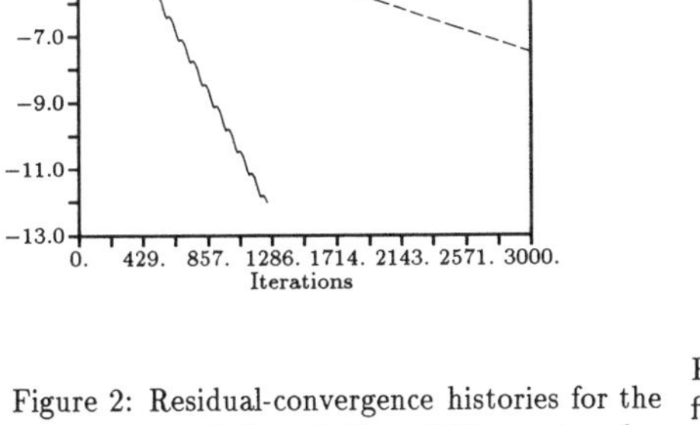

Figure 2: Residual-convergence histories for the computation of the solution of Figure 1 and a similar computation without preconditioning.

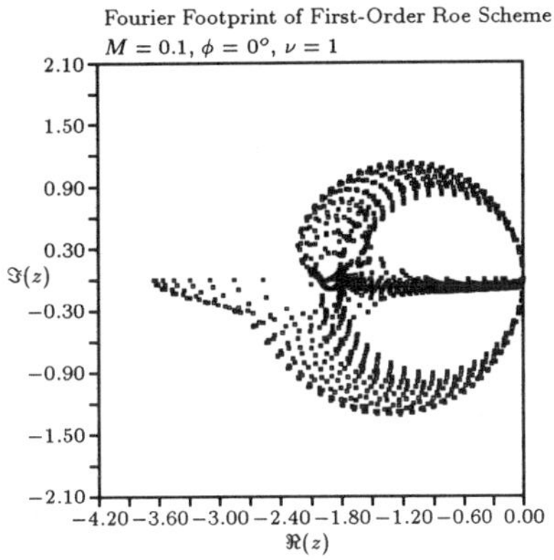

Figure 3: Fourier footprint of the first-order upwind approximation of the spatial Euler operator, for $M = 0.1$, and flow angle 0°.

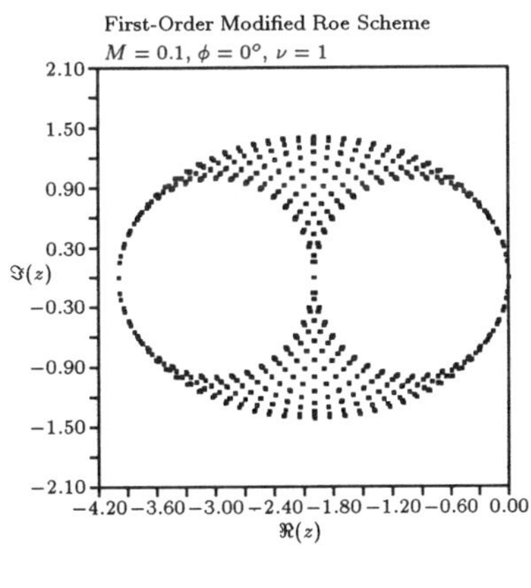

Figure 4: Fourier footprint of the preconditioned first-order upwind approximation of the spatial Euler operator, for $M = 0.1$, and flow angle 0°.

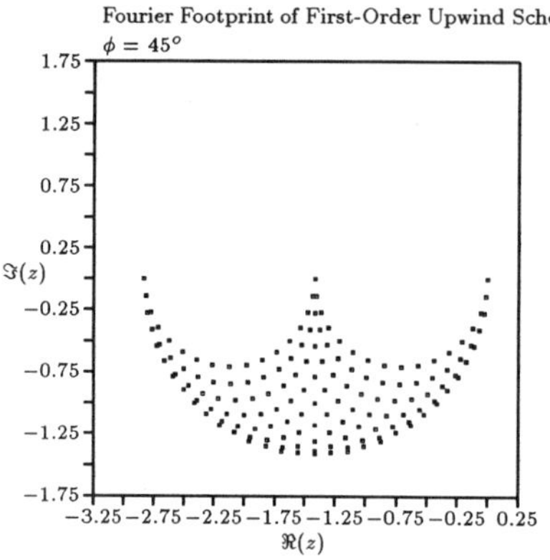

Figure 5: Fourier footprint of the first-order upwind approximation of the two-dimensional convection equation; convection angle 45°.

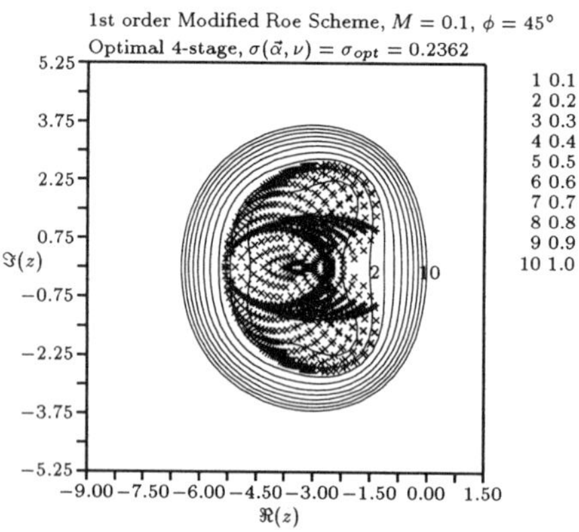

Figure 6: High-frequency Fourier footprint of the preconditioned first-order upwind Euler operator plotted on top of the level lines of the amplification factor of the associated optimal 4-stage scheme. Flow angle 45°.

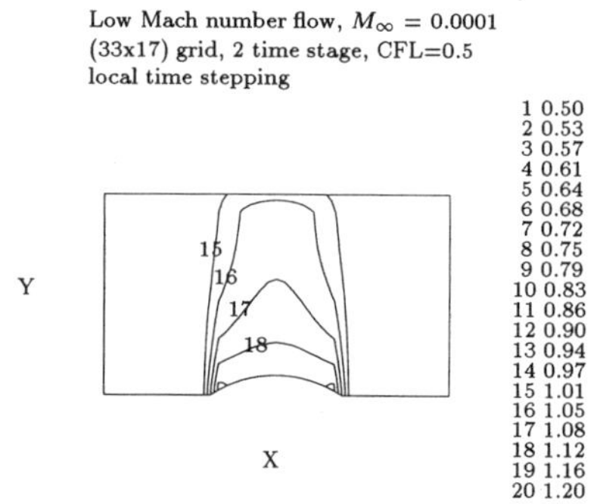

Figure 7: Mach-number contours of steady solution of the Euler equations for flow over a circular bump in a semi-infinite plane, with $M_\infty = 0.0001$. First-order upwind differencing, no preconditioning. Level values are in units of 10^{-4}.

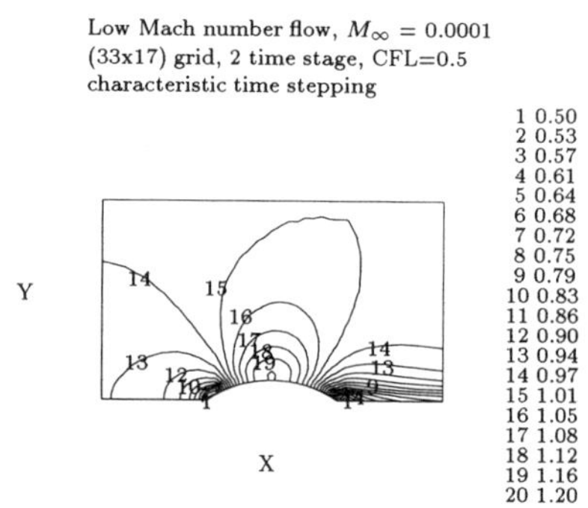

Figure 8: Same as Figure 8, except preconditioning was applied.

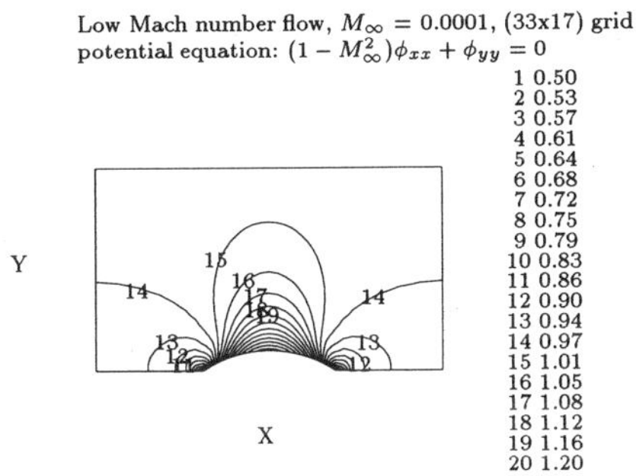

Figure 9: Mach-number contours of the steady potential flow over a circular bump in a semi-infinite plane, with $M_\infty = 0.0001$.

On Mode III Propagation of Adiabatic Shear Bands

T. W. Wright[1] and **J. W. Walter** [1]

Abstract

Local solutions are found for the tip of a shear band as it propagates edgewise into relatively undisturbed material. Only the case of antiplane motion is considered.

1 Introduction

An adiabatic shear band is a thin layer of material where the deformation has become extremely localized due to a material instability that results from plastic heating and subsequent thermal softening of the flow stress. Bands tend to form in regions of extremely high strain rate such as those that might occur in impact problems. Typical lateral dimensions of a shear band are of the order of millimeters, with thicknesses of the order of 10's of microns, so the aspect ratios are of the order of a few 100's.

The dynamics of shear bands are becoming fairly well understood in a one-dimensional setting, and this seems sufficient for explaining most features of the simple Hopkinson torsion bar experiment, where a short tubular specimen (gage length about 2mm, wall thickness about 0.1mm) is placed between two solid bars and subjected to a torsional pulse. The pulse travels as an elastic wave down one bar and strikes the specimen, which deforms plastically, reflecting part of the pulse and transmitting the rest into the second elastic bar. Nominal or applied plastic strain rates in the specimen are usually of the order of 1000 s^{-1}, but the maximum strain rate in the shear band may be several orders of magnitude higher. For details of a typical experiment, see Marchand and Duffy [1988].

Shear bands also form in bulk material, however; there it is believed that they behave more like cracks, propagating edgewise into relatively undisturbed material, although the material deformation only localizes, rather than separates, as in a crack. If the deformation is such that material velocity occurs only in the z-direction and depends only on x, y, and t, it is called antiplane, and in the parlance of fracture mechanics, is said to exhibit Mode III motion. (Modes I and II refer to in-plane motions). Such deformation is the simplest to describe mathematically, and therefore, only Mode III motions will be considered in this paper.

2 Equations

Consider a rigid/plastic material that exhibits thermal softening and strain rate sensitivity, but not work hardening. The conservation laws and a simple flow law for such a material may be written as follows:

$$\begin{array}{lrcl} \text{Momentum:} & \rho\dot{\mathbf{v}} & = & \mathrm{div}\mathcal{S} - \nabla p \\ \text{Energy:} & \rho c\dot{\vartheta} & = & k\nabla^2\vartheta + \mathcal{S} : \mathcal{D} \\ \text{Flow Law:} & \mathcal{S} & = & \kappa_0(1 - a\vartheta)(1 + 2bI)^m \mathcal{D}/I \end{array} \qquad (1)$$

[1]Army Research Laboratory, Aberdeen Proving Ground, Maryland 21005

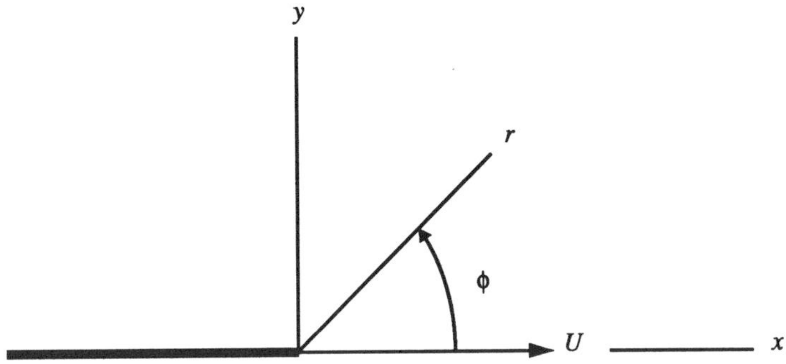

Figure 1: A propagating shear band lies on the negative x-axis with Cartesian and polar coordinates attached.

In Equation (1) $\mathbf{v}$ is particle velocity, $\mathcal{S}$ is the deviatoric stress tensor, p is pressure, and ϑ is temperature. The stretching tensor, $\mathcal{D}$, is defined by $\frac{1}{2}(\nabla\mathbf{v} + \nabla\mathbf{v}^T)$, and $I = (\frac{1}{2}\mathrm{tr}\mathcal{D}^2)^{\frac{1}{2}}$. The physical constants ρ, c, k, κ_0, a, b, and m are density, heat capacity, thermal conductivity, strength, softening coefficient, time constant, and strain rate sensitivity, respectively. The flow law in this form is due to Batra [1988] and is a generalization of Litonski's law [1977].

For antiplane motions $\mathbf{v} = (0, 0, w)$, the pressure may be assumed to vanish, and only two components of the deviatoric stress are required. For the flow law assumed above, it works out that the tractions on the xy-plane are given by $\mathcal{S}\mathbf{e}_3 = s\mathbf{e}_w$, where s is the magnitude of the shear traction, and $\mathbf{e}_w$ is a unit vector in the direction of ∇w. If, in addition, it is assumed that the shear band propagates edgewise at a constant speed, U, as shown in Figure 1, that the motion, as seen from the tip of the propagating band, is essentially steady, and that the strain rate is large enough so that $(b\dot{\gamma}_0)^m \gg 1$, then in nondimensional form and with $g = 1 - a\vartheta$, the reduced equations become

$$\begin{aligned} -\alpha w_x &= \nabla\cdot(s\nabla w/\dot{\gamma}) \\ \beta g_x &= -\epsilon\nabla^2 g + s\dot{\gamma} \\ s &= g\dot{\gamma}^m \\ \dot{\gamma} &= |\nabla w| \end{aligned} \tag{2}$$

where α, β, and ϵ are nondimensional parameters

$$\alpha = \frac{\rho U^2}{S_0} \quad \beta = \frac{\rho c}{a S_0}, \quad \text{and } \epsilon = \frac{k\dot{\gamma}_0}{a S_0 U^2} \tag{3}$$

The following straightforward system of nondimensionalization has been used:

$$\begin{aligned} t &= \dot{\gamma}_0 \bar{t}, & x &= \frac{\bar{x}}{L}, & y &= \frac{\bar{y}}{L} \\ w &= \frac{\bar{w}}{U}, & \vartheta &= \frac{\rho c \bar{\vartheta}}{S_0}, & s &= \frac{\bar{s}}{S_0} \end{aligned} \tag{4}$$

where $\dot{\gamma}_0$ and L are the reference strain rate and length respectively, with the relationship $U = L\dot{\gamma}_0$, and the reference stress is given by $S_0 = \kappa_0(b\dot{\gamma}_0)^m$. In Equation (4) the overbar identifies the dimensional variables.

Typical values of material constants in a strong steel are approximately $\rho = 8\text{x}10^3 \frac{kg}{m^3}$, $c = 5\text{x}10^2 \frac{J}{kg-K}$, $k = 50 \frac{W}{m-K}$, $a = 8\text{x}10^{-4} \frac{1}{K}$, and $S_0 = 0.5$GPa. Then for $\dot{\gamma}_0 = 10^4$/s and $U = 250$m/s, the nondimensional parameters are $\alpha = 1$, $\beta = 10$, and $\epsilon = 2\text{x}10^{-5}$. Clearly heat conduction can be ignored except in regions with large and rapidly changing temperature gradients such as the flanks of a fully formed shear band.

3 Solution

As shown in Figure 1, the band propagates into relatively undisturbed material, so for the near tip region that lies in the annulus $r_1 \leq r \leq r_2$, $-\pi < \phi < \pi$ (where both r_1 and r_2 are small), heat conduction will be ignored. A near field similarity solution will be assumed such that each dependent variable has the form of the dimensional polar radius, $\bar{r}$, raised to some power, times a function that depends only on the polar angle, ϕ. After converting Equations (2) to polar coordinates and performing some other algebraic manipulations, it comes out that with

$$\begin{aligned} w &= \left\{\frac{2m}{1+m}\cdot\frac{\rho c}{a\kappa_0}\right\}^{\frac{1}{(1+m)}} \left\{\frac{\bar{r}}{U\bar{b}}\right\}^{\frac{m}{(1+m)}} W(\phi) \\ g &= \frac{\rho U^2}{\kappa_0}\cdot\left\{\frac{2m}{1+m}\cdot\frac{\rho c}{a\kappa_0}\right\}^{\frac{(1-m)}{(1+m)}} \left\{\frac{\bar{r}}{U\bar{b}}\right\}^{\frac{2m}{(1+m)}} G(\phi) \end{aligned} \tag{5}$$

the nondimensional functions $W(\phi)$ and $G(\phi)$ must satisfy the coupled ordinary differential equations

$$\begin{aligned} \left(\frac{GW'}{\Gamma^{1-m}}\right)' + \frac{1+2m}{1+m}\cdot\frac{m}{1+m}\cdot\frac{GW}{\Gamma^{1-m}} + \frac{m}{1+m}W\cos\phi - W'\sin\phi &= 0 \\ \sin\phi\cdot G' + \frac{2m}{1+m}(\Gamma^{1+m} - \cos\phi)G &= 0 \end{aligned} \tag{6}$$

where $\Gamma^2 = (W')^2 + (\frac{m}{1+m}W)^2$ and $(\cdot)' = \frac{d}{d\phi}(\cdot)$, with the initial conditions

$$W(0) = 0, \quad W'(0) = 1, \quad \text{and } G(0) = G_0 \tag{7}$$

The exponents of $\bar{r}$ in Equations (5) are demanded by the equations, but the scaling coefficients were chosen so as to adjust the magnitude of $\Gamma(0)$, thus simplifying the second equation in (6) and eliminating α and β as explicit coefficients in the equations. The first initial condition is required by antiplane symmetry, but the second is demanded by the second equation in (6); otherwise, G_0 could not be chosen arbitrarily. Note that Equations (6) will apply to *any and all materials with the same strain rate sensitivity* since for the scaling shown in Equations (5), all material properties except m are contained only in the leading coefficients and not in the differential equations themselves.

Since $\sin\phi$ vanishes at $\phi = 0$, the numerical integration of the second equation in (6) requires special treatment near the initial values. By linearizing Equations (6) about the initial values, it can be shown that the behavior of the solution starts as follows:

$$\begin{aligned} W &= \phi + \frac{\phi^3 \ln\phi}{3m(1+m)G_0} - \frac{1}{6}\phi^3\left\{\left(\frac{m}{1+m}\right)^2 + \frac{2}{3}\cdot\frac{1}{m(1+m)G_0}\right\} \\ W' &= 1 + \frac{\phi^2\ln\phi}{m(1+m)G_0} - \frac{1}{2}\left(\frac{m}{1+m}\right)^2\phi^2 \\ G &= G_0 - \frac{\phi^2\ln\phi}{1+m} + \frac{1-mG_0}{2(1+m)}\cdot\phi^2 \end{aligned} \tag{8}$$

After these equations have been used to find a consistent set of initial values for some small value of ϕ, further numerical integration with *Mathematica*™ is straightforward.

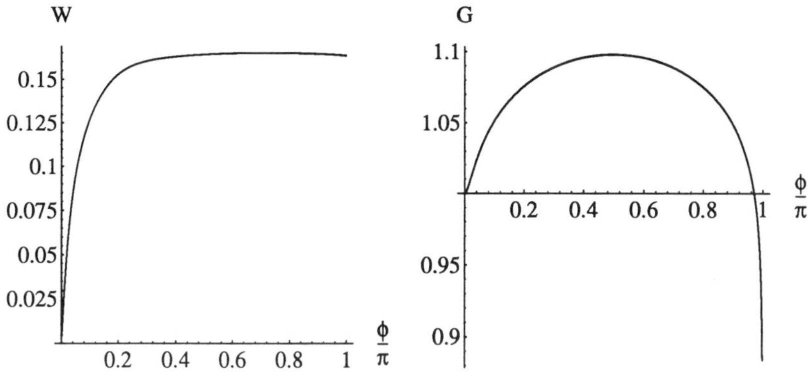

Figure 2: Plots of nondimensional functions W and G for $G_0 = 1$ and $m = 0.02$.

Typical solutions for W and G are shown in Figure 2 for $m = 0.02$ and $G_0 = 1.0$. The general appearance of the curves does not change for other values of G_0. W increases from zero initially, rising to a plateau that extends virtually all the way to $\phi = \pi$, and G rises at first (indicating cooler temperatures) as ϕ increases from zero, then plunges as $\phi \to \pi$. Both W and G shift upwards with increasing G_0. Increasing the parameter m has the effect of broadening the initial region of lower G (higher temperature), increasing G near $\phi = \frac{\pi}{2}$, and broadening the region of decreasing G near π. Also as m increases, W takes longer to rise to its plateau, which in turn has a higher level.

Nondimensional stresses that correspond to Equations (5) are given by

$$
\begin{aligned}
s_{13} &= \frac{\rho U^2}{S_0}\left\{\frac{2m}{1+m}\cdot\frac{\rho c}{a\kappa_0}\right\}^{\frac{1}{(1+m)}}\left(\frac{\bar{r}}{Ub}\right)^{\frac{m}{(1+m)}}\frac{G}{\Gamma^{1-m}}\left\{\frac{m}{1+m}W\cos\phi - W'\sin\phi\right\} \\
s_{23} &= \frac{\rho U^2}{S_0}\left\{\frac{2m}{1+m}\cdot\frac{\rho c}{a\kappa_0}\right\}^{\frac{1}{(1+m)}}\left(\frac{\bar{r}}{Ub}\right)^{\frac{m}{(1+m)}}\frac{G}{\Gamma^{1-m}}\left\{W'\cos\phi + \frac{m}{m+1}W\sin\phi\right\}
\end{aligned}
\tag{9}
$$

and are plotted in Figure 3. The shear stress that drives the shear band, s_{23}, has its highest values ahead of the propagating band, but it falls off immediately after a leading wedge and ends up on a plateau that extends all the way to $\phi = \pi$. On the other hand, the transverse component of shear stress, s_{23}, vanishes ahead of the band, rises to a plateau, and falls sharply again as $\phi \to \pi$.

Near to $\phi = \pi$, where the fully developed shear band lies, the inertial similarity solution can no longer be valid because of high temperatures and steep temperature gradients near the band. However, to take advantage of the lamellar nature of shear bands, it seems reasonable to assume that the band profile itself is essentially one-dimensional with slowly changing parameters along the band. Glimm, Plohr, and Sharp [1993] showed that for the constitutive law $\sigma = \kappa_0 g(\vartheta)(1 + bv_y)^m$, where m is small, the asymptotic structure of the temperature and velocity fields near a shear band may be written

$$
\begin{aligned}
\vartheta &= \vartheta_c - \{2mg(\vartheta_c)/[-g'(\vartheta_c)]\}\cdot\ln\{\cosh(y/\delta)\}\cdot[1 + \mathcal{O}(m)] \\
v &= v_c - \delta\lambda\tanh(y/\delta)[1 + \mathcal{O}(m)]
\end{aligned}
\tag{10}
$$

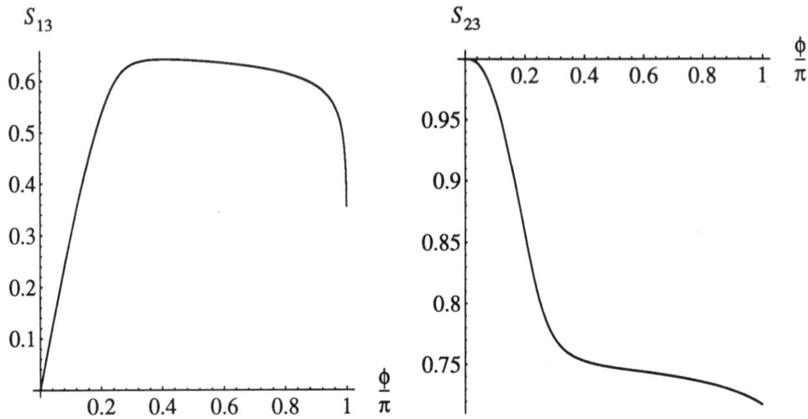

Figure 3: Plots of nondimensional shear tractions S_{13} and S_{23} for $G_0 = 1$ and $m = 0.02$.

In Equation (10) ϑ_c and v_c are temperature and velocity in the center of the shear band, respectively, (in the present case v_0), λ is the strain rate in the center, and δ is a characteristic length given by

$$\delta = \left\{ \frac{2kmg(\vartheta_c)}{-g'(\vartheta_c)\sigma\lambda} \right\}^{\frac{1}{2}} \tag{11}$$

where σ is the driving shear stress, which is constant through the band. From Equation (10) it may be seen that the velocity jump across the band, $\triangle v$, is related to δ and λ by

$$\frac{1}{2}\triangle v = \delta\lambda \tag{12}$$

Since the stress is also given in the present case by the constitutive law

$$\sigma = \kappa_0(1 - a\vartheta_c)(b\lambda)^m \tag{13}$$

Equations (11)–(13) may be solved for the central temperature, strain rate, and characteristic length, with the shear stress, σ, and the velocity jump, $\triangle v$, as driving parameters.

$$\begin{aligned} 1 - a\vartheta_c &= \frac{\sigma}{\kappa_0} \left\{ \frac{2km}{\kappa_0 ab(\frac{1}{2}\triangle v)^2} \right\}^{\frac{m}{(1-m)}} \\ \lambda &= \frac{1}{b} \left\{ \frac{\kappa_0 ab(\frac{1}{2}\triangle v)^2}{2km} \right\}^{\frac{1}{(1-m)}} \\ \delta &= \left\{ \frac{2km}{\kappa_0 ab^m} \right\}^{\frac{1}{(1-m)}} (\tfrac{1}{2}\triangle v)^{-\frac{1+m}{(1-m)}}. \end{aligned} \tag{14}$$

Now with σ and $\frac{1}{2}\triangle v$ replaced by s_{23} and w from the inertial similarity solution, respectively, the structure of the fully developed shear band is completely determined along its length.

Figure 4 shows a typical cross section of the temperature field (actually $1 - a\vartheta$) near the fully developed shear band. The upper curve is the inertial solution, the lower curve is the core solution, and the dashed connecting curve is an approximation (not actually computed) of the smoothing

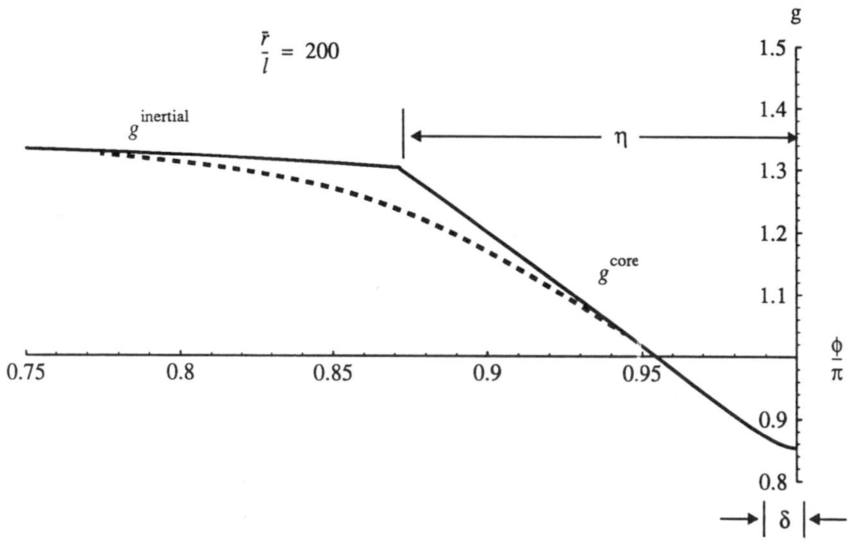

Figure 4: Solution profile for softening function g at $\frac{\bar{r}}{\ell} = 200$ and near to $\phi = \pi$.

effect of heat conduction in the intermediate region where the material could be treated as a rigid, semi-infinite heat conductor with a known heat source (temperature gradient) on its inner boundary.

If δ is regarded as the mechanical width of the shear band, then the intersection of the core and inertial solutions could be taken as a rough measure of the thermal width, η. The width δ marks the transition from high strain rate deformation of material to nearly rigid translation with a constant temperature gradient, as may be seen by examining Equation (10) for $y \gg \delta$. The thermal width, η, is found by setting $g^{(core)} = g^{(inertial)}$ and plotting the resulting curve, which is given intrinsically in polar coordinates by

$$
\begin{aligned}
1 \;=\; & \left\{\tfrac{1+m}{\bar{r}/\ell}\right\}^{\frac{m}{1-m}} \Gamma^{-1+m} W^{-\frac{2m}{1-m}} (W' \cos\phi + \tfrac{m}{1+m} W \sin\phi)\cdot \\
& \left\{1 + 2m \ln(\cosh\left\{(\tfrac{\bar{r}/\ell}{1+m}\cdot W^{1+m})^{\frac{1}{1-m}} \sin\phi\right\})\right\}
\end{aligned}
\tag{15}
$$

and is shown in Figure 5. In Equation (15) the length scale ℓ is given by $\ell = \frac{k}{\rho c U}$. For the numbers assumed previously, ℓ is 50 nm, so the thermal width comes out to be in the range of 3 to 4 μm. For the purposes of comparison the mechanical width, δ, is nearly an order of magnitude smaller. Owing to the simplistic constitutive law in Equation (1) these numbers are probably too small by a factor of five or ten compared to observed band configurations, but the general configuration predicted is entirely reasonable.

Since U and G_0 have not yet been determined, the end result is a two-parameter family of solutions for a propagating anti-plane sheer band where the undetermined parameters must be found by making the present solutions compatible with a complete boundary value problem.

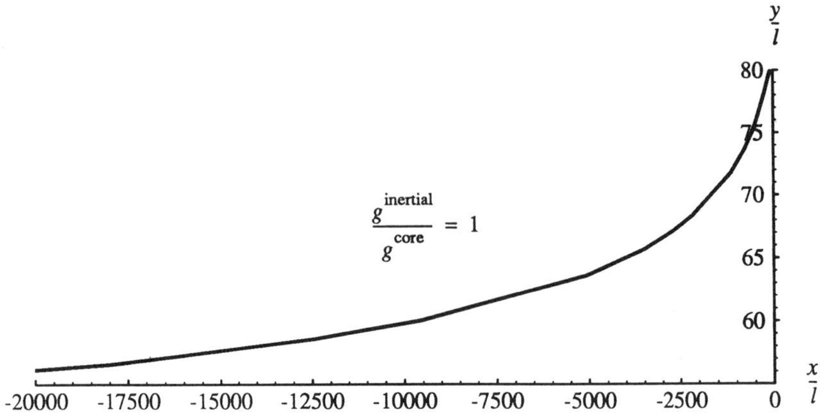

Figure 5: Thermal width as given implicitly by Equation (15).

4 References

[1977] J. Litonski, Plastic Flow of a Tube under Adiabatic Torsion, Bull. de l'Acad. Polonaise des Sciences, v. 25, pp 7-14.

[1988] R. C. Batra, Steady State Penetration of Thermoviscoplastic Targets, Comp. Mech. v. 3, pp 1 - 12.

[1988] A. Marchand and J. Duffy, An Experimental Study of the Formation of Adiabatic Shear Bands in a Structural Steel, J. Mech. Phys. Solids v. 36, pp 251-283.

[1993] J.G. Glimm, B.J. Plohr, and D.H. Sharp, A Conservative Formulation for Large-Deformation Plasticity, Appl. Mech. Rev. v. 46, pp 519-526.

ON NONLINEAR STABILITY OF CONTACT DISCONTINUITIES

ZHOUPING XIN

ABSTRACT. The large time asymptotic behavior toward contact discontinuities for the 1-dimensional compressible Euler equations with uniform viscosity is investigated in this note. It is found that for the polytropic gases, although a contact discontinuity is not an asymptotic attractor for the viscous system, a viscous wave pattern which approximates the contact discontinuity on any finite time interval can be constructed. This is shown to be asymptotically stable with small generic perturbations for the viscous system, provided that the strength of the contact discontinuity is suitably small. The asymptotic form of the viscous solutions can be obtained *a priori* by the distribution of the initial excessive mass. A generic perturbation of the contact discontinuity introduces a shift in the center of the diffusive contact wave, and nonlinear diffusion waves in the two sound wave families. A coupled linear diffusion wave located essentially in the region of the diffusive contact wave is needed to obtain the asymptotic ansatz. The stability is obtained by using a carefully designed weighted energy estimate.

1. INTRODUCTION

Consider the 1-dimensional compressible Euler equations in Lagrangian coordinates:

$$\begin{cases} v_t - u_x = 0, \\ u_t + p_x = 0, \qquad x \in \mathbb{R}^1, \quad t > 0 \\ E_t + (up)_x = 0, \end{cases} \tag{1}$$

where v, u, E, and p are specific volume, velocity, total energy, and pressure respectively. Denote by e the internal energy. Then $E = e + \frac{1}{2}u^2$. We will only consider the polytropic gases, i.e., we assume that the equation of state is given by

$$p(v,e) = \frac{e}{v}.$$

Courant Institute, 251 Mercer Street, New York, NY 10012. Email: xinz@cims.nyu.edu
The research is supported in part by a Sloan Foundation Fellowship, NSF grant DMS-93-03887, DOE Grant DE-FG02-88ER-25053.

A *contact discontinuity* solution to (1) is given by $\widetilde{U} = (\tilde{v}, \tilde{u}, \tilde{E})$ such that

$$\tilde{u} \equiv 0\,, \qquad \tilde{p} \equiv 1\,, \qquad \tilde{v}(x,t) = \begin{cases} v_\ell, & x < 0 \\ v_r, & x > 0 \end{cases} \tag{2}$$

where v_ℓ and v_r are positive constants, and $v_\ell \neq v_r$.

We will study the large time asymptotic behavior of solutions toward the inviscid discontinuity given in (2) for the corresponding viscous system of the form

$$\begin{cases} v_t - u_x = \varepsilon v_{xx}, \\ u_t + p_x = \varepsilon u_{xx}, \qquad x \in \mathbb{R}^1, \quad t > 0 \\ E_t + (up)_x = \varepsilon E_{xx}, \end{cases} \tag{3}$$

with initial data

$$(u, v, E)(x, 0) = (v_0(x), u_0(x), E_0(x)) \equiv U_0(x), \qquad x \in \mathbb{R}^1, \tag{4}$$

where ε is a positive constant. Thus let U_0 be a generic small perturbation of the contact discontinuity $\widetilde{U}$ given in (2). The main questions we would like to answer are:

(1) Is there a global (in time) solution to (3) and (4) for fixed $\varepsilon > 0$?
(2) When the solution $U(x,t) \equiv (v, u, E)(x,t)$ to (3) and (4) exists, what is the precise large time asymptotic behavior of $U(x,t)$? and can $\widetilde{U}(x,t)$ be an asymptotic state for the solutions of the viscous problem (3) and (4)?

A simple calculation for the linear heat equation shows that an inviscid contact discontinuity is only a *metastable* state for the viscous problem. Thus, despite the nonlinear modes in (3), one would expect that a similar phenomenon arises for our problem here. Indeed, our main result in this note, roughly speaking, states that *although the inviscid contact discontinuity $\widetilde{U}(x,t)$ is not an asymptotic state for (3) and (4), one can construct a smooth viscous contact wave, $\overline{U}(x,t)$, with the following properties*

(i) $\overline{U}(x,t)$ solves (3) exactly;
(ii) the estimate

$$\|\overline{U}(\cdot,t) - \tilde{U}(\cdot,t)\|_{L_p(\mathbb{R}^1)} \leq c(\varepsilon t)^{\frac{1}{2p}}$$

holds for a constant $c > 0$, $p \geq 1$;

(iii) $\overline{U}(x,t)$ is nonlinearly stable for the problem (3) and (4). Furthermore, the asymptotic structure of the solution to (3) and (4) is described by a shifted viscous contact wave superimposed with nonlinear diffusion waves in sound wave families, and the shift and the masses carried by the nonlinear diffusion waves can be determined *a priori* from the distribution of the initial excessive mass.

For more details, see Section 2.

Since (3) has uniform viscosity, it can be regarded only as a model for the compressible Navier-Stokes equations. However, we expect that similar results hold for the Navier-Stokes equation, at least for the polytropic gases. We should also remark that the asymptotic behavior toward contact discontinuity of (3) itself is *interesting* since (3) governs the leading order behavior of the Lax-Friedrichs discretization of the Euler equations (1).

There have been intensive studies on the large time asymptotic behavior for solutions to systems of viscous conservation laws including (3), see [3,4,6,7]. Almost all the results deal with the nonlinear stability of constant states, shock profiles, and rarefaction waves. The main difficulty in analyzing contact waves lies in the fact that contact waves are less stable [2]. Since a generic perturbation of the linear contact wave introduces nonlinear waves in the two sound wave families, the propagation and interaction of these nonlinear waves with the linear contact wave is the major issue here.

The rest of the paper is organized as follows. In Section 2, we first construct the asymptotic ansatz as a superposition of a shifted viscous contact wave, two nonlinear diffusion waves in the sound families, and the coupled linear diffusion wave. The viscous constant wave, which approximates the inviscid contact discontinuity, is constructed by diffusing out the discontinuity in the contact manifold. The coupled linear diffusion wave is introduced to take into account the wave interactions. The pointwise estimate for the coupled linear diffusion wave is presented, whose proof will appear elsewhere. The general solution of (3) and (4) can be decomposed as sum of this ansatz with a remainder. We then state our precise stability theorem. The sketch of the proof of the stability theorem by weighted energy method is presented in Section 3.

2. Asymptotic Ansatz and the Stability Theorem

To describe the asymptotic ansatz, we first introduce the *viscous contact wave*, $\overline{U}(x,t)$, as the solution to the problem:

$$\overline{u}(x,t) \equiv 0, \quad p(\overline{v},\overline{e}) = \overline{p} \equiv 1, \quad \begin{cases} \overline{v}_t = \varepsilon \overline{v}_{xx}, \\ \overline{v}(x,-1) = v_\ell + (v_r - v_\ell)H(x), \end{cases} \tag{5}$$

here $H(x)$ is the Heaviside function. It is clear from (5) that $\overline{U}(x,t)$ has the following properties.

Lemma 2.1. *The viscous contact wave $\overline{U}(x,t)$ in (5) satisfies:*

(*i*) *$\overline{U}(x,t)$ solves system (3) exactly;*

(*ii*) $\|\overline{U}(\cdot,t) - \tilde{U}(\cdot,t)\|_{Lp} = 0(1)(\varepsilon(1+t)^{\frac{1}{2p}}$ *for all* $p \geq 1$;

(*iii*) $\overline{v}_x(x,t) < 0 (> 0)$ *if* $v_r < v_\ell (v_r > v_\ell)$.

Thus $\overline{U}(x,t)$ gives the desired viscous approximation to the inviscid contact discontinuity. A generic perturbation of the linear contact wave should also introduce waves in the sound wave families. These are the nonlinear diffusion waves introduced by Liu [3], which can be explicitly written as

$$\theta_i(x,t) = \frac{1}{\sqrt{t+1}} c\left(\frac{x - \lambda_1(v_{io})(t+1)}{\sqrt{t+1}}\right), \quad i = 1,3, \tag{6}$$

with

$$c(y) = \frac{\alpha_i^{-1}(\exp\{\alpha_i m_i/2\} - 1)\exp\{-y^2\}}{\sqrt{\pi} + (\exp(\alpha_i m_i/2) - 1)\int_y^\infty \exp\{-\xi^2\}d\xi},$$

where $\lambda_1, \lambda_2 = 0, \lambda_3$ are the eigenvalues of the Jacobian matrix of the flux in (1) with corresponding left and right eigenvectors r_i and ℓ_i $(i = 1, 2, 3)$ respectively, $v_{1o} = v_\ell$ and $v_{3o} = v_r$, $\alpha_i = \nabla\lambda_i(v_{io}) \cdot r_i(v_{io})$, and $m_i = \int \theta_i(x,t)dx$ is the mass carried by θ_i. It should be noted that $\theta_i(x,t)$ in (6) solves the convected Burger's equation (with speed $\lambda_i(v_{io})$). We set

$$\theta(x,t) = \theta_1(x,t)r_1(v_\ell) + \theta_3(x,t)r_3(v_r). \tag{7}$$

To handle the interactions among decoupled diffusive sound waves and the viscous contact wave; we need to introduce a *coupled diffusion wave*, $D(x,t)$, as the solution to the following problem:

$$\begin{cases} d_{1t} - d_{2x} - \varepsilon d_{1xx} = (\tilde{s}_1)_x \\ d_{2t} + \left(\frac{1}{\overline{v}}(d_3 - d_1)\right)_x - \varepsilon d_{2xx} = (\tilde{s}_2)_x - (\nabla^2 p(\overline{U})(\theta, D))_x \\ d_{3t} + d_{2x} - \varepsilon d_{3xx} \\ \qquad = (\tilde{s}_3)_x - \left(2\frac{d_2}{\overline{v}}(\theta_1 + \theta_3) + \left(\sqrt{\frac{2}{v_r}}\theta_3 - \sqrt{\frac{2}{v_\ell}}\theta_1\right)\frac{1}{\overline{v}}(d_3 - d_1)\right)_x \\ D(x,0) = D_0(x) \equiv 0, \end{cases} \tag{8}$$

where

$$(\tilde{s}_1, \tilde{s}_2, \tilde{s}_3) = (\frac{1}{2}(\alpha_1\theta_1^2 + d_3\theta_3^2), -\frac{1}{6}(\alpha_1\theta_1^2 r_{12}(v_\ell) + \alpha_3\theta_3^2 r_{32}(v_r), \frac{5}{6}(\alpha_1\theta_1^2 + \alpha_3\theta_3^2),$$

is the major source term due to wave interactions. By a delicate approximate parametrix method, we can obtain the following optimal global pointwise estimate for the solution of (8).

Proposition 2.2. *Let $D(x,t)$ be the coupled diffusion wave given by (8). Denote by*

$$R(\overline{U}) = (r_1(\overline{U}), r_2(\overline{U}), r_3(\overline{U}))$$

the matrix of the right eigenvectors of the Jacobian matrix of the fluxes in (1). Set

$$D \equiv R(\overline{U})_x\xi + R(\overline{U})\xi_x = (R(\overline{U})\xi)_x. \tag{9}$$

Then for suitably chosen N, we have

$$(10)\quad (|\xi|,|\xi_x|)=\begin{cases} O(1)\exp\{(-c(|x|+t)\}, & |x|\geq c'(t+1)\\ O(1)\left(\dfrac{\log(1+t)}{(1+t)^{1/4}},\dfrac{\log(1+t)}{(1+t)^{3/4}}\right) & \text{on } I_i(t)\\ O(1)\left(\frac{\log(1+t)}{(\min\limits_i|x-\lambda_i t|+\sqrt{t})^{1/2}},\frac{\log(1+t)}{\sqrt{1+t}(\min\limits_i|x-\lambda_i t|+\sqrt{t})^{1/2}}\right) & \text{otherwise}\end{cases}$$

where

$$I_i(t)=\{x\cdot|x-\lambda_i(v_{io})(t+1)|\leq CN\sqrt{t}\log(1+t)\}.$$

The proof of this proposition is based on an approximate parametrix method making use of scalar duals and crucial structures of the underlying waves. The details are very technical and will appear elsewhere.

We now decompose the initial excessive mass as follows:

$$(11)\qquad \int_{-\infty}^{+\infty}(v_0-\overline{u})(x,0)dx=x_0(v_r-v_\ell)r_2(\overline{u})+m_1r_1(v_\ell)+m_3r_3(v_r).$$

Then for $\delta\equiv|v_r-v_\ell|$ small, (11) determines x_0, m_1, and m_2 uniquely. With x_0, m_1 and m_3 so determined, we define the asymptotic ansatz for the solution of (3) and (4) as

$$(12)\qquad U^a(x,t)=\overline{U}(x+x_0,t)+\theta(x,t)+D(x,t).$$

It follows from the definition that

$$\int_{-\infty}^{\infty}(U_0(x)-U^a(x,0))dx=0$$

since

$$\int_{-\infty}^{+\infty}(\overline{U}(x+x_0,t)-\overline{U}(x,t)dx=x_0(v_r-v_\ell)r_2(\overline{U}).$$

Due to the conservation laws for θ, D and $\overline{U}$, we have for all t

$$(13)\qquad \int_{-\infty}^{\infty}(V(x,t)-U^a(x,t))dx=0$$

provided $U(x,t)$ is a solution to (3) and (4).

Our main theorem then states that there exists a unique solution to the problems (3) and (4) in the neighborhood $U^a(x,t)$ and whose asymptotic behavior is completely determined by $U^a(x,t)$. More precisely, we have

Theorem 2.3. *Let $U^a(x,t)$ be the function constructed above, and ε be a fixed positive constant. Then there exist positive constants δ_1 and δ_2 such that if*

(*i*) $U_0-\overline{U}\in H^1(\mathbb{R}^1)$,

(*ii*) $\delta=|v_r-v_\ell|\leq\delta_1$,

(*iii*) $\left|\int_{\mathbb{R}^1}(U_0-\overline{U})(x)dx\right|+\int_{\mathbb{R}^1}(1+x^2)|U_0-\overline{U}|^2dx\leq\delta_2$,

then there exists a unique smooth global solution, $U(x,t)$, to the Cauchy problem (3) and (4) with the following properties:

(*a*) $U-\overline{U}\in C([0,\infty):\ H^1(\mathbb{R}^1))$,

(*b*) $U-U^a\in C([0,\infty);H^1(\mathbb{R}^1))\cap L^2([0,\infty),H^1(\mathbb{R}^1))$,

(*c*) $\lim_{t\to+\infty}\|U(\cdot,t)-\overline{U}(\cdot+x_0,t)\|_{L^p}=0$ *for all* $p\geq 2$,

(*d*) $\lim_{t\to+\infty}\int_t^{+\infty}\int_{\mathbb{R}}|U(x,t)-U^a(x,t)|^2dxdt=0$.

A few remarks are in order.

(1) The coupled diffusion wave defined by (8) is not only a technical device. Indeed, although D decays faster than the nonlinear decoupled diffusion waves pointwise, we have

$$\int_0^{\infty}\int_{-\infty}^{+\infty}|D(x,t)|^2dxdt=+\infty,$$

and estimate *(d)* in the Theorem 2.3 cannot hold true without the coupled diffusion waves.

(2) In Theorem 2.3, the decay estimates to the viscous contact wave are given in average norms, based on a simple weighted energy analysis outlined in Section 3. However, using an approximate parametrix method as for the coupled diffusion wave given in (8) and an elaborate iteration procedure, we can also obtain a detailed pointwise decay estimate for viscous contact waves. This will be reported in a forthcoming paper [5].

(3) Though Theorem 2.3 deals with system (3) which is of uniform viscosity, one would expect that similar results apply to the 1-dimensional Navier-Stokes equations for the polytropic gases. This is currently under study [1].

3. Weights and Stability Analysis

By assuming Proposition 2.2, whose proof is the most technical part of this study, the stability theorem can be obtained rather easily by a weighted energy method outlined below.

Let $U(x,t)$ be a smooth solution to the Cauchy problem (3) and (4). We decompose U as

$$U(x,t)=U^a(x,t)+V(x,t), \tag{14}$$

where $U^a(x,t)$ is the asymptotic ansatz constructed in Section 2. A simple computation, using the construction of U^a, shows that U^a solves

$$\begin{cases} v_t^a-u_x^a-\varepsilon v_{xx}^a=0\\ u_t^a+(p(U^a))_x-\varepsilon u_{xx}^a=(\xi_1)_x\\ E_t^a+(u^ap(U^a))_x-\varepsilon E_{xx}^a=(\xi_2)_x,\end{cases} \tag{15}$$

here the typical terms in the errors ξ_1 and ξ_2 are

$$|D|^2, \quad |\theta|^2|D|, \quad |\theta|\,|D|, \text{ and } |\overline{U}+\theta|\,|\theta|\,|D|,$$

which can be estimated by Proposition 2.2. It follows from (13) that

$$\int_{\mathbb{R}^1} V(x,t)dx = 0 \quad \text{for all } t \geq 0$$

Thus setting

$$V(x,t) = \Phi_x(x,t), \tag{16}$$

one can reformulate the Cauchy problem (3) and (4) as

$$\begin{cases} \phi_{1t} - \phi_{2x} - \varepsilon\phi_{1xx} = 0 \\ \phi_{2t} + p'(U^a)\Phi_x - \varepsilon\phi_{2xx} = -\widetilde{\xi}_1 \\ \phi_{3t} + p(U^a)\phi_{2x} + u^a p'(U^a)\Phi_x - \varepsilon\phi_{3xx} = -\widetilde{\xi}_2, \end{cases} \tag{17}$$

with initial data

$$\Phi(x,0) = \Phi_0(x) \in H^2(\mathbb{R}^1) \tag{18}$$

where $\widetilde{\xi}_i = \xi_i + O(1)|\Phi_x|^2 + O(1)(|D|\,|\Phi_x|)$ are modular small terms. We need to show that the Cauchy problem (17) and (18) has a unique global solution which decays to zero uniformly as time approaches infinity.

Let A^a be the convection matrix in (17), and $R^a = (r_1^a, r_2^a, r_3^a)$ and $L^a = (\ell_1^a, \ell_2^a, \ell_3^a)^t$ be the right and left eigenvector matrices respectively, so that

$$L^a A^a R^a = \Lambda^a = diag\{\lambda_1^a, 0, \lambda_3^a\}.$$

In terms of the characteristic variables

$$\Psi = L^a \Phi, \tag{19}$$

system (17) becomes

$$\Psi_t + \Lambda^a \Psi_x - \varepsilon \Psi_{xx} = \{\dots\}, \tag{20}$$

where $\{\dots\}$ denotes the coupling terms, nonlinear terms and error terms. To handle the wave interactions and obtain the desired basic energy estimate, we define the weight functions for each characteristic field as follows. Set

$$W(x,t) = \text{diag}\{W_1, W_2, W_3\}(x,t), \tag{21}$$

and

$$W_i(x,t) = \prod_{i=1}^{3} w_{i,j}(x,t)$$

where w_{ij} is designed to take into account the effect of the j-wave on the i-th field. More precisely, we define

(1) $w_{i,i}, (i = 1,2,3)$ by

$$w_{2,2} \equiv 1$$
$$w_{i,i} = \exp\left\{-\frac{\alpha_i}{2}\int_{-\infty}^{x}\theta_i(y,t)dy\right\}, \quad i = 1,3,$$

for self-interaction;
(2) $w_{i,j}, (i \neq j, j = 1,3)$, satisfying

$$\begin{cases} (\lambda_j - \lambda_i)w_{i,jx} = q_i(1+t)^\beta|\theta_{j,x}|w_{i,j}, & \lambda_k = \lambda_k(v_{k_0}), \\ \lim\limits_{x\to\infty} w_{i,j} = 1 \end{cases}$$

for the coupling of nonlinear diffusive waves with other fields; and
(3) $w_{j,2}, (j = 1,3)$, by

$$\begin{cases} w_{1,2} = (\lambda(\overline{v}(x,t)))^q, \\ w_{3,2} = (\lambda(\overline{v}(x,t)))^{-q} \\ \lambda(\overline{v}) = -\lambda_1(\overline{v}) = \lambda_3(\overline{v}) \end{cases}$$

for the effect of the contact wave on the sound wave families. Here $|q|$ is a large integer to be chosen (the sign of q depends on the sign of $v_r - v_\ell$). With this construction, one can show:

Lemma 3.1. *For δ suitably small, one can always choose the constants q_i, β, and q in (21) such that*

$$\begin{cases} -W_{2t} - \varepsilon W_{2xx} \geq -O(1)\dfrac{\theta_o}{(1+t)^\gamma} + q_2(1+t)^\beta \sum\limits_{i\neq 2}|\theta_{ix}| \\ -W_{it} - (\lambda_i^a W_i)_x - \varepsilon W_{ixx} \geq -O(1)\dfrac{\theta_0}{(1+t)^\gamma} \\ \qquad + q_1(1+t)^\beta \sum\limits_{\substack{j\neq i \\ j\neq 2}}|\theta_{jx}| + q_3|\lambda_x(\overline{U})|, \quad i = 1,3, \end{cases} \tag{22}$$

for some constant $\gamma > 1$, where $\theta_0 = |m_1| + |m_3|$.

With the weights so determined, one can take the inner product of (20) with $\Psi^t W$ and integrate the resulting equation over $\mathbb{R}^1$ to obtain, after applying (22),

$$\begin{aligned} &\frac{1}{2}\frac{d}{dt}\int \Psi^t W\Psi dx + O(1)\int q_3|\lambda_x(\overline{u})|(\Psi_1^2 + \Psi_3^2)dx + \int \Psi_x^t W\Psi_x dx + \\ &\quad + q_1(1+t)^\beta \int (|\theta_{1x}|\Psi_3^2 + (|\theta_{1x}| + |\theta_{3x}|)\Psi_2^2 + |\theta_{3x}|\Psi_1^2)dx \\ &\quad \leq \int \Psi^t W\{\dots\}dx + O(1)\frac{\theta_0}{(1+t)^\gamma}\int \Psi^t W\Psi dx. \end{aligned} \tag{23}$$

By analyzing the coupling terms in $\{\dots\}$ carefully and choosing the appropriate constants q_i, q and β, one can obtain the following desired energy estimate.

Proposition 3.2. *There exists a positive constant $O(1)$ independent of time t such that*

$$\begin{aligned}(24)\qquad &\|\Psi(\cdot,t)\|^2+\int_0^t\int|\lambda_x(\overline{u})|(\Psi_1^2+\Psi_3^2)(x,\tau)dxd\tau+\int_0^t\int|\Psi_x|^2(x,\tau)dxd\tau\\ &+\int_0^t\int(1+\tau)^\beta\sum_{j\neq i}|\theta_{ix}|\Psi_j^2dxd\tau\le O(1)\{\|\Psi(x,0)\|^2+\theta_0\}\end{aligned}$$

provided δ_i and $\|\psi(x,t)\|_{L^\infty}$ are suitable small.

Based on the *a priori* estimate (24), the proof of Theorem 2.3 can be completed in a standard way ([3,4,6]).

Acknowledgment. I am grateful to Professor D. Hoff for many useful comments, suggestions and fruitful discussions on aspects of this work.

References

1. D. Hoff and Z. Xin, in preparation.
2. T.-P. Liu, *Linear and nonlinear large time behavior of general system of hyperbolic conservation laws*, Comm. Pure Appl. Math. **30** (1977), 767–796
3. T.-P. Liu, *Nonlinear Stability of Shock Waves for Viscous Conservation Laws*, Memoirs AMS, **56**, #308 (1985).
4. T.-P. Liu and Zhouping Xin, *Nonlinear stability of rarefaction waves for compressible Navier-Stokes Equations*, Comm. Math. Phys., Vol. 118 # (1988), 451–465.
5. T.-P. Liu and Zhouping Xin, *Pointwise decay to contact discontinuity for systems of viscous conservation laws*, in preparation.
6. A. Szepessy and Zhouping Xin, *Nonlinear stability of viscous shock waves*, Arch. Rat. Mech. Anal., **122** (1993), 53–103.
7. A. Szepessy and K. Zumbrum, *Stability of viscous rarefaction wave*, this proceeding.

SOLUTIONS TO THE EULER EQUATIONS WITH LARGE DATA

BLAKE TEMPLE[1] AND ROBIN YOUNG[2]

ABSTRACT. We consider the large time existence problem for 3×3 systems of conservation laws with large total variation and small oscillation. Under these conditions it is known that general solutions can blow up in finite time due to destabilizing geometric wave interaction effects. We discuss a general framework for rigorously addressing this problem, and provide a sufficient condition for existence of solutions. We then restrict to a class of systems which includes the Euler equations, having simplified wave interactions. We obtain an explicit path integral formula for approximate solutions. We then find time–dependent bounds for the total variation. The variation grows exponentially in time, and known examples show that our estimates are sharp.

1. INTRODUCTION

We consider the behavior of solutions to 3×3 systems of strictly hyperbolic conservation laws

$$u_t + f(u)_x = 0, \qquad u(x,0) = u_0(x), \tag{1}$$

in one space dimension x, with initial data having total variation V_0 and sup-norm $\epsilon > 0$. We address the problem of long time existence for small ϵ and large V_0. In general the time of existence of solutions is highly dependent on geometric features of the conservation law, determined by the Lie brackets of eigenvectors of the Jacobian matrix Df. For gas dynamics, symmetries arise as a consequence of the Second Law of Thermodynamics, and our primary interest here is to explore the role that such symmetries play in the large-time existence problem for the Euler equations.

Our analysis uses Glimm's method, for which compactness is based on a bound for the total variation of approximate solutions [1]. Glimm obtained this bound as long as V_0 is small, or if ϵV_0 is small when there is a full set of Riemann coordinates

[1]Department of Mathematics, Univ. of California, Davis CA 95616.
Supported in part by NSF Applied Mathematics Grant Number DMS–92–06631, in part by ONR, US NAVY grant number N00014-94-1-0691, by a Guggenheim Fellowship, and by the Institute of Theoretical Dynamics, UC-Davis.
[2]Department of Applied Mathematics, SUNY at Stony Brook, Stony Brook NY 11794.
Partially supported by DOE grant number DE-FG02-88ER25053 at the Courant Institute, and by NSF grant number DMS-9201581 and DOE grant number DE-FG02-90ER25084.

(so that Lie brackets vanish identically.) In either of these cases, solutions decay due to shock formation and interaction [2, 4], but new methods are required when the total variation is large and the Lie brackets are non-zero.

There are two different types of nonlinearity in systems of three or more conservation laws. The first is the well-known dependence of wave-speed on the solution at each point, known as genuine nonlinearity [3]. This leads to the formation of shock waves and the subsequent decay of solutions in each family. Weak solutions must be introduced to continue the solution beyond the time of shock wave formation. The other nonlinearity is the geometric nonlinear coupling of different wave families [7]. In systems of 3 equations, this coupling is quadratic, and leads to the generation of new waves of quadratic strength when waves of different families interact. We emphasize that genuine nonlinearity is stabilizing, while geometric nonlinearity is destabilizing.

In recent years some progress has been made in understanding the nature of the geometric nonlinearity, both through the method of weakly nonlinear geometric optics, and directly through Glimm's method. The most dramatic results show the resonant blow-up of solutions to general 3×3 systems. This clearly indicates that extra assumptions must be made to ensure nonlinear stability of solutions.

The method of reorderings was developed in [7] to account for quadratic effects of wave interactions, which arise due to geometric coupling between wave fields. Using this framework, a potential for the oscillation was built by including quadratic terms which correspond to future interactions. After interaction, the class of future interactions is smaller, so that the potential decreases, up to errors which are cubic. This led to a separation of the total variation and sup-norms, but those results still required that the total variation lie below some threshold. In order to remove this restriction, we must account for cumulative quadratic effects. Here we describe an extension of the method and derive an algebraic condition which is sufficient to guarantee existence of solutions.

The crucial modification is the definition of a new error potential, analogous to Glimm's, but including all cumulative quadratic effects. The defining property of this functional expresses the fact that the error in a series of interactions should be a sum of individual errors. This is a statement of 'conservation of interaction errors.' We then isolate a continuity property, needed to take higher order errors into account. The important difference between our error potential and Glimm's is that we include the contribution of each interacting pair using wave strengths at the time of interaction rather than initially, thus including all quadratic effects of earlier interactions.

Since our error functional takes quadratic terms into account, continuity implies that the error potential decreases if the sup-norm ϵ is small enough, rather than the total variation. Indeed, we show that continuity of the error potential is a sufficient condition for existence of solutions. We remark that boundedness of the functional is a consequence of continuity. Since solutions to general systems may be unbounded, the error potential cannot be continuous in general.

To obtain precise results for gas dynamics, we restrict consideration to a subclass of systems that includes the 3×3 Euler equations. We assume that the second (entropy) field possesses a Riemann coordinate, and for convenience assume that it has constant wave-speed. This assumption greatly simplifies reordered sequences by simplifying pairwise wave interactions, and allows us to model entropy fluctuations as a background source for scattering sound waves, which does not change across reorderings.

In our quadratic model, sound waves add linearly, so that each sound wave can be treated separately. We thus examine the scattering pattern of a single sound wave passing through the fluctuating entropy field. We describe the scattering process in terms of paths of scattered waves, each given an appropriate strength. Combining these linearly, we get a path integral formula for reordered sequences. We thus have an algebraic expression for the piecewise constant approximate solutions in Glimm's scheme which includes all quadratic effects of wave interactions.

Our final task is to use the path integral formula to find bounds for the reordered sequences, and thus show that the functionals are continuous. By bounding the strengths of scattered waves and carefully counting paths, we find time-dependent bounds, with exponential growth for the total variation. We do not expect solutions of gas dynamics to have exponential growth, as the fact that the entropy is convex (which we have not used) introduces extra symmetries, leading to extra cancellation due to different paths, see [6]. On the other hand, there are known examples which satisfy our hypotheses and exhibit exponential growth [8], so that our estimates are sharp.

To find the bounds, we assume a bound for 2-wave strengths, and count the number of paths. In this paper, we present a simplified argument which yields the bound as long as the entropy remains Lipschitz. A more sophisticated path count allows this condition to be dropped, and yields new physics. This paper is a summary of results in [5], and the interested reader is referred to that paper for details.

2. Reorderings

We briefly recall the method of reorderings in Glimm's scheme, and describe the strengths of reordered waves [7]. The Glimm approximations consist of piecewise constant states separated by elementary waves, which are shocks, rarefactions or contact discontinuities. Our notation is as follows. For a fixed mesh size Δx, we consider the sequence of constant states $\{u_0, \ldots, u_n\}$ separated by the wave sequence $\gamma = (\gamma_1, \ldots, \gamma_n)$ of index $(c_1, \ldots, c_n)$. This means that the wave separating constant states u_{i-1} and u_i is a wave of strength γ_i from the c_i-th family. Here a positive strength means that the wave is a rarefaction, while a negative strength indicates a shock, and we do not distinguish between a wave and its strength. The total variation and sup-norm of the solution can then be measured by

$$V(\gamma) = \sum |\gamma_i| \quad \text{and} \quad S(\gamma) = \max \Big| \sum_{c_i = r} \gamma_i \Big|, \tag{2}$$

respectively, the last sum being taken over sets of consecutive waves of each family.

One of the main ideas in Glimm's scheme is to use induction to reduce complicated large-time wave interaction patterns to the interaction of two Riemann solutions. If α_j and β_j, $j = 1 \ldots 3$, represent the two Riemann solutions, then the resulting Riemann solution is represented by

$$\epsilon_i = \alpha_i + \beta_i + \sum_{j>k} \alpha_j \beta_k \Lambda_i^{jk} + O(1) SD(\alpha, \beta), \tag{3}$$

where the Λ_i^{jk} are (constant) interaction coefficients measuring the strength of geometric nonlinearity, defined by the Lie bracket of eigenvector fields. The local interaction error D is defined by

$$D(\alpha, \beta) = \sum_{App} |\alpha_j \beta_k|, \tag{4}$$

the sum being over all approaching pairs of waves. Our main concern is the appropriate global extension of this error potential.

The method of reorderings is a tool for accounting for the quadratic effects of interactions in Glimm's scheme, through changes in wave sequences. Each pairwise interaction of waves is modeled by transposing the wave pair (in case they are from different families), or combining them into one (if they are of the same family). The only quadratic effect occurs when waves of different families cross, in which case new waves are generated. Instead of introducing a new wave into the model, we adjust the strength of the nearest wave of that family by the appropriate amount; this exactly models Glimm's scheme, in which the number of waves is conserved.

A surjective map $\tau : \{1, \ldots, n\} \to \{1, \ldots, m\}$ defines an action on sequences of n waves through permutation of subscripts. That is, τ acts on the wave sequence and its index by shifting the wave γ_i from position i to position $\tau(i)$. As long as domains and ranges match up correctly, we can define compositions of maps, and so we can factor maps (non-uniquely) into products of elementary maps. These are the transposition or merge of a single adjacent pair of waves.

To extract more information from the map τ, we define the *crossing set* C_τ by

$$C_\tau = \{ (j, k) \mid j < k \text{ and } \tau(j) > \tau(k) \}, \tag{5}$$

which identifies those pairs of waves which cross under the action of τ, and similarly define the *merge set*

$$M_\tau = \{ (j, k) \mid j < k \text{ and } \tau(j) = \tau(k) \}, \tag{6}$$

and the union of these as the interaction set I_τ. Using these sets, we characterize the set of 'physical' admissible reorderings, as follows. In order for a pair of waves to cross, that is to pass through each other, the wave on the left should travel faster than that on the right, so we require that $c_j > c_k$ for all pairs $(j, k) \in C_\tau$. Similarly, only waves in the same family may merge to form one wave, so we require that if $(j, k) \in M_\tau$ then $c_j = c_k$. In particular, this condition implies that the reordered index of a sequence is single-valued. Note that the class of admissible reorderings depends only on the index of the wave sequence, so that perturbing wave strengths does not change the class of reorderings.

In factoring reorderings into interactions of adjacent wave pairs, we allow only admissible factors, and treat different factorizations as different reorderings. Each factorization of $\tau = \lambda_t \ldots \lambda_1$ determines a time-like ordering on the set I_τ of interactions which make up the reordering, by ordering pairs corresponding to each factor from right to left. Similarly, to each interacting pair we can associate a 'sub-ordering', namely the largest ordering $\mu = \lambda_m \ldots \lambda_1$ under which that pair does not interact.

To accurately model the evolution of the approximate solutions, we must consider changes in the strengths of waves as well as their relative positions. We define the action of a reordering on wave strengths inductively, so we need only describe the action of a single join or transposition of waves. Since our estimates for local interactions are exact to second order, we treat all cubic terms as errors, and ignore them in the model. Interacting waves of the same family add linearly with cubic errors, so if the join ϕ maps p and $p+1$ onto p, then the new wave γ'_p has strength $\gamma'_p = \gamma_p + \gamma_{p+1}$, while no other strengths change (although their positions may shift).

It remains to model the effect of a single transposition, which is to produce a wave of the third family whose strength is the weighted product of incident wave strengths. We do not allow a new wave to be created, but rather adjust the strength of the nearest wave of the appropriate family. This is analogous to sampling in the Glimm scheme, since only 3 waves leave each interaction diamond. The wave whose strength is to be adjusted when a particular pair interacts is determined by an interaction map. These maps are inductively defined on the set C_τ of crossing pairs, and the image $\imath^\tau(j,k)$ of a crossing pair is that wave of the third family whose strength is to be changed. We thus describe the effect of the transposition κ of γ_k and γ_{k+1} on the wave γ_i, where $i = \imath^\kappa(k, k+1)$, as

$$\gamma'_i = \gamma_i + \Lambda_{c_i} \gamma_k \gamma_{k+1}, \tag{7}$$

while other strengths are unchanged.

Having described the effect of elementary interactions, we use the factorization to inductively define the reordered wave sequence, so that $\tau\gamma = \lambda_t(\ldots\lambda_1(\gamma))$. We have thus modeled the evolution of approximate solutions, including cumulative quadratic effects.

For the class of solutions we consider, quadratic terms dominate higher order effects, so that control of these terms allows us to deduce existence of solutions. Each space-like curve in Glimm's scheme has an associated reordering, so that the waves crossing a curve can be approximated by a reordered sequence. We use these sequences to estimate norms of the solution at large times. In particular, we define the functional F for the (future) total variation to be

$$F(\gamma) = \sup_\tau V(\tau\gamma), \tag{8}$$

where γ is the sequence of waves initially, $\tau\gamma$ is the reordered sequence, and the supremum is over admissible reorderings. This functional is analogous to the functional P for the sup-norm, introduced in [7]. Since our bounds are time–dependent, we further restrict the class of reorderings to those whose interactions take place within (some large) time T.

3. Error Potential

Glimm's scheme is based on the construction of a decreasing global error potential. This is a measure of interaction errors for future wave interactions, and is defined by

$$G = \sum_{App} |\alpha||\beta|, \tag{9}$$

where α and β are wave strengths, and the sum is over all approaching wave pairs. The key idea is that after an interaction, the local interaction error D is balanced by the removal of terms from G, as fewer waves approach. Thus although the total variation (and other norms) may grow, some combination $V + cG$ is decreasing, so that the total variation is bounded. The change in G across a diamond is

$$\Delta G \leq -D + O(1)VD, \tag{10}$$

where D is the local error (removed from G), and the second term is due to higher order errors. Thus in order for G to decrease, we require the variation V to be small.

Our goal is to write down a similar error potential which takes into account quadratic effects, and thus allow larger total variation. The functional G includes all future interactions, but with first order accuracy in wave strengths. We define a functional Q which includes all quadratic terms by means of reorderings. That is, we consider approaching wave pairs, but instead of using initial wave strengths, we use the wave strengths at the time of interaction. Suppose we are given a wave sequence γ and reordering τ, together with an interacting pair (j, k). If μ is the subordering associated to this pair, then we let γ_j^μ be the wave in $\mu\gamma$ corresponding to γ_j; this is the strength of γ_j when that wave interacts with γ_k. We define the error potential by

$$Q(\gamma, \tau) = \sum_{I_\tau} |\gamma_j^\mu||\gamma_k^\mu|, \quad \text{and} \quad Q = \sup_\tau Q(\gamma, \tau), \tag{11}$$

where μ is different for each interacting pair $(j, k) \in I_\tau$. We have the functional relation

$$Q(\gamma, \sigma\tau) = Q(\tau\gamma, \sigma) + Q(\gamma, \tau) \tag{12}$$

for the composition of reorderings, which is the statement that the error potential decreases by the local error after interaction. This functional relation thus expresses the 'conservation of interaction error.' Again, since our bounds are time-dependent, we restrict ourselves to those reorderings spanning time T.

To deal with higher order effects, we consider continuity of the functional. Suppose we are given sequences γ and γ' with the same index (and reorderings), but which differ in a single wave strength, say $\gamma_q \neq \gamma'_q$. We say that Q is continuous, if

$$Q(\gamma,\tau) - Q(\gamma',\tau) \leq K_Q|\gamma_q - \gamma'_q|, \tag{13}$$

where K_Q does not depend on the reordering τ or mesh-dependent parameters, although it may depend on time and the initial data.

Across a diamond, the local interaction error is $O(1)SD$, so that continuity gives

$$\Delta Q \leq -D + O(1)SK_Q D, \tag{14}$$

Thus for a decreasing error potential, we require SK_Q to be small. The following theorem holds for general systems, and says that the large–time behavior of solutions can be determined by the quadratic model.

Theorem 1. *Suppose that the functional Q is continuous, with K_Q depending on time T and norms of the initial data. Then given V_0, there is an $\epsilon > 0$ such that if the initial data satisfies*

$$\|u_0\|_\infty \leq \epsilon \quad \text{and} \quad TV(u_0) \leq V_0, \tag{15}$$

then a weak solution with bounded oscillation and total variation exists up to time T. Moreover, the initial bound for the functional F yields a bound for the total variation of the solution.

Here ϵ is defined in terms of K_Q and V_0, so any restrictions made in finding K_Q apply to the solution. Continuity and boundedness of the functionals are essentially equivalent, so only one of these needs to be checked. Since there are known counterexamples, the functionals F and Q will not have time–independent bounds. Since F and Q are defined in similar ways, $Q \approx F^2$, we make the following conjecture.

Conjecture. *For general systems, boundedness of the functional F independent of mesh parameters is necessary and sufficient for existence of solutions. Any assumptions on the initial data and time of existence used in bounding F translate into conditions on solutions to the Cauchy problem.*

4. Path Integral Formula

Although we have found an algebraic criterion for existence of solutions, this is difficult to analyze without extra assumptions. In general, reordered sequences become unmanageable, and we know that the functionals are unbounded. Motivated by gas dynamics, we make the assumption that our system possesses a Riemann coordinate w. This means that ∇w is a left eigenvector,

$$\ell \cdot Df = \lambda \ell, \quad \text{where} \quad \ell = \nabla w, \tag{16}$$

so that w does not change across rarefactions of other wave families. We also assume that the family corresponding to w has constant wave-speed λ. This is true for gas dynamics if the Lagrange formalism is used, and allows us to derive a scalar transport equation for smooth solutions, namely

$$w_t + \lambda(u) w_x = 0. \tag{17}$$

Note that this is not the same as assuming the existence of an entropy–flux pair.

As a consequence of our assumptions, the second (entropy) field is degenerate in two ways : first, since we have a Riemann coordinate, the coefficient Λ_2 vanishes, so that (at the quadratic level) no new wave is generated when two sound waves interact. Also, since the wave-speed is constant, 2-waves do not interact, as only one 2-wave enters or leaves each diamond. These observations allow us to treat the entropy as a constant background source at the quadratic level.

With these assumptions, we can now describe the effects of a reordering on a wave sequence precisely. We know that 2-waves do not change, so we need only consider changes in sound waves due to interactions. Moreover, sound waves of different families produce no quadratic effects, and waves of the same family add linearly, so that we can consider each sound wave separately. We thus consider the effect of passing a single sound wave through a static pattern of entropy fluctuations. If a sound wave ϵ crosses an entropy jump γ_{k_1}, a sound wave of strength $\epsilon^{(1)} = \Lambda\epsilon\gamma_{k_1}$ of the opposite family is reflected. This new wave then interacts with other entropy waves, say γ_{k_2}, to produce a wave $\epsilon^{(2)} = \Lambda'\epsilon^{(1)}\gamma_{k_2}$ in the same family as ϵ. This process continues inductively, and after r interactions, say r even, the wave generated by this series of interactions has strength

$$\epsilon^{(r)} = \epsilon^{(r-1)}\gamma_{k_r}\Lambda' = \epsilon\gamma_{k_1}\ldots\gamma_{k_r}(\Lambda\Lambda')^{r/2}. \tag{18}$$

This scattering process occurs for each distinct pattern of reflections of the wave as it passes through the field of entropy fluctuations. We shall refer to one of these interaction patterns as a path.

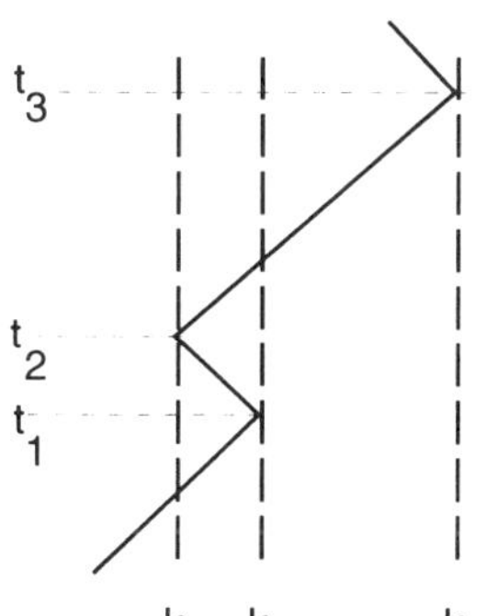

FIGURE 1. A single path

Formally, a path is a set of crossing wave pairs $\{(j_p, k_p)\} \subset C_\tau$, where each k_p represents a 2-wave, and satisfying $j_{p+1} = \imath^\tau(j_p, k_p)$. We say that it begins at j_1 and ends at $\tau\imath^\tau(j_d, k_d)$, where (j_d, k_d) is the last interacting pair in the path. The position j_{p+1} represents the position of the scattered wave after the first p interactions, and that scattered wave has strength given as above. Since all sound waves combine linearly, we can add up each of these contributions to get a discretized path integral formula.

Theorem 2. *Given the wave sequence γ and reordering τ, let $\delta = \tau(\gamma)$ be the reordered wave sequence. Then the wave δ_i is given by*

$$\delta_i = \sum_j \gamma_j \sum_d \Lambda(i,d) \sum_{\Pi_d(j,i,\tau)} \gamma_{k_1} \dots \gamma_{k_d}. \tag{19}$$

where $\Pi_d(j,i,\tau)$ is the collection of all d-paths in τ beginning at j and ending at $i = \tau\imath^\tau(j_d, k_d)$, the γ_k's are 2-waves, and Λ is the appropriate weight.

5. Functional Bounds

Having found a formula describing reordered sequences, we look for bounds which will allow us to conclude existence of solutions. It is evident from the path integral formula that the large-time behavior of the solution is highly dependent on the entropy of the solution. Our approach is to bound the 2-wave strengths by some number β, and then count the number of possible paths. This leads to time-dependent bounds for the reordered sequence, and we have the following theorem.

Theorem 3. *If the system of conservation laws admits a 2-Riemann coordinate, and τ is a reordering corresponding to a set of diamonds spanning time T, then the error potential is continuous,*

$$Q(\gamma,\tau) - Q(\gamma',\tau) \le |\gamma_q - \gamma_q'| V(\gamma)^2 O(T), \tag{20}$$

when the sequences differ in the single wave strength γ_q. Moreover, the total variation of the reordered sequence $\tau\gamma$ grows at most exponentially in time,

$$V(\tau\gamma) \le V(\gamma) e^{kT}, \tag{21}$$

where k depends on the flux and the entropy of the initial data.

Since our bounds are independent of mesh parameters, we can now invoke Theorem 1 to obtain existence of solutions, with total variation growing exponentially in time.

We briefly describe the idea of proof of the total variation bound, which in turn yields a bound for Q. Using the path integral formula, we have

$$V(\tau\gamma) = \sum |\delta_i| \le V(\gamma) \max_j |\sum \Lambda(i,d) \gamma_{k_1} \dots \gamma_{k_d}|, \tag{22}$$

where the sum is over d, i and $\Pi_d(j,i,\tau)$. We now suppose that each $|\gamma_{k_p}| \le \beta$, and use the bound $|\Lambda(i,d)| \le \Lambda^d$, to get

$$V(\tau\gamma) \le V(\gamma) \sum_d (\beta\Lambda)^d \# \Pi_d(J,\cdot,\tau), \tag{23}$$

where $\#\Pi_d(J,\cdot,\tau) = \max_j \sum_i \#\Pi_d(j,i,\tau)$ is the maximum number of d-paths starting at a fixed point.

We now count the number of paths as follows : a path is determined by its starting position, together with the 2-waves $k_1,\ldots,k_d$ with which the scattered wave interacts. Now only one 2-wave passes through each diamond, and each sound wave passes through only one diamond per time step. Therefore the path can be determined by a knowledge of its starting position and a list of (discrete) interaction times. That is, k_p is determined by position j_p and the time t_p at which the p-th interaction occurs, see Figure 1. In particular, the number of paths starting at a fixed point is bounded by the number of lists of interaction times $t_1 < \cdots < t_d$. Since the path spans a time T, we must have $t_d < T/\Delta t$, so that the number of paths is

$$\#\Pi_d(J,\cdot,\tau) \le \binom{T/\Delta t}{d}. \tag{24}$$

Substituting this path count in and using the binomial theorem, gives

$$V(\tau\gamma) \le V(\gamma)(1+\beta\Lambda)^{T/\Delta t} \le V(\gamma)e^{\beta\Lambda T/\Delta t}. \tag{25}$$

As stated here, this bound explodes as the mesh size Δt decreases to 0, unless the entropy remains Lipschitz continuous, which is a major assumption. However, a more sophisticated estimate based on these ideas yields a time–dependent bound with no extra assumptions on the solution. For details of the estimate together with a physical interpretation, the reader is referred to [5].

References

1. J. Glimm, *Solutions in the large for nonlinear hyperbolic systems of equations*, Comm. Pure Appl. Math. **18** (1965), 697–715.
2. J. Glimm and P.D. Lax, *Decay of solutions of systems of nonlinear hyperbolic conservation laws*, Memoirs Amer. Math. Soc. **101** (1970).
3. P.D. Lax, *Hyperbolic systems of conservation laws, II*, Comm. Pure Appl. Math. **10** (1957), 537–566.
4. Tai-Ping Liu, *Decay to n-waves of solutions of general systems of nonlinear hyperbolic conservation laws*, Comm. Pure Appl. Math. **30** (1977), 585–610.
5. B. Temple and R. Young, *Solutions to the Euler equations with large data*, In preparation.
6. Tong Yang, *A functional integral approach to shock wave solutions of Euler equations with spherical symmetry*, Ph.D. thesis, U.C. Davis, 1993.
7. Robin Young, *Sup-norm stability for Glimm's scheme*, Comm. Pure Appl. Math. **46** (1993), 903–948.
8. ———, *Exact solutions to degenerate conservation laws*, Preprint, 1994.

FLUID MIXING INDUCED BY RANDOM FIELDS

QIANG ZHANG[1]

ABSTRACT. We present a qualitative theory for the mixing of fluids induced by a random velocity field. The theory provides a quantitative prediction for the growth of the mixing region. The theoretical predictions are in excellent agreement with the results of numerical computations. There are three distinct regimes for the asymptotic scaling behavior of the mixing layer, depending on the asymptotic behavior of the random velocity field. The asymptotic diffusion is Fickian when the correlation function of the random field decays rapidly at large length scales. Otherwise the asymptotic diffusion is non-Fickian. The scaling behavior of the mixing layer is determined over all length scales. Our results show that, in general, the scaling exponent of the mixing layer is non-Fickian on all finite length scales. In Lagrangian picture, due to the non-linearity of the effective dynamical equation derived from the Taylor diffusion theory, the mixing layer is not a fractal even if the random velocity field is a fractal.

1. INTRODUCTION

As passive tracers flow through random velocity field, a mixing layer develops between the tagged and untagged regions of the fluid. The mixing region expands as the time evolves. The determination of the growth of the mixing region induced by a random velocity field has importance in oil recovery processes, ground water ecology and fully developed turbulence. The works on the turbulent fluid mixing and on fluid in porous media can be traced from [12, 2, 11]. The questions which we considered here are: Which effective equation governs the dynamics of the mixing regime? Can we predict the size of the mixing regime from the statistical properties of the random velocity field? How fast the mixing regime will grow with time? We provide quantitative answers for these questions.

The mixing length $l(t)$ is a quantity which characterizes the size of the mixing regime. If $l(t)$ is proportional to $t^{1/2}$, the scaling is called Fickian. Otherwise the scaling is called non-Fickian. If $l(t)$ is proportional to t^γ with γ being constant, then the mixing regime is termed as a fractal. Otherwise it is termed as multi-length-scale-fractal, due to the fact the scaling exponent changes with the length scales.

[1]Department of Applied Mathematics, SUNY at Stony Brook. Partially supported by DOE grant number DE-FG02-90ER25084.

Any positive, piecewise differentiable function $f(t)$ can be expressed in a multi-fractal form, $f(t) = a(t)t^{b(t)}$, where $b(t) = \frac{dlnf(t)}{dlnt}$ and $lna(t) = lnf(t) - lnt\frac{dlnf(t)}{dlnt}$. In the $ln(l(t))$-$ln(t)$ plane, $b(t)$ is the slope at lnt. Therefore, the mixing length $l(t)$ can be expressed as

$$l(t) = c(t)t^{\gamma(t)}. \tag{1}$$

Here $\gamma(t)$ is the exponent of the mixing length $l(t)$. Similarly, one can define ψ and β as the scaling exponents of the dispersivity and the correlation function of the fluctuation of the velocity field respectively.

The asymptotic scaling law among the asymptotic exponents γ_∞, ψ_∞ and β_∞ are given by [17, 15]

$$\gamma_\infty = max\{\frac{1}{2}, 1 + \frac{\beta_\infty}{2}, 1 + \frac{\alpha\beta_\infty}{2}\} \tag{2}$$

and

$$\psi_\infty = max\{0, 1 + \beta_\infty, 1 + \alpha\beta_\infty\}. \tag{3}$$

Here α is a parameter characterizes the rate of removal of the infrared cutoff.

Equations (2) and (3) show that there are three regimes of distinct scaling behavior. These three regimes are connected by two critical points: $\beta_\infty = -1$ and $\beta_\infty = 0$. The regime $\beta_\infty < -1$ is called Fickian dispersion regime. This regime is characterized by the property that the correlation function is integrable over one spatial dimension. Therefore a finite asymptotic limit exists for the dispersion coefficient. The second regime lies in the range $-1 < \beta_\infty < 0$. In this regime, the asymptotic dispersion is non-Fickian and the dispersion coefficient diverges at large time. The third regime is specified by $0 < \beta_\infty$. In this regime, the scaling relationship is non-unique before specifying how the infrared cutoff is removed from the system. In other words the scaling relation contains a family of solutions which depends on the rate of removal of the infrared cutoff. As α tends to infinity, which corresponds to the fastest removal of the infrared cutoff, the scaling exponent tends to infinity as well. The scaling relations in first and second regimes have also been obtained from perturbative methods for a fractal velocity field [7], and for a general velocity field [14], and, have been validated by numerical computations [4, 5, 6].

The study of mixing induced by random velocity has been carried out in both Lagrangian picture and Eulerian picture. Both pictures give the same asymptotic scaling law given by (2) and (3).

2. Eulerian Picture

In Eulerian picture, the motion of the tracer is governed by a linear transport equation,

$$s_t + \vec{v} \cdot \nabla s = 0. \tag{4}$$

where s is the saturation value of the fluid, $s = 1$ for tagged fluid and $s = 0$ for untagged fluid. The shape of the tracer interface at $t = 0$ is given by $s(0, \vec{x})$. $\vec{v}$ is the velocity field. We assume that the velocity field is stationary, random and statistically invariant under a spatial translation. For incompressible fluid, the velocity field $\vec{v}$ is divergent free, i.e. $\nabla \cdot \vec{v} = 0$. We separate the velocity $\vec{v}$ into two parts, an ensemble averaged velocity $\vec{v}_0 =< \vec{v} >$ and a fluctuation $\delta\vec{v} = \vec{v} - < \vec{v} >$. Then the mean value of $\delta\vec{v}$ vanishes. After changing the independent variable to $\vec{\eta} = \vec{x} - \vec{v}_0 t$, i.e. going to the frame moving with the average position of the mixing region, (1) can be written as

$$s_t = -\delta\vec{v} \cdot \nabla s \tag{5}$$

where $\delta\vec{v} = \delta\vec{v}(\vec{\eta} + \vec{v}_0 t)$, $s = s(t, \vec{\eta})$ and the operator ∇ is defined with respect to the variable $\vec{\eta}$.

Applying renormalization group perturbation theory, one has the following the effective dynamical equation for the ensemble averaged saturation [14]

$$\frac{\partial}{\partial t} < s(\vec{\eta}, t) >= \nabla \cdot \int_0^t Q(\vec{v}_0 t')dt' \cdot \nabla < s(\vec{\eta}, t) > . \tag{6}$$

Here $Q(\vec{x} - \vec{x}') =< \delta\vec{v}(\vec{x})\delta\vec{v}(\vec{x}') >$ is the covariance tensor of the fluctuations of the velocity field.

For simplicity, we assume that the initial interface between tagged and untagged fluids a plane located $x_1 = 0$, namely $s(\vec{x}, 0) = \theta(\vec{\eta})$, and $\vec{v}_0$ is in the x_1 direction. Then (6) can be solved analytically and the result is

$$< s(t, \vec{\eta}) >= \frac{1}{2} erfc(\frac{x_1 - v_0 t}{2(w(t))^{1/2}}), \tag{7}$$

where $erfc(\cdot)$ is the complimentary error function and

$$w(t) = \int_0^t \int_0^\zeta q(v_0\xi)d\zeta d\xi = \int_0^t (t - \xi)q(v_0\xi)d\xi. \tag{8}$$

Here $v_0\xi$ represents the distance over which the mixing interface swept during the time period ξ and $q =< \delta v_1(x)\delta v_1(x') >$ is the longitudinal velocity correlation function. The effective dispersion coefficient is given by

$$c(t) = \int_0^t q(v_0\xi)d\xi. \tag{9}$$

The mixing length l corresponds to the distance over which $< s >$ varies from a value close to 0 to a value close to 1. Equation (7) shows that $l(t)$ has the scaling [14]

$$l(t) = 2(w(t))^{1/2} = 2[\int_0^t (t - \xi)q(\xi)d\xi]^{1/2}. \tag{10}$$

From Eqs. (8), (10) and the definition (1), the scaling exponent $\gamma(t)$, and the coefficient $c(t)$ can be determined analytically [14]

$$\gamma(t) = \frac{d(ln(l(t)))}{d(ln(t))} = \frac{t}{l}\frac{dl(t)}{dt} = \frac{1}{2}[1 - \frac{\int_0^t \xi q(\xi)d\xi}{t\int_0^t q(\xi)d\xi}]^{-1}, \tag{11}$$

and

$$ln(c(t)) = ln(2) + \frac{1}{2}ln[\int_0^t (t - \xi)q(\xi)d\xi] - \frac{1}{2}[1 - \frac{\int_0^t \xi q(\xi)d\xi}{t\int_0^t q(\xi)d\xi}]^{-1} ln(t).$$

Several important conclusions can be obtained from (9)-(11).

1) In general, the growth of the mixing layer is non Fickian on all finite length scales (unless the random velocity is white noise).

2) If the correlation function of the random velocity is non-singular at short length scales, the mixing length l grows linearly with time t, i.e. $\gamma(0) = 1$, for $t << 1$.

3) The longitudinal asymptotic diffusion is Fickian if the correlation function of the random velocity field is integrable along the longitudinal direction. Otherwise it is asymptotically non-Fickian. Therefore $\gamma_\infty = max\{\frac{1}{2}, 1 + \frac{\beta_\infty}{2}\}$ for $\beta_\infty < 0$. $\beta_\infty = -1$ is a critical point which joints the asymptotic Fickian diffusion branch ($\beta_\infty < -1$) with the asymptotic non-Fickian diffusion ($\beta_\infty > -1$) branch. At the critical point, the system has the slowest rate for approaching its asymptotic exponent.

4) Over the range where the correlation function is positive, the effective diffusion coefficient is an increasing function of time. This phenomenon is known as "scale effect" and has been observed in experimental studies. Therefore our theory has explained the "scale effect".

5) If the exponent $\beta(t)$ of the velocity correlation function $q(v_0 t)$ varies slowly over curtain length scales, then $\gamma(t)$ approaches to a instantaneous fractal exponent $max\{\frac{1}{2}, 1 + \frac{\beta(t)}{2}\}$.

6) The exponent of the mixing length at length scale l depends on the velocity field correlation at all length scales smaller than l. The scaling behavior of fluid at short length scales is not enough to predict the scaling behavior of fluid at intermediate and large length scales.

7) A quantitative prediction of the mixing induced by a random field can be achieved from information of the field at each distinct length scale. Such information is necessary for the solution of the scale up problem.

The proofs for these conclusions can be found in [Zha1].

The theoretical predictions given by (10) are in excellent agreement with the solutions of full numerical simulations for both asymptotic Fickian and non-Fickian random fields over the entire time range of the numerical simulations [8, 9].

3. Lagrangian Picture

In the Lagrangian framework, the dispersion induced by a random velocity field can be formulated in terms of Taylor's dispersion theory. Driven by a velocity field $\vec{v}(\vec{x}, t)$, a particle located initially at $\vec{a}$ will be found at the location $\vec{x}_{ttl}(t)$ at time t. Here $\vec{x}_{ttl}(t)$ is the total particle displacement and is determined by $\vec{x}_{ttl}(t) = \vec{a} + \int_0^t \vec{v}(\vec{x}_{ttl}(t'), t')dt'$. Without lost of generality, we choose $\vec{a} = 0$. We express the velocity as a sum of three parts $\vec{v} = \vec{v}_0 + \delta\vec{v} + \vec{v}_d$. Here $\vec{v}_0$ is the ensemble averaged velocity, and $\delta\vec{v}$ is the fluctuation around the ensemble averaged velocity, $\vec{v}_d$ is a velocity associated with the Brownian motion. We assume that $\vec{v}_0$ is a constant vector and both $\delta\vec{v}$ and $\vec{v}_d$ are Gaussian with mean zero. Similarly, $\vec{x}_{ttl}$ can be expressed as $\vec{x}_{ttl} = \vec{x}_m + \vec{x}' + \vec{x}_d$. It follows that $\vec{x}_m(t) = \vec{v}_0 t$ and $\vec{x}'(t) = \int_0^t \delta\vec{v}(\vec{x}_{ttl}(t'), t')dt'$. Let X be the second moment of the fluctuation of the particle displacement. Then X is given by $X = <\vec{x}'(t)\vec{x}'(t)> = \int_0^t (t - t') < \delta\vec{v}(\vec{x}_{ttl}(t'), t')\delta\vec{v}(0, 0) > dt'$. Here

$< \cdot >$ denotes ensemble average and we have assumed that the random velocity is stationary and statistically homogeneous.

$$\frac{d^2X}{dt^2} = \frac{2}{(2\pi)^m}\int\int < \tilde{\delta\vec{v}}(\vec{k}',0)\tilde{\delta\vec{v}}^*(\vec{k},t)e^{i\vec{k}\cdot\vec{x}'} > exp(-i\vec{k}\cdot\vec{v}_0 t - d_l k^2 t)d\vec{k}d\vec{k}'. \quad (12)$$

Here $*$ denotes the complex conjugate. m is the spatial dimension. d_l is the local dispersion coefficient. $\tilde{\delta\vec{v}}$ is the Fourier transform of $\delta\vec{v}$.

There is a closure problem in (12); it appears in the factor $< \tilde{\delta\vec{v}}\tilde{\delta\vec{v}}e^{i\vec{k}\cdot\vec{x}'} >$. This closure problem can be resolved approximately by applying Corrsin's hypothesis [12, 13]. This approximation is based on two assumptions: $\tilde{\delta\vec{v}}\tilde{\delta\vec{v}}$ and $e^{i\vec{k}\cdot\vec{x}'}$ are uncorrelated at large time, and $\vec{x}'$ is Gaussian. Based on the first assumption, one replaces $< \tilde{\delta\vec{v}}\tilde{\delta\vec{v}}e^{i\vec{k}\cdot\vec{x}'} >$ by $< \tilde{\delta\vec{v}}\tilde{\delta\vec{v}} >< e^{i\vec{k}\cdot\vec{x}'} >$. Based on the second assumption, one further replaces $< e^{i\vec{k}\cdot\vec{x}'} >$ by $e^{-\frac{1}{2}\vec{k}\cdot X\cdot\vec{k}}$. Therefore under Corrsin's hypothesis, we have an approximate governing equation.

$$\frac{d^2X}{dt^2} = \frac{2}{(2\pi)^m/2}\int\int < \tilde{\delta\vec{v}}(\vec{k}',0)\tilde{\delta\vec{v}}^*(\vec{k},t) > e^{-\frac{1}{2}\vec{k}\cdot X\cdot\vec{k}}exp(i\vec{k}\cdot\vec{v}_0 t - d_l k^2 t)d\vec{k}d\vec{k}'. \quad (13)$$

For simplicity, we choose the initial conditions $X|_{t=0} = 0$, $dX/dt|_{t=0} = 0$ and $m = 3$. We assume that the velocity correlation tensor has the form $< \tilde{\delta\vec{v}}\tilde{\delta\vec{v}} >= \tilde{g}(\vec{k})G(\vec{k})$, where $\tilde{g}(\vec{k})$ is a scalar function, $G(\vec{k}) = (I - \frac{\vec{k}\vec{k}^T}{|\vec{k}|^2})\frac{\vec{v}_0\vec{v}_0^T}{|\vec{v}_0|^2}(I - \frac{\vec{k}\vec{k}^T}{|\vec{k}|^2})$, and $I - |\vec{k}|^{-2}\vec{k}\vec{k}^T$ is the projection operator onto the space of incompressible velocity fields. An example for a velocity field with such a correlation tensor is the incompressible velocity field obtained from the linearized solution of Darcy's law [3, 13]. In this case, $\tilde{g}(\vec{k})$ is the covariance function of the permeability field in Fourier space.

The macroscopic dispersion tensor, D, induced by the random velocity field is related to X by $D = \frac{1}{2}\frac{dX}{dt}$, and the total dispersion tensor, i.e. the contribution from both the random velocity field and molecular dispersion, is given by $D + d_l I$.

Equation (13) is a non-linear differential equation. In addition it involves an integration over Fourier space. Equation (13) shows that the dispersion induced by the random velocity field is non-local. Dagan [3] has linearized (13) by setting X to zero on the right hand side of (13). The dispersion is local for the linearized equation. Neuman and Zhang [13] considered a specific correlation function, which decays exponentially with distance. For such systems, the asymptotic dispersion is Fickian and X is asymptotically proportional to t for large time t. They quasi-linearize (13) by replacing X on the right hand side of (13) by its asymptotic form $2d_0 t$. Here d_0 is a (constant) asymptotic dispersion tensor and is determined numerically by an iterative procedure.

A recent analysis of Corrsin's hypothesis shows that for a weak random field, Corrsin's hypothesis may be applicable not only to large time scales but also to finite time scales [16]. The governing equation obtained from Corrsin's hypothesis is the leading term in a renormalization group perturbation expansion [16].

For an isotropic correlation function, both X and D become diagonal matrices. There are only two distinct diagonal elements corresponding to the longitudinal and transverse directions respectively. Let f_L and f_T be the longitudinal and transverse elements of X, κ_L and κ_T be the longitudinal and transverse elements of D. Then the scaling exponent of f_i and κ_i at time t are defined as $\gamma_i(t) = \frac{1}{2}\frac{d\ln f_i(t)}{d\ln t}$ and $\psi_i(t) = \frac{d\ln\kappa_i(t)}{d\ln t}$, for $i = L$ and T.

To illustrate the scale dependence of the exponent γ, we consider the mixing induced by a random velocity field with a fractal correlation function

$$g(r) = c|\vec{r}|^\beta, \tag{14}$$

with $-1 < \beta < 0$. We show that even for a fractal correlation function, X and D determined by (13) and (14) are not fractals. Both X and D have different growth rates at different length scales.

For the fractal correlation function given by (14), (13) can be expressed in terms of parabolic cylinder function,

$$\frac{d^2}{dt^2}\begin{pmatrix} f_L \\ f_T \end{pmatrix} = \frac{cv_0^2}{4\pi^{1/2}}\int_0^1 \Gamma(\beta+2)2^{(\beta+3)/2}a^{-\beta}e^{-a^2h^2/8}$$

$$\times[D_{-\beta-1}(\frac{ah}{\sqrt{2}}) + D_{-\beta-1}(-\frac{ah}{\sqrt{2}})]\begin{pmatrix} 2(1-\tau^2)^2 \\ \tau^2(1-\tau^2) \end{pmatrix} d\tau. \tag{15}$$

Here $a = [\frac{1}{2}f_T + \frac{1}{2}(f_L - f_T)\tau^2 + d_l t]^{-1/2}$ and $h = \tau v_0 t$. The parabolic cylinder function is defined as

$$D_p(z) = \frac{e^{-x^2/4}}{\Gamma(-p)}\int_0^\infty e^{-(2xz+x^2)/2}x^{-p-1}dx,$$

Both the small and large time asymptotic expressions can be obtained from (15). For large time, the asymptotic dominant solution of (15) is given by [15]

$$f_L^{dom} = 16c[(\beta+1)(\beta+2)(\beta+3)(\beta+5)]^{-1}(v_0t)^{\beta+2},$$

$$f_T^{dom} = 2c[(\beta+2)(\beta+3)(\beta+5)]^{-1}(v_0t)^{\beta+2},$$

$$\kappa_L^{dom} = 8cv_0[(\beta+1)(\beta+3)(\beta+5)]^{-1}(v_0t)^{\beta+1}$$

and

$$\kappa_T^{dom} = cv_0[(\beta+3)(\beta+5)]^{-1}(v_0t)^{\beta+1}.$$

The dispersion induced by the random field with a fractal correlation function given by (14) is non-Fickian at asymptotic large length scales, (it diverges), while the molecular dispersion is Fickian. Therefore at large length scales the dispersion induced by the random field dominates. This is why the molecular dispersion does not appear in the dominant asymptotic solutions. The situation is quite different at short length scales [15, 16]. In the absence of molecular dispersion, i.e. $d_l = 0$, we have

$$f_i = B_i J_1{}^{2/(2-\beta)}t^{4/(2-\beta)} + high\ order\ terms$$

and

$$\kappa_i = B_i\frac{2}{2-\beta}J_1{}^{2/(2-\beta)}t^{(2+\beta)/(2-\beta)} + high\ order\ terms$$

for $t << 1$ and $i = L, T$. Here

$$J_1 = \frac{cv_0^2(2-\beta)^2}{2\sqrt{\pi}(2+\beta)}\Gamma(\frac{\beta+3}{2})2^{\beta/2}.$$

B_L and B_T are determined by

$$\begin{pmatrix} B_L \\ B_T \end{pmatrix} = \int_0^1 [B_T + (B_L - B_T)\tau^2]^{\beta/2} \begin{pmatrix} 2(1-\tau^2)^2 \\ \tau^2(1-\tau^2) \end{pmatrix} d\tau.$$

In the presence of molecular dispersion, i.e. $d_l \neq 0$, we have

$$f_i = B_i J_2 t^{(4+\beta)/2} + high\ order\ terms$$

and

$$\kappa_i = B_i \frac{4+\beta}{4} J_2 t^{(2+\beta)/2} + high\ order\ terms$$

for $t << 1$ and $i = L, T$. Here

$$B_L = 8, B_T = 1 \quad and \quad J_2 = \frac{cv_0^2}{\sqrt{\pi}} \frac{2^{\beta+4}\Gamma(\frac{\beta+3}{2})}{15(\beta+2)(\beta+4)} d_l^{\beta/2}.$$

Therefore γ_L and γ_T will evolve from $2/(2-\beta)$ (when $d_l = 0$) or from $(4+\beta)/4$ (when $d_l \neq 0$) at short time scales to $(2+\beta)/2$ at asymptotically large time scale. Similarly, ψ_L and ψ_T will evolve from $(2+\beta)/(2-\beta)$ (when $d_l = 0$) or from $(2+\beta)/2$ (when $d_l \neq 0$) at short time scales to $1+\beta$ at asymptotically large time scale. This implies that even for a random velocity field with a fractal correlation function, the mixing induced by such fractal random velocity field is not a fractal; it is a multi-length-scale-fractal. The length scale dependence of the scaling exponents of X and D comes from the fact that the right hand side of (13) involves X. If one sets X on the right hand side of (13) to zero, as in the linear theory of Dagan, then the mixing becomes a fractal also, and the results coincide with the large time dominant asymptotic solution.

It is interesting to observe that the results from the linear theory agree with the exact solution of (15) at asymptotic large time scales, rather than at short time scales. As explained in [15], it is the ratio $\|X\|/t^2$, rather than X alone, which determines the asymptotic behavior of the solution. The linear theory corresponds to the limit $\|X\|/t^2 \to 0$. For $\beta < 0$, this ratio tends to zero as t tends to infinity, and tends to infinity as t tends to zero. Therefore the results from the linear theory agree with the exact solution of (15) at asymptotically large time scales. When the asymptotic dispersion is Fickian, such agreement no longer helds, because the linear approximation does not dominate in the asymptotically large time scales.

The solution of (15) depends on the amplitude c and the mean velocity v_0. We introduce the following set of dimensionless variables: $\bar{t} = c^{1/\beta}v_0 t$, $\bar{d}_l = c^{1/\beta}v_0^{-1}d_l$, $\bar{\kappa}_i = c^{1/\beta}v_0^{-1}\kappa_i$, $\bar{f}_i = c^{2/\beta}f_i$, $\bar{\gamma}_i = \gamma_i$ and $\bar{\psi}_i = \psi_i$, for $i = L, T$. These dimensionless quantities are independent of c and v_0. For any given c and v_0, the original dimensional variables can be easily obtained from the dimensionless variables by inverting the relation.

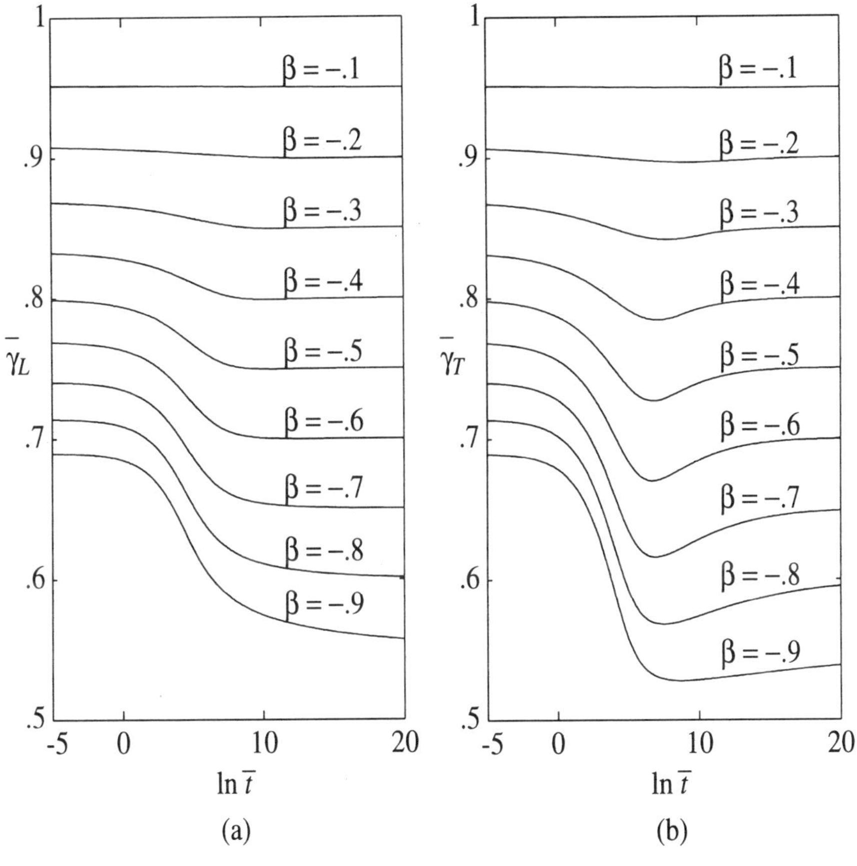

FIGURE 1. The longitudinal and transverse multi-length-scale-fractal mixing exponents $\bar{\gamma}_L$ and $\bar{\gamma}_T$. The figure shows that even for a fractal correlation function, the longitudinal and transverse mixing exponents are not constants. The exponents $\bar{\gamma}_L$ and $\bar{\gamma}_T$ vary from $2/(2-\beta)$ at short time scales to $1+\beta/2$ at large time scales.

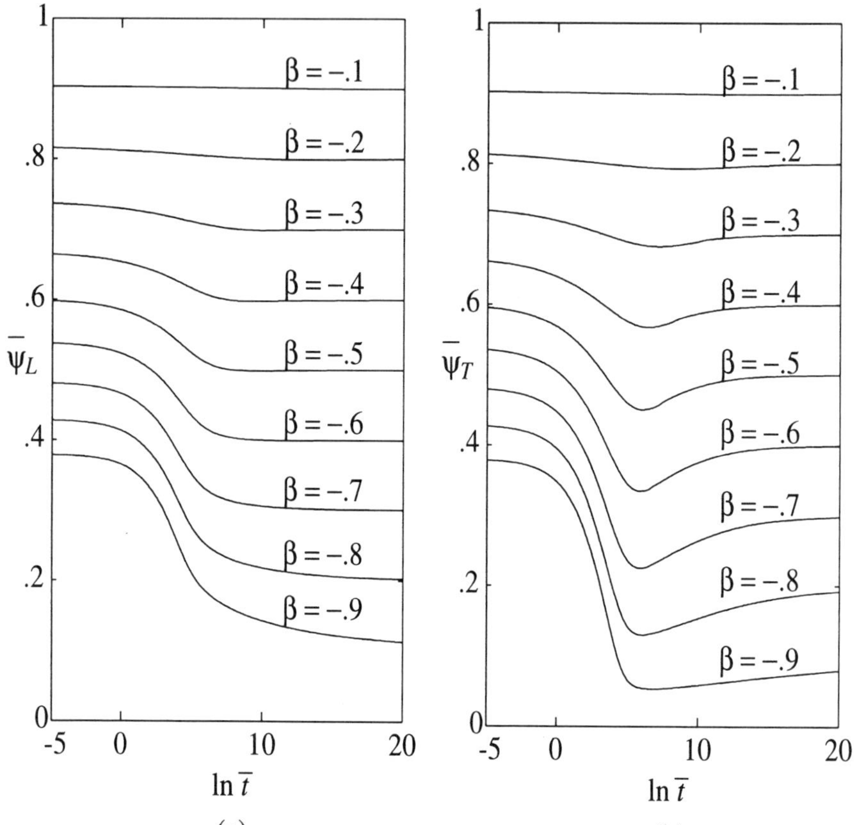

FIGURE 2. The multi-length-scale-fractal exponent of the longitudinal and transverse dispersion coefficients, $\bar{\psi}_L$ and $\bar{\psi}_T$. As in Fig. 1, the exponents for the dispersion coefficients also depend on the time scales. The exponents $\bar{\psi}_L$ and $\bar{\psi}_T$. vary from $(2+\beta)/(2-\beta)$ at short time scales to $1+\beta$ at large time scales.

We present the solutions of (15) for $\bar{d}_l = 0$. In Figure 1 and 2, we plot $\bar{\gamma}_L$ and $\bar{\gamma}_T$, the multi-length-scale-fractal mixing exponents. In Figure 2, we plot $\bar{\psi}_L$ and $\bar{\psi}_T$, the multi-length-scale-fractal exponents of the longitudinal and transverse dispersion coefficients. $\bar{\gamma}_L$ and $\bar{\gamma}_T$ vary from $2/(2-\beta)$ at short time scales to $1+\beta/2$ at large time scales. $\bar{\psi}_L$ and $\bar{\psi}_T$ vary from $(2+\beta)/(2-\beta)$ at short time scales to $1+\beta$ at large time scales. Therefore they are multi-length-scale-fractals.

REFERENCES

1. M. Avellaneda, and A. Majda, *Mathematical Models with Exact Renormalization for Turbulent Transport*, Commun. Math. Phys. **131**, 381 (1990).
2. G. Dagan, G. *Flow and Transport in Porous Formations:* Springer-Verlag, New York, 1989.
3. G. Dagan, *Solute transport in heterogeneous porous formations*, J. Fluid Mech., **145**, 151 (1984)
4. F. Furtado, J. Glimm, B. Lindquist and F. Pereira, *Multi-length scale computations of the mixing length growth in tracer flow*proceedings of the Emerging Technologies Conference, Kovarik F. ed.:251, Houston TX, July 1990.
5. F. Furtado, J. Glimm, B. Lindquist and F. Pereira, *Characterization of mixing length growth for flow in heterogeneous porous media* SPE 21233, Proceedings of the 11th SPE Symposium on Reservoir Simulation, Feb. 1991.
6. F. Furtado, J. Glimm, B. Lindquist, F. Pereira and Q. Zhang, *Time Dependent Anomalous Diffusion for Flow in Multi-fractal Porous Media*, Proceedings of the Workshop on Numerical Methods for the Simulation of Multiphase and Complex Flow. Lecture Notes in Physics, Verheggan T. M. ed., Springer Verlag, New York, (1991).
7. J. Glimm and D. H. Sharp, *A random field model for anomalous diffusion in heterogeneous porous media*, J. Stat. Phys. **62**, 415 (1991).
8. J. Glimm, W. B. Lindquist, F. Pereira and R. Peierls, *The Multi-fractal Hypothesis and Anomalous Diffusion*, Revista Brasileira de Matematica Aplicade e Computacional, **11**, 189 (1992).
9. J. Glimm, W. B. Lindquist, F. Pereira and Q. Zhang, *A Theory of Macrodispersion for the Scale-Up Problem*, Transport in Porous Media, **13**, 97 (1993).
10. T.A. Hewett, *Fractal Distributions of Reservoir Heterogeneity and Their Influence on Fluid Transport*, SPE 15386, 1986.
11. *Lake, L. and Carroll, H. eds., Reservoir Characterization*, Academic Press, 1986.
12. W. D. McComb, *The Physics of Turbulence, Oxford*, Oxford University Press, 1990.
13. S. P. Neuman and Y. K. Zhang, *A Quasi-Linear Theory of Non-Fickian Subsurface Dispersion*, Water Resources Res. **26** 887 (1990).
14. Q. Zhang, *A Multi-length Scale Theory of the Anomalous Mixing Length Growth For Tracer Flow In Heterogeneous Porous Media*, J. Stat. Phys. **66**, 485 (1992).
15. Q. Zhang, *The Asymptotic Scaling Behavior of Mixing Induced by a Random Velocity Field*, Advances in Applied Mathematics, **16**, 23 (1995).
16. Q. Zhang, *The Transient Behavior of Mixing Induced by a Random Velocity Field*, Water Resources Research, **31**, 577 (1995).
17. Q. Zhang and J. Glimm, *Inertial Range Scaling of Laminar Shear Flow as a Model of Turbulent Transport*, Comm. in Math. Phys. **146**, 217 (1992).

The Initial Boundary Value Problem for Hyperbolic Systems of Conservation Laws

Fumioki Asakura[1]

1. Introduction

Let us consider a general n-system of conservation laws:

$$U_t + F(U)_x = 0. \tag{1.1}$$

We assume that (1.1) is *hyperbolic* with *non-zero characteristic speeds* and each characteristic field is *genuinely nonlinear* in the sense of Lax.

We study the large–time behavior of solutions to the initial boundary value problem of (1.1) with prescribed initial values:

$$U(x,0) = U_0(x) \quad \text{for} \quad x > 0 \tag{1.2}$$

and boundary conditions:

$$F(U(0,t)) - H(t) \in \mathcal{M} \quad \text{for} \quad t > 0. \tag{1.3}$$

Here $U_0(x)$ and $H(t)$ are given vector functions and $\mathcal{M}$ is a fixed subspace of R^n whose dimension is $n - n_+$ (n_+: the number of positive eigenvalues of $F'(U)$). Let $\mathcal{E}^+(U)$ be the direct sum of the eigenspaces corresponding to the positive eigenvalues of $F'(U)$. We further assume that the boundary conditions satisfy the *Lopatinski* condition for all $U \in \Omega$:

$$\mathcal{E}^+(U) \cap \mathcal{M} = \{0\} \tag{1.4}$$

(cf. Kreiss [5]). The existence of global solutions to this general initial boundary value problem was established by M. Sablé-Tougeron in [12]. She showed by using the Glimm method that a weak solution $U(x,t) \in L^\infty(R_+; BV(R_+)) \cap \mathrm{Lip}\,(R_+; L^1_{loc}(R_+))$ exists for data having sufficiently small total variation.

The aim of this article is to give a brief account of [1] in which the author proves that these solutions are strongly stable in the following sense provided the data are flat at infinity:

$$U_0(x) = U^+_\infty, \quad |x| \geq M, \quad \text{and} \quad H(t) = H_\infty, \quad t \geq T. \tag{1.5}$$

Let Ω_j denote the *the j-primary region* defined by Liu [8].

Theorem 1.1. *Suppose that the initial and boundary data are small in total variation and the solution $U_\infty(x,t)$ to the Riemann-Goursat problem with data U^+_∞ and H_∞ consists of centered waves $\alpha_{n-n_++1}, \ldots, \alpha_n$. Then it follows that, as $t \to \infty$:*

(1) *$U(x,t)$ approaches U_0 in the wake region at the rate $t^{-3/2}$.*

---

[1]Faculty of Economics Otemon-Gakuin University, 2–1–15 Nishiai, Ibaraki-city, Osaka 567, JAPAN

(2) *If α_j is a rarefaction wave, the amount of j-shock wave in Ω_j approaches zero uniformly at the rate of $t^{-1/2}$, and $U(x,t)$ approaches the rarefaction wave α_j uniformly in Ω_j also at the rate of $t^{-1/2}$.*
(3) *If α_j is a shock wave, then there exists a j-shock wave approaching the shock wave α_j both in strength and speed at the rate of $t^{-3/2}$. Moreover, the total variation of the solution in Ω_j outside this shock wave approaches zero uniformly at the rate of $t^{-3/2}$. Similarly, $U(x,t)$ approaches U_{j-1} (or U_j) uniformly at the rate of $t^{-3/2}$ to the left (or right, respectively) of the relatively strong j-shock wave in Ω_j.*

2. The Glimm Difference Scheme

Let h, k be mesh lengths satisfying the stability condition and $\theta = \{\theta_n\}$ be a sequence of random numbers uniformly distributed in $(0,1)$. For integers m, n chosen so that $m+n$ is even, $m \geq 0$, and $n \geq 0$, we set $A_{m,n} = ((m+2\theta_n)h, nk)$ which will be our interior mesh points. For $n =$ odd, we define $A_{0,n} = (0, nk)$ which are called boundary mesh points. The quadrant $R_+ \times R_+$ is divided by *interaction diamonds* $\Delta_{m,n}$ centered at (mh, nk). A curve which consists of segments starting at boundary mesh points and joining vertices of diamonds from W to N or S is called an *approximate space-like curve*.

In the interior, approximate solutions $U_{h,\theta}(x,t)$ are constructed in the same manner as Glimm [3] by solving the Riemann problem for (1.1), the solution of which consists of $n+1$ constant regions connected by rarefaction waves and shock waves (Lax [6]).

In order to construct the Glimm approximate solutions at the boundary $x=0$, we have to solve the following boundary value problem. Let U_R and H_0 be constant vectors such that $U_R \in \mathcal{U}$ and the distance $|BF(U_R) - BH_0|$ is sufficiently small. Find a constant vector U_0 satisfying

$$B(F(U_0) - H_0) = 0 \tag{2.1}$$

and a weak solution to the Riemann-Goursat problem:

$$\begin{aligned} U_t + F(U)_x = 0, &\quad (x,t) \in R_+ \times R_+, \\ U(x,0) = U_R, &\quad x > 0, \\ U(0,t) = U_0, &\quad t > 0. \end{aligned} \tag{2.2}$$

The solution is: for given $U_R \in \mathcal{U}$ to find U_0 satisfying (2.1) and connected to U_R by rarefaction waves and shock waves with positive characteristic speeds. Since the Lopatinski condition (1.4) implies that the tangent space of $\mathcal{B}(H_0) : BF(U) = BH_0$ and $\mathcal{E}^+(U)$ span R^n, we obtain the solution by applying the implicit function theorem.

Theorem 2.1. *There exists an open set $\mathcal{U}$ satisfying $\overline{\mathcal{U}} \subset\subset \widetilde{\mathcal{U}}$ and a positive constant d such that if $U_R \in \mathcal{U}, H_0 \in R^n$, and the distance $|BF(U_R) - BH_0|$ is less than d, a constant vector U_0 satisfying* (2.1) *is determined uniquely and the Riemann-Goursat problem* (2.2) *has a solution. This solution consists of* $(n_+ + 1)$ *constant* *regions connected by rarefaction waves and shock waves whose characteristic speeds are positive. Moreover the solution of this form is unique provided the intermediate constant vectors are restricted to a certain neighborhood of* $\mathcal{U}$.

Since the above solution can be obviously continued to the solutions in the interior, we can construct approximate solutions $U_{h,\theta}(x,t)$ for all $(x,t) \in R_+ \times R_+$.

3. Interaction Estimates

Let $\Delta_{m,n}$, denoted simply by Δ, be an interior interaction diamond. The elementary waves issuing from $((m-1)h,(n-1)k)$ and $((m+1)h,(n-1)k)$ and entering Δ are denoted by $\beta = (\beta_1, \beta_2, \ldots, \beta_n)$ and $\gamma = (\gamma_1, \gamma_2, \ldots, \gamma_n)$ respectively. Also the elementary waves issuing from (mh, nk) are denoted by $\alpha = (\alpha_1, \alpha_2, \ldots, \alpha_n)$.

Theorem 3.1 (Glimm [3], Liu [10]). *Suppose that* U_L *and* U_R *are near* $\overline{U}$. *Then it follows that*

$$\alpha_j = \beta_j + \gamma_j + O(1)Q(\Delta) \tag{3.1}$$

for $1 \le j \le n$. *Here* $O(1)$ *depends only on* U_L, U_R, *and the system.*

The quantity $Q(\Delta)$ is defined as the following:

$$Q(\Delta) = \sum_{i>j} |\beta_i||\gamma_j| + \sum_{1\le i\le n} Q_S(\beta_i, \gamma_i), \tag{3.2}$$

where

$$Q_S(\beta,\gamma) = \begin{cases} 0 & \beta \ge 0, \gamma \ge 0 \\ |\beta|^3 & |\beta| \le |\gamma|, \gamma \ge 0, \beta < 0 \\ |\beta||\gamma|^2 & |\beta| \le |\gamma|, \gamma < 0 \\ |\gamma||\beta|^2 & |\gamma| \le |\beta|, \beta < 0 \\ |\gamma|^3 & |\gamma| \le |\beta|, \beta \ge 0; \gamma < 0 \end{cases} \tag{3.3}$$

or

$$= \begin{cases} 0 & \beta \ge 0, \gamma \ge 0 \\ |\beta\gamma|(|\beta| + |\gamma|) & \beta < 0 \ \text{ or } \ \gamma < 0 \end{cases} \tag{3.4}$$

(cf. Liu [10] Theorem 2.2).

Let $\Delta = \Delta_{0,2l}$ be an interaction diamond at the boundary; $(2l+1)k, (2l-1)k$ are denoted by t_1, t_0 respectively. Let $\beta = (\beta_1, \beta_2, \ldots, \beta_n)$ connect U_0 and U_n where U_0 is on $\mathcal{B}(H(t_0))$, and $\alpha = (0, \ldots, 0, \alpha_{n-n_++1}, \ldots, \alpha_n)$ connect certain $U_{n+1} \in \mathcal{B}(H(t_1))$ and U_n. We define quantities $Q(\Delta)$ at the boundary (c.f. Chern [2]).

$$Q(\Delta) = \sum \{|\beta_j \beta_k|;\ 1 \le k \le n - n_+,\ 1 \le j \le n,\ j \ne k\}. \tag{3.5}$$

Theorem 3.2. *Suppose that* U_0 *and* U_n *are near* $\overline{U}$ *and* $|t_1 - t_0|$ *is sufficiently small. Then it follows that*

$$\alpha_j = \beta_j + O(1) \sum_{1 \le l \le n-n_+} |\beta_l| + O(1)\Delta H + O(1)Q(\Delta) \tag{3.6}$$

for $j \geq n - n_+ + 1$. *Here* $\Delta H = |H(t_1) - H(t_0)|$, $Q(\Delta)$ *are defined by* (3.5) *and* $O(1)$ *depends only on* U_0, U_n, *and the system.*

Let $W(J)$ denote the set of waves crossing J and $W_A(J)$ the set of waves in $W(J)$ approaching the boundary. We define $L(J) = \sum_{\alpha_j \in W(J)} |\alpha_j|$ which is equivalent to the total variation of $U_{h,\theta}(x,t)$ on J. Let β, γ be two sets of waves in $U_{h,\theta}$ crossing J. Suppose that β lie to the left of γ on J. We say that β_i and γ_j are approaching, if either $i > j$, or $i = j$ and at least one of them is a shock. The potential function $Q(J)$ is defined for any J by

$$Q(J) = Q_{OD}(J) + \sum_{1 \leq j \leq n} Q^j_{OS}(J) + KQ_A(J) + K^2 Q_{AA}(J), \tag{3.7}$$

where

$$\begin{aligned} Q_{OD}(J) = \sum \{ |\beta_j \gamma_k| ;\ & \beta_j, \gamma_k \in W(J), \beta_j \text{ is any } j\text{-wave lying toward} \\ & \text{the left of a } k\text{-wave } \gamma_k \text{ with } j > k \}, \end{aligned} \tag{3.8}$$

$$\begin{aligned} Q^j_{OS}(J) = \sum \{ |\alpha|^3 ;\ & \alpha \text{ is any } j\text{-shock wave in } W(J) \} + 4 \sum \{ |\beta + \gamma| |\beta\gamma| ; \\ & \beta, \gamma \text{ are any distinct } j\text{-shock waves in } W(J) \} + 8 \sum \{ |\epsilon\zeta\eta| ; \\ & \epsilon, \zeta, \eta \text{ are any distinct } j\text{-shock waves in } W(J) \} \end{aligned} \tag{3.9}$$

or

$$\begin{aligned} Q^j_{OS}(J) = \sum \{ (|\beta| + |\gamma|) |\beta\gamma| ;\ & \beta, \gamma \text{ are any } j\text{-waves in } W(J); \text{ not both of} \\ & \beta, \gamma \text{ are } j\text{-rarefaction waves} \}; \end{aligned} \tag{3.10}$$

$$Q_A(J) = \sum \{ |\beta_j \gamma_k| ;\ \beta_j \in W_A(J), \gamma_k \in W(J), \beta_j \neq \gamma_k \}, \tag{3.11}$$

$$Q_{AA}(J) = \sum \{ |\beta_j \gamma_k| ;\ \beta_j, \gamma_k \in W_A(J) \}, \quad \text{(cf. Chern [2])}. \tag{3.12}$$

Lemma 3.1. *Let* J_1, J_2 *be two consecutive approximate space-like curves such that* $J_2 > J_1$ *and* Δ *the diamond between them. Assume that* $L(J_1)$ *is sufficiently small. Then we have*

$$Q(\Delta) \leq 2(Q(J_1) - Q(J_2)) + 2\Delta H. \tag{3.13}$$

This lemma implies the following estimate: the total amount of interaction in the Glimm approximations is uniformly bounded.

$$\sum_{\Delta} Q(\Delta) \leq 2Q(O) + 2 \sum_{n \geq 0} \Delta H. \tag{3.14}$$

4. Asymptotic Stability

Liu [8] shows that the global interaction estimate (3.14) implies the weak stability theorem: that is, Theorem 1.1 holds without algebraic rates. Here we shall give the rate of the convergence.

Let $\chi^R_n = \{(x_r(t), t); t \geq 0\}$ be the generalized n-characteristic issued from $(M, 0)$ such that $U(x,t) = U^+_\infty$ for $x > x_r(t)$. The j-characteristics through $\{(x_r(t), t)\}$ are denoted by $\chi_j^2(t)$. For $1 \le j \le n - n_+$, all $\chi_j^2(t)$ hit the boundary $x = 0$ before time $t_0 = O(1)t$; the j-generalized characteristic through $(0, t_0)$ is denote by $\chi_j^1(t)$, $j \ge n - n_+ + 1$. Since the characteristic roots are distinct, there exists $t_1 > t_0$ such that all $\chi_i^2(t)$ and $\chi_j^1(t)$ satisfying $i < j$ intersect before $t_1 = O(1)t_0$. We set $t_1 = t_1(t)$ and $t^* = t_1(t_1(t))$. Clearly there exists a constant C, $C > 1$ such that t^* satisfies

$$t^* = Ct. \tag{4.1}$$

For each $s \ge t_1(t), t$, and $j \ge n - n_+ + 1$, we denote by $X_j^{\pm}(s;t)$ the amount of j-wave at time s in the interior of the region between $\chi_j^1(t)$ and $\chi_j^2(t)$ and Str.$\chi_j^i(s;t)$ the strength of $\chi_j^i(t)$ at time s. Let $X_k^{\pm}(t)$ denote respectively the amount of k-rarefaction and k-shock at time t, and $X_k(t) = X_k^+(t) + |X_k^-(t)|$; the limit of $Q(J)$ is denoted by $Q(t)$. Let $S_j(t)$ denote the magnitude of the j-shock waves approaching S_j.

Theorem 1.1 is proved by using the next lemmas which are originally due to Liu [10].

Lemma 4.1. *For any $s \ge t_1(t)$ and any (x_j, t_j) in the region between $\chi_j^2(t)$ and $\chi_{j+1}^1(t)$, we have*

$$\begin{aligned} &|U(x_j,t_j) - U_j| \\ &= O(1)Q(t) + O(1)\sum_{S_j \ne 0}(X_j(t) - |S_j(t)|) + O(1)\sum_{S_j = 0}|X_j^-(t)|^3, \end{aligned} \tag{4.2}$$

$$\begin{aligned} &X_j^+(s;t) + X_j^-(s;t) + \text{Str. } \chi_j^1(s;t) + \text{Str. } \chi_j^2(s;t) \\ &= S_j + O(1)Q(t) + O(1)\sum_{S_j \ne 0}(X_j(t) - |S_j(t)|) + O(1)\sum_{S_j = 0}|X_j^-(t)|^3. \end{aligned} \tag{4.3}$$

Lemma 4.2. *For $T \ge T_0$ and $m = 1, 2, \dots$, we have*

$$Q(C^mT) = Q_{OS}(C^mT) + \sum_{1 \le k \le m-1}(O(1)\epsilon)^k Q_{OS}(C^{m-k}T) + (O(1)\epsilon)^m Q(T). \tag{4.4}$$

Here the constant C is given by (4.1) and ϵ is the total variation of the initial and boundary data.

References

1. F. Asakura, Asymptotic stability of solutions to the initial boundary value problem for the hyperbolic system of conservation laws, preprint (1994).
2. I. L. Chern, Stability theorem and truncation error analysis for the Glimm scheme and for a front tracking method for flows with strong discontinuities. Comm. Pure Appl. Math., 42 (1989), 815–844.
3. J. Glimm, Solutions in the large for nonlinear hyperbolic systems of equations. Comm. Pure Appl. Math., 18 (1965), 697–715.
4. J. Glimm and P. D. Lax, Decay of solutions of systems of nonlinear hyperbolic conservation laws. Amer. Math. Soc. Memoir, No. 101. A.M.S. Providence, 1970.

5. H. O. Kreiss, Initial boundary value problems for hyperbolic systems, Comm. Pure Appl. Math., 23 (1970), 277–298.
6. P. D. Lax, Hyperbolic systems of conservation laws II. Comm. Pure Appl. Math., 10 (1957), 537–566.
7. T. P. Liu, Initial boundary value problem for gas dynamics, Arch. Rat. Mech. Anal., 64 (1977), 147–177.
8. T. P. Liu, Large time behavior of initial and initial-boundary-value problems of general systems of hyperbolic conservation laws. Comm. Math. Phys., 55 (1977), 163–177.
9. T. P. Liu, Decay to N-waves of solutions of general systems of nonlinear hyperbolic conservation laws. Comm. Pure Appl. Math., 30 (1977), 585–610.
10. T. P. Liu, Linear and nonlinear large time behavior of general systems of hyperbolic conservation laws. Comm. Pure Appl. Math., 30 (1977), 767–796.
11. J. Smoller, *Shock Waves and Reaction-Diffusion Equations*, Springer Verlag, New York, 1983.
12. M. Sablé-Tougeron, Méthode de Glimm et problème mixte, Ann. Inst. Henri Poincaré, Analyse non linéaire, 10 (1993), 423–443.

Viscosity Solutions and Uniqueness for Systems of Conservation Laws

Alberto Bressan*

1 - Introduction

Consider the Cauchy problem

$$u_t + \big[F(u)\big]_x = 0, \tag{1.1}$$

$$u(0,x) = \bar{u}(x), \tag{1.2}$$

for a nonlinear $n \times n$ system of conservation laws in one space dimension. As usual, we assume that the system is strictly hyperbolic and that each characteristic field is either linearly degenerate or genuinely nonlinear.

The existence of a globally defined weak solution was proved in the fundamental paper of Glimm [11]. Since then, the uniqueness of this solution has remained a major research problem. Results in this direction can be found in [7, 9, 13, 15, 18] and in several other papers. In many of these works, uniqueness is obtained using the Holmgren principle, within a class of sufficiently regular functions which satisfy an additional entropy-admissibility condition.

In this note, we propose an alternative approach to the problem of uniqueness and continuous dependence. Instead of a single Cauchy problem, we begin by looking at an entire flow of solutions of (1.1). Our first goal is to prove that there exists exactly one Lipschitz continuous flow which is compatible with the standard self-similar solutions [14] of the Riemann problems.

Having established the existence of this unique flow, the proof of uniqueness for the solution of a specific Cauchy problem then becomes a much easier task. Indeed, it then suffices to show that any admissible solution u of (1.1)-(1.2) actually coincides with the unique trajectory of the flow starting at $\bar{u}$. This is conveniently done by introducing a concept of "viscosity solution", which precisely characterizes the trajectories of the flow.

* S.I.S.S.A., Via Beirut 4, Trieste 34014, ITALY. E-mail: bressan@tsmi19.sissa.it

Our program thus consists of three main steps:

STEP 1: Show that there exists a nontrivial domain $\mathcal{D} \subset \mathbf{L}^1(\mathbb{R};\ \mathbb{R}^n)$, containing all functions $\bar{u}$ with sufficiently small total variation, and a continuous semigroup $S : \mathcal{D} \times [0, \infty[\ \mapsto \mathcal{D}$ with the following properties.

(P1) There exists a constant L such that, for all $t \geq 0$ and $\bar{u}, \bar{v} \in \mathcal{D}$, one has

$$\left\|S_t\bar{u} - S_t\bar{v}\right\|_{\mathbf{L}^1} \leq L \cdot \|\bar{u} - \bar{v}\|_{\mathbf{L}^1} .$$

(P2) If $\bar{u} \in \mathcal{D}$ is piecewise constant, then, for $t > 0$ sufficiently small, $S_t\bar{u}$ coincides with the solution of (1.1)-(1.2) obtained by piecing together the standard self-similar solutions of the corresponding Riemann problems.

A map S with the above properties will be called a *Standard Riemann Semigroup.*

In the current literature, three methods for constructing global weak solutions are well known, namely:

(i) The deterministic version of Glimm's scheme [16],

(ii) The vanishing viscosity method [10],

(iii) Wave-front tracking [2, 8, 19].

All of the above constructions rely on a compactness argument: a-priori bounds on the total variation, or a compensated compactness lemma, guarantee that some subsequence of approximate solutions actually converge to a globally defined weak solution. Unfortunately, these techniques do not provide any information on the dependence of the limit solution on the initial data. New constructive methods must therefore be devised. Very recently, the existence of a uniformly Lipschitz semigroup of solutions was proved in [1, 3, 6] for various classes of hyperbolic systems.

We remark that, for all $n \times n$ systems, the variational analysis in [4] suggests the precise form of a metric, equivalent to the usual $\mathbf{L}^1$ distance, which appears to be contractive w.r.t. a semigroup of solutions of (1.1). A rigorous construction of such a semigroup, however, has not yet been carried out for this general case.

STEP 2: Show that, for a suitable domain $\mathcal{D}$, the semigroup S with properties (P1)-(P2) is unique.

Notice that here we are concerned not with the uniqueness of the solution to a single Cauchy problem, but with the uniqueness of an entire flow. This step can thus be accomplished rather easily, for general $n \times n$ systems. The proof relies on the existence of a dense set of piecewise constant approximate solutions, obtained by wave-front tracking.

From the proof it also follows that, by properties (P1)-(P2), all trajectories of the semigroup are weak solutions of the system of conservation laws. If the system admits a convex entropy, these trajectories are entropy-admissible as well.

STEP 3: Isolate some property which characterizes the trajectories $t \mapsto S_t\bar{u}$ of the semigroup. Use this characterization in order to prove the uniqueness of the solution to the Cauchy problem (1.1)-(1.2).

To single out this characteristic property, the key idea is to compare a candidate function $u = u(t,x)$ on one hand with the self-similar solution $U^\sharp$ of a standard Riemann problem, and on the other hand with the solution $U^\flat$ of a linear hyperbolic system with constant coefficients. At each point (t_0, x_0), the maps $U^\sharp, U^\flat$ play the role of approximate parametrices for (1.1). A function u, which is suitably close (in a sense specified later) to the corresponding functions $U^\sharp, U^\flat$ on a neighborhood of each point (t_0, x_0), will be called a *viscosity solution* of the system of conservation laws (1.1). The main result in [5] states that a continuous function $u : [0,T] \mapsto \mathcal{D}$ is a trajectory of the semigroup if and only if it is a viscosity solution.

At this stage, in order to prove that the solution of (1.1)-(1.2) constructed by a given algorithm is unique and depends continuously on the initial data, it suffices to check that it is a viscosity solution. In [5] this is proved to be the case for solutions obtained by wave-front tracking and those generated by the Glimm scheme.

In the remaining sections of the paper, the above steps will be examined in somewhat greater detail.

2 - Existence of the semigroup

In connection with the system (1.1), given a piecewise constant function $v \in \mathbf{L}^1(\mathbb{R};\mathbb{R}^n)$ with small total variation, call $V(v)$ and $Q(v)$ respectively the total strength of waves and the wave interaction potential of v, as defined in [11]. For given constants $C_0, \delta_0 > 0$, consider the closed domain

$$\mathcal{D} \doteq cl\Big\{v \in \mathbf{L}^1(\mathbb{R};\ \mathbb{R}^n), \quad v \text{ is piecewise constant}, \ \ V(v) + C_0 Q(v) < \delta_0\Big\}, \tag{2.1}$$

where cl denotes closure in the $\mathbf{L}^1$ topology. Concerning the existence of a Standard Riemann Semigroup, the presently known results are collected in the following theorem.

Theorem 1. *Let the system (1.1) be strictly hyperbolic and assume that each characteristic field is either linearly degenerate or genuinely nonlinear. Then, for some constants* $C_0, \delta_0 > 0$ *in (2.1), a continuous semigroup* $S : \mathcal{D} \times [0, \infty[\mapsto \mathcal{D}$ *with properties (P1)-(P2) exists in each of the following cases:*

(1) *Every characteristic field is linearly degenerate.*

(2) *All characteristic curves are straight lines.*

(3) *There exists a coordinate system of Riemann invariants, and, in addition, all shock and rarefaction curves coincide.*

(4) $n = 2$.

Indeed, in case **(1)**, the semigroup was constructed in [1] by deriving a-priori estimates on a dense set of globally defined smooth solutions. The paper [3] deals with the cases **(2)**-**(3)**, relying on periodic mollifications and on piecewise constant approximations, respectively. The fact that every 2×2 system of conservation laws generates a uniformly Lipschitz semigroup was proved in [6]. A special case of **(4)**, motivated by gas dynamics, was previously considered in [17].

We conjecture that the result remains valid for general $n \times n$ systems. For an intuitive description of the basic ideas involved in the construction of this semigroup, see the introduction to [6].

3 - Uniqueness of the semigroup

Throughout the following, it is assumed that the constants $C_0, \delta_0 > 0$ are suitably chosen, so that the domain $\mathcal{D}$ in (2.1) is positively invariant. We begin by recalling a property of approximate solutions obtained by wave-front tracking [2, 5].

Lemma 1. *Let $S : \mathcal{D} \times [0, \infty[\mapsto \mathcal{D}$ be a Standard Riemann Semigroup for (1.1). Then for every $\bar{u} \in \mathcal{D}$ and $\varepsilon > 0$ there exists a piecewise constant approximate solution $u^\varepsilon = u^\varepsilon(t,x)$, with jumps occurring only along finitely many lines in the t-x plane, having the properties:*

$$\left\|u^\varepsilon(0,\cdot) - \bar{u}\right\|_{\mathbf{L}^1} \leq \varepsilon, \qquad u^\varepsilon(t,\cdot) \in \mathcal{D} \qquad \forall t \geq 0, \tag{3.1}$$

$$\limsup_{h \to 0+} \frac{\left\|u^\varepsilon(t+h,\cdot) - S_h u^\varepsilon(t,\cdot)\right\|_{\mathbf{L}^1}}{h} \leq \varepsilon \qquad \forall t \geq 0. \tag{3.2}$$

Observe that, since $u^\varepsilon(t,\cdot)$ is piecewise constant, for $h > 0$ small the function $S_h u^\varepsilon(t,\cdot)$ is entirely determined by the assumption (P2).

From (3.1)-(3.2) and property (P1) of the semigroup, for every $\tau > 0$ it follows

$$\begin{aligned}\left\|u^\varepsilon(\tau,\cdot) - S_\tau \bar{u}\right\|_{\mathbf{L}^1} &\leq \left\|S_\tau u^\varepsilon(0,\cdot) - S_\tau \bar{u}\right\|_{\mathbf{L}^1} + \int_0^\tau L \cdot \left(\limsup_{h \to 0+} \frac{\left\|u(t+h,\cdot) - S_h u(t,\cdot)\right\|_{\mathbf{L}^1}}{h} \right) dt \\ &\leq L\varepsilon + L\tau\varepsilon.\end{aligned} \tag{3.3}$$

Now let $\bar{u} \in \mathcal{D}$ and let $u = u(t,x)$ be a weak solution of (1.1)-(1.2) obtained by the wave-front tracking method [2]. In other words, assume

$$u(t,\cdot) = \lim_{\varepsilon_n \to 0+} u^{\varepsilon_n}(t,\cdot) \qquad \forall t \geq 0,$$

for a sequence of piecewise constant approximations u^{ε_n} satisfying (3.1)-(3.2). The bound (3.3) then implies

$$u(t,\cdot) = S_t \bar{u} \qquad \forall t \geq 0. \tag{3.4}$$

Observing that (3.4) must hold for every semigroup S satisfying (P1)-(P2) and every limit solution u, one obtains:

Theorem 2. *On the domain $\mathcal{D}$, if a Standard Riemann Semigroup generated by (1.1) exists, then it is necessarily unique. Its trajectories coincide with the the weak (entropy-admissible) solutions obtained by the wave-front tracking algorithm in [2]. In particular, this algorithm produces unique solutions, continuously depending on the initial data.*

Remark 1. In the classical approach, a "good" solution of (1.1)-(1.2) is usually meant to be a function $u = u(t,x)$ which satisfies the Cauchy problem in the distributional sense, plus some additional admissibility conditions. In particular, if the system admits a convex entropy η with entropy flux q, one requires that $\big[\eta(u)\big]_t + \big[q(u)\big]_x \leq 0$ in the sense of distributions.

Here, by a "good" solution we mean in the first place a "semigroup solution", i.e. a function $u = u(t,\cdot)$, with $u(0,\cdot) = \bar{u}$, which can be embedded in a continuous flow satisfying (P1)-(P2). By Theorem 2, this embedding condition alone suffices to completely characterize the solution. In other words, since the set of piecewise constant functions is dense, the entire flow is uniquely determined as soon as we prescribe how each Riemann problem is solved.

Remark 2. Throughout this paper, we exclusively consider the self-similar solutions of the Riemann problems constructed by Lax [14]. This choice guarantees that every trajectory of the semigroup is a weak, entropy-admissible solution of (1.1). Alternatively, however, one could also consider the flow generated by different "Riemann solvers". This may provide an appropriate framework for defining weak solutions to hyperbolic systems in non-conservative form.

4 - Viscosity Solutions

Let the system (1.1) admit a Standard Riemann Semigroup S. In order to characterize its trajectories, we introduce two types of local approximate parametrices for (1.1). One is derived from the self-similar solution of a Riemann problem, the other is obtained by "freezing" the coefficients of the corresponding quasilinear hyperbolic system in a neighborhood of a given point.

Let $u : [0,T] \times I\!\!R \mapsto I\!\!R^n$ be a locally integrable function, and fix any point (τ, ξ) in the domain of u. If $u(\tau,\cdot) \in \mathcal{D}$, the bound on the total variation implies the existence of the limits

$$u^- = \lim_{x\to\xi-} u(\tau,x), \qquad\qquad u^+ = \lim_{x\to\xi+} u(\tau,x).$$

Call $\omega = \omega(t,x)$ the self-similar solution of the Riemann problem

$$u_t + \big[F(u)\big]_x = 0, \qquad\qquad u(0,x) = \begin{cases} u^- & \text{if} \quad x < 0, \\ u^+ & \text{if} \quad x > 0. \end{cases} \tag{4.1}$$

Let $\hat{\lambda}$ be an upper bound for all characteristic speeds. For $t > \tau$, define

$$U^\sharp_{(u;\tau,\xi)}(t,x) \doteq \begin{cases} \omega(t-\tau,\ x-\xi) & \text{if} \quad |x-\xi| \leq \hat{\lambda}(t-\tau), \\ u(\tau,x) & \text{if} \quad |x-\xi| > \hat{\lambda}(t-\tau). \end{cases} \tag{4.2}$$

The function $t \mapsto U^\sharp_{(u;\tau,\xi)}(t,\cdot)$ is then Lipschitz continuous with values in $\mathbf{L}^1$, and approaches $u(\tau,\cdot)$ as $t \to \tau+$. Notice that this last property would fail, had we defined simply $U^\sharp(t,x) = \omega(t-\tau, x-\xi)$ for all t,x.

Next, call $\widetilde{A} \doteq DF(u(\tau,\xi))$ the Jacobian matrix of F computed at the point $u(\tau,\xi)$. For $t>\tau$, define $U^\flat_{(u;\tau,\xi)}(t,x)$ to be the solution of the linear hyperbolic Cauchy problem with constant coefficients

$$w_t + \widetilde{A} w_x = 0, \qquad w(\tau,x) = u(\tau,x). \tag{4.3}$$

In the following, by $\mathrm{TV}\{u(\tau);\ I\}$ we denote the total variation of the function $u(\tau,\cdot)$ over the set I.

Definition 1. Let $u : [0,T] \mapsto \mathcal{D}$ be continuous w.r.t. the $\mathbf{L}^1$-norm. We say that u is a *viscosity solution* of the system of conservation laws (1.1) if there exists a constant $C>0$ such that, at each point $(\tau,\xi) \in [0,T[\times I\!R$, for every $\rho,\varepsilon>0$ sufficiently small one has

$$\frac{1}{\varepsilon}\int_{\xi-\rho+\varepsilon\hat\lambda}^{\xi+\rho-\varepsilon\hat\lambda} \left|u(\tau+\varepsilon,x) - U^\sharp_{(u;\tau,\xi)}(\tau+\varepsilon,x)\right| dx \ \le C\cdot \mathrm{TV}\{u(\tau);\ ]\xi-\rho,\ \xi[\ \cup\ ]\xi,\ \xi+\rho[\}, \tag{4.4}$$

$$\frac{1}{\varepsilon}\int_{\xi-\rho+\varepsilon\hat\lambda\cdot}^{\xi+\rho-\varepsilon\hat\lambda} \left|u(\tau+\varepsilon,x) - U^\flat_{(u;\tau,\xi)}(\tau+\varepsilon,x)\right| dx \le C\cdot \Big(\mathrm{TV}\{u(\tau);\ ]\xi-\rho,\ \xi+\rho[\}\Big)^2. \tag{4.5}$$

If also $u(0,\cdot)=\bar u$, then we say that u is a viscosity solution of the Cauchy problem (1.1)-(1.2).

Remark 3. In the above definition, both estimates are required to hold at every point (τ,ξ). In practice, however, the behavior of a solution u is determined by (4.4) at points of discontinuity, and by (4.5) at points where u is continuous.

Indeed, if the function $u(\tau,\cdot)$ is continuous at $x=\xi$, then its total variation on $]\xi-\rho,\ \xi+\rho[$ is just the sum of the total variation on $]\xi-\rho,\ \xi[$ and on $]\xi,\ \xi+\rho[$. Hence (4.5) is a much sharper estimate than (4.4), because of the square on the right hand side. On the other hand, if $u(\tau,\cdot)$ has a jump at $x=\xi$, then (4.5) provides little information on the local behavior of u near (τ,ξ), because the right hand side does not approach zero as $\rho\to 0$.

The next result, proved in [5], provides a complete characterization of semigroup trajectories.

Theorem 3. *Assume that the hyperbolic system (1.1) admits a Standard Riemann Semigroup S on a domain $\mathcal{D}$ of the form (2.1). Then a continuous map $u:[0,T]\mapsto\mathcal{D}$ is a viscosity solution of*

(1.1)-(1.2) if and only if

$$u(t,\cdot) = S_t\bar{u} \qquad \forall t \in [0,T]. \tag{4.6}$$

Remark 4. It remains an open problem to prove that the limits as $\varepsilon \to 0$ of the viscous approximations

$$u_t + \big[F(u)\big]_x = \varepsilon u_{xx}$$

actually coincide with semigroup trajectories. In this regard we observe that, if the initial data $u(0,\cdot)$ is piecewise constant, then property (P2) implies that the semigroup solution $u(t,\cdot) = S_t\bar{u}$ is piecewise smooth for small $t > 0$. By [12], this solution can be obtained as a limit of viscous approximations. If the vanishing viscosity method produces solutions with bounded variation which depend Lipschitz continuously on the initial data, then, by Theorem 2, these solutions must coincide with the semigroup trajectories. In particular, they must be "viscosity solutions" according to the above definition.

What makes Definition 1 particularly convenient to work with, is that it only involves the $\mathbf{L}^1$ distance and the total variation. Concerning these two quantities, several estimates are already available in the literature. Often the bounds (4.4)-(4.5) can thus be established by well known techniques. In particular, using the theory of generalized characteristics, one can prove:

Theorem 4. *Let the system (1.1) admit a Standard Riemann Semigroup* $S : \mathcal{D} \times [0,\infty[\mapsto \mathcal{D}$. *Then, for every initial condition* $\bar{u} \in \mathcal{D}$, *the solution obtained by the Glimm scheme with uniformly distributed sampling is a viscosity solution; hence it is unique and coincides with the semigroup trajectory* $t \mapsto S_t\bar{u}$.

References

[1] A. Bressan, Contractive metrics for nonlinear hyperbolic systems, *Indiana Univ. Math. J.* **37** (1988), 409-421.

[2] A. Bressan, Global solutions of systems of conservation laws by wave-front tracking, *J. Math. Anal. Appl.* **170** (1992), 414-432.

[3] A. Bressan, A contractive metric for systems of conservation laws with coinciding shock and rarefaction waves, *J. Differential Equations* **106** (1993), 332-366.

[4] A. Bressan, A locally contractive metric for systems of conservation laws, *Ann. Scuola Norm. Sup. Pisa* **IV-22** (1995), 109-135.

[5] A. Bressan, The unique limit of the Glimm scheme, *Arch. Rat. Mech. Anal.*, to appear.

[6] A. Bressan and R. M. Colombo, The semigroup generated by 2×2 systems of conservation laws, *Arch. Rational Mech. Anal.*, to appear.

[7] C. Dafermos and X. Geng, Generalized characteristics, uniqueness and regularity of solutions in a hyperbolic system of conservation laws, *Ann. Inst. Henri Poincaré - Nonlinear Analysis* **8** (1991), 231-269.

[8] R. DiPerna, Global existence of solutions to nonlinear hyperbolic systems of conservation laws, *J. Differential Equations* **20** (1976), 187-212.

[9] R. DiPerna, Uniqueness of solutions to hyperbolic conservation laws, *Indiana Univ. Math. J.* **28** (1979), 137-188.

[10] R. DiPerna, Convergence of approximate solutions to conservation laws, *Arch. Rat. Mech. Anal.* **82** (1983), 27-70.

[11] J. Glimm, Solutions in the large for nonlinear hyperbolic systems of equations, *Comm. Pure Appl. Math.* **18** (1965), 697-715.

[12] J. Goodman and Z. Xin, Viscous limits for piecewise smooth solutions to systems of conservation laws, *Arch. Rat. Mech. Anal.* **121** (1992), 235-265.

[13] A. Heibig, Existence and uniqueness of solutions for some hyperbolic systems of conservation laws, *Arch. Rat. Mech. Anal.* **126** (1994), 79-101.

[14] P. Lax, Hyperbolic systems of conservation laws II, *Comm. Pure Appl. Math.* **10** (1957), 537-566.

[15] T. P. Liu, Uniqueness of weak solutions of the Cauchy problem for general 2×2 conservation laws, *J. Differential Equations* **20** (1976), 369-388.

[16] T. P. Liu, The deterministic version of the Glimm scheme, *Comm. Math. Phys.* **57** (1977), 135-148.

[17] V. J. Ljapidevskii, On correctness classes for nonlinear hyperbolic systems, *Soviet Math. Dokl.* **16** (1975), 1505-1509.

[18] O. A. Oleinik, Discontinuous solutions of nonlinear differential equations, *Usp. Mat. Nauk* **12** (1957), 3-73. English Transl. *Amer. Math. Soc. Transl.* Ser. 2, Vol.**26**, 95-172.

[19] N. H. Risebro, A front-tracking alternative to the random choice method, *Proc. Amer. Math. Soc.* **117** (1993), 1125-1139.

A High-order Godunov Scheme for Ideal Magnetohydrodynamical Equations Based on a Nonlinear Riemann Solver

Wenlong Dai[1] and *Paul R. Woodward*[1]

Abstract

A high-order Godunov scheme is developed for multi-dimensional ideal magneto-hydrodynamical (MHD) equations based on a nonlinear Riemann solver and the dimension splitting technique. The scheme is second-order accurate in both space and time. Its correctness and robustness are demonstrated through numerical examples.

1 Introduction

Godunov's approach [1] of using Riemann problems to obtain robust, upstream centered difference schemes has been very effective for compressible gas dynamics. Since the early MUSCL scheme [2], several other schemes involving high order constructions with characteristic formulations have been devised (for example, see [3-5]). The principal advantages of these schemes are the robust operation in the presence of very strong waves and thin shock fronts with little attendant noise generation.

For several years efforts have been underway to develop Godunov schemes for magnetohydrodynamical (MHD) equations. Brio and Wu [6] developed a high-resolution method for MHD equations based on Roe's approach. Zachary et al. [7,8] proposed a high order method for MHD equations using an extended Engquist-Osher flux. A major difficulty in developing Godunov schemes for MHD equations has been the development of a Riemann solver and dealing with the nonstrict hyperbolicity of the system.

The scheme described here is built upon a nonlinear MHD Riemann solver [9] and the dimensional splitting technique [10]. The procedure for multidimensional MHD simulations is described and the correctness and robustness of the scheme is demonstrated in this paper. The plan of the paper is as follows. In the second section the numerical scheme is described. The numerical examples are presented in the third section. Conclusions and brief discussions may be found at the end of this paper.

2 Numerical Schemes

The ideal MHD equations have the form:

$$\partial_t \rho + \nabla \cdot (\rho \mathbf{u}) = 0, \tag{1}$$

$$\rho(\partial_t \mathbf{u} + \mathbf{u} \cdot \nabla \mathbf{u}) = - \nabla p + \frac{1}{4\pi}(\nabla \times \mathbf{B}) \times \mathbf{B}, \tag{2}$$

[1]University of Minnesota, 1100 Washington Avenue South, Minneapolis, MN 55415

$$\partial_t(\frac{1}{2}\rho\mathbf{u}^2 + \rho\varepsilon + \frac{1}{8\pi}\mathbf{B}^2) = -\nabla\cdot\mathbf{q}, \tag{3}$$

$$\partial_t\mathbf{B} = \nabla\times(\mathbf{u}\times\mathbf{B}). \tag{4}$$

Here ρ, $\mathbf{u}$, p, $\mathbf{B}$, and ε are the mass density, flow velocity, thermal pressure, magnetic field and the specific internal energy density, and

$$\mathbf{q} \equiv \rho\mathbf{u}(\frac{1}{2}\mathbf{u}^2 + \varepsilon + p/\rho) + \frac{1}{4\pi}\mathbf{B}\times(\mathbf{u}\times\mathbf{B}).$$

The field $\mathbf{B}$ is subject to the divergence-free condition. We assume a γ-law for the equation of state: $p = (\gamma - 1)\rho\varepsilon$, with γ being the ratio of specific heats.

Since we will construct a scheme which uses dimensional splitting, we write the equations in the one-dimensional situation (e.g., in the x-direction):

$$\partial_t\mathbf{U} + \partial_m\mathbf{F}(\mathbf{U}) = 0, \tag{5}$$

Here we have introduced a Lagrangian mass coordinate defined by $dm \equiv \rho dx$. $\mathbf{U}$ and $\mathbf{F}$ are given by

$$\mathbf{U} \equiv \begin{pmatrix} V \\ u_x \\ u_y \\ u_z \\ VB_y \\ VB_z \\ E \end{pmatrix}, \quad \mathbf{F}(\mathbf{U}) \equiv \begin{pmatrix} -u_x \\ P \\ \Lambda_y \\ \Lambda_z \\ -B_x u_y \\ -B_x u_z \\ Pu_x + \Lambda_y u_y + \Lambda_z u_z \end{pmatrix},$$

where $V \equiv 1/\rho$.

Considering a numerical zone $[x_i,\, x_{i+1}]$, we write Eq.(5) in a difference form:

$$\tilde{\mathbf{U}}(\triangle t)_i = \tilde{\mathbf{U}}(0)_i + \frac{\triangle t}{\triangle m_i}(\overline{\mathbf{F}}_i - \overline{\mathbf{F}}_{i+1}). \tag{6}$$

Here $\triangle t$ is the time step and $\triangle m_i$ is the mass contained in the zone, $\tilde{\mathbf{U}}(t)_i$ is the average of $\mathbf{U}$ over the zone at time t, and $\overline{\mathbf{F}}_i$ is the time-averaged flux at the interface x_i during the time step, i.e.,

$$\tilde{\mathbf{U}}(\triangle t)_i \equiv \frac{1}{\triangle x_i}\int_{x_i}^{x_{i+1}} \mathbf{U}(x,t)dx, \tag{7}$$

$$\overline{\mathbf{F}}_i \equiv \frac{1}{\triangle t}\int_{\triangle t} \mathbf{F}(\mathbf{U}(x_i,t))dt$$

with $\triangle x_i$ being the width of the zone. Thus one of the key issues for this type of scheme is to obtain the time-averaged flux at the interfaces of numerical zones. The flux needed in our scheme is approximately calculated through a nonlinear and iterative Riemann solver presented in [9]. The initial guess needed in the Riemann solver comes from a linear Riemann solver which is based on characteristic formulations. The characteristic formulations for MHD equations have been discussed in previous papers [6-8]. The differentials of Riemann invariants of the system are

$$dR_0 = d(pV^\gamma), \tag{8}$$

$$dR_{f\pm} = (C_f^2 - C_a^2)(dP \pm C_f du_x) + \rho\Lambda_y(d\Lambda_y \pm C_f du_y) + \rho\Lambda_z(d\Lambda_z \pm C_f du_z), \tag{9}$$

$$dR_{s\pm} = (C_s^2 - C_a^2)(dP \pm C_s du_x) + \rho\Lambda_y(d\Lambda_y \pm C_s du_y) + \rho\Lambda_z(d\Lambda_z \pm C_s du_z), \tag{10}$$

$$dR_{a\pm} = \pm C_a(B_z du_y - B_y du_z) + (B_z d\Lambda_y - B_y d\Lambda_z). \tag{11}$$

Here dR_0 is for the entropy wave, $dR_{f\pm}$ for two fast waves, $dR_{s\pm}$ for two slow waves, and $dR_{a\pm}$ for two Alfven waves. The plus (or minus) sign stands for a wave propagating in the positive (or negative) x-direction.

Our one-dimensional code starts from zone-averages of a set of variables (ρ, p, u_x, u_y, u_z, B_y, B_z), which are defined through Eq.(7). For each of these variables, a cubic polynomial is used to find its value at an interface x_i from its zone-averages on zones (x_{i-2}, x_{i-1}), (x_{i-1}, x_i), (x_i, x_{i+1}) and (x_{i+1}, x_{i+2}). After we get the values of the variables at interfaces, the monotonicity constraint originally suggested by van Leer [5] is applied to these values at interfaces. As we know, interpolated structures are not always monotone increasing (decreasing) even though they have been constructed from monotone data. The over- and undershoots in the interpolated internal zone structures eventually give rise to over- and undershoots in the zone-averaged data. Van Leer realized that an advection scheme may be made to preserve the monotonicity of its initial data if any non-monotone interpolated zone structures are flattened so that they become monotone. This leads to van Leer's monotonicity constraint: no values interpolated within a zone shall lie outside the range defined by the zone averages for this zone and its two neighbors.

After the monotonicity constraint is applied, a parabola defined by a zone-average and the values at two interfaces of the zone is used to interpolate the structure inside the zone for the calculation of domain-averages for all kinds of waves. The flow is considered continuous inside each zone, but may change significantly across a zone interface. After we find the various domain-averages on fast, Alfven and slow domains, we use Riemann invariants to find effective left and right states for the Riemann problem arising from the interface x_i.

Once we have found the effective left and right states, we use the Riemann solver in [9] to find the time-averaged flux needed in the Godunov scheme Eq.(6), and update conserved quantities, i.e., mass, momentum, energy and magnetic flux. After zone-averages are updated on the Lagrangian grid, we map the conserved quantities onto the Eulerian grid. The mapping procedure is only a transformation between two grids, and keeps the conservation of the mass, momentum, energy and magnetic flux.

Mathematically, the vanishing of the longitudinal or tangential component of a local magnetic field may cause a singularity because of the vanishing expansion coefficients when the Riemann invariants are used. If any component of the magnetic field is close to zero, our code uses a small value for it in the calculation for the time-averaged flux to avoid the singularity. Since the value is small but is within the accuracy of digits of a machine, the resulting magnetic pressure is very small compared to the local thermal pressure.

We should mention that in the one-dimensional code only two transverse components of the magnetic field are updated, while the longitudinal component remains unchanged. The longitudinal component will be updated in other passes of the multidimensional scheme.

The multidimensional scheme is based on the dimensional splitting technique. Each time step of a higher dimensional problem is broken down into one-dimensional passes in which derivatives in all other directions are set to zero. Each pass (e.g., x-pass) of the scheme updates all the variables except B_x which is updated in other passes. The value of B_x at an interface is used in the

Riemann problem arising from that interface. We introduce neither any source term arising from the projection of the set of multidimensional equations to the set of one-dimensional equations nor any additional steps to formally ensure the divergence-free condition.

The divergence-free condition is imposed on an initial magnetic field in our numerical scheme. Although this condition is analytically maintained, it remains true only up to a certain accuracy for a difference scheme. Our numerical scheme does not involve derivatives of any variable. If calculated, the divergence could be O(1) near a discontinuity because of the approximation of a polynomial which is used to calculate the divergence, and the divergence is meaningless inside a shock in our formulation. It is the conservation of the magnetic flux that is important for a high-order difference scheme. Each one-dimensional pass in our scheme conserves the magnetic flux exactly. We will show the flux conservation across a shock through numerical examples.

3 Numerical Examples

The scheme has been tested on variety of problems. A few of them are presented here. These calculations are chosen to illustrate the correctness and robustness of the scheme. In these calculations, $\gamma = 5/3$, and only one iteration in the Riemann solver is used.

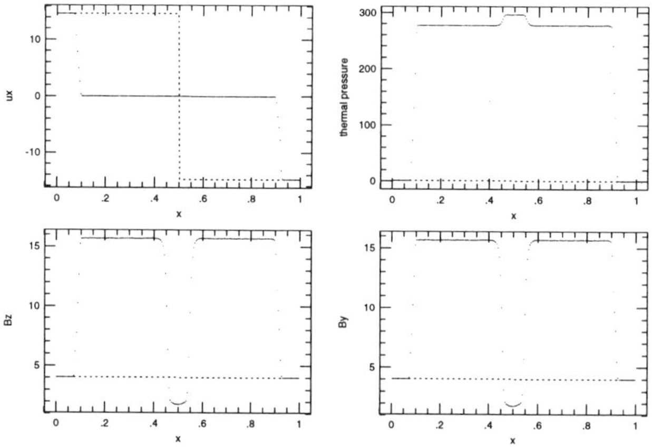

Figure 1: A one-dimensional test problem for the numerical scheme. The dashed and dotted lines are profiles at t = 0.0 and 0.08 respectively.

The first numerical example is a one-dimensional shock tube problem. Two hundred numerical zones, 4 for B_x are used. The initial condition is shown by the dashed lines in Fig.1. The solid lines are the profiles at t = 0.08, which show the generation of two slow shocks and two fast shocks with Mach numbers 8.9 and 10.1. The exact solution of the corresponding Riemann problem may be found through enough iterations in the Riemann solver, and the result given in the figure has excellent agreement with the exact solution.

The second example is the propagation of a fast wave with unit wavelength in the (x, y)-domain at an angle of 30° with respect to the x-axis. The simulation is performed on a $(x_{max}$- $0)\times(y_{max}$ - $0)$ domain with $x_{max} = y_{max} = 1/\cos 30^o$ and 256×256 uniform zones. Fig.2 shows the profiles along the line y = 128 zones at t = 0.5, 1.0 (solid lines). The dotted lines in the figure are the references obtained from purely one-dimensional simulations.

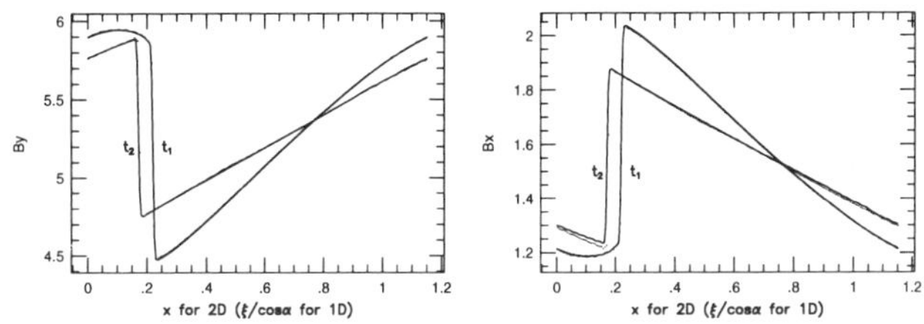

Figure 2: The comparison between one- and two-dimensional simulations for the steepening of a fast wave. Solid (or dotted) lines are from the two- (or one-) dimensional simulation. The lines indicated by t_1 and t_2 are the profiles after the wave has travelled about one wavelength and two wavelengths respectively.

The third calculation is for the Orszag-Tang vortex system which was introduced by Orszag and Tang [11] and was studied by Picone and Dahlburg [12]. The problem was also used to test the numerical scheme by Zachary et al. [8]. We choose this problem to test our simple approach for the nonstrict hyperbolicity of the equations when several waves propagating at different angles interact. The initial condition we used are

$$\rho(x, y, 0) = \rho_0,$$

$$p(x, y, 0) = p_0,$$

$$\mathbf{u}(x, y, 0) = -\sin(2\pi y)\mathbf{i} + \sin(2\pi x)\mathbf{j},$$

$$\mathbf{B}(x, y, 0) = -\sin(2\pi y)\mathbf{i} + \sin(4\pi x)\mathbf{j}.$$

Here $\mathbf{i}$ and $\mathbf{j}$ are unit vectors in the x- and y-directions. The parameters ρ_0 and p_0 are such chosen that the initial average Mach number M = 1.0, and the initial average β (the ratio of thermal pressure and magnetic pressure) $\beta = 10/3$. Here $\beta = 8\pi p_0/ < \mathbf{B}^2 >$ and $M^2 = \rho_0 < \mathbf{u}^2 > /\gamma p_0$ with the symbol $<>$ being the average. 256×256 zones in a unit square are used. Fig.3 shows the contours of the thermal pressure at at $t = 0.5$. The contours appear to match those described by Picone and Dahlburg [12] and described by Zachary et al. [8].

4 Conclusions and Discussions

In this paper a high order Godunov scheme has been developed for multidimensional MHD equations. The scheme is based on an iterative and nonlinear Riemann solver, which takes into account all the discontinuities in ideal MHD equations. The paper shows that the scheme offers the principle advantages of a high-order Godunov scheme. The scheme may be used to study MHD flows with strong discontinuities.

In this paper, we have not given the results for Brio and Wu's problem [6] which involves compound waves, although our simulation for the problem gave the same structures. Compound waves are phenomena of dissipative MHD equations. It should be stressed that a Godunov scheme, without any numerical viscosity added, already contains error terms which produce numerical diffusion in this type of scheme. As long as the internal structure of a discontinuity is not important for the dynamics, the scheme may correctly simulate resistive MHD phenomena, although it is based on the formulation of ideal MHD equations. The dissipative terms neglected in Eqs.(1-4) determine the internal structure of a discontinuity.

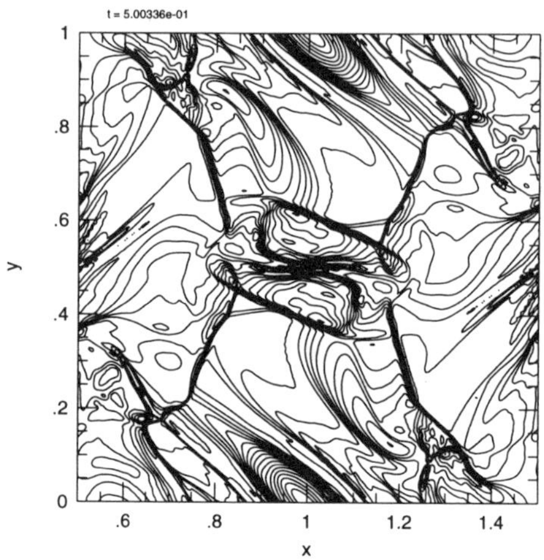

Figure 3: The contours of the thermal pressure in the Orszag-Tang vortex system at $t = 0.5$.

5 Acknowledgments

This work was supported by the U.S. National Science Foundation under grant NSF-ASC-9309829, by the U.S. Department of Energy through grant DE-FG02-87ER25035, by the Minnesota Supercomputer Institute, and by the Army Research Office contract number DAALO3-89-C-0038 with the Army High Performance Computing Research Center at the University of Minnesota.

References

[1] S. K. Godunov, *Difference methods for the numerical calculation of the equations of fluid dynamics*, Math. Sb. **47**, 271 (1959).

[2] B. Van Leer, *Toward the ultimate conservative difference scheme, V. a second-order sequel to Godunov's method*, J. Comput. Phys. **32**, 101 (1979).

[3] P. L. Roe, *Approximate Riemann solvers, parameters vectors, and difference schemes*, J. Comput. Phys. **43**, 358 (1981).

[4] A. Harten, *High resolution schemes for hyperbolic conservation laws*, J Comput. Phys. **49**, 357 (1983).

[5] P. Colella and P. R. Woodward, *The piecewise parabolic method (PPM) for gas-dynamical simulations*, J. Comput. Phys. **54**, 174 (1984).

[6] M. Brio and C. C. Wu, *An upwind differencing scheme for the equations of ideal magnetohydrodynamics*, J. Comput. Phys. **75**, 400 (1988).

[7] A. L. Zachary and P. Colella, *A high-order Godunov method for the equations of ideal magnetohydrodynamics*, J. Comput. Phys. **99**, 341 (1992).

[8] A. L. Zachary, A. Malagoli and P. Colella, *A higher-order Godunov method for multidimensional ideal magnetohydrodynamics*, SIAM J. Sci. Comput. **15**, 263 (1994).

[9] W. Dai and P. R. Woodward, *An approximate Riemann solver for ideal magnetohydrodynamics*, J. Comput. Phys. **111**, 354 (1994).

[10] G. Strang, *On construction and comparison of differential schemes*, SIAM J. of Num. Anal. **5**, 506 (1968).

[11] S.A. Orszag and C.M. Tang, *Small-scale structure of two-dimensional magnetohydrodynamical turbulence*, J. Fluid Mech. **90**, 129 (1979).

[12] J.M. Picone and R.B. Dahlburg, *Evolution of the Orszag-Tang vortex system in a compressible medium. II. Supersonic flow*, Phys. Fluid B **3**, 29 (1991).

Singularities in Contact Curves

Aparecido J. de Souza[1], Dan Marchesin[2]

ABSTRACT. We analyze the behavior of the linearly degenerate charcateristic field of a system of three conservation laws. The system describes three phase flow in porous media; it can model polymer injection of thermal processes used to improve oil recovery. Generically, there exists a curve consisting of points where the line field is singular. For polymer injection, we obtain the curve of singular points explicitly and we show that the line field has a saddle behavior. The curve of singular points coincides with the secondary bifurcation locus of Hugoniot branches associated with contact discontinuities.
Keywords: linearly degenerate field, contact curve, singular point.

1. Introduction.

We are interested in studying the behavior of the linearly degenerate line field that occurs in generic systems of conservation laws in a neighborhood of a curve of singular points. We consider systems of the following type:

$$
\begin{aligned}
u_t + f(u,v,z)_x &= 0 \quad x \in I\!R \\
v_t + g(u,v,z)_x &= 0 \quad t \in I\!R^+ \\
((\alpha u + \beta v + \gamma)z)_t + ((\alpha f(u,v,z) + \beta g(u,v,z) + \delta)z)_x &= 0,
\end{aligned}
\tag{1.1}
$$

where f and g are $\mathcal{C}^2$ functions defined in a domain $\Omega \in I\!R^3$, with $\alpha u + \beta v + \gamma \neq 0$ in Ω. Here α, β, γ and δ are real constants. In multiphase flow in porous media, when $\alpha = 1$ and $\beta = \gamma = \delta = 0$, system (1.1) models the injection of a polymer in the water phase [9]. For a model of injection of hot-water in petroleum reservoirs the constants α, β γ and δ are related to the thermal capacities of the phases, [7]. In both cases the system generalizes the systems studied by Isaacson&Temple and Keyfitz&Kranzer, [2, 4].

The system (1.1) possess a characteristic linearly degenerate field. The novel feature of this model is that under the implicit theorem function assumptions there is a curve where the linearly degenerate line field has saddle singularities. The curve of singularities is also the set of states where branches of Hugoniot curves associated with contact discontinuities have secondary bifurcation, [1].

In one way of interpreting Lax's Theorem [8], assuming the implicit theorem function hypotheses, the construction of Riemann solution of a 3×3 strictly hyperbolic system with a fixed left state U_L utilizes a *smooth* surface $\mathcal{M}$ in the state space Ω. This surface consists of the intermediate states U_M that can be connected to U_L by means of two waves (with lower and middle speeds) separated by a constant state. These waves satisfy an entropy condition and the compatibility condition

[1]Departamento de Matemática e Estatística, Universidade Federal da Paraíba, 58109-970, Campina Grande, PB, Brazil, desouza@brufpb2.bitnet

[2]Instituto de Matemática Pura e Aplicada, 22460-320, Rio de Janeiro, RJ, Brazil, marchesi@fluid.impa.br

that the speed of the waves increase from left to right in physical space (x,t). In our case, due to the presence of the curve of singular points, the construction of the Riemann solution is slightly more complicated. The surface $\mathcal{M}$ is *not* smooth any more since wave curves change direction abruptly. The relevant fact is that the surface $\mathcal{M}$ maintains transversality to the *upper* wave curves, thus permitting the construction of the Riemann solution.

For $\alpha = 1$, $\beta = \gamma = \delta = 0$, and f, g a pair of given functions used in Petroleum Engineering, we perform explicit calculations and characterize the curve of singular points of the linearly degenerate line field. For these parameters the Riemann solution of (1.1) near the curve of singular points is presented in [5].

2. Preliminaries.

The system (1.1) can be written in compact notation as

(2.1)
$$\mathcal{H}(U)_t + \mathcal{F}(U)_x = 0\,, \quad x \in I\!R, \quad t \in I\!R^+, \text{ where } U = (u,v,z)\,,$$
$$\mathcal{H}(U) = (u, v, (\alpha u + \beta v + \gamma)z) \text{ and } \mathcal{F}(U) = (f(U), g(U), (\alpha f(U) + \beta g(U) + \delta)z)\,.$$

The *Riemann problem* for (2.1) is the initial value problem with piecewise constant initial data given by:

$$U(x,t=0) = \begin{cases} U_L, & \text{for } x < 0 \\ U_R, & \text{for } x > 0 \end{cases} \quad . \tag{2.2}$$

For differentiable solutions U with values in Ω, system (2.1) can be written as

$$U_t + A(U) \cdot U_x = 0 \quad ,$$

where $A(U)$ is the matrix

$$A = (D\mathcal{H})^{-1} D\mathcal{F} = \begin{pmatrix} f_u & f_v & f_z \\ g_u & g_v & g_z \\ 0 & 0 & \lambda^z \end{pmatrix}, \quad \text{and } \lambda^z = \frac{\alpha f + \beta g + \delta}{\alpha u + \beta v + \gamma}. \tag{2.3}$$

A *rarefaction wave* connecting U_- to U_+ is a continuous piecewise smooth solution of system (2.1) with initial data $U_L = U_-$ and $U_R = U_+$, depending on x/t. For such a wave, the values of the function $U(x/t)$ must lie on the portion of an integral curve through U_- associated with one of the characteristic line fields, along which the corresponding characteristic speed is monotonically increasing. Such portion of the integral curve is called the *rarefaction curve* through U_-. Isolated critical points of the characteristic speed $\lambda(U)$ along the corresponding integral curve are called *inflection points*. However, a characteristic field of A associated with $\lambda(U)$ is called *linearly degenerate* if $\lambda(U)$ is constant along its corresponding integral curve.

Now we define the function $G : \Omega \times I\!R \times \Omega \to I\!R^3$ by

$$G(U, \sigma, U_-) = \sigma[\mathcal{H}(U) - \mathcal{H}(U_-)] - [\mathcal{F}(U) - \mathcal{F}(U_-)] \quad . \tag{2.4}$$

A *shock wave* connecting U_- to U_+ is a discontinuous (weak) solution of system (2.1), with initial data $U_L = U_-$ and $U_R = U_+$, consisting only of the two constant states U_- and U_+ related by the Rankine-Hugoniot condition

$$G(U_+, \sigma, U_-) = 0 \quad . \tag{2.5}$$

Notice that if $U_+ \neq U_-$, σ is uniquely determined by the equation (2.5), so we also write $\sigma = \sigma(U_-; U_+)$. The constant σ is called the *shock speed* between U_- and U_+. Actually, the states U_- and U_+ have to satisfy entropy conditions not discussed here.

The *Hugoniot curve* for the fixed left state U_-, denoted by $H(U_-)$, is the projection on Ω of the following set

$$\{(U, \sigma) \in \Omega \times I\!R \,|\, G(U, \sigma, U_-) = 0\} \quad .$$

For system (2.1), generically a Hugoniot curve consists of three branches emanating from U_- in the direction of the right eigenvectors of $A(U_-)$, together with detached branches. Following [1], the state U_- is called a primary bifurcation point of $H(U_-)$. Away from U_- we adopt the following definition. A *right-secondary bifurcation point* of a Hugoniot curve $H(U_-)$ is a state $U_0 \in H(U_-)$ distinct from U_- where the 3×4 matrix

$$\left(\frac{\partial G(U, \sigma, U_-)}{\partial (U, \sigma)} \right)_{|(U = U_0, \sigma = \sigma_0)}$$

is rank deficient. Notice that in such case $\sigma_0 = \lambda(U_0)$, where $\lambda(U_0)$ is a right eigenvalue of the matrix $A(U_0)$ defined in (2.3).

The *right-secondary bifurcation locus* is the set of states U_0 for which a right-secondary bifurcation point in $H(U_-)$ occurs for U_- in Ω, [3]. The *left-secondary bifurcation locus* is the set of states U_- for which $H(U_-)$ has a right-secondary bifurcation at some state U_0.

If for a shock wave the shock speed σ coincides with the characteristic speed λ on both sides of the discontinuity, the shock wave is called a *contact discontinuity*. A Hugoniot branch along which the shock speed σ is constant and coincides with a characteristic speed λ is called a *contact Hugoniot branch.*

According to [10], contact Hugoniot branches coincide with integral curves of linearly degenerate characteristic fields. Thus, we use the nomenclature *contact curves* for both types of waves.

A *Riemann solution* of system (2.1) with initial data U_L and U_R given in (2.2) is a sequence of rarefaction and shock waves separated by sectors of constant states. The leftmost constant state is U_L and the rightmost constant state is U_R.

3. The linearly degenerate field.

In this section we analyze the linearly degenerate characteristic field of system (2.1).

It is clear that one of characteristic speeds of matrix A in (2.3) is λ^z. Let us denote the eigenvector associated with $\lambda^z(U)$ by $X(U)$. It is easy to see that the characteristic speed λ^z is linearly degenerate, i.e., $\nabla\lambda^z(U)\cdot X(U)\equiv 0$. From (2.3), the line field $X(U)$ turns out to be:

$$X = ((\lambda^z - g_v)f_z + f_v g_z, (\lambda^z - f_u)g_z + g_u f_z, (\lambda^z - f_u)(\lambda^z - g_v) - f_v g_u). \tag{3.1}$$

It is easy to show that a sufficient condition for the line field X to possess a singular point U_0, where $X(U_0)=0$, is that any two of its three components vanish at U_0.

Now we connect the existence of singular points of the line field X with the existence of secondary bifurcation points of Hugoniot contact branches. First, let us characterize generic Hugoniot curves for system (2.1).

Theorem 3.1: Let U_- be a state in Ω. Then the Hugoniot curve $H(U_-)$ consists of two parts. The first part consists of branches lying on the plane $z=z_-$. The second part of $H(U_-)$ consists of contact branches generically transversal to planes $z = constant$.

Proof. Using the notation $f(U)$ for f, $f(U_-)$ for f_-, $g(U)$ for g and $g(U_-)$ for g_- in (2.5), we see that the Rankine-Hugoniot relation is defined by the system of three equations with four variables (u,v,z,σ):

$$\begin{gathered}
\sigma[u-u_-] = [f-f_-]\\
\sigma[v-v_-] = [g-g_-]\\
\sigma[(\alpha u+\beta v+\gamma)z - (\alpha u_- + \beta v_- + \gamma)z_-] =\\
[(\alpha f+\beta g+\delta)z - (\alpha f_- + \beta g_- + \delta)z_-].
\end{gathered}$$

If $z=z_-$, the three relations above reduce to the Rankine-Hugoniot relation for the subsystem

$$\begin{aligned}
u_t + f(u,v;z_-)_x &= 0\\
v_t + g(u,v;z_-)_x &= 0 \quad ,
\end{aligned}$$

which proves the first part of the Theorem.

Now let us consider $z\neq z_-$. A straightforward calculation gives us:

$$\begin{aligned}
\sigma(U_-;U) &= \lambda^z(U_-),\\
\lambda^z(U_-)[u-u_-] &= [f-f_-],\\
\lambda^z(U_-)[v-v_-] &= [g-g_-] \quad .
\end{aligned}$$

Thus such Hugoniot branches consist of states U which can be connected to U_- by contact discontinuities with speed $\lambda^z(U_-)$ and the proof is complete.

Finally, we have the following:

Proposition 3.2: Let U_- and U_0 be states in Ω with $z_0\neq z_-$, and $U_0\in H(U_-)$. Then U_0 is a singular point of the line field X, if and only if, U_0 is a right-secondary bifurcation point of a contact branch of $H(U_-)$.

Proof. According to the definition of right-secondary bifurcation point (see §2), we have to consider the function $G:\Omega\times\mathbb{R}\times\Omega\to\mathbb{R}^3$ defined by (2.4). Since we are interested only in secondary bifurcations of contact Hugoniot branches we fix $\sigma_0 = \sigma(U_-; U_0) = \lambda^z(U_0)$. By a direct calculation, we obtain that

$$Rank\left(\frac{\partial G(U,\sigma,U_-)}{\partial(U,\sigma)}\right)_{|(U=U_0,\sigma=\sigma_0)} = 2 \quad \Leftrightarrow \quad X(U_0) = 0\,,$$

and the proof is complete.

Since level surfaces of λ^z are foliated by integral curves of X, one can fix a state U_0 and analyze the line field X restricted to $S_0 = \{U \in \Omega \,|\, \lambda^z(U) = \lambda^z(U_0)\}$. Generically, there are only two possibilities for the singular point U_0. Either U_0 is a center corresponding to an isolated point of a Hugoniot contact branch, or U_0 is a point corresponding to a saddle (a right-secondary bifurcation point) where the two Hugoniot contact branches are separatrices.

If we assume that the implicit function theorem hypotheses hold in a neighborhood of a state $U_0 \in \Omega$ where X vanishes, then there exists a curve of singular points of the line field X. If this curve of singular points consists of saddle singularities of X (on the level surfaces of λ^z), then such a curve coincides with the right-secondary bifurcation locus of the Hugoniot contact branches with saddle structure.

Remark: Calculations similar to the ones we have shown for three equations can be performed for $n+1$ equations, replacing the first two flux functions of (1.1) by n flux functions and replacing the last equation of (1.1) by a linear combination of the previous n flux functions, [6].

4. The curve of singularities for the polymer model.

Here we consider the case of system (1.1), when $\alpha = 1$, $\beta = \gamma = \delta = 0$. We take the flux functions as $f(u,v,z) = \mu(z)u^2/D(u,v,z)$ and $g(u,v,z) = bv^2/D(u,v,z)$, with $D(u,v,z) = \mu(z)u^2 + bv^2 + c(1-u-v)^2$. Here b and c are positive real constants and $\mu(z)$ is a prescribed C^2 real function such that $\mu(z) > 0$ and $\frac{d\mu(z)}{dz} < 0$. The dependent variables take values in the prism $\Omega = \{(u,v,z) \in I\!R^3 \,| 0 \le u, v, u+v \le 1 \text{ and } 0 \le z \ll 1\}$. In the context of flow in porous media, u, v, $1-u-v$ may be thought of as the saturations of each of the three phases, water, oil and gas. The concentration of a polymer transported in the water phase is z; the presence of the polymer affects the viscosity of the water phase. With this interpretation, this is a model for polymer injection, a method used in Petroleum Engineering to improve oil recovery in reservoirs.

Remark: In [5], the function $\mu(z)$ considered is $\frac{1}{a(z)}$. In that case $\mu(z)$ is the inverse of the viscosity parameter of the water phase.

Proposition 4.1: The line field X has a curve of singularities interior to the domain Ω.

Proof. By a direct calculation using the explicit formulas for the model above and eliminating denominators, the line field (3.1), scaled by D^3, can be written as:

$$\begin{aligned} X_1(u,v,z) &= u^2[(\mu' D - \mu D_z)(\mu u - 2bv) + \mu' b v^2 D_v] \\ X_2(u,v,z) &= buv^2[\mu D_z - \mu' u D_u] \\ X_3(u,v,z) &= \mu u[(\mu u - 2bv)(uD_u - D) - bv^2 D_v] \end{aligned} \tag{4.1}$$

A straightforward calculation shows that the line field X vanishes at an interior curve Γ with $\frac{\partial \Gamma}{\partial z} \neq 0$. This curve may be parameterized by z as:

$$\Gamma(z) = \left(\frac{2bc}{(b+c)\mu(z) + 2bc}, \quad \frac{c\mu(z)}{(b+c)\mu(z) + 2bc}, \quad z \right) . \tag{4.2}$$

Let us study the degenerate characteristic field X near the curve of singularities Γ. Let $U_0 = (u_0, v_0, z_0)$ be an arbitrary state on Γ and consider the line field in a neighborhood of U_0.

A direct calculation shows that the plane π_0 tangent to the surface S_0, where λ^z is constant, at the point U_0 is given by:

$$(z - z_0) = K(u - u_0), \tag{4.3}$$

where $K = -\mu_0/(\mu_0' u_0)$ and $\mu_0 = \mu(z_0)$.

We focus our attention on a neighborhood of U_0. Let us consider the line field X restricted to π_0. Consider the basis $\{\mathbf{f_1}, \mathbf{f_2}\}$ which spans the plane π_0, with $\mathbf{f_1} = \mathbf{e_1} + K\mathbf{e_3}$ and $\mathbf{f_2} = \mathbf{e_2}$, where $\{\mathbf{e_i}\}_{i=1}^{3}$ is the standard basis for $\mathbb{R}^3$. Substituting (4.3) in (4.1), the line field X can be written (for $p = (u, v, z_0 + K(u - u_0))$ on π_0) as

$$X = X_1(p)\,\mathbf{f_1} + X_2(p)\,\mathbf{f_2} + (X_3(p) - KX_1(p))\,\mathbf{e_3}. \tag{4.4}$$

Notice that the coefficients of $\mathbf{f_1}$ and $\mathbf{f_2}$ vanish at (u_0, v_0, z_0) while the coefficient of $\mathbf{e_3}$ vanishes quadratically at this point. Let $\tilde{X}(u,v) = X_1(p)\mathbf{f_1} + X_2(p)\mathbf{f_2}$ be the projection of X onto π_0 along $\mathbf{e_3}$.

Linearizing $\tilde{X}$ near $U_0 \equiv p_0$ we obtain:

$$\frac{\partial(X_1, X_2)}{\partial(u,v)} = \begin{pmatrix} 2bc\mu_0' u_0^2 v_0^2 & \frac{2b\mu_0' u_0^2 v_0^2}{c}(c^2 - b^2) \\ -2bc\mu_0' u_0^2 v_0^2 & -2bc\mu_0' u_0^2 v_0^2 \end{pmatrix} ,$$

which yields

$$det\left(\frac{\partial(X_1,X_2)}{\partial(u,v)} \right)_{|U_0} = -(2b^2\mu_0' u_0^2 v_0^2)^2 < 0$$

This means that for any value z_0 the state U_0 on the singular curve Γ corresponds to a saddle point of the line field $\tilde{X}$ defined on the tangent plane π_0. Since Γ is transversal to any plane π_0, varying U_0 along Γ, we see that the orbits of the line field X topologically are a family of saddles centered on the curve Γ. See figure (4.1).

Conclusion. The classical Theorem of Lax proves the existence of Riemann solutions provided the contact line field has a trivial flow structure. In certain classes of models, we have shown that the contact line field has a saddle structure. Nevertheless, we will show in future work that the Theorem of Lax can be reinterpreted and extended to prove the existence of Riemann solutions for such models.

Acknowledgments. We thank the Brazilian research foundation "Conselho Nacional de Desenvolvimento Científico e Tecnológico-CNPq", as well as the Instituto

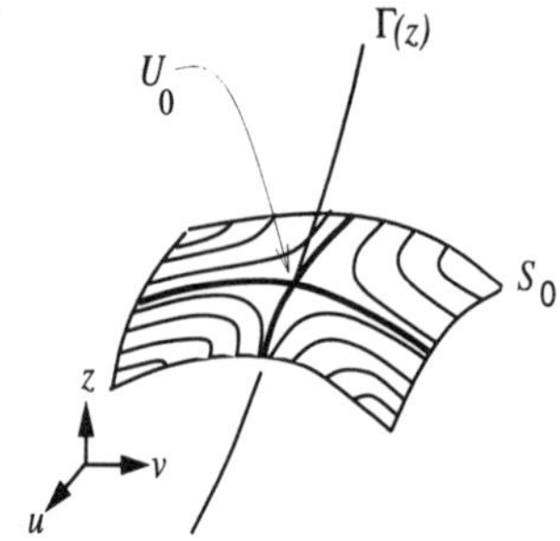

Fig. 4.1: The curve of singular points Γ and the line field X on S_0, for the polymer model.

de Matemática Pura e Aplicada -IMPA, and the Universidade Federal da Paraíba for support. This work was supported in part by the Conselho Nacional de Desenvolvimento Tecnológico e Científico under Grant CNPq/NSF 910087/92-0, under Grant CNPq 52.0609/93-0 and the Financiadora de Estudos e Projetos under Grant 65920311-00.

References.

[1] M. Golubitsky, D. G. Shaeffer, *Singularities and Groups in Bifurcation Theory*, vol. I, Springer Verlag, 1985.

[2] E. Isaacson, *Global solution of a Riemann problem for a non-strictly hyperbolic system of conservation laws arising in enhanced oil recovery*, preprint, Rockefeller University, New York, (1981).

[3] E. Isaacson, D. Marchesin, F. Palmeira, B. Plohr, *A global formalism for nonlinear waves in conservation laws*, J. Comm. Math. Phys.,vol. 146, (1992), 505–552.

[4] B. Keyfitz, H. Kranzer, *A system of non-strictly hyperbolic conservation laws arising in elasticity theory*, Arch. Rat. Mech. Anal., 72 (1980), 219–241.

[5] D. Marchesin, A. J. de Souza, *Fundamental solutions of a system of conservation laws with a singular characteristic field*, Mat. Contemporânea, (1994), to appear.

[6] D. Marchesin, A. J. de Souza, *On a class of systems of conservation laws with singularities on contact curves*, in preparation.

[7] J. C. da Mota, The Riemann problem for a simple thermal model for two phase flow in porous media, Mat. Aplic. Comp., vol. 11, #2, (1992), 117–145.

[8] J. Smoller, *Shock Waves and Reaction Diffusion Equations*, Springer Verlag, 1983.

[9] A. J. Souza, *Wave structure for a non-strictly hyperbolic system of three conservation laws*, Advances in Hyperbolic Diff. Equations, to appear.

[10] J. B. Temple, *Systems of Conservation Laws with Invariant submanifolds*, Trans. Amer. Math. Soc., vol. 280, #2, (1980), 781–795.

Wave Processes in a Three Phase System of a Collapsing Bubble Near a Compliant Wall

Christian Dickopp[1], Josef Ballmann[1]

With the intention to find the still unknown mechanism of cavitation damaging the basic problem of a single bubble near to a compliant wall is simulated by the coupling of two numerical methods in a time accurate way:

- The two phase flow of a gas or vapour bubble in a liquid is described by the system of the Navier-Stokes equations completed with the material laws of both fluids and jump conditions at the phase interface. The solution of these equations is approximated by a finite element method with a flux corrected transport approach following Löhner, et al.
- For the simulation of the behavior of the adjacent elastic solid, a Godunov-type-method solves the system of the elastodynamic equations with the assumption of a linear stress strain relation.

Both methods are coupled in a time correct manner by an alternate calculation of the numerical approaches using a predictor corrector strategy.

1 Introduction

Cavitation damage results from the collapse of bubbles in liquids at high pressure near a solid surface and is an important problem in many technical applications of liquid flows. Although theoretical and experimental investigations have been done for about 150 years the mechanism of this kind of erosive material damage is still unknown. Probably pressure waves and jets which are generated in the liquid phase load the solid surface in such a way that the stresses inside the solid cause a break of this material locally.

With the intention to find an explanation for the cavitation damage the dynamical basic problem of a single gas or vapour bubble collapsing near a solid surface is observed (Figure 1).

In this three phase problem the motion of the two fluids is described by the Navier-Stokes-equations completed by material laws whereas the elastodynamical equations with a linear stress-strain relation are solved to simulate the behavior of the solid. The problem contains two phase interfaces across which jump conditions for contact discontinuities have to be satisfied. Because of the high importance of the interface between the two fluids, this contact discontinuity is fitted during the calculation in a time accurate way by an adaption of the grid in such a manner that the discontinuity appears as a closed curve in the net.

[1]Lehr- und Forschungsgebiet für Mechanik, RWTH Aachen, University of Technology, Templergraben 64, D-52056 Aachen, Germany. dickopp@riemann.lufmech.rwth-aachen.de

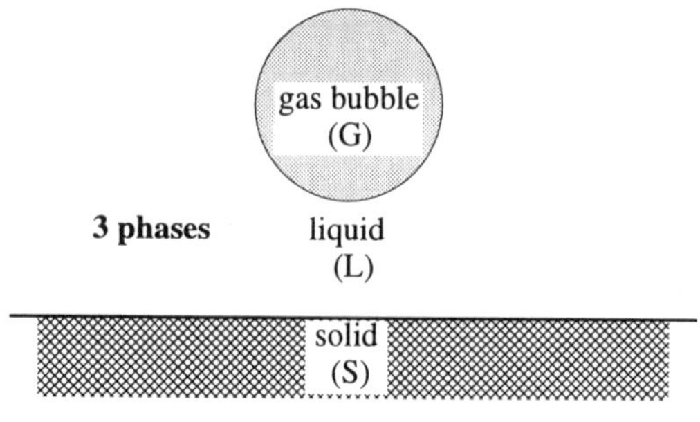

Figure 1: Schematic presentation of the problem

2 The Mathematical Formulation

The motion of the gas and the liquid is described by the system of the conservation equations of mass, momentum and energy for compressible, viscous fluids in two space dimensions on a moving grid with the velocity vector $\underline{v}_{grid}$:

$$\underline{U}_{,t} + \operatorname{div} F(\underline{U}) = \underline{0}$$

where $\underline{U}$ denotes the vector of conserved variables and $F(\underline{U})$ the flux function:

$$\underline{U} = \begin{pmatrix} \rho \\ \rho \underline{v} \\ \rho(\frac{v^2}{2} + e) \end{pmatrix}, \qquad \underline{F}(\underline{U}) = \begin{pmatrix} \rho(\underline{v} + \underline{v}_{grid}) \\ \rho \underline{v} \circ (\underline{v} + \underline{v}_{grid}) + pI - \tau \\ \rho(\frac{v^2}{2} + e)(\underline{v} + \underline{v}_{grid}) + p\underline{v} - \tau \underline{v} + \underline{q} \end{pmatrix}.$$

To complete this system of equations a Newtonian formulation of the stress tensor, τ, with the Stokes hypothesis, the Fourier ansatz for the heat flux q, and an equation of state for both fluids are used. The pressure in liquids such as water can be approximated by the TAIT-equation of state:

$$\frac{p + B}{p_0 + B} = \left(\frac{\rho}{\rho_0}\right)^n,$$

$$p_0 = 7025 \text{ Pa}, \qquad \rho_0 = 998.2 \,\frac{\text{kg}}{\text{m}^3}, \qquad B = 3.01 \times 10^8 \text{ Pa}, \qquad n = 7.15.$$

The behavior of the gas can optionally be assumed as thermally perfect or alternatively according to the van-der-Waals-equation, the Virial-equation or the Beattie-Bridgeman equation of state. The according relations between the internal energy on the one side and the temperature and the density on the other side result from thermodynamical laws.

Across the gas liquid interface the following jump conditions for contact discontinuities resulting from the Navier-Stokes equations have to be satisfied. $< x > = x_L - x_G$ denotes the jump in the variable x between the value on the liquid side and that on the gas side of the discontinuity, $\underline{n}$ is the unit normal vector on the discontinuity, indices: G - gas, L - liquid, S - solid:

$$<\rho>= \text{ arbitrary }, \quad \frac{2\sigma}{R} + < p-(\tau\underline{n})\underline{n} >= 0, \quad \frac{2\sigma\underline{v}\underline{n}}{R} + < (p-\tau\underline{n}\underline{n})\underline{v}\underline{n} + \underline{q}\underline{n} >= 0.$$

σ denotes the coefficient of suface tension, R is the local radius of curvature.

Optionally the mass flux across the gas liquid interface Γ_{GL} caused by condensation and evaporation is also taken into account by a source term on the right hand side of the continuity equation:

$$\rho_{,t} + \operatorname{div}(\rho\underline{v}) = \int_{\Gamma_{GL}} \dot{m}\, ds$$

where $\dot{m}$ denotes the mass flux per second obtained from thermodynamical relations.

Because the fluid interface Γ_{GL} is a contact discontinuity the additional condition $v_L = v_G = v_{grid}$ has to be incorporated in the solution method. The hyperbolic parts of the resulting system of equations for the two phase flow are the dominant ones.

The velocity $\underline{v}_S$ and the stresses σ inside the compliant solid are obtained as the solution of the following system of elastodynamical equations in combination with a Hookean law for the material in cylindrial coordinates r, θ and z:

$$\frac{\partial \underline{w}}{\partial t} = \frac{\partial \underline{f}}{\partial r} + \frac{\partial \underline{g}}{\partial z} + \underline{h}$$

with

$$\underline{w} = \begin{pmatrix} \rho v_{S1} \\ \rho v_{S2} \\ \frac{1}{\rho c_1^2}\sigma_r \\ \frac{1}{\rho c_1^2}\sigma_z \\ \frac{1}{\rho c_1^2}\sigma_\theta \\ \frac{1}{\rho c_2^2}\sigma_{rz} \end{pmatrix}, \quad \underline{f} = \begin{pmatrix} \sigma_r \\ \sigma_{rz} \\ v_{S1} \\ \alpha v_{S1} \\ \alpha v_{S2} \\ v_{S2} \end{pmatrix}, \quad \underline{g} = \begin{pmatrix} \sigma_{rz} \\ \sigma_z \\ \alpha v_{S2} \\ v_{S2} \\ \alpha v_{S2} \\ v_{S1} \end{pmatrix}, \quad \underline{h} = \begin{pmatrix} \frac{\sigma_r - \sigma_\theta}{r} \\ \frac{\sigma_{rz}}{r} \\ \alpha\frac{v_{S1}}{r} \\ \alpha\frac{v_{S1}}{r} \\ \frac{v_{S1}}{r} \\ 0 \end{pmatrix}.$$

where

$\rho =$ density of the solid, $c_1 =$ pressure wave velocity, $c_2 =$ shear wave velocity,

$$\alpha = 1 - 2(\frac{c_2}{c_1})^2.$$

The two systems of equations, i.e. the fluid dynamical equations for the gas and the liquid on the one hand and the elastodynamical equations for the solid on the other hand, are mathematically coupled by the conditions across the liquid solid interface Γ_{LS} that form a second contact discontinuity of the observed problem with the normal vector $\underline{n}$:

$$-p_L + \tau_L\underline{n}\underline{n} = \sigma_S\underline{n}\underline{n}, \qquad \tau_L\underline{n}\overline{\underline{n}}^T = \sigma_S\underline{n}\overline{\underline{n}}^T, \qquad \underline{v}_L = \underline{v}_S.$$

3 The Numerical Method

The solution of the fluid dynamical equations is approximated by a finite element method with a flux corrected transport approach following Löhner, et al. ([1]), a Taylor-Galerkin method with second order of accuracy in space and time. Across discontinuities the order is reduced to one. This method has been improved for the treatment of the liquid phase because of the very sensitive TAIT-equation of state used for fluids like water.

The foundation of the discretization is a Galerkin formulation of the conservation laws with the test functions ϕ_N for the variables $\underline{U}$ according to each grid point N:

$$\int_\Omega \phi_N \phi_M \, d\Omega \, \underline{U}_{M,t} + \int_\Omega \phi_N \phi_{M,i} \, d\Omega \, F_i(\underline{U})_M = \underline{0}.$$

The jump conditions across the gas liquid interface Γ_{GL} are incorporated in this discretization by boundary integrals over Γ_{GL}:

$$\begin{aligned}
\textstyle\int_\Omega \phi_N \phi_M \, d\Omega \, \rho_{M,t} + \int_\Omega \phi_N \phi_{M,i} \, d\Omega \, (\rho(\underline{v}_i + \underline{v}_{grid\,i}))_M &= \textstyle\int_{\Gamma_{GL}} \phi_N^* \dot{m} \, ds, \\
\textstyle\int_\Omega \phi_N \phi_M \, d\Omega \, (\rho \underline{v})_{M,t} & \\
+ \textstyle\int_\Omega \phi_N \phi_{M,i} \, d\Omega \, (\rho \underline{v}(\underline{v}_i + \underline{v}_{grid\,i}) + pI - \tau e_i)_M & \\
+ \textstyle\int_{\Gamma_{GL}} \phi_N^* \phi_M^* \, ds(\frac{2\sigma}{R}\underline{n} + < p - (\tau \underline{n})\underline{n} > \underline{n})_M &= \underline{0}, \\
\textstyle\int_\Omega \phi_N \phi_M \, d\Omega \, (\rho(\frac{\underline{v}^2}{2} + e))_{M,t} & \\
+ \textstyle\int_\Omega \phi_N \phi_{M,i} \, d\Omega \, (\rho(\frac{\underline{v}^2}{2} + e)(\underline{v}_i + \underline{v}_{grid\,i}) + (p - \tau \underline{v})\underline{v}_i)_M & \\
+ \textstyle\int_{\Gamma_{GL}} \phi_N^* \phi_M^* \, ds(\frac{2\sigma \underline{v}\underline{n}}{R} + < (p - \tau \underline{n}\underline{n})\underline{v}\underline{n} + \underline{q}\underline{n} >)_M &= 0.
\end{aligned}$$

The jump conditions at the liquid solid interface Γ_{LS}, are satisfied in a similar way. To fulfil the contact condition $\underline{v}_L = \underline{v}_G$ at the fluid interface Γ_{GL} the momenta, $\rho_G \underline{v}_G$ and $\rho_L \underline{v}_L$, are averaged for the points on Γ_{LG} in such a way that the total momentum is conserved.

To fit the contact discontinuity of the gas liquid interface a remeshing of the unstructered grid is necesarry in every time step. With respect to the performance and to the run time of the program, an algebraic grid generator was preferred that forms an unstructured grid

For the simulation of the behavior of the elastic solid in the neightborhood of the collapsing bubble, a second order Godunov type method solves the system of elastodynamic equations ([2]). Both methods are coupled in a time correct manner by the alternate application of the programs using a predictor corrector strategy. The liquid solid interface is treated as an outer boundary for the fluid solver and the Godunov-type-method. The boundary values change with time in both methods. The alternate use of the programs in a predictor corrector sense increases the time accuracy up to two for the values on the liquid solid interface and therewith the same order as in the fluid or solid regions.

4 Results

The numerical calculations are started with homogeneous but different density, temperature and pressure distributions in both fluids and an unloaded solid. At the liquid points on the gas-liquid interface, the same pressure as in the gas was assumed. All three phases and the according grids are at rest in the beginning.

The results printed on the next two pages show the development of an underpressure region around the bubble in the liquid and especially between the bubble and the solid following the initiation of the collapse. The solid prohibits a simple radial acceleration of the liquid toward the bubble as it appears in the upper part of the flow region. So the pressure gradient accelerates the liquid in the direction parallel to the wall from both sides and furthermore to the collapsing bubble. This seems to be the mechanism of formation of a liquid jet from the wall towards the bubble that has been already recognized in experiments for certain distances between the bubble and the solid.

During the further collapse, pressure waves towards the solid can be generated if the pressure of the gas inside the bubble achieves or overshoots the pressure in the liquid. Another effect is the formation of a second jet already observed in the simulations of other authors as well as in experimental researches. The resulting dynamical changes of the loading of the solid surface and the resulting high shear stresses in the solid might be a reason for cavitation damage. This research work, which is still in progress, aims to contribute new clarifying arguments.

References

[1] Löhner, R., Morgan, K., Peraire, J., Vahdati, M., "Finite Element Flux Corrected Transport (FEM-FCT) for the Euler and Navier-Stokes equations", Finite Elements in Fluids - Volume 7, 105 pp., (1987) John Wiley & Sons Ltd.

[2] Lin, X., Ballmann, J., "A Riemann Solver and a Second Order Godunov Method for Elastic Plastic Wave Propagation in Solids", Int. J. Impact Engng - Vol. 13, p. 473-483, (1993).

[3] Westenberger, H., "Zur Dynamik des Impulstransportes in blasigen, kompressiblen Flüssigkeiten", Dr. rer. nat. thesis, RWTH Aachen, (1987)

timestep = 100, time = 8.13 10^{-6} sec | **timestep = 200, time = 1.35 10^{-5} sec**

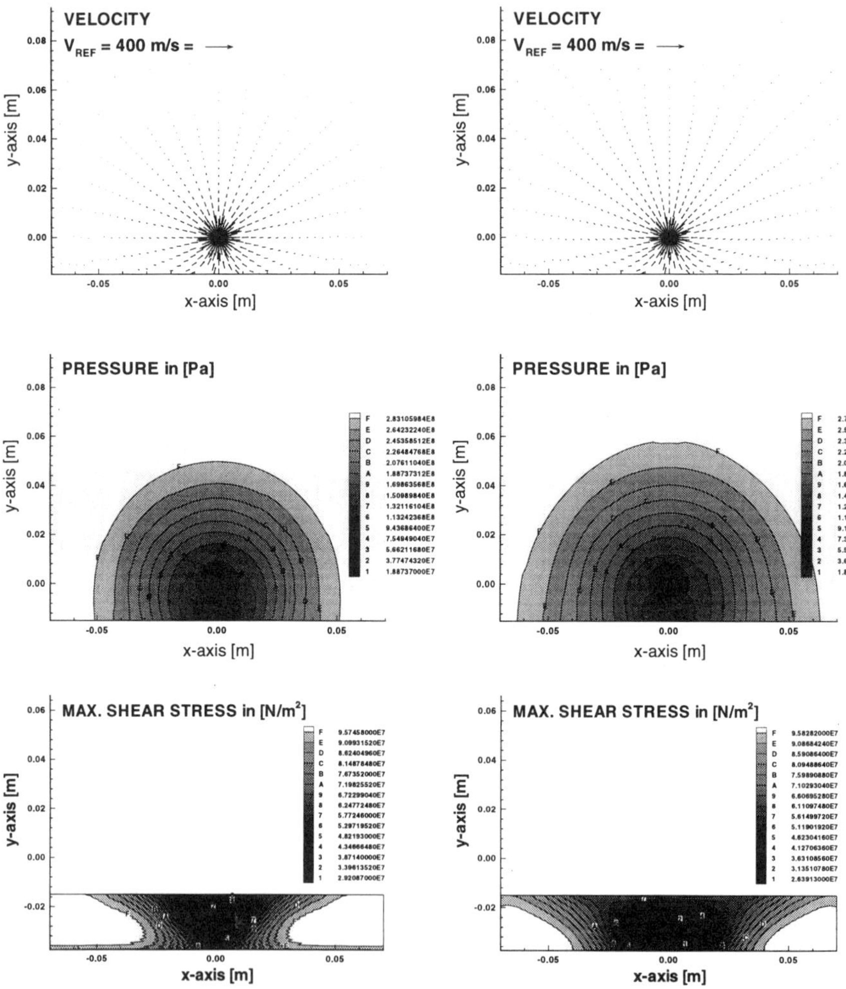

Structure and Stability of Radiative Shock Waves

D. Folini[1] and R. Walder[2]

ABSTRACT. Radiative shock waves are a widespread phenomenon in astrophysics. To model them we numerically solve the time dependent Euler equations with a source term to account for radiative energy losses. For this special source term, which can be very stiff, only very little is known about the solutions of these equations. We have solved them in spherical symmetry for a few astrophysical settings. We find that the solution often shows an oscillatory instability due to the presence of the source term.

This is the first time, at least in an astrophysical context, that the interaction zone of two radiative colliding flows could be resolved in such detail. To achieve the necessary resolution we use the adaptive mesh refinement algorithm of Berger [1].

Keywords: radiative shock waves, Euler equations, stiff source terms, adaptive meshes

1. Astrophysical background

From an astronomical point of view, radiative shock waves are important as they can be found in many different astronomical objects. Prominent examples are supernova remnants or active galactic nuclei. However, supersonically colliding flows can also be found in many less spectacular objects. Depending on the particular object, the radiation from such shock waves can contribute significantly to the emitted spectrum. Therefore, a proper understanding of radiative shock waves is an important step towards a better understanding of the entire object.

However, our interest in this context lies in the mathematical formulation of the problem and in the numerical treatment. The astrophysical background is of no more importance for what follows. We first present the mathematical formulation and the numerical method. Then we discuss the general structure of the solution and, finally, consider its stability.

2. Mathematical formulation and numerical method

In the following we are looking for solutions of the Euler equations in spherical symmetry with a source term, $u_t + f(u)_r = \Psi(u)$. We consider a particular source term of the form $\dot{E} = -N^2\Lambda(T)$ to account for radiative losses of the shock heated matter. N denotes the particle density of the gas and the radiative loss function $\Lambda(T)$ describes the temperature dependence of the source term, according to $\Lambda(T) =$

[1]Seminar für Angewandte Mathematik ETH-Zentrum 8092 Zürich, Switzerland, folini@sam.math.ethz.ch

[2]Inst. für Astron., ETH-Zentrum 8092 Zürich, Switzerland, walder@astro.phys.ethz.ch

$\Lambda_{0i}T^{\beta_i}, T \in [T_i, T_{i+1})$. In Fig. 1 it can be seen that there are temperature ranges where the radiated power increases with decreasing temperature. These regions have significant influence on the stability properties of the solution. A second source term accounts for radiative heating, ensuring that $T \geq T_0$, where T_0 is a certain cutoff temperature. We use the caloric equation of state, $e = p/(\gamma - 1)$, with $\gamma = 5/3$.

The equations are designed on a Cartesian mesh and solved by a MUSCL-type integrator (Colella and Glaz [2]). A Strang splitting accounts for the source terms (geometrical and physical ones), and the method is second order accurate in space and time. To achieve the necessary spatial and temporal resolution, we use the adaptive mesh refinement algorithm (AMR in the following) of Berger [1].

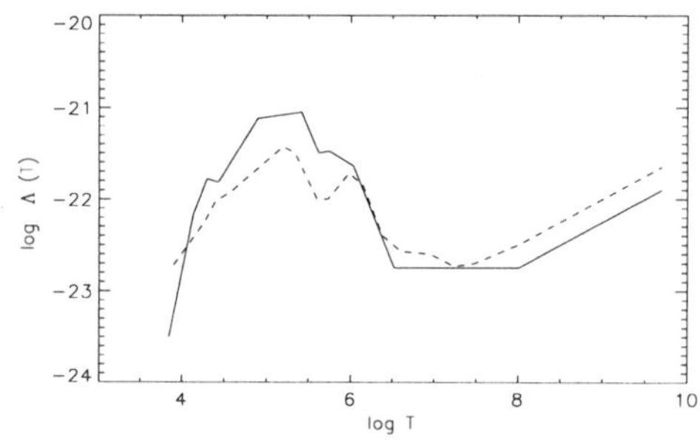

FIGURE 1. Two radiative loss functions, $\Lambda(T) = \Lambda_{0i}T^{\beta_i}, T \in [T_i, T_{i+1})$. The two curves correspond to different physical assumptions. The stability of the solution is influenced by the negative slopes.

3. SOLUTION

In this section we describe the basic flow pattern in the radiative case for two particular examples. Our results should be regarded as two (physical) examples which illustrate some basic properties of the solution.

3.1. Example settings. In spherical symmetry, we investigate the effect of a highly supersonic stellar wind (flow 1) crashing into a slower wind or the interstellar medium (flow 2). The boundary conditions are supersonic inflow on the left side of the computational domain and free outflow on the right side. Initially, the domain has been filled with material of flow 2. Then flow 1 starts to flow into the domain from the left side.

Setting 1: A fast, tenuous flow (flow 1) crashes into a slow, dense flow (flow 2). The flow parameters are $N_1(r = 10^{16}cm) \simeq 30\,part./cm^3$, $v_1 = 2.5 \cdot 10^3\,km/s$, $(Mach_1 \simeq 170, T = 1.5 \cdot 10^4\,K)$; $N_2(r = 10^{16}cm) \simeq 2 \cdot 10^4\,part./cm^3$, $v_2 = 20\,km/s$, $(Mach_2 \simeq 1.4, T = 1.5 \cdot 10^4\,K)$. $\Lambda(T)$ corresponds to the dashed line in Fig. 1.

Setting 2: A fast, tenuous flow (flow 1) crashes into an ambient medium of constant density at rest (flow 2). The flow parameters are $N_1(r = 10^{19}cm) \simeq$

$10^{-2}\,part./cm^3$, $v_1 = 1.6 \cdot 10^3\,km/s$, $(Mach_1 \simeq 110, T = 1.5 \cdot 10^4\,K)$; $N_2 = 1.5\,part./cm^3$, $v_2 = 0$. $\Lambda(T)$ corresponds to the solid line in Fig. 1.

3.2. Basic flow pattern.

3.2.1. *Adiabatic case.* This case is quite well known. For both settings the interaction zone of the two flows consists of two shock heated regions which are separated by a contact discontinuity. Setting 1 and 2 can be regarded as natural extensions of the Riemann problem to spherical symmetry. In both cases the self-similar solutions are known. For setting 1 the solution of Chevalier and Imamura [4] applies whereas for setting 2 one has a classical Sedov-Taylor solution (Taylor [6], Sedov [5]). For setting 1 the speed of the leading shock is constant in time whereas for setting 2 this speed decreases as $t^{3/5}$.

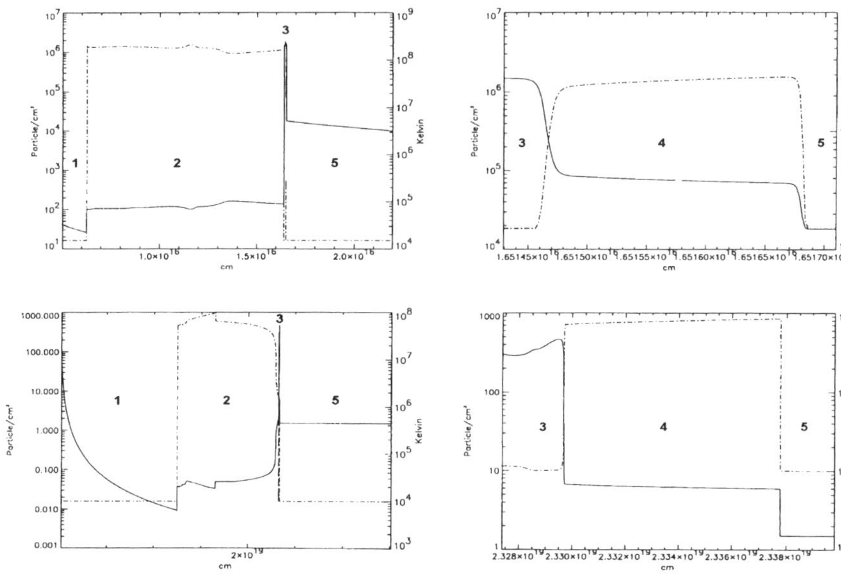

FIGURE 2. The basic radiative flow pattern for setting 1 *(top)* and setting 2 *(bottom)*. **Left:** Entire interaction zone. The basic flow pattern contains the following parts. *1:* flow 1, *2:* reverse shell (shocked flow 1), *3:* cold layer, *5:* flow 2. **Right:** Leading part of the interaction zone at a time where the leading shell is growing (cf. Sec. 4). *3:* part of cold layer, *4:* leading shell (shocked flow 2), *5:* flow 2. The left scale denotes the density in part./cm^3 (solid line), the right scale the temperature in degree Kelvin (dash-dotted line).

3.2.2. *Radiative case.* The flow pattern we find from numerical simulations are shown in Fig. 2. Both patterns contain a region of shock heated flow 1 material (reverse shell), a region of cooled flow 2 material (cold layer), and a region of shock heated flow 2 material (leading shell). These three regions are the basic components of the radiative pattern, instead of the two regions in the adiabatic case.

Additional regions are found for setting 2, and the two patterns differ globally. The structure of the leading part, however, is the same for both patterns: cold layer (3), hot leading shell (4), flow 2 (5) (see Fig. 2). Nevertheless, the stability properties of the leading shell are very different.

For setting 2 the pattern basically agrees with the one derived by Bertschinger [3]. Under certain assumptions he has derived a piecewise self-similar solution. It is important to note that his solution shows that the interface between the cold layer (3) and the hot leading shell (4) is, in fact, a boundary layer. The size of this boundary layer is several orders of magnitude smaller than the leading shell.

4. Stability of the Leading Shell

4.1. Changing timescales. A typical timescale of the source term, $\dot{E} = -N^2\Lambda(T)$, is given by $\tau_{cool} = T/|dT/dt|$. For the model we use we have $\tau_{cool} \propto T^{1-\beta_i}/N$, $T \in [T_i, T_{i+1})$. τ_{flow} is a typical flow timescale. As we are looking at the stability properties of the leading shell, it is reasonable to take $\tau_{flow} = l/v$, where l is the shell size and v is the post shock flow velocity in the comoving frame of the leading shock. Interestingly, for the source term we are considering, we find that the timescales in the leading shell steadily change between the following three cases.

- $\tau_{cool} << \tau_{flow}$: The solution approximately corresponds to the solution of the ODE part of the equations.
- $\tau_{cool} >> \tau_{flow}$: The solution approximately corresponds to the solution of the Euler equations alone.
- $\tau_{cool} \approx \tau_{flow}$: The source part and the transport part work on the same timescale. An *a priori* approximation to the solution cannot be given.

4.2. Linear stability analysis. Bertschinger [3] has performed a linear stability analysis for the piecewise self-similar solution quoted above. He finds that the leading shell can show an oscillatory instability. Taking temperature independent β values in the source term, he has determined several different modes of oscillation and the corresponding critical values β_{crit}. For $\beta < \beta_{crit}$ the solution becomes unstable. For radial perturbations the highest critical value he finds is $\beta_{crit} \simeq 0.8$.

4.3. Numerical results.

4.3.1. *Oscillatory instability.* With the numerical solution the non linear regime can also be covered. Over a wide temperature range we have β values in our radiative loss function which, according to Bertschinger [3], should lead to an unstable solution. A comparison, however, is difficult as we do not have only one fixed β value for the entire temperature range.

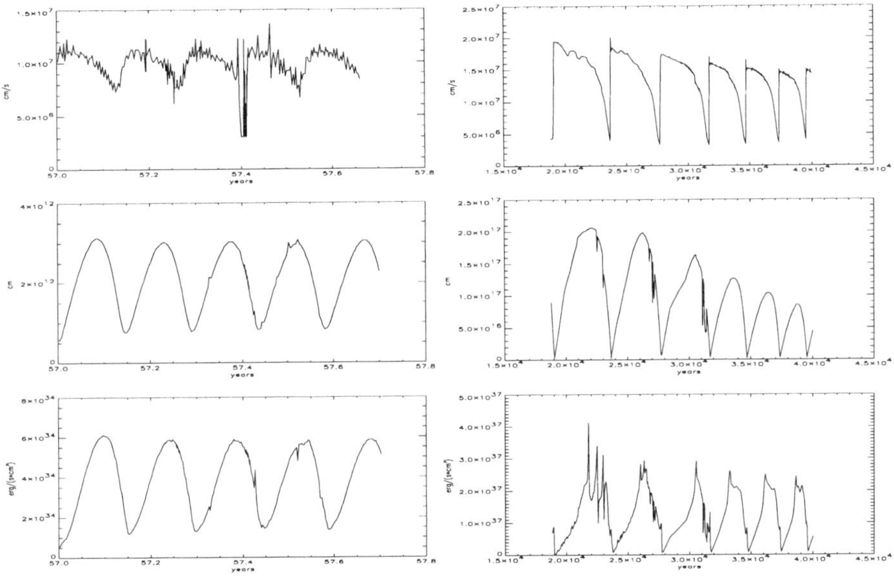

FIGURE 3. Oscillatory instability of the leading shell. *Top*: Velocity of leading shock. *Middle*: Size of leading shell. *Bottom*: Energy loss per time of leading shell due to the source term. **Left column**: 'Smooth instability' at the example of setting 1. The strong changes in velocity around 57.4 years are due to misleading determinations of the shock position and are not physical. **Right column**: 'Catastrophic instability' at the example of setting 2. The graphs show the qualitatively different behavior of the two types of instabilities.

We find indeed the hot leading shell to be unstable for post shock temperatures larger than a few times 10^5 K. The instability is oscillatory. The shell grows and shrinks in size. The velocity of the leading shock oscillates around some mean velocity. The energy loss of the shell per time increases from time to time. Due to the limited space, a detailed description of the working mechanism of the instability cannot be given here.

4.3.2. *Two different types.* Two different types can be distinguished, a 'smooth type' and a 'catastrophic type'. The smooth type is illustrated by setting 1 whereas setting 2 illustrates the catastrophic type. It is likely that not only the value of β but also the size of the temperature range over which this value is constant is important for the type of the instability. The following particularities should be noted.

'Smooth instability' (Left column of Fig. 3.)

- The shell size changes by about a factor 6. This amplitude is too big to be treated by linear stability analysis, but it is small compared to the amplitude of the catastrophic type.
- The velocity of the leading shock, the shell size, and the energy output change smoothly in time.
- The energy output varies in the same way as the shell size.

Note: Period and amplitude of the oscillation are constant due to the particular setting.

'Catastrophic instability' (Right column of Fig. 3.)

- The amplitude of the oscillation is very big: The size of the shell changes by more than a factor 100.
- The velocity of the leading shock, the shell size, and the energy output sometimes change abruptly.
- The energy output varies differently in time than the shell size.

Note: The decrease of the period and amplitude is due to the particular setting.

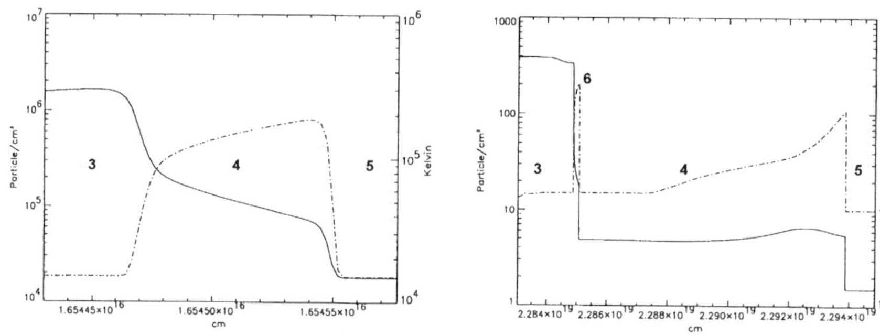

FIGURE 4. The different nature of the two types of instability. Shown is the leading shell while it is decreasing in size. The solid line denotes the density, the dash-dotted line the temperature. The numbers denote the cold layer (3), the leading shell (4) and flow 2 (5). **Left:** Smooth instability. The leading shell stays hot while decreasing in size. τ_{cool} is only slightly smaller than τ_{flow}. **Right:** The leading shell has already cooled significantly while its size has only decreased by a factor 2. Here $\tau_{cool} << \tau_{flow}$. Number 6 denotes a secondary shock which is forming when the shell nearly completely collapses because the momentum balance is completely out of equilibrium.

5. A NOTE ON NUMERICAL DIFFICULTIES

To resolve the leading shell an AMR, is necessary. (Setting 2: total computational domain: 10^{20} cm, leading shell: 10^{14} cm. With AMR this requires about 3000 grid points.)

Despite the use of AMR, the boundary layer between the leading shell and the cold layer cannot spatially be resolved. For points within the boundary layer between the cold layer and the leading shell the value of τ_{cool} typically is on the order of Δt_{coarse} to $0.1 \cdot \Delta t_{coarse}$ where, Δt_{coarse} is a typical timestep on the coarsest grid. Therefore, the timestep on the finest grid (about $10^{-6} \cdot \Delta t_{coarse}$) is sufficiently small for a stable computation of the energy loss. Nevertheless, we find a grid-dependence of the propagation speed of the boundary layer. When changing both the space and time discretization by a factor of about 500, the speed of the boundary layer changes by 3-5 %. We interpret this as a direct consequence of the insufficient spatial resolution of this boundary layer.

6. Conclusion

We have investigated the structure and stability of radiative colliding flows by numerically solving the Euler equations with a source term in spherical symmetry. The use of an adaptive mesh refinement algorithm allows a previously unachieved spatial resolution. For two astrophysical examples the results have been presented. We find that the leading shell of the interaction zone is unstable for wide temperature ranges due to the source term, $\dot{E} = -N^2\Lambda(T)$. Two different types of oscillatory instabilities should be distinguished, a smooth and a catastrophic type. A systematic study, however, has still to be performed. On the other hand, more physics (heat conduction, magnetic fields) should be included to achieve a more realistic model. Moreover, a multidimensional model is necessary.

References

1. M.J. Berger, *Adaptive mesh refinement for hyperbolic equations*, Lectures in applied Mathematics **22** (1985), 31–40.
2. P. Colella and H.M. Glaz, *Efficient solution algorithms for the riemann problem for real gases*, Journal of Computational Physics **59** (1984), 264–289.
3. E.Bertschinger, *On the structure and stability of radiative shock waves*, Astrophysical Journal **304** (1986), 154–177.
4. R.A.Chevalier and J.N.Imamura, *Self-similar solutions for the interaction regions of colliding winds*, Astrophysical Journal **270** (1983), 554-563.
5. L.I. Sedov, *Similarity and dimensional methods in mechanics*, New York: Academic Press, 1959.
6. G.I. Taylor, *The formation of a blast wave by a very intense explosion. I.Theoretical discussion.*, Proceedings Royal Society A **201** (1950), 159–174.

The Riemann Problem for Polymer Flooding with Hysteresis

Khaled Furati*

Abstract

The analytic solution to the Riemann problem for a polymer flooding model with hysteresis is described. The model is a non-strictly hyperbolic system of conservation laws with a history-dependent flux function. Because the solution of the model without hysteresis shows cases with non-monotone saturation profile, the introduction of hysteresis necessarily produces structurally different solutions.

1 Introduction

In this paper we describe the global analytic solution to the Riemann problem for the mathematical model

$$\partial_t s + \partial_x F = 0, \qquad (1)$$
$$\partial_t (cs) + \partial_x (cF) = 0,$$

with the flux function, F, of the form

$$F = \begin{cases} f(s,c) & \text{if } s = \pi,\ \partial_t s > 0, \quad (\textit{primary flow}), \\ h(s,\pi,c) & \text{otherwise}, \qquad\qquad (\textit{secondary flow}), \end{cases} \qquad (2)$$

and initial conditions

$$(s,\pi,c)(x,0) = \begin{cases} (s^L, \pi^L, c^L) & x \leq 0, \\ (s^R, \pi^R, c^R) & x > 0, \end{cases} \qquad (3)$$

where, $t \in \mathcal{R}^+$, $x \in \mathcal{R}$. The variables $s, \pi, c \in I$ and $s \leq \pi$, where I is the interval $[0,1]$. The system (1) is used in enhanced oil recovery to model polymer flooding in which oil is displaced by water containing dissolved polymer. The variable, s, is the saturation of the aqueous phase and c is the concentration of polymer. The flux function, F, is the fractional flow function for the aqueous phase and π is the hysteresis parameter. For the derivation of (1) and physical assumptions see [12].

This form of the flux function models the history dependence of the flow. The function $F = f(s,c)$ describes the flow when the saturation increases monotonically from the irreducible aqueous phase saturation, s_{ra}, to the maximum saturation, $1 - s_{rl}$, where s_{rl} is the residual liquid phase saturation. We will consider this imbibition of the aqueous phase as the *primary flow*. When this imbibition is reversed at $s = \pi$, the fractional flow function will be less and is given by $F = h(s,\pi,c)$. This drainage of the aqueous phase can be continued until s reaches a minimum value beyond which the phase can no longer flow. We will denote this irreducible saturation by s_π. We will assume

*Department of Mathematics, Duke University, Durham, NC 27706. fura@math.duke.edu

that as long as $s < \pi$, the flow (imbibition or drainage) is described by $h(s, \pi, c)$. We will call this flow a *secondary flow*. When s increases again beyond π, the fractional flow is given by $f(s, c)$.

Buckley and Leverett [1] constructed the analytical solution for immiscible waterflooding. Pope [12] generalized the fractional flow theory to more complicated floodings and used the method of characteristics to construct the corresponding solutions. The polymer flooding model without hysteresis was analyzed and solved by Isaacson [5] and Johansen and Winther [6]. The problem is similar to the Riemann problem for vibration of an elastic string considered by Keyfitz and Kranzer [7].

Hysteresis phenomena in multi-phase flows in porous media have been recognized by many researchers. For example, Colonna et al. [2], Killough [8], Gladfelter and Gupta [4] and Lenhard [9]. Marchesin et al. [10] used a double-valued flux function to model a two-component two-phase flow with hysteresis and constructed the complete solution of the associated Riemann problem graphically. Here, we consider a three-component two-phase flow with a more complicated treatment of hysteresis. The treatment of the history dependence in this paper is similar to that of motion in an elastic-plastic bar studied by Trangenstein and Pember in [13].

The outline of the paper is as follows. In section 2 we present the exact mathematical assumptions on the flux function and analyze the hyperbolicity of the model. In section 3 we list the different wave families. In section 4 we describe briefly the construction of the Riemann problem solutions.

2 The Mathematical description of the Model

In this section we present the mathematical assumptions and properties of the flux functions and analyze the hyperbolicity of the system of conservation laws (1). Let $I = [0, 1]$, $I_s = [s_{ra}, 1 - s_{rl}]$ and $I_\pi = [s_\pi, \pi]$, where $s_\pi = \max\{s : h(s, \pi, c) = 0\}$. We will assume that the flux functions, f and h, satisfy the following properties as shown in Figure 1:

1. $f(s, c)$ is a C^2 function in $I_s \times I$, and $f(s_{ra}, c) = 0$ and $f(1 - s_{rl}, c) = 1$, $\forall c \in I$;

2. $\partial_s f(s, c) > 0$ and $\partial_c f(s, c) < 0$ in $I_s \times I$;

3. For each $c \in I$, there is exactly one inflection point, $s^I = s^I(c) \in (s_{ra}, 1 - s_{rl})$, such that $\partial_{ss} f(s, c) > 0$ for $s < s^I$ and $\partial_{ss} f(s, c) < 0$ for $s > s^I$;

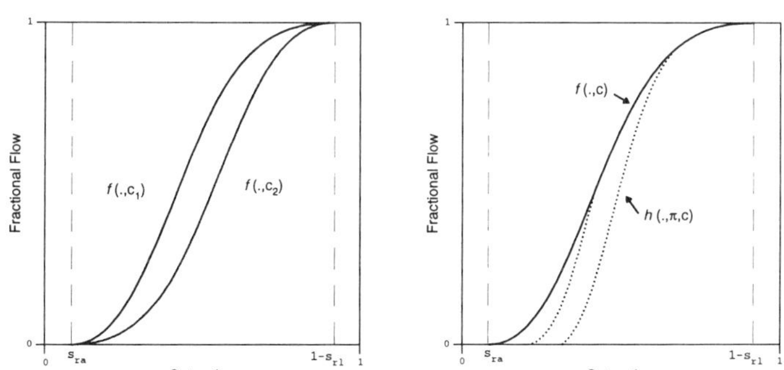

Figure 1: Fractional flow curves, $c_2 > c_1$.

4. $h(s,\pi,c)$ is a C^2 function in $s \in I_\pi$, C^1 in $\pi \in I_s$ and C^2 in $c \in I$;

5. $h(s_\pi,\pi,c) = 0$ and $h(s,s,c) = f(s,c)$ for each $\pi, s \in I_s$, and $c \in I$;

6. $\partial_s h(s,\pi,c) > 0$, $\partial_\pi h(s,\pi,c) < 0$, and $\partial_c h(s,\pi,c) < 0$ in $I_\pi \times I_s \times I$;

7. $\partial_{ss} h(s_\pi,\pi,c) > 0$ for each $c \in I$ and $\pi \in I_s$, and h has at most one inflection point $s^I(\pi,c)$, for $s \in I_\pi$;

8. $\partial_s f(\pi,c) \leq \partial_s^- h(\pi,\pi,c)$ for each $c \in I$, where ∂_s^- is the left derivative.

Property 6 implies that the secondary fractional flow curves for a fixed c are disjoint for distinct values of π. It also implies that, for a fixed π, the curves are disjoint for distinct values of c. For more details on these properties, and how they follow from the properties of the relative permeabilities, see [3].

For the primary flow, system (1) can be written in the quasilinear form

$$\partial_t \begin{bmatrix} s \\ c \end{bmatrix} + A\,\partial_x \begin{bmatrix} s \\ c \end{bmatrix} = 0, \qquad A = \begin{bmatrix} \partial_s f & \partial_c f \\ 0 & f/s \end{bmatrix}. \tag{4}$$

Since the eigenvalues of A are real, the system of conservation laws is hyperbolic, and has the two characteristic speeds and the corresponding eigenvectors

$$\lambda_b = \partial_s f, \quad r_b = \begin{bmatrix} 1 \\ 0 \end{bmatrix}, \quad \lambda_p = \frac{f}{s}, \quad r_p = \begin{bmatrix} \partial_c f \\ \lambda_p - \lambda_b \end{bmatrix}. \tag{5}$$

The characteristic speed, λ_b, is the Buckley-Leverett rarefaction wave speed, which is equal to the slope of the fractional flow function, f. The second characteristic speed is the aqueous particle velocity, which is equal to the slope of the chord from the origin to the fractional flow function.

If the flow is secondary, $F = h(s,\pi,c)$, then the system of conservation laws (1) can be closed by adding the constraint $\partial_t \pi = 0$, and written in the following quasilinear form

$$\partial_t \begin{bmatrix} s \\ c \\ \pi \end{bmatrix} + \tilde{A}\,\partial_x \begin{bmatrix} s \\ c \\ \pi \end{bmatrix} = 0, \qquad \tilde{A} = \begin{bmatrix} \partial_s h & \partial_c h & \partial_\pi h \\ 0 & h/s & 0 \\ 0 & 0 & 0 \end{bmatrix}. \tag{6}$$

The characteristic speeds in this case are

$$\lambda_{bs} = \partial_s h, \qquad \lambda_{ps} = h/s, \qquad \lambda_\pi = 0. \tag{7}$$

Note that all the characteristic speeds are non-negative. This implies that all the wave families are either stationary or traveling to the right.

For the border states, $u = (s,s,c)$, we define the Buckley-Leverett characteristic speeds by

$$\lambda_b(s,s,c) = \partial_s f(s,c), \qquad \lambda_{bs}(s,s,c) = \partial_s^- h(s,s,c). \tag{8}$$

The type of the flow (secondary or primary) determines which velocity is to be used in determining the Riemann problem solutions.

Note that, for each $c \in I$, there exists at most one $s^{tan}(c)$ such that $\lambda_p(u^{tan}) = \lambda_b(u^{tan})$, $u^{tan}(c) = (s^{tan}(c), c)$. Similarly, for each $c \in I$ and $\pi \in I_s$, there exists at most one $s^{tans}(\pi,c)$ such that $\lambda_{ps}(u^{tans}) = \lambda_{bs}(u^{tans})$, $u^{tans}(\pi,c) = (s^{tans}(\pi,c),\pi,c)$. We will assume that the state $u^{tan}(c)$ exists for all $c \in I$. Moreover, it is easy to show that for each $c \in I$ and $\pi \in I_s$, if $\pi < s^{tan}(c)$ then $\lambda_{ps}(u) < \lambda_{bs}(u)$ for all $u = (s,\pi,c)$ with $s_\pi < s \leq \pi$. If $\pi \geq s^{tan}(c)$, then either the state $u^{tans}(\pi,c)$ exists or the state (π,π,c) is a transitional state across which the relation between the two speeds is reversed. The three types of the transitional states are shown in Figure 2.

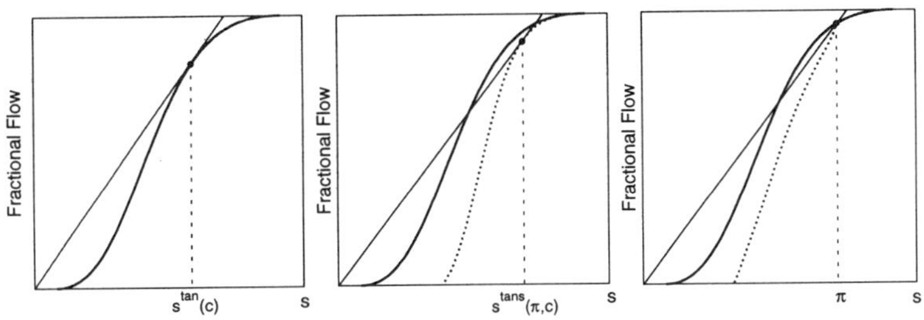

Figure 2: Transitional states.

3 Wave Families

In this section we list all wave families associated with the Riemann problem (1)–(3). From the solutions of the Rankine-Hugoniot conditions and the integral curves of the eigenvectors, the model has the following wave families.

b-waves and bs-waves

These two families correspond to the characteristic speeds λ_b and λ_{bs}, respectively, and consist of rarefactions and shocks. Due to the inflection points, the two families are in general not genuinely nonlinear. Note that c is a Riemann invariant for both families and π is a Riemann invariant of the bs-family.

Thus, $u^L = (s^L, c^L)$ and $u^R = (s^R, c^R)$, $s^L > s^R$, can be connected by a b-rarefaction wave if $c^L = c^R$ and λ_b is increasing from s^L to s^R. They can be connected by a b-shock with speed $\sigma = [f]/[s]$ if $c^L = c^R$ and the Oleinik entropy condition for scalar equations [11] is satisfied.

The two states $u^L = (s^L, \pi^L, c^L)$ and $u^R = (s^R, \pi^R, c^R)$ can be connected by a bs-rarefaction if $c^L = c^R$, $\pi^L = \pi^R$ and λ_{bs} is increasing from s^L to s^R. They can be connected by a bs-shock with speed $\sigma = [h]/[s]$ if $c^L = c^R$, $\pi^L = \pi^R$, and the Oleinik condition is satisfied.

p-waves and ps-waves

These two families correspond to the characteristic speeds λ_p and λ_{ps}, respectively. The two families are linearly degenerate since λ_p and λ_{ps} are Riemann invariants. Thus, the two families consist of contact discontinuities that travel with speeds equal to the characteristic speed.

The wave curves for the p-family are the curves $c = c(s)$ along which λ_p is constant. The slope of such curves depends on the relation between λ_p and λ_b since $dc/ds = (\lambda_p - \lambda_b)/f_c$. The same argument is true for the ps-family curves $c = c(s, \pi)$.

The states, $u^L = (s^L, c^L)$ and $u^R = (s^R, c^R)$, $s^L > s^R$, can be connected by a contact discontinuity if $\lambda_p(u^L) = \lambda_p(u^R)$ and they lie on the same side of the transitional state. The last condition is needed to fulfill Lax's entropy condition which requires that exactly one of the b-characteristics leave the discontinuity. With $\pi^L = \pi^R$, similar results are true for the ps-family.

st-waves

This family consists of stationary contact discontinuities satisfying Rankine-Hugoniot conditions with speed $\sigma = 0$. Across this discontinuity $[h] = [c] = 0$.

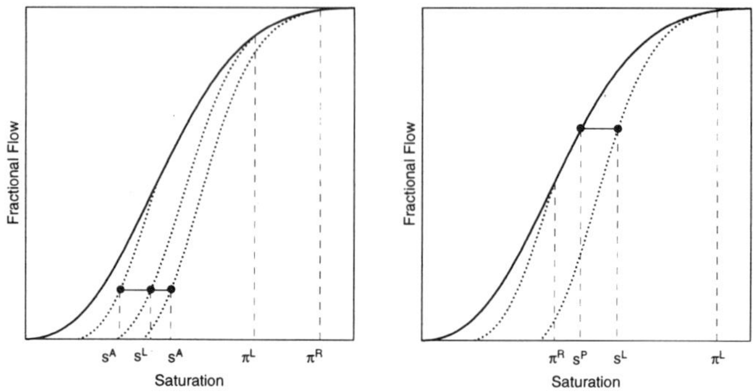

Figure 3: The states u^A and u^P.

4 Global Solution of the Riemann Problem

The global solution of the Riemann problem can be constructed graphically using the wave families mentioned above. The uniqueness of the solutions can easily be verified by imposing the compatibility condition which requires the initial speed of each wave to be greater than or equal to the final speed of the preceding wave.

Note that for any given two states $u^L = (s^L, \pi^L, c^L)$ and $u^R = (s^R, \pi^R, c^R)$, there are two cases according to the relation between $h^I = h(\pi^R, \pi^R, c^L)$ and $h^L = h(u^L)$, as shown in Figure 3. If $h^L \leq h^I$ then u^L can be connected to $u^A = (s^A, \pi^R, c^L)$ by a stationary wave, otherwise u^L can be connected to $u^P = (s^P, s^P, c^L)$, $s^P > \pi^R$ by a stationary wave. Thus, the problem can be reduced to solving for either $\pi^L = \pi^R$ or $s^L = \pi^L > \pi^R$ for any arbitrary u^R.

When $c^L = c^R$, the solution is obtained by constructing the convex or concave hulls of the flux functions and consists of combinations of b-waves and bs-waves. An example is shown in Figure 4. On the other hand, when $c^L \neq c^R$, the solutions consist also of p-waves and ps-waves. The Riemann problem in this case has many different cases. To account for all possible cases, a list of *if-else* statements is used. The cases are based on the relations between c^L and c^R, λ^L_{ps} and λ^L_{bs}, π^R and

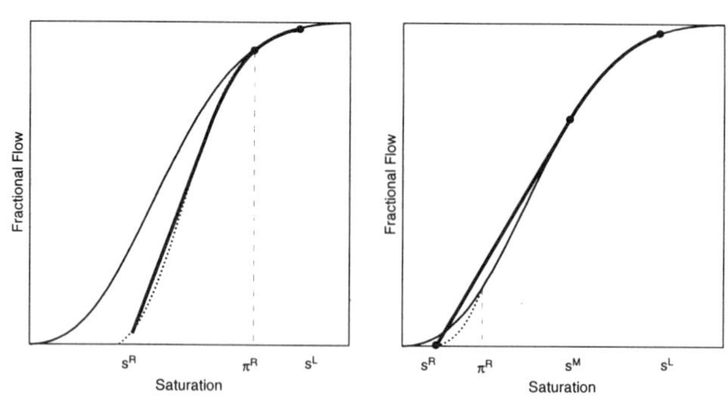

Figure 4: Solutions to hysteretic Buckley-Leverett equation with $s^L = \pi^L > \pi^R$.

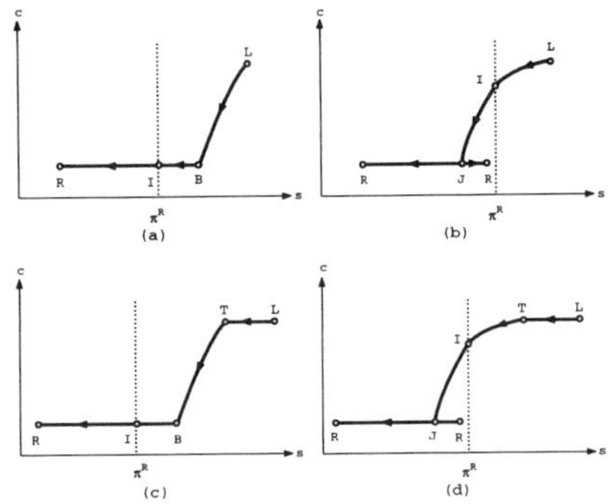

Figure 5: Solutions to hysteretic polymer flooding with $c^L > c^R$ and $s^L = \pi^L > \pi^R$.

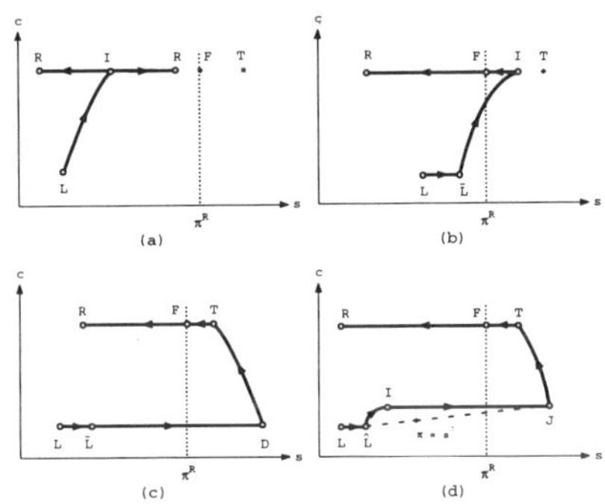

Figure 6: Solutions to hysteretic polymer flooding with $c^L < c^R$, $\pi^L = \pi^R$ and $\pi^R < s^{tan}(c^R)$.

$s^{tan}(c)$, etc. Examples of such solutions are shown in Figures 5–7. The complete set of solutions are described in [3].

Acknowledgment

I would like to thank professor John Trangenstein, my PhD advisor, for introducing me to this field and suggesting this problem.

References

[1] S. E. Buckley and M. C. Leverett, *Mechanisms of fluid displacement in sands*, Transactions of the American Institute of Mining and Metallurgical Engineers **146** (1942), 107–116.

[2] J. Colonna, F. Brissaud, and J. L. Millet, *Evolution of capillarity and relative permeability hysteresis*, Society of Petroleum Engineers Journal Transactions **253** (1972), 28–38.

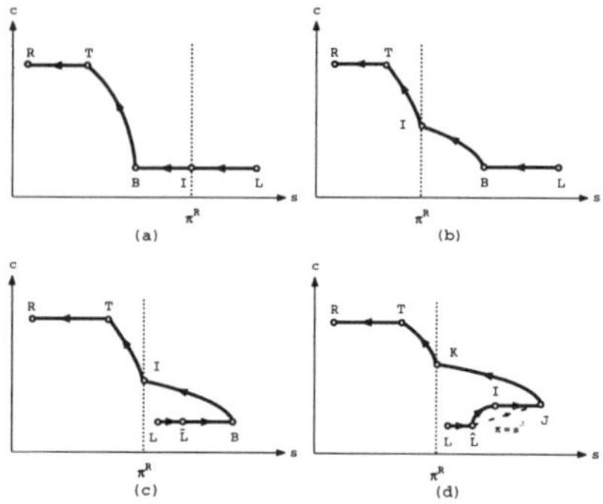

Figure 7: Solutions to hysteretic polymer flooding with $c^L < c^R$, $s^L = \pi^L > \pi^R$ and $\pi^R \geq s^{tan}(c^R)$.

[3] K. Furati, *The solution of the Riemann problem for a hyperbolic system of conservation laws modeling polymer flooding with hysteresis*, Ph.D. thesis, Duke University, Durham, NC, 1995.

[4] R. E. Gladfelter and S. P. Gupta, *Effect of fractional flow hysteresis on recovery of tertiary oil*, Society of Petroleum Engineers Journal **20** (1980), 508–520.

[5] E. L. Isaacson, *Global solution of a Riemann problem for a non-strictly hyperbolic system of conservation laws arising in enhanced oil recovery*, Unpublished.

[6] T. Johansen and R. Winther, *The solution of the Riemann problem for a hyperbolic system of conservation laws modeling polymer flooding*, SIAM Journal of Mathematical Analysis **19** (1988), no. 3, 541–566.

[7] B. L. Keyfitz and H. C. Kranzer, *A system of non-strictly hyperbolic conservation laws arising in elasticity*, Archive for Rational Mechanics and Analysis **72** (1980), 220–241.

[8] J. E. Killough, *Reservoir simulation with history-dependent saturation functions*, Society of Petroleum Engineers Journal **16** (1976), 37–48.

[9] R. J. Lenhard and J. C. Parker, *A model for hysteretic constitutive relations governing multiphase flow, 2. permeability-saturation relations*, Water Resources Research **23** (1987), no. 12, 2197–2206.

[10] D. Marchesin, H. B. Medeiros, and P. J. Paes-Leme, *A model for two phase flow with hysteresis*, Contemporary Mathematics **60** (1987), 89–107.

[11] O. A. Oleinik, *Uniqueness and a stability of the generalized solution of the Cauchy problem for a quasilinear equation*, Uspekhi Mat. Nauk. (N.S.) **14** (1959), 165–170, English translation in Amer. Math. Soc. Transl. Ser. 2, 33(1964), 285–290.

[12] G. A. Pope, *The application of fractional flow theory to enhanced oil recovery*, Society of Petroleum Engineers Journal **20** (1980), 191–205.

[13] J. A. Trangenstein and R. B. Pember, *The Riemann problem for longitudinal motion in an elastic-plastic bar*, SIAM J. Sci. Stat. Comput **12** (1991), no. 1, 180–207.

Tracking of Shear Bands in an Antiplane Shear Model

F. Xabier Garaizar[1,2,3] and John Trangenstein[4,5]

1. Introduction

In this notice we describe a numerical algorithm for the solution of a granular flow problem in two dimensions where shear bands are present. The main difficulty in this problem is the resolution of the motion near the shear band. In this region, traditional methods break down and a new approach, incorporating ideas from front-tracking techniques, is used.

We consider elasto-plastic deformations in an antiplane shear motion. That is, while the deformation is homogeneous in the z-direction (the state variables only depend on x and y) the motion is only in the z-direction. In other words, the only non-trivial component of the velocity vector is in the z-direction.

When the material reaches a critical value of loading, shear banding occurs. While the overall deformation is described by a hyperbolic system of equations, hyperbolicity is lost at the points when shear banding occurs. At these points, the deformation is described by jump relations in a manner similar to shock-hugoniots. From a numerical point of view, shear bands are treated as internal boundaries.

An added problem is that the endpoints of the band can propagate at a rate that can be larger than the elastic speed. Effectively this means that the growth of the band is not limited by the CFL condition and the endpoints of a band can propagate across several cells during a single time-step. We update the cells where the shear band is likely to grow, and those where it already exists, at fractional time-steps allowing for the proper growth of the band.

2. Equations

We study a problem that models antiplane shearing as described in [7]. This motion is described by three unknowns: the z-component of the velocity, which we write as a scalar velocity v, and the T_{31} and T_{32} components of the Cauchy stress tensor $\mathbf{T}$. We define a dimensionless stress vector $\vec{\tau} = (\tau_1, \tau_2)^T$ by identifying $\tau_1 = T_{31}$ and $\tau_2 = T_{32}$.

The equations governing the motion are: the momentum equation

$$\partial_t v = c^2 \text{ div } \vec{\tau} \, , \tag{1}$$

where c is the speed of elastic waves, and the constitutive relations. These relations take different forms for elastic and plastic deformations :

$$\partial_t \vec{\tau} = \text{grad } v \quad \text{if elastic} \tag{2}$$

$$\left(\mathbf{I} + \frac{1}{h(|\vec{\tau}|) + |\vec{\tau}|^2 \cos\alpha}\mathbf{R}\right) \partial_t \vec{\tau} = \text{grad } v \quad \text{if plastic.} \tag{3}$$

---

[1]Center for Research in Scientific Computation and Department of Mathematics, Box 8205, North Carolina State Univ., Raleigh, NC 27695, Email: garaizar@math.ncsu.edu

[2]Supported by NSF Grant DMS 9201115, which includes funds from AFOSR.

[3]Current Address: Department of Mathematics, Box 90320, Duke University, Durham, NC 27706-0320

[4]Department of Mathematics, Box 90320, Duke University, Durham, NC 27706-0320, Email: johnt@math.duke.edu

[5]Supported by NSF Grants DMS 9201034, DMS 9201361,DMS 9407029, and DOE Grant DE-F605-92-ER25145.

Here $\mathbf{R}$ is the rotation matrix $\begin{pmatrix} \cos\alpha & \sin\alpha \\ -\sin\alpha & \cos\alpha \end{pmatrix}$ and $h(|\vec{\tau}|)$ is the hardening modulus, which we assume satisfies $h' < 0$ and $h(1) = 0$. The parameter α in the rotation matrix $\mathbf{R}$ characterizes the degree of non-associativity in the model.

The change from elastic to plastic behavior is described in terms of a yield condition of von Mises type. We say that the material is deforming plastically if $\partial_t|\vec{\tau}(t,x,y)| > 0$ and $|\vec{\tau}(t,x,y)| = \max_{0\leq s\leq t}(|\vec{\tau}(s,x,y)|)$, otherwise we say that the material is deforming elastically. For later use we define a measure of the accumulated stress as $\gamma(t,x,y) = \max_{0\leq s\leq t}(|\vec{\tau}(s,x,y)|)$.

For most of what follows it will be more convenient to write system (1)-(3) in quasilinear form:

$$\partial_t v = c^2 \operatorname{div} \vec{\tau} \tag{4}$$

$$\partial_t \vec{\tau} = \mathbf{A}(\vec{\tau}) \operatorname{grad} v \tag{5}$$

where $\mathbf{A}(\vec{\tau}) = \mathbf{I}$ if the deformation is elastic and $\mathbf{A}(\vec{\tau}) = \mathbf{I} - (h + |\vec{\tau}|^2\cos(\alpha))^{-1}\mathbf{R}\vec{\tau}\vec{\tau}^T$ if the deformation is plastic. As noted in [7] the formation of shear bands is related to the ill-posedness of the problem, which occurs when speeds of planar waves, in some direction, become complex. For a given direction $\vec{\xi}$, the equations for planar waves propagating in that direction (i.e. $(x,y) = s\vec{\xi}$), simplify to:

$$\partial_t v = c^2\, \partial_s(\vec{\xi}\cdot\vec{\tau}) \tag{6}$$

$$\partial_t \vec{\tau} = \vec{b}\, \partial_s\, v \tag{7}$$

where $\vec{b}$ is defined as $\vec{b}(\vec{\xi}) = \vec{\xi} - (h + |\vec{\tau}|^2\cos(\alpha))^{-1}\mathbf{R}\vec{\tau}(\vec{\xi}\cdot\vec{\tau})$ if the deformation is plastic and $\vec{b}(\vec{\xi}) = \vec{\xi}$ if elastic. For this one-dimensional system, the eigenvalues λ are 0 and $\pm c\sqrt{\vec{\xi}\cdot\vec{b}}$ with corresponding eigenvectors

$$\begin{pmatrix} 0 \\ \vec{\xi}^{\perp} \end{pmatrix} \quad \text{and} \quad \begin{pmatrix} -\lambda \\ \vec{b} \end{pmatrix} \tag{8}$$

These eigenvalues will be real provided that $h(|\vec{\tau}|) > |\vec{\tau}|^2\sin^2(\frac{\alpha}{2})$. Otherwise there is a direction that minimizes $\vec{\xi}\cdot\vec{b}$ and for which the eigenvalues are imaginary. This direction of extreme ill-posedness is the direction bisecting the angle made by $\vec{\tau}$ and $\mathbf{R}\vec{\tau}$. That is, if $\vec{\tau} = |\vec{\tau}|(\cos\theta, \sin\theta)^T$, the extreme direction is $\vec{\xi} = (\cos\frac{\theta-\alpha}{2}, \sin\frac{\theta-\alpha}{2})^T$. For more details, see [7].

3. Shear Band

As explained above, shear bands originate at points of ill-posedness. The shear bands are idealized as stationary points of discontinuity in a weak solution of (4)–(5). If $\vec{n}$ is the vector normal to the band, the normal component of the stress ($\sigma = \vec{\tau}\cdot\vec{n}$) is continuous across the band, while the tangential component ($\omega = \vec{\tau}\cdot\vec{n}^{\perp}$) may suffer a jump. The variables σ_b and ω_b inside the band are related to the outside variables by a differential-jump relation:

$$\partial_t \vec{\tau}_b = \mathbf{A}(\vec{\tau}_b)\begin{pmatrix} [v]/\delta \\ 0 \end{pmatrix} \tag{9}$$

where $\vec{\tau}_b = (\sigma_b, \omega_b)^T$, $[v]$ is the jump in the velocity across the band and δ is a "grain diameter" parameter. This equation provides the evolution of the stress inside the shear band.

4. Numerical Algorithms

Let γ_c be the root to the equation $h(\gamma) = \gamma^2 \sin^2(\frac{\alpha}{2})$. For those points in the domain where $\gamma(t,x,y) < \gamma_c$, the system is hyperbolic and the eigenvalue-eigenvector study of section 2 is valid. The algorithm is organized in two blocks: the first one corresponds to the integration of the cells where the system is still hyperbolic ($\gamma < \gamma_c$). This is done via a second order algorithm based on Godunov methods. The second block consists of updating the cells intersected by the shear bands ($\gamma \geq \gamma_c$) and surrounding cells, which are expected to be affected by the presence of the shear band. This part of the algorithm is first-order accurate.

4.1. Calculation before the formation of the shear band. We momentarily assume that the states at the cell boundaries ($x = x_i = i\Delta x$ or $y = y_j = j\Delta y$) at the half timestep $t = t_n + \frac{1}{2}\Delta t$ are known. We refer to those states by $U^{n+\frac{1}{2}}_{i+\frac{1}{2},j}$ and $U^{n+\frac{1}{2}}_{i,j+\frac{1}{2}}$, where $U = (v, \tau_1, \tau_2, \gamma)^T$. The momentum equation (1) is in conservation form and we can write the usual Godunov type algorithm to update the velocity at a new time:

$$v^{n+1}_{i,j} = v^n_{i,j} + \frac{\Delta t}{\Delta x \Delta y}\left\{\Delta y\left[(\tau_1)^{n+\frac{1}{2}}_{i-\frac{1}{2},j} - (\tau_1)^{n+\frac{1}{2}}_{i+\frac{1}{2},j}\right] + \Delta x\left[(\tau_2)^{n+\frac{1}{2}}_{i,j-\frac{1}{2}} - (\tau_2)^{n+\frac{1}{2}}_{i,j+\frac{1}{2}}\right]\right\}. \tag{10}$$

Since the constitutive relations are in conservation form only while the material deforms elastically (but not for plastic deformations), the update of the stress variables cannot be done via a Godunov type algorithm. We integrate (5) along a particle path (similar to [9]), and since the motion is only in the z-direction (particles do not move in the xy-plane), this integration reduces to a simple integration in time for constant x and y. To perform this integration first we approximate $\operatorname{grad} v$ in (5) by

$$\operatorname{grad} v \approx \frac{[\vec{\tau}]}{[x,y]}(\equiv)_d \begin{pmatrix} \dfrac{v^{n+\frac{1}{2}}_{i+\frac{1}{2},j} - v^{n+\frac{1}{2}}_{i-\frac{1}{2},j}}{\Delta x} \\ \dfrac{v^{n+\frac{1}{2}}_{i,j+\frac{1}{2}} - v^{n+\frac{1}{2}}_{i,j-\frac{1}{2}}}{\Delta y} \end{pmatrix} \tag{11}$$

and then we integrate numerically

$$\partial_t \vec{\tau} = \mathbf{A}(\vec{\tau})\frac{[\vec{\tau}]}{[x,y]} \tag{12}$$

from $t = t_n$ to $t = t_n + \Delta t$. We notice that for elastic deformations, the integration of (12) is equivalent to an algorithm of Godunov type as that of (10).

The states $U^{n+\frac{1}{2}}_{i+\frac{1}{2},j}$ and $U^{n+\frac{1}{2}}_{i,j+\frac{1}{2}}$ are calculated via a one dimensional Riemann Solver [3, 4] and characteristic tracing algorithm [2, 8, 9] that incorporates transverse flow corrections as to achieve second order accuracy in space and time as well as stability.

4.2. Integration algorithm near the shear band. We define, geometrically, a shear band as an oriented curve in the xy-plane, whose intersection with a given cell is either empty or a segment of a straight line dividing the cell in two disjoint regions. That is we do not allow the shear band to end in the interior of a cell. These segments are called bonds and the intersection of the shear band with grid lines are called points.

The shear band is treated as an internal boundary where the stress evolves according to a particular evolution equation (9) as established in [7]. The approach used here follows traditional front tracking techniques [1], [5] with modifications proper to this problem: the tracked front is neither a hyperbolic wave nor does it divide the domain into unconnected domains. At the shear band the approximated solution is first order accurate. We organize the description of the method as follows: (a) averaging of states near the band, (b) update of states at the band, (c) recalculation of fluxes at and near the band and (d) integration on subcells and cells near the band.

For each cell intersected by a shear band, the eight surrounding cells will need to be updated again to account for the influence of the shear band. Let the (i_b, j_b) cell be crossed by a shear band. We will refer to this cell as C_b and the bond (portion of the shear band in the cell) as I_b. For latėr use we define the box centered at C_b, $B = [i_b - 1, i_b + 1] \times [j_b - 1, j_b + 1]$, and $B_b = B - (i_b, j_b)$. Associated with the bond I_b we have information proper of the band:
- the states U_b at the shear band.
- the states U_b^L and U_b^R at the sides of the band at time $t_n + \Delta t/2$,
- the states U^L and U^R on the subcells of cell C_b

and that of surrounding cells, $U_{i,j}$ for $(i, j) \in B$. The normal to the band, $\vec{n}$, is used to determine if a cell, or portion of a cell, is to the left or right of the band. For the cells next to the tips of the bands, and only for the purpose of assigning left and right areas to these cells, we extrapolate the band at its endpoints by one or two cells in the direction of the last bond.

(a) Averaging of states. In what follows, we will refer to a left or right state with the index $S = L, R$. Since the side states U^S correspond to subcells whose area can be very small, we average these states according to a weighted interpolation [1] :

$$\hat{U}^S = \frac{A^S_{i_b,j_b}}{\Delta x \Delta y} U^S + (1 - \frac{A^S_{i_b,j_b}}{\Delta x \Delta y}) \frac{\sum_{(i,j)\in B_b} A^S_{ij} U^S_{ij}}{\sum_{(i,j)\in B_b} A^S_{ij}},$$

where $A^S_{i,j}$ is the area of the portion of the cell $C_{i,j}$ to the S side of the band.

(b) Updating of states at the band. We denote by σ_b^n, ω_b^n, γ_b^n the stress variables inside the band at $t = t_n$. We integrate (9) from t_n to $t_n + \Delta t$, where $[v] = \tilde{v}_b^R - \tilde{v}_b^L$ and $\tilde{v}_b^S$ are the velocities of the states at the sides of the band at time t_n. These velocities are obtained as the result of tracing one parameter wave curves, with σ as the parameter, from the side states $\hat{U}^S$ to $\sigma = \sigma_b^n$. The variables at the end of this integration are σ_b^{n+1}, ω_b^{n+1}, γ_b^{n+1}. For the next step we need the value of σ_b at a half-time step. We approximate this value by $\sigma_b^{n+1/2} = (\sigma_b^n + \sigma_b^{n+1})/2$ and identify $\sigma_b^R = \sigma_b^L = \sigma_b^{n+1/2}$.

(c) Calculation of fluxes. By tracing the one parameter wave curves, with σ as the parameter, from the side states $\hat{U}^S$ to $\sigma = \sigma_b^S$ we obtain an approximation of the states U_b^S at the sides of the band at $t = t_n + \Delta t/2$. We use these states to find the fluxes at the band. To calculate the fluxes at cell edges in B, to the left and right of the band, we repeat the calculation of $U^{n+\frac{1}{2}}_{i+\frac{1}{2},j}$ and $U^{n+\frac{1}{2}}_{i,j+\frac{1}{2}}$ as in section 4.1 (though this time with only first order accuracy) where we have replaced U_{i_b,j_b} by $\hat{U}^S$ and U_{ij} by U^S_{ij} for $(i, j) \in B$.

(d) Integration near the band. If we denote by $d^S_{i+1/2,j}$ (resp. $d^S_{i,j+1/2}$) the length of the $(i + 1/2, j)$ (resp. $(i, j + 1/2)$) boundary on the S side of the band and d_b the length of the bond segment, the momentum equation is integrated on the S subcell to

$$\begin{aligned} A^S_{i_b,j_b}\left[(v^S)^{n+1}_{i_b,j_b} - (v^S)^n_{i_b,j_b}\right] \;=\; \Delta t \Big[& d^S_{i_b-\frac{1}{2},j_b}(\tau_1^S)^{n+\frac{1}{2}}_{i_b-\frac{1}{2},j_b} - d^S_{i_b+\frac{1}{2},j_b}(\tau_1^S)^{n+\frac{1}{2}}_{i_b+\frac{1}{2},j_b} + \\ & d^S_{i_b,j_b-\frac{1}{2}}(\tau_2^S)^{n+\frac{1}{2}}_{i_b,j_b-\frac{1}{2}} - d^S_{i_b,j_b+\frac{1}{2}}\tau_2^S)^{n+\frac{1}{2}}_{i_b,j_b+\frac{1}{2}} + (-1)^S d_b \sigma_b^S \Big], \end{aligned} \tag{13}$$

where $(-1)^L = -1$ and $(-1)^R = 1$).

If we write M^S for the RHS of (13) and $M_{i,j}$ as the corresponding quantity in (10), we would like to write

$$(v^S)^{n+1}_{i_b,j_b} = (v^S)^n_{i_b,j_b} + \frac{M^S}{A^S_{i_b,j_b}}. \tag{14}$$

But, if $A^S_{i_b,j_b}$ is too small, (14) would violate the CFL stability condition. Instead, we write

$$(v^S)^{n+1}_{i_b,j_b} = (v^S)^n_{i_b,j_b} + \frac{M^S}{\Delta x \Delta y}. \tag{15}$$

This expression for $(v^S)^{n+1}_{i_b,j_b}$ satisfies the CFL condition, but when we compute the integral of the momentum equation in the region $C^S_b \times [t_n, t_n + \Delta t]$ and compare with (13) we observe that there is an excess of mass flux that is given by

$$M^S - A^S_{i_b,j_b}((v^S)^{n+1}_{i_b,j_b} - (v^S)^n_{i_b,j_b}) = M^S(1 - \frac{A^S_{i_b,j_b}}{\Delta x \Delta y}). \tag{16}$$

In order to be conservative in the region, we need to put the above excess "mass-flux" back into the algorithm. There are several techniques to carry out this mass redistribution. We partially follow the ideas of [1] with some variations in the sense that we perform a weighted area redistribution. The fact that the front is not hyperbolic, precludes a wave decomposition at the front, and the fact that the material presents a sharp discontinuity in the displacement indicates that transfer of "mass-flux" across the discontinuity is not proper. We now update the cells in B by

$$v^{n+1}_{i,j} = v^n_{i,j} + \frac{M_{i,j}}{\Delta x \Delta y} + \sum_S \sum_F \left(M^S(1 - \frac{A^S_{i_b,j_b}}{\Delta x \Delta y}) \frac{1}{\sum_{B_b} A^S_{m,n}} \right) \tag{17}$$

if the shear band does not intersect $C_{i,j}$, and otherwise by

$$(v^S)^{n+1}_{i,j} = (v^S)^n_{i,j} + \frac{M^S_{i,j}}{\Delta x \Delta y} + \sum_F \left(M^S(1 - \frac{A^S_{i_b,j_b}}{\Delta x \Delta y}) \frac{1}{\sum_{B_b} A^S_{m,n}} \right). \tag{18}$$

The sum $\sum_F$ is over all the band bonds that affect the (i,j) cell. A similar process is done for the stress update.

4.3. Initialization of the shear band cell. When the total stress γ reaches the critical value γ_c at some cell (i_b, j_b) and time t_0, a shear band is introduced in that cell. This can be due either to a shear band growing into a new cell or by the formation of a new shear band. Both cases are treated similarly. The direction of extreme ill-posedness for system (4)–(5) is taken to be the normal direction to the shear band being initialized at that point. As seen above, this normal is $\vec{n} = (\cos\frac{\theta-\alpha}{2}, \sin\frac{\theta-\alpha}{2})^T$, where θ is the angle that the stress vector at that point makes with the positive x-axis. The exact placement of the band in the cell is done by: (a) continuation of a previous band, if present, from the boundary cell, or (b) locating a local maximum of $\gamma(t_0, x, y)$ in the cell (i_b, j_b).

It may occur that during a single timestep, the states at several cells enter the region of ill-posedness. These cells may be associated with different shear bands or one single band growing rapidly since there is no a-priori limit in the rate of growth of the band. This potential difficulty is solved by integrating the stress at the "new" shear band cells at time subintervals. We explain the process: After a regular time step $[t_n, t_n + \Delta t]$, we sweep over all the cells which were not intersected by shear bands at $t = t_n$. We mark those cells, $C_1, \ldots, C_m$, where the stress has gone beyond the critical yield value $|\vec{\tau}| = \gamma_c$.

We write $\gamma(t,k)$ for the solution of (12) at each C_k in $[t_n, t_n + \Delta t]$. For each k we find the first t_k such that $\gamma(t_k, k) = \gamma_c$ and label $t^*_{k_b} = \min_{k=1,\dots m} t_k$. We redo the integration up to $t^*_{k_b}$ at the C_k cells and the cells intersected by the shear band. This will give us the first cell C_{k_b} in which to place a shear band as discussed above. We now repeat the above process from time $t = t^*_{k_b}$, adding, if necessary, new cells to the collection of marked cells until we exhaust the time interval. Although the process is potentially computationally intensive it is performed only during a small portion of the simulation.

With this process we take into account the effect of newly formed shear bands on their neighboring cells, either by accelerating the loading rate (and possibly introducing new band cells) or by unloading the nearby cells (and eliminating extraneous shear bands).

5. EXAMPLES

In this test run we consider a uniformly loaded sample with constant stress throughout the domain $\tau_1 = .820352871$, $\tau_2 = .409702741$ and $\gamma = .945303919$. The initial velocity is linear in the x-direction and constant in the y-direction, $v(0,x,y) = .05(x - .5)$, for $0 < x, y < 1$. If the material is homogeneous, the deformation, at all times, will consist of a velocity which is constant in time and stress that, although changing in time, remains constant in space (see [7]) . We introduce an inhomogeneity in the sample at the origin: $\alpha = .475$ everywhere except at the cell at the center of the domain and cells immediatly above and below for which we set $\alpha = .5$. At short time after initialization, a shear band is formed in those cells. This band grows diagonally through the domain. We show two snapshots of the contours for the velocity (to the right) and γ (to the left), when the shear band has grown half-way through the domain (Figure 1) and after it has reached the boundaries of the domain (Figure 2).

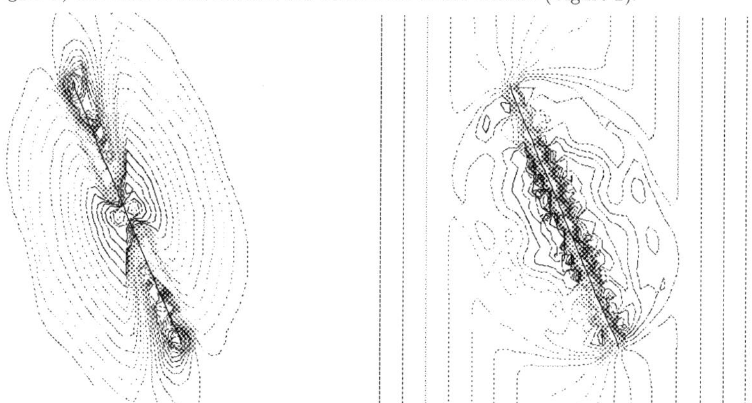

FIGURE 1. γ and v contour lines at intermediate time.

Some features to be noticed are: (a) the localization of high stress in the direction, and preceding, the shear band, (b) the unloading of the material in the direction orthogonal to the shear band and (c) the localization of the variation (jump) in the velocity at the shear band. All these phenomena agree with analytical and/or numerical studies performed for one-dimensional deformations ([4], [6], [7]).

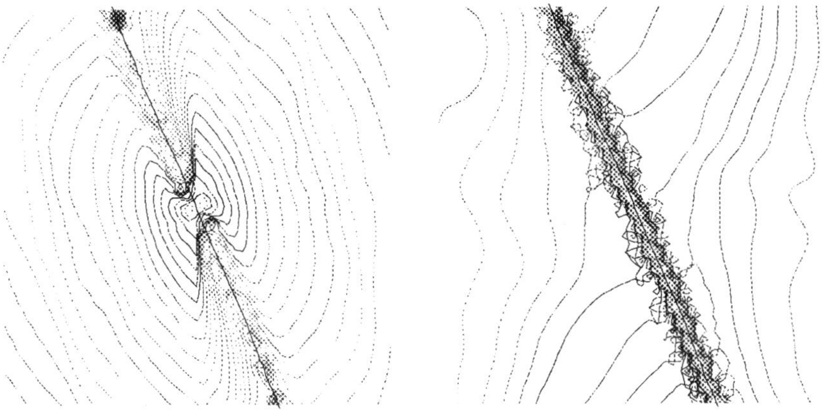

FIGURE 2. γ and v contour lines at larger time.

REFERENCES

1. I-L. Chern and P. Colella, *A conservative front tracking method for hyperbolic conservation laws*, Tech. Report UCRL-97258, Lawrence Livermore National Laboratory, Lawrence Livermore National Laboratory, Berkeley, CA 94720, July 1987.
2. P. Colella, *Multidimensional upwind methods for hyperbolic conservation laws*, Journal of Computational Physics **87** (1990), 171–200.
3. X. Garaizar, *Numerical computations for antiplane shear in a granular flow model*, Quarterly of Applied Mathematics **52** (1994), 289–309.
4. X. Garaizar and D. Schaeffer, *Numerical computations for shear bands in an antiplane shear model*, J. Mech. Phys. Solids **42** (1994), no. 1, 21–50.
5. J. Glimm, E. Isaacson, D. Marchesin, and O. McBryan, *Front tracking for hyperbolic systems*, Advances in Applied Mathematics **2** (1985), 91–119.
6. D.G. Schaeffer M. Shearer, F.X. Garaizar and J. Trangenstein, *Numerical computations for shear bands in an antiplane shear model*, Annual ARO Conference on Applied Mathematics and Computing, vol. ARO 94-1, 1994, pp. 15–28.
7. D. G. Schaeffer, *A mathematical model for localization in granular flow*, Proceedings of the Royal Society of London **436** (1992), 217–250.
8. John A. Trangenstein, *A second-order godunov algorithm for two-dimensional solid mechanics*, Computational Mechanics **13** (1994), 343–359.
9. John A. Trangenstein and P. Colella, *A higher-order godunov method for modeling finite deformation in elastic-plastic solids*, Comm. Pure Appl. Math. **44** (1991), 41–100.

Shock-Wave Solutions of the Relativistic Euler Equations on Non-Flat Spacetimes

Jeffrey M. Groah[1]

1. Introduction

The Einstein equation $G = kT$, where G is the Einstein curvature tensor, k a constant, and T the energy momentum tensor, describes how matter determines curvature. The Einstein curvature tensor is specifically concocted so that conservation of energy and momentum holds; *i.e.*, $\operatorname{div} T = 0$. This equation is known as the *relativistic Euler equation* and describes how curvature moves matter.

In an article published in 1993 by Blake Temple and Joel Smoller [3] it was shown that if the total variation of the initial data is bounded, then shock wave solutions to the relativistic Euler equations for a flat two dimensional spacetime with equation of state $p = \sigma^2\rho$ exist for all time, and the total variation of the solution remains uniformly bounded. Curved spacetimes raise several complications, not the least of which are the bending of space to form black holes, or the spacetime ceasing to exist in finite time due to its crashing to a point. Neither of these two questions arise in a classical setting, and show that Smoller and Temple's conclusion is not expected to be true for curved spacetimes in general.

From Einstein's equation it is clear that there are two necessary ingredients to study the relativistic Euler equations. First, curvature arises from a metric, or inner product, so we need to decide what class of metrics to consider. The other ingredient is a class of energy-momentum tensors.

Since our approach to the problem is from the very large scale structure of spacetime it is appropriate to assume that the fluid is perfect. This leads to the energy-momentum tensor $T^{ij} = (p + \rho c^2)u^i u^j + pg^{ij}$ where p is the pressure, ρ the energy density, g the metric, and $U = u^i \frac{\partial}{\partial x^i}$ the four-velocity which is a unit vector (*i.e.*, $g(U, U) = -1$.) Note that in an n-dimensional spacetime, U has $n - 1$ independent components, so there are $n + 1$ unknowns. However, $\operatorname{div} T = 0$ represents only n equations. With the equation of state $p(\rho)$ it is possible to reduce this to n unknowns. For present purposes it is assumed that $p = \sigma^2\rho$.

This equation of state can be justified by noting that for a polytropic equation of state ($p = C\rho^\gamma, \gamma > 1$) the sound speed exceeds the speed of light for sufficiently large ρ, except in the limiting case $\gamma = 1$. Thus it is only in the case $\gamma = 1$ that we do not have to modify the equation of state at large densities.

The equation of state $p = \sigma^2\rho$ does arise theoretically. Anile [1] shows that $p = c^2\rho/3$ arises in connection with a gas in local thermodynamical equilibrium with radiation when the radiation density greatly exceeds the total gas energy density, as well as for a fluid of massless neutrinos, or a fluid of ultrarelativistic electron-positron pairs. He also notes that in the early universe, $p = c^2\rho/3$ would apply.

[1]Mathematics Department, University of California at Davis, Davis, CA 95616, Email: groah@ucdmath.ucdavis.edu

Weinberg [4] states that the equation of state $p = c^2\rho/3$ holds for a hot, extremely relativistic gas.

Even with the correct number of unknowns, the number of equations is still too large to solve. With enough symmetry it is possible to reduce the number of equations to two. Here, particles will be assumed to move only along one spatial coordinate: x^1. This requires some symmetry in the manifold so that gravitation won't divert the fluid out of the x^1 coordinate direction. The metric must be chosen so that the acceleration in the coordinate directions other than the time coordinate x^0, and x^1, is zero (*i.e.*, $g(\nabla_U U, \frac{\partial}{\partial x^k}) = 0$, $k \geq 2$.) If the metric has the block structure,

$$g = \begin{pmatrix} \tilde{g} & 0 \\ 0 & \hat{g} \end{pmatrix} \tag{1}$$

(where $\tilde{g}$ is a 2×2 block) then this condition implies that the entries in the upper block depend only on x^0 and x^1.

It is also necessary to assume that in the initial data, the components of particle velocity ($u^k = 0, k \geq 2$) in the directions we don't want particles to move, as well as the spatial pressure gradient in those directions, is zero (*i.e.*, $g(\text{grad}_\perp(p), \frac{\partial}{\partial x^k}) = 0, k \geq 2$.) The spatial pressure gradient is the component of the pressure gradient orthogonal to particle four-velocity, $0 = g(\text{grad}_\perp(p), U)$ and so $\text{grad}(p) = -U(p)U + \text{grad}_\perp(p)$. In addition, in order to properly assign initial data, we assume that the manifold under consideration is globally hyperbolic, inextendable, and either non-compact or compact with Euler characteristic zero.

The last assumption we will make here is that the choice of coordinate along which particles are permitted to move was made wisely enough to avoid the possibility of the fluid focusing at a point, as fluids do in spherically symmetric solutions of the classical Euler equations. This is not only a coordinate assumption but also an assumption about curvature and hence about the metric. To see this, we can allow lots of curvature in the directions particles do not move, but not too much in the direction of motion.

If the spatial part of the manifold is compact, then a focusing point must always exist. However, even if a space has no curvature, such a point can still exist in certain coordinate systems (such as spherical coordinates in $\mathbb{R}^n$). So what is the condition on the coordinates and the metric? For the present purposes, restrict attention to the spatial part of the metric. The coordinate system generates a coordinate basis for the tangent space to a manifold. The basis vector for the direction particles are allowed to move generates a vector field on the manifold. To avoid focusing, the divergence of this vector field must be prevented from becoming undefined. This means that

$$\text{div}\left(\frac{\partial}{\partial x^1}\right) = \frac{\partial}{\partial x^k}(\delta_1^k) + \Gamma^i_{ik}\delta_1^k = \Gamma^i_{i1} = \frac{1}{2}\frac{\partial}{\partial x^1}(\ln(-g)) \tag{2}$$

must never become undefined at any point in the manifold (where g is the determinant of the entries in the metric.) Thus, in general, it is best to assume that

$\ln(-g)_{,1}$ as well as $\ln(-g)_{,0}$ is never undefined. The latter condition implies that the spacetime won't crash to a point in finite time in a big crunch.

Given all of these assumptions about the metric, the relativistic Euler equations are still formidable. They have a source term, so an operator splitting technique is used to prove existence of weak solutions.

2. The Operator Splitting Technique

It is possible to put the equations into the form

$$U_t + F(U)_x = G(x,t,U) \tag{3}$$

Approximate solutions are constructed by first, setting the source term to zero (*i.e.*, $U_t + F(U)_x = 0$) and then Riemann problems are solved on the discretized data. Then, the ODE's $U_t = G(x,t,U)$ are solved with initial values being the solutions to the Riemann problems. To show that the approximates are well-defined and to show convergence it is necessary to obtain uniform bounds on the sum of the wave strengths. This is done in two parts. First, the sum of the wave strengths in the Riemann problem part is shown to be nonincreasing, and then the increase of the wave strengths after the ODE step is shown to be bounded. In order to obtain uniform bounds it is necessary to dampen the source term in most circumstances. The rate at which the source is dampened determines how the wave strengths grow. It is possible to obtain linear or exponential growth bounds and even uniform bounds depending on metric assumptions.

The bound on the strength of waves from the Riemann problem part follows closely the paper, "Global Solutions of the Relativistic Euler Equations, " by Temple and Smoller [3]. This paper is itself a generalization of Nishida's 'big data' global existence theorem for weak solutions of the classical compressible Euler equations [2]. The idea is to show that the shock curves based at different points in the plane of Riemann invariants are translations or reflections of each other.

The bound on the increase in wave strength on the acceleration step is obtained by solving the ODE's explicitly for $\ln(\rho_{,0})$ as a function of $v = \frac{dx^1}{dt}$ (the velocity) and the metric and then determining the appropriate bound on the metric, which will give the desired bound on wave strengths. Standard arguments are used to show convergence of the approximates to a weak solution.

3. Conformally Flat Spaces

For the present purposes we restrict the class of metrics further to the class of conformally flat metrics in order to exploit some additional structure. The conformally flat metrics can have the form $g = e^{2\phi}\eta$ where η is the flat space Minkowski metric, and ϕ is a function of x^0 and x^1 alone. Note that with the change of variable $\frac{d\tau}{dt} = e^{\phi(t)}$ the Robertson-Walker metric $ds^2 = -dt^2 + e^{2\phi}(dx^2 + dy^2 + dz^2)$ can be put into conformally flat form. The power of conformally flat metrics is seen in the following proposition.

Proposition 3.0.1. *Suppose T is a symmetric (2,0) tensor, $\bar{T} = e^{(n+2)\phi}T$ where n is the dimension of the manifold, and the manifold is endowed with a metric $\tilde{g}$ conformal to another metric g (i.e., $\tilde{g} = e^{2\phi}g$.) Then $\tilde{\operatorname{div}}(T) = 0$ if and only if*

$$\operatorname{div}(\bar{T}) = e^{(n+2)\phi}\operatorname{grad}(\phi)\operatorname{tr}(T) \tag{4}$$

Here grad and tr are taken with respect to the metric g. A similar proposition appears in Anile's book on relativistic fluids [1] of which this is a generalization. The proposition in Anile's book applies only when the metric is conformal to the Minkowski metric in a four dimensional spacetime and when the energy-momentum tensor is trace-free, thus killing off the source term on the right hand side. The advantage of this proposition is that if g is the Minkowski metric, then $\operatorname{div}(\bar{T}) = \bar{T}^{ij}_{,i} + \Gamma^i_{ik}\bar{T}^{kj} + \Gamma^j_{ik}\bar{T}^{ik}$ takes a particularly simple form since the Christoffel symbols Γ^j_{ik} vanish. Here $_{,i}$ represents partial differentiation with respect to the x^i coordinate direction. The equations are especially simple if $\operatorname{grad}(\phi)\operatorname{tr}(T) = 0$. $\operatorname{grad}(\phi) = 0$ is just the case that both metrics are the Minkowski metric, and $\operatorname{tr}(T) = 0$ is equivalent to the equation of state $p = \frac{c^2}{(n-1)}\rho$.

Note especially that if $\operatorname{grad}(\phi)\operatorname{tr}(T) = 0$, then the source term is zero. Moreover, with the change of variable to conformal density $\bar{\rho} = e^{n\phi}\rho$ the equations reduce to the same equations generated in the Minkowski metric case.

More generally, an existence theorem for $\operatorname{div}(\bar{T}) = e^{(n+2)\phi}\operatorname{grad}(\phi)\operatorname{tr}(T)$ with one dimensional fluid flow in an n-dimensional spacetime with conformally flat metric has been worked out as follows.

Proposition 3.0.2. *Suppose the spacetime M is conformally flat (i.e., $g = e^{2\phi}\eta$ where η is the Minkowski metric), ϕ is a function of x^0 and x^1 alone, fluid focusing in the direction of motion does not occur, and the initial data is appropriately constructed for one dimensional fluid flow. If $\{\sigma^2(n-1) - c^2\}\{|\phi_{,0}| + |\phi_{,1}|\}$ is bounded, and*

$$\{\sigma^2(n-1) - c^2\}\{|\phi_{,0}| + |\phi_{,1}|\} \leq \frac{C}{|x^0|^p + |x^1|^q} \tag{5}$$

where p and q are greater than 1, and C is a constant independent of σ, the sound speed, then weak solutions of

$$\operatorname{div}(\bar{T}) = e^{(n+2)\phi}\operatorname{grad}(\phi)\operatorname{tr}(T) \tag{6}$$

exist for all time, and the total variation of a solution is bounded by a constant plus the total variation of its initial data.

$$TV(U) \leq TV(U_0) + C. \tag{7}$$

Other assumptions about the metric could be made which give linear or exponential growth bounds. The present case however, represents spaces which are eventually flat (*i.e.*, Minkowskian) and which become flat at a fast enough rate.

The case $\operatorname{grad}(\phi)\operatorname{tr}(T) = 0$ is worth isolating. As already noted, when $0 = \operatorname{grad}(\phi)\operatorname{tr}(T)$, the source term vanishes. Given this, the above theorem yields global bounds on the total variation of $\ln(\bar{\rho})$. This means that the conformal density can be bounded clear up to the collapse (or birth) of the spacetime, and hence a computational scheme based on this method may be used to model a spacetime all the way from a big bang to a big crunch.

References

1. A. M. Anile, *Relativistic fluids and magneto-fluids*, Cambridge University Press, 32 East 57th Street, New York, NY 10022, USA, 1989.
2. T. Nishida, *Global solution for an initial boundary value problem of a quasilinear hyperbolic system*, Proc. Jap. Acad. **44** (1968), 642–646.
3. J. Smoller and J. B. Temple, *Global solutions of the relativistic euler equations*, Communications in Mathematical Physics **156** (1993), 67–99.
4. S. Weinberg, *Gravitation and cosmology: Principles and applications of the general theory of relativity*, Wiley, New York, 1972.

Gaussian-Based Moment-Method Closures for the Solution of the Boltzmann Equation

Clinton P. T. Groth[1], Tamas I. Gombosi[2], Philip L. Roe[1], and Shawn L. Brown[1]

ABSTRACT. A new hierarchy of higher-order moment-method closures is described that provide approximate non-equilibrium solutions to the Boltzmann equation. Members of this hierarchy include 14-, 21-, 26-, and 35-moment solutions. Each of the proposed models are based on a Chapman-Enskog like perturbative expansion for the particle velocity distribution function expressed in terms of an anisotropic Gaussian weight function, and fourth-order velocity moments are included explicitly in the expansions. This permits the inclusion of hydrodynamic stresses in a non-perturbative fashion. The Gaussian-based expansions can also be shown to be positive valued for all velocity space and lead to sets of quasi-linear symmetric moment equations that are stable and hyperbolic for significant ranges of physical conditions. Positivity and hyperbolicity are important properties for they ensure finite propagation speeds, well-posedness, and prevent the breakdown of the closure models. Previous approaches based on perturbations to a Maxwellian cannot be easily endowed with these properties. Relevant theory for the closures is discussed. Particular attention is given to the 35-moment model which appears, in many respects, to be a very natural closure.

1. INTRODUCTION

Accurate prediction of flows of non-equilibrium rarefied gases and plasmas requires a consideration of the microscopic behaviour of the fluid particles. This is because the macroscopic descriptions provided by conventional hydrodynamic theories (derived on the basis of continuum assumptions) become inaccurate as inter-particle collisions occur less frequently and particle mean free paths become comparable with physical length scales. Classical kinetic theory provides a convenient and useful mathematical description of rarefied gases beyond the hydrodynamic limit. In this modeling approach, a statistical representation of particle motions is adopted in which a *particle phase-space distribution function*, F, is used to prescribe the probable particle velocities, and the time evolution of the distribution is governed by the Boltzmann equation [1]. Neglecting external forces, the Boltzmann equation for a monatomic gas can be formulated in terms of the random particle velocity c_i, the average or bulk particle velocity u_i, position x_i, and time t and expressed as

$$\frac{\partial F}{\partial t} + (u_i + c_i)\frac{\partial F}{\partial x_i} - \left[\frac{\partial u_i}{\partial t} + (u_j + c_j)\frac{\partial u_i}{\partial x_j}\right]\frac{\partial F}{\partial c_i} = \frac{\delta F}{\delta t}, \tag{1}$$

where $\delta F/\delta t$ is the *collision operator* representing the time rate of change of the particle distribution function produced by collisional processes. Refer to reference [8] for further details.

Equation (1) is a non-linear integro-differential equation in terms of seven independent variables. Closed-form analytic solutions are virtually unobtainable. *Extended-continuum* approaches, such as the methods of Chapman and Enskog [4, 7] and Grad [10], provide one means of constructing approximate solutions. These methods would seem particularly suited for *transitional* rarefied flows

[1]W. M. Keck Foundation Laboratory for Computational Fluid Dynamics, Department of Aerospace Engineering, University of Michigan, Ann Arbor, MI 48109-2118, Email: groth@engin.umich.edu, philroe@engin.umich.edu, brownsl@engin.umich.edu

[2]Space Physics Research Laboratory, Department of Atmospheric, Oceanic and Space Sciences, University of Michigan, Ann Arbor, MI 48109–2143 tamas@engin.umich.edu

Research was supported in part by the NASA grant NAGW-1366 and by NSF-NASA-AFOSR interagency grant NSF-9318181. The first author was also suuported by a NSERC Postdoctoral Fellowship. The fourth author was also supported by the US Air Force Palace Knight Program.

having still relatively high collision rates (cases for which the computational costs of *direct* solution techniques can be prohibitive) for they replace the problem of directly solving the Boltzmann equation with one of solving a system of generalized transport equations for various hydrodynamic quantities. The velocity distribution function is related to the fluid variables by means of a perturbative expansion technique. Unfortunately, most closures of this type are based on an expansion about a local Maxwellian distribution and, as such, require that the gas be close to local thermodynamic equilibrium (LTE). For rarefied flows exhibiting significant departures from LTE, several important weaknesses limit the applicability of the methods. Problems such as non-positivity of the distribution function, non-realizable velocity moments, unstable and/or non-hyperbolic transport equations (the latter leads to non-finite wave speeds for disturbances), ill-posed initial boundary value problems (IBVPs), and general breakdown of the closure model are encountered. It is for these reasons that extended continuum approaches have not received more widespread use.

Having introduced the limitations of traditional solution techniques for the Boltzmann equation, the remainder of the paper is concerned with the description of a new hierarchy of closures that avoid these deficiencies. The proposed hierarchy is obtained by expanding about a generalized anisotropic Gaussian distribution and including fourth-order terms in the expansion. This permits the inclusion of fluid stresses in a non-perturbative fashion. It also results in approximate distributions that are positive valued and quasi-linear moment equations that are both stable and hyperbolic for significant ranges of physical conditions, thereby preventing the breakdown of the transport equations. The study was inspired by the recent theoretical work of Levermore concerning moment closure hierarchies [15]. In fact, the proposed closure technique may be regarded as a practical compromise between the non-perturbative method of Levermore and classical methods. For this reason, a brief summary of his method is first given in Section 2. This includes a description of the anisotropic Gaussian closure (this solution is used as the weight function in the proposed expansion technique). The hierarchy of Gaussian-based moment closures, including 14-, 21-, 26-, and 35-moment models, are subsequently presented in Section 3. Conclusions follow in Section 4.

2. Moment Closure Hierarchy of Levermore

2.1. Non-Perturbative Closure Technique. The deleterious features of previous methods have motivated Levermore [15] to develop a new hierarchy of closed systems of moment equations having many desirable properties. Each member of this hierarchy is hyperbolic, has an entropy, and possesses realizability of its predicted moments. Levermore's closure technique has similarities to the moment method of Grad [10]. A finite number, N, of functions, $M_\alpha^N = M_\alpha^N(\mathbf{c})$, $\alpha \in \{1, 2, \ldots, N\}$, are selected to form a basis or subspace for the closure (traditionally these are polynomial functions); moments of the Boltzmann equation are taken to arrive at a set of generalized transport equations for the moment densities $\sigma_\alpha = m < M_\alpha^N F^N >= m \int_{-\infty}^{\infty} M_\alpha^N F^N d^3c$ where m is the molecular weight (for polynomial functions, the moments correspond to various hydrodynamic quantities); and formal closure of the resulting system is achieved by expressing the remaining unknown moments in terms of the densities σ_α. However, instead of basing the closure on a perturbative expansion for the velocity distribution about a Maxwellian, the distribution function is assumed to have the form

$$F^N = \exp\left(-\Lambda_\alpha^N M_\alpha^N\right) , \qquad (2)$$

where Λ_α^N are coefficients specified by satisfying the prescribed velocity moments (i.e., $\Lambda_\alpha = \Lambda_\alpha(\sigma_\alpha)$). The functions, M_α^N, comprising the subspace for the closure are selected such that $\exp(-\Lambda_\alpha^N M_\alpha^N)$ is finite and the subspace contains $\{1, c_i, c^2\}$ and is invariant under translational and orthogonal transformations (requirements for frame indifference). If these conditions are met, then Levermore has shown that positivity of the distribution function is ensured, the resulting transport equations are hyperbolic, and all predicted moments are realizable. Not only does the expression of Eq. (2) guarantee positivity of the approximate distribution function, it can also be shown that it is the unique distribution that maximizes the entropy, $s = -k < F \ln F > /\rho$, subject to the conditions that the moments $\sigma_\alpha = m < M_\alpha^N F^N >$ are fixed [16]. Here, k is Boltzmann's constant and $\rho = m < F >$ is the mass density. Additional mathematical properties of Levermore's non-perturbative closure technique are given in reference [15]. It is interesting that Cercignani and Tironi [3] have also postulated a similar hierarchy but, beyond suggesting a form for the closures, do not appear to have carried out a detailed analysis concerning their use.

For polynomial basis functions, the first six subspaces of Levermore's hierarchy are given by $M_\alpha^5 \in \{1, c_i, c^2\}$, $M_\alpha^{10} \in \{1, c_i, c_ic_j\}$, $M_\alpha^{14} \in \{1, c_i, c_ic_j, c_ic^2, c^4\}$, $M_\alpha^{21} \in \{1, c_i, c_ic_j, c_ic_jc_k, c^4\}$, $M_\alpha^{26} \in \{1, c_i, c_ic_j, c_ic_jc_k, c_ic_jc^2\}$, $M_\alpha^{35} \in \{1, c_i, c_ic_j, c_ic_jc_k, c_ic_jc_kc_l\}$. They correspond to the 5-, 10-, 14-, 21-, 26-, and 35-moment closures, respectively. The first, or 5-moment closure, leads to the Euler transport equations for which the particle phase-space distribution function is approximated by an isotropic Maxwellian distribution. This closure is for the most part well understood. The second or 10-moment closure is of more interest. Although it is not a complete approximation because heat conduction is neglected, the closure leads to a non-perturbative, non-isotropic, and positive Gaussian particle distribution and furnishes an explicit set of transport equations for the hydrodynamic quantities ρ, u_i, and P_{ij}. The anisotropic Gaussian is a vital ingredient in the present work and a description of the model now follows.

2.2. Anisotropic Gaussian Closure. As indicated above, the ten basis functions for the Gaussian closure are $\{1, c_i, c_ic_j\}$. The coefficients of the associated particle phase-space distribution function, F^{10}, are related to the velocity moments by requiring that

$$m < F^{10} >= \rho\,, \quad m < c_iF^{10} >= 0\,, \quad m < c_ic_jF^{10} >= P_{ij}\,, \tag{3}$$

where P_{ij} is the symmetric generalized pressure tensor (the usual hydrodynamic pressure, p, is related to the trace of the pressure tensor and given by $p = P_{ii}/3$). These constraints yield an analytic closed-form expression for the distribution function in terms of the hydrodynamic quantities ρ, u_i, and P_{ij}. The 10-moment distribution has the form

$$F^{10} = \frac{\rho}{m(2\pi)^{3/2}\Delta^{1/2}} \exp\left(-\frac{1}{2}\Theta_{\alpha\beta}^{-1}c_\alpha c_\beta\right)\,, \tag{4}$$

where Θ_{ij} is a symmetric tensor related to P_{ij} by the expression $\Theta_{ij} = P_{ij}/\rho$ and $\Delta = \det\Theta$.

The velocity distribution of Eq. (4) appears to have been first derived in earlier work by Schlüter [12, 17] and then in subsequent but independent research by Holway [13]. Holway referred to the solution as the *ellipsoidal distribution function* (EDF). It can be regarded as a generalization of the bi- and tri-Maxwellian velocity distribution functions that does not require the identification of the planes of principal stress [5, 12, 17]. Levermore[15] has shown that for $\rho > 0$, $\Delta > 0$ and Θ is a positive definite matrix. Hence, Θ^{-1} is also positive definite and $\Theta_{ij}^{-1}c_ic_j > 0$ for all $\mathbf{c}$ other than $c_i = 0$, from which it readily follows that the Gaussian distribution function is finite and positive valued for all physically realistic values of $\mathbf{P}$ and ρ. The region of positivity in parameter space (the space defined by the hydrodynamic variables) is defined by the conditions that both $\rho > 0$ and $\Delta > 0$. It is significant that there are effectively no unphysical limitations on the magnitude of the individual elements of the stress tensor $\tau_{ij} = p\delta_{ij} - P_{ij}$ for positivity of F^{10}. In addition to its positivity properties, the form of the distribution function of Eq. (4) is also pleasing from a mathematical standpoint as it is possible to evaluate all velocity moments analytically (see reference [9]). In particular, odd order random velocity moments of F^{10} are zero.

Generalized transport equations for the Gaussian closure can be obtained by taking velocity moments of Eq. (1) with $M = \{m, mc_i, mc_ic_j\}$. This yields a set of ten partial-differential equations (PDEs) for the hydrodynamic quantities ρ, u_i, and P_{ij}. For Cartesian coordinates (x, y, z), the transport equations can be expressed in vector form and written as

$$\partial_t \mathbf{W}^{10} + \mathcal{A}^{10}\partial_x \mathbf{W}^{10} + \mathcal{B}^{10}\partial_y \mathbf{W}^{10} + \mathcal{C}^{10}\partial_z \mathbf{W}^{10} = \mathcal{S}^{10}\,, \tag{5}$$

where $\mathbf{W}^{10} = [\rho, u_i, P_{ij}]^{\mathrm{T}}$ is the solution vector, $\mathcal{A}^{10}$, $\mathcal{B}^{10}$, and $\mathcal{C}^{10}$ are flux coefficient matrices, and $\mathcal{S}^{10}$ is the source vector containing terms that represent changes in the moment densities brought on by particle collisions. The solution of Eq. (5) provides a complete description of F^{10}.

A key feature of Levermore's closures is that the positivity of the velocity distributions ensures realizability of the predicted moments (i.e., solutions of Eq. (5) are physically meaningful and consistent with a positive valued distribution function). It also ensures that the transport equations are hyperbolic[1]. For the 10-moment closure, this second feature can be seen by considering the eigenstructure of the coefficient matrices $\mathcal{A}^{10}$, $\mathcal{B}^{10}$, and $\mathcal{C}^{10}$. Quasi-linear PDEs of the form given in Eq. (5) are said to be strictly hyperbolic if the eigenvalues of these matrices are all real and the corresponding eigenvectors form a complete non-degenerate set. This is a revised version of the more restrictive demand that the eigenvalues be both real and distinct [14]. For the x-direction, a characteristic analysis reveals that the eigenvalues, λ, of $\mathcal{A}^{10}$ are the roots of

$$\det\left(\mathcal{A}^{10} - \lambda\mathbf{I}\right) = Z^4\left(Z^2 - a_x^2\right)^2\left(Z^2 - 3a_x^2\right) = 0\,, \tag{6}$$

where $Z = u - \lambda$ and $a_x = \sqrt{P_{xx}/\rho}$. Solution of Eq. (6) yields the eigenvalues

$$\lambda_k = u + \eta_k a_x\,, \quad \text{for } \eta_1 = -\sqrt{3}\,, \quad \eta_{2,3} = -1\,, \quad \eta_{4,5,6,7} = 0\,, \quad \eta_{8,9} = 1\,, \quad \eta_{10} = \sqrt{3}\,. \tag{7}$$

Although some eigenvalues are repeated (there are 5 unique values and 10 in total), a complete set of right eigenvectors, $\boldsymbol{r}$, satisfying $\mathcal{A}^{10}\boldsymbol{r} = \lambda\boldsymbol{r}$ can also be defined. However, due to space limitations, these are not given here. A detailed investigation of the wave structure, dispersive properties, and computed shock structure for the 10-moment closure is given elsewhere [2, 9].

It is evident from Eq. (7) that the eigenvalues of $\mathcal{A}^{10}$ are real for $P_{xx}/\rho > 0$. Moreover, a complete set of eigenvectors can be found. As the eigensystems of $\mathcal{B}^{10}$ and $\mathcal{C}^{10}$ have much the same structure, it follows that the moment equations of the Gaussian closure are of the hyperbolic type. The fact that the choice of the x-direction for performing the analysis was arbitrary means that the 10-moment equations are hyperbolic provided that $P_{xx}/\rho > 0$, for any orientation of the Cartesian coordinates. This amounts to requiring that Θ is positive definite. Hence, the equations are hyperbolic whenever the moments of F^{10} are realizable. This result was indeed predicted by Levermore on the basis of general theorems that make use of the fact that a reformulation of the closure equations in terms of the distribution coefficients, Λ_α, leads to a symmetric set of PDEs [15]. Evidently, a strong relationship between hyperbolicity, positivity, and realizability exists.

2.3. Higher-Order Closures. Despite the promising features of the anisotropic Gaussian closure described above, for higher-order approximations (i.e., the 14-, 21-, 26-, and 35-moment closures), a significant limitation of Levermore's closure procedure becomes apparent. The difficulty is that explicit closed-form expressions for the coefficients, Λ_α, of the distribution function in terms of the hydrodynamic quantities, σ_α, prove to be impossible to obtain (i.e., $\Lambda_\alpha = \Lambda_\alpha(\sigma_\alpha)$ defines a transformation and the inverse transformation $\sigma_\alpha = \sigma_\alpha(\Lambda_\alpha)$ is not easily determined). Although numerical techniques can be employed to solve PDEs for Λ_α and subsequently evaluate the moment densities (this approach is advocated by Levermore), transport equations for the hydrodynamic quantities, which are traditionally of interest, cannot be formulated explicitly without performing a linearization of various non-equilibrium quantities as is done in molecular extended thermodynamics [16]. This is a drawback because transport equations for the moment densities are highly desirable.

[1] The hyperbolic and symmetric properties of the generalized transport equations automatically imply the existence of an entropy balance or dissipation law as discussed by Dafermos [6]. This is another desirable feature of Levermore's closure hierarchy.

3. Perturbative Anisotropic Gaussian-Based Closure Hierarchy

In this section, a hierarchy of higher-order closures beyond the 10-moment level are formulated by using the polynomial basis functions of Levermore's hierarchy as the subspaces for the closures and by approximating the particle phase-space distribution functions with Chapman-Enskog type expansions about the Gaussian F^{10} having the form

$$F = F^{10}\,[\,1 + A + B_\alpha\, c_\alpha + C_{\alpha\beta}\, c_\alpha c_\beta + D_{\alpha\beta\gamma}\, c_\alpha c_\beta c_\gamma + E_{\alpha\beta\gamma\delta}\, c_\alpha c_\beta c_\gamma c_\delta + \ldots]\,, \tag{8}$$

where A, B_α, $C_{\alpha\beta}$, $D_{\alpha\beta\gamma}$, $E_{\alpha\beta\gamma\delta}$, etc. are coefficients that are independent of the random velocity $\mathbf{c}$. The proposed perturbative approach is, in essence, a marriage of the Chapman-Enskog and Levermore approaches with a view to obtaining closed forms for the generalized transport equations while still retaining many of the desirable features of Levermore's non-perturbative models. It differs from traditional moment methods in several important ways:

(1) Most conventional methods are based on perturbative expansions for the distribution functions in terms of a local Maxwellian whereas the expansion of Eq. (8) is based on the Gaussian F^{10}. The latter is fundamentally a non-equilibrium solution that includes the effects of non-zero hydrodynamic shear stress in a non-perturbative manner.

(2) For any non-equilibrium state, there are always regions of phase space for which the approximate distributions of virtually all perturbative methods are negative and hence non-physical. In contrast, the Gaussian-based expansions are expressed in terms of the polynomial basis functions used in Levermore's closure hierarchy and are therefore positive definite for all particle velocities within finite regions of parameter space in the vicinity of F^{10}.

(3) The Gaussian-based expansion of Eq. (8) can be viewed as first-order approximations to Levermore's non-perturbative solutions of Eq. (2) and therefore should lead to well-posed sets of hyperbolic transport equations, at least in the Gaussian limit.

The use of an anisotropic Gaussian weight function is not entirely new. Both Schlüter and Hertweck [12, 17] have made use of the anisotropic 10-moment solution when formulating systems of transport equations for describing non-equilibrium plasmas. However, their formulations do not make use of Levermore's subspaces of basis functions, which are key ingredients in the present models.

In what follows, 14-, 21-, 26-, and 35-moment closures are developed. Although each is discussed, the 35-moment closure is considered first and in the most detail for it appears to result in the most pleasing set of transport equations. The other models are derived directly from the 35-moment approximation by means of various contraction procedures.

3.1. 35-Moment Closure.

The 35-moment closure is derived by employing the basis functions $\{1, c_i, c_ic_j, c_ic_jc_k, c_ic_jc_kc_l\}$ and assuming the velocity distribution, F^{35}, has the form of Eq. (8) with all non-contracted terms up to fourth-order included in the expansion. The coefficients A, B_α, $C_{\alpha\beta}$, $D_{\alpha\beta\gamma}$, and $E_{\alpha\beta\gamma\delta}$ are specified by relating them to the first 35 velocity moments

$$m < F^{35} >= \rho\,, \quad m < c_i F^{35} >= 0\,, \quad m < c_ic_j F^{35} >= P_{ij}\,, \tag{9}$$

$$m < c_ic_jc_k F^{35} >= Q_{ijk}\,, \quad m < c_ic_jc_kc_l F^{35} >= R_{ijkl}\,. \tag{10}$$

These constraints result in the following expression for F^{35}:

$$F^{35} = F^{10}\left[1 + D_{\alpha\beta\gamma}\, c_\alpha \left(c_\beta c_\gamma - 3\frac{P_{\beta\gamma}}{\rho}\right) + E_{\alpha\beta\gamma\delta}\left(c_\alpha c_\beta c_\gamma c_\delta - 6\frac{P_{\gamma\delta}}{\rho} c_\alpha c_\beta + 3\frac{P_{\alpha\beta}P_{\gamma\delta}}{\rho^2}\right)\right], \tag{11}$$

where $D_{\alpha\beta\gamma}$ and $E_{\alpha\beta\gamma\delta}$ are related to the heat flow and fourth-moment tensors Q_{ijk} and R_{ijkl} as follows:

$$Q_{ijk} = \frac{6}{\rho^2} P_{i\alpha}P_{j\beta}P_{k\gamma}D_{\alpha\beta\gamma}\,, \quad R_{ijkl} = \frac{1}{\rho}\,\{P_{ij}P_{kl}\}^{(3)}_{[ijkl]} + \frac{24}{\rho^3} P_{i\alpha}P_{j\beta}P_{k\gamma}P_{l\delta}E_{\alpha\beta\gamma\delta}\,, \tag{12}$$

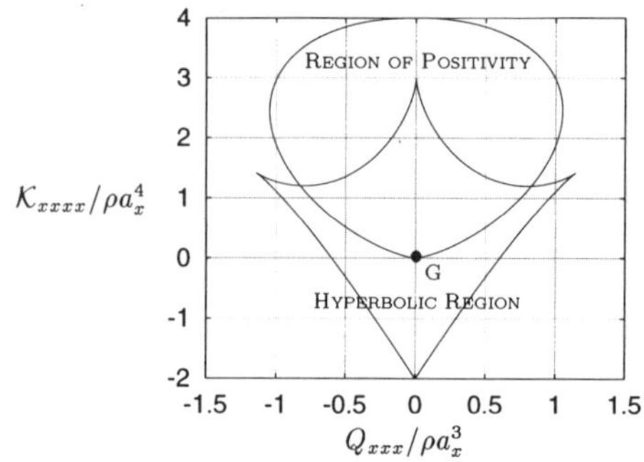

FIGURE 1. Regions of positivity for 35-moment distribution function and hyperbolicity for 35-moment transport equations. Point G represents the Gaussian limit.

and where the notation $\{\}^{(N)}_{[ijk..pq]}$ represents a symmetric tensor obtained by taking the sum of all N independent permutations of the indices. Equations (11) and (12) provide a description of the approximate particle phase-space distribution function F^{35} in terms of the 35 hydrodynamic quantities ρ, u_i, P_{ij}, Q_{ijk}, and R_{ijkl}. The time evolution of these quantities can be described by a set of transport equations. The latter may be derived by taking velocity moments of the Boltzmann equation and can be written as

$$\partial_t \mathbf{W}^{35} + \boldsymbol{\mathcal{A}}^{35}\partial_x \mathbf{W}^{35} + \boldsymbol{\mathcal{B}}^{35}\partial_y \mathbf{W}^{35} + \boldsymbol{\mathcal{C}}^{35}\partial_z \mathbf{W}^{35} = \boldsymbol{\mathcal{S}}^{35}, \tag{13}$$

where the solution column vector is $\mathbf{W}^{35} = [\rho, u_i, P_{ij}, Q_{ijk}, \mathcal{K}_{ijkl}]^{\mathrm{T}}$ and the resulting closing relation for the fifth-order velocity moment is a natural one having the form

$$\begin{aligned} S_{ijklm} = m < c_i c_j c_k c_l c_m F^{35} > \; = \; & \frac{1}{\rho}\left[P_{ij}Q_{klm} + P_{ik}Q_{jlm} + P_{il}Q_{jkm} + P_{im}Q_{jkl} + P_{jk}Q_{ilm}\right. \\ & \left. + P_{jl}Q_{ikm} + P_{jm}Q_{ikl} + P_{kl}Q_{ijm} + P_{km}Q_{ijl} + P_{lm}Q_{ijk}\right]. \end{aligned} \tag{14}$$

Here, the perturbative tensor $\mathcal{K}_{ijkl} = R_{ijkl} - \{P_{ij}P_{kl}\}^{(3)}_{[ijkl]}/\rho$ has been introduced.

The Gaussian-based expansion of Eq. (11) affords arbitrarily large values for the fluid stresses provided that $\Delta > 0$. This is a definite improvement over Maxwellian-based expansions which only make allowances for small deviations in the stresses from zero. The expression for F^{35} is also positive valued and hence possesses realizable velocity moments for well-defined regions of parameter space close to solutions provided by the Gaussian closure (i.e., near $Q_{ijk} = 0$ and $\mathcal{K}_{ijkl} = 0$). Furthermore, and possibly more importantly, it results in a hyperbolic set of moment equations in the vicinity of this Gaussian limit. Space limitations do not permit a thorough examination of these properties herein; however, for planar propagations in the x-direction, it can be shown that Eq. (13) is hyperbolic (i.e., the eigenvalues of $\boldsymbol{\mathcal{A}}^{35}$ are real and its eigenvectors distinct) and Eq. (11) is positive valued for the regions in parameter space depicted in Fig. 1. Note that it is significant that the regions shown in the figure overlap. A more complete derivation of the 35-moment closure, including a detailed investigation of the positivity of F^{35} and hyperbolicity, wave structure, and dispersive properties of the 35-moment transport equations, is given in references [9] and [11].

3.2. 26-, 21-, and 14-Moment Closures. The 26-, 21-, and 14-moment Gaussian-based closures can be derived from the preceding 35-moment model by contraction. For the 26-moment closure, it is assumed that $E_{\alpha\beta\gamma\delta} = E_{\alpha\beta}\delta_{\gamma\delta} + E_{\alpha\gamma}\delta_{\beta\delta} + E_{\alpha\delta}\delta_{\beta\gamma}$ where $E_{\alpha\beta}$ is a second-order symmetric tensor and the resulting approximate distribution function is

$$F^{26} = F^{10}\left[1 + D_{\alpha\beta\gamma}\, c_\alpha \left(c_\beta c_\gamma - 3\frac{P_{\beta\gamma}}{\rho}\right) + 3E_{\alpha\beta}\left(c_\alpha c_\beta c^2 - 2\frac{P_{\gamma\gamma}}{\rho}c_\alpha c_\beta - 4\frac{P_{\beta\gamma}}{\rho}c_\alpha c_\gamma + \frac{P_{\gamma\gamma}P_{\alpha\beta}}{\rho^2} + 2\frac{P_{\alpha\gamma}P_{\beta\gamma}}{\rho^2}\right)\right], \tag{15}$$

where $D_{\alpha\beta\gamma}$ is as defined in Eq. (12), $r_{ij} = m < c_i c_j c^2 F^{26} >= \kappa_{ij} + (P_{ij}P_{\alpha\alpha} + 2P_{i\alpha}P_{j\alpha})/\rho$, and $\kappa_{ij} = 24(P_{i\alpha}P_{i\beta}P_{k\gamma}P_{\gamma k}E_{\alpha\beta} + 2P_{i\alpha}P_{j\beta}P_{k\beta}P_{k\gamma}E_{\alpha\gamma})/\rho^2$. In the case of the 21-moment closure, it is assumed that $E_{\alpha\beta\gamma\delta} = (\delta_{\alpha\beta}\delta_{\gamma\delta} + \delta_{\alpha\gamma}\delta_{\beta\delta} + \delta_{\alpha\delta}\delta_{\beta\gamma})E$ where E is a scalar and F^{21} takes the form

$$F^{21} = F^{10}\left[1 + D_{\alpha\beta\gamma}\, c_\alpha \left(c_\beta c_\gamma - 3\frac{P_{\beta\gamma}}{\rho}\right) + 3E\left(c^4 - 2\frac{P_{\alpha\alpha}}{\rho}c^2 - 4\frac{P_{\alpha\beta}}{\rho}c_\alpha c_\beta + \frac{P_{\alpha\alpha}P_{\beta\beta}}{\rho^2} + 2\frac{P_{\alpha\beta}P_{\beta\alpha}}{\rho^2}\right)\right], \tag{16}$$

where, for this closure, $E = \rho^2[15\rho r - (P_{\alpha\alpha}P_{\beta\beta} + 2P_{\alpha\beta}P_{\beta\alpha})]/24[P_{\alpha\gamma}P_{\gamma\alpha}P_{\beta\delta}P_{\delta\beta} + 2P_{\alpha\gamma}P_{\alpha\delta}P_{\beta\gamma}P_{\beta\delta}]$ and $r = m < c^4 F^{21} > /15$. Finally, the 14-moment model can be obtained by further contraction of the 35-moment closure by assuming that $D_{\alpha\beta\gamma} = (\Theta^{-1}_{\alpha\delta}D_\delta\delta_{\beta\gamma} + \Theta^{-1}_{\beta\delta}D_\delta\delta_{\alpha\gamma} + \Theta^{-1}_{\gamma\delta}D_\delta\delta_{\alpha\beta})/15$. The resulting distribution function is

$$F^{14} = F^{10}\left[1 + \frac{D_\beta}{5}\left[\Theta^{-1}_{\alpha\beta}\left(c_\alpha c^2 - \frac{P_{\gamma\gamma}}{\rho}c_\alpha\right) - 2c_\beta\right] + 3E\left(c^4 - 2\frac{P_{\alpha\alpha}}{\rho}c^2 - 4\frac{P_{\alpha\beta}}{\rho}c_\alpha c_\beta + \frac{P_{\alpha\alpha}P_{\beta\beta}}{\rho^2} + 2\frac{P_{\alpha\beta}P_{\beta\alpha}}{\rho^2}\right)\right], \tag{17}$$

where, in order to satisfy the constraints imposed by the velocity moments $h_i = m < c_i c^2 F^{14} > /2$, the coefficients D_α are required to have the form $h_i = (P_{j\alpha}P_{\alpha j}D_i + 2P_{i\alpha}P_{j\alpha}D_j)/5\rho$. For each of the contracted closures, the transport equations for the hydrodynamic quantities have forms similar to those of Eqs. (5) and (13). The solution vectors in these cases are $\mathbf{W}^{26} = [\rho, u_i, P_{ij}, Q_{ijk}, \kappa_{ij}]^{\mathrm{T}}$, $\mathbf{W}^{21} = [\rho, u_i, P_{ij}, Q_{ijk}, \kappa]^{\mathrm{T}}$, and $\mathbf{W}^{14} = [\rho, u_i, P_{ij}, h_i, \kappa]^{\mathrm{T}}$.

Positivity of F^{26}, F^{21}, and F^{14} can be demonstrated for solutions near the Gaussian limit. Unfortunately, hyperbolicity of the corresponding transport equations is more difficult to show. Unlike the 35-moment closure which results in a closed form for the moment equations and a strikingly elegant yet simple wave structure [9, 11], explicit forms for the 26- and 14-moment equations are impossible to obtain in the general case and furthermore, the eigenstructures of the 21-moment equations, as well as reduced forms of the 26- and 14-moment equations, are found to be highly coupled, complicated, and rather messy, even if portions of the regions of hyperbolicity can be mapped. It seems that the contraction procedures have spoiled the natural closure properties of the 35-moment approximation. It is felt that further study of various simplifications of the 35-moment closure is warranted.

4. Summary and Conclusions

A family of Gaussian-based moment closures and generalized transport models has been described that provides approximate solutions to the Boltzmann equation for rarefied flows in the transitional regime, beyond but near the hydrodynamic (continuum) limit. The use of the 10-moment Gaussian weight function accounts for arbitrary pressure anisotropies and stability and hyperbolicity of the moment equations and positivity of the approximate particle phase-space distribution functions are possible for well-defined regions of parameter space. These features provide extended ranges of validity for the closure models and make them well-suited for today's sophisticated numerical algorithms for solving hyperbolic conservation laws.

Included in the proposed hierarchy are the 35-, 26-, 21-, and 14-moment models. The 35-moment closure model appears to be a natural one for it provides a closing relation that leads to a closed form for the transport equations and eigensystem, dispersion, and positivity analyses reveal a number of desirable features (see references [9] and [11] for details). Unfortunately, the 26-, 21-, and 14-moment closures are not quite as promising. Although the number of hydrodynamic quantities are reduced, it appears that the contraction procedures used to derive these models have disrupted much of the rather elegant yet simple structure of the 35-moment closure. Further study of possible simplifications of the 35-moment closure is on-going.

References

1. L. Boltzmann, *Weitere Studien Über das Wärmegleichgewicht unter Gasmolekülen*, Sitz. Math.-Naturwiss. Cl. Akad. Wiss. Wien **66** (1872), 275–370.
2. S. L. Brown, P. L. Roe, and C. P. T. Groth, *Numerical Solution of a 10-Moment Model for Nonequilibrium Gasdynamics*, submitted for presentation at the AIAA 12th Computational Fluid Dynamics Conference, September 1994.
3. C. Cercignani and G. Torino, *Non-linear Heat Transfer Between Two Parallel Plates According to a Model with Correct Prandtl Number*, Rarefied Gas Dynamics (New York) (C. L. Brundin, ed.), vol. I, Academic Press, 1967, pp. 441–453.
4. S. Chapman, *On the Kinetic Theory of a Gas. Part II. - A Composite Monoatomic Gas: Diffusion, Viscosity, and Thermal Conduction*, Phil. Trans. Royal Soc. London A **217** (1916), 115–116.
5. R. Chodura and F. Pohl, *Hydrodynamic Equations for Plasmas in Strong Magnetic Fields — II Transport Equations Including Collisions*, Plasma Phys. **13** (1971), 645–658.
6. C. M. Dafermos, *Hyperbolic Systems of Conservation Laws*, Systems of Nonlinear Partial Differential Equations (Dordrecht, Holland) (J. M. Ball, ed.), D. Reidel Publishing Company, 1983, pp. 25–70.
7. D. Enskog, *Kinetishe Theorie der Vorgänge in Massing Verdumten Gasen*, Ph.D. thesis, University of Upsala, 1917.
8. T. I. Gombosi, *Gaskinetic Theory*, Cambridge University Press, Cambridge, 1994.
9. T. I. Gombosi, C. P. T. Groth, P. L. Roe, and S. L. Brown, *35-Moment Closure for Rarefied Gases: Derivation, Transport Equations, and Wave Structure*, submitted to the Physics of Fluids, October 1994.
10. H. Grad, *On the Kinetic Theory of Rarefied Gases*, Commun. Pure Appl. Math. **2** (1949), 331–407.
11. C. P. T. Groth, P. L. Roe, T. I. Gombosi, and S. L. Brown, *On the Nonstationary Wave Structure of a 35-Moment Closure for Rarefied Gas Dynamics*, submitted for presentation at the AIAA 26th Fluid Dynamics Conference, September 1994.
12. F. Hertweck, *Allgemeine 13-Momenten-Näherung zur Fokker-Planck-Gleichung eines Plasmas*, Zeit. Naturfor. **20a** (1965), 1243–1255.
13. L. H. Holway, *Approximation Procedures for Kinetic Theory*, Ph.D. thesis, Harvard University, 1963.
14. P. D. Lax, *Hyperbolic Systems of Conservation Laws and the Mathematical Theory of Shock Waves*, CMBS–NSF Regional Conference Series in Applied Mathematics, vol. 11, SIAM, Philadelphia, Pa., 1973.
15. C. D. Levermore, *Moment Closure Hierarchies for Kinetic Theory*, in preparation, June 1993.
16. I. Müller and T. Ruggeri, *Extended Thermodynamics*, Springer-Verlag, New York, 1993.
17. V. Oraevskii, R. Chodura, and W. Feneberg, *Hydrodynamic Equations for Plasmas in Strong Magnetic Fields — I Collisionless Approximation*, Plasma Phys. **10** (1968), 819–828.

Irregular Reflection of Weak Shocks

Moysey Brio[1] and John K. Hunter[2]

1 Introduction

In this paper, we present numerical solutions of an asymptotic equation modeling irregular weak shock reflection. The solutions show a von Neumann reflection, in agreement with numerical results of Colella and Henderson for the full Euler equations [3]. The von Neumann reflection must contain a continuous supersonic wave. We suggest that this wave is generated by the reflection of incoming characteristics off the sonic line.

The asymptotic equation for irregular weak shock reflection is the two-dimensional Burgers equation [1, 2, 4, 5, 6, 7],

$$u_t + \left(\frac{1}{2}u^2\right)_x + v_y = 0, \qquad v_x - u_y = 0. \tag{1.1}$$

Here, u and v are scaled x and y velocity components. This equation holds in the half-space $y > 0$. The appropriate initial and boundary conditions are

$$u|_{t=0} = \begin{cases} 0 & x > ay \\ 1 & x < ay \end{cases}, \qquad v|_{t=0} = \begin{cases} 0 & x > ay \\ -a & x < ay \end{cases}; \tag{1.2}$$

$$v|_{y=0} = 0. \tag{1.3}$$

Thus, the initial location of the incident shock is $x = ay$, and the jump in u across the shock is normalized to one. Equations (1.1)–(1.3) are self-similar.

2 Numerical solutions

The numerical scheme used here is similar to the one described in [1]. The issue of constructing a numerical scheme is rather subtle because the problem is a characteristic initial value problem. Surprisingly, standard numerical schemes for solving (1.1) do not seem to be very robust. As a result, there

[1]Department of Mathematics, University of Arizona, Tucson AZ

[2]Department of Mathematics and Institute of Theoretical Dynamics, University of California, Davis CA

are some numerical oscillations in the solutions shown below. We made careful convergence tests to check that these oscillations do not invalidate the numerical solutions. A second numerical difficulty is that the initial data (1.2) contain a discontinuity in v along the negative x-axis. Instead of imposing (1.2) directly, we used as initial data the linearized solution of (1.1)–(1.3) at a small positive starting time. We then computed the numerical solution until it established self-similarity, typically five to ten times longer than the starting time.

Theoretical arguments [1, 4, 7] show that regular reflection is impossible for $a < \sqrt{2}$. On the scales resolved in our computation, the irregular reflection looked qualitatively similar for various values of a ranging from 0.2 to 1.2. A more detailed description of the numerical results will be given elsewhere.

A typical numerical solution for $a = 0.5$ is shown here. The computation was done on a 2400 by 400 grid with $\Delta x = 4.5 \times 10^{-4}, \Delta y = 1.5 \times 10^{-3}$, and $\Delta t = \Delta y^2$. Figure 1 shows the v-contour plot after 80000 iterations. Numerical plots of the minus and plus families of characteristics, defined in (3.2) below, are shown in Figure 2 and Figure 3. A blow up of the v-contours near the point where the reflected wave hits the incident shock is shown in Figure 4. We also plot the sonic line (3.3). This picture shows the gradual steepening of the reflected wave into a shock over a distance of about fifty x-grid points. The thickness of the reflected wave varies from about ten y-grid points near the incident shock to about five y-grid points across the reflected shock.

3 Theoretical discussion

The main qualitative feature of our numerical solutions is that the slope of the incident shock varies continuously along its whole length and the reflected wave is continuous immediately behind the incident shock. The continuity of the incident shock slope is consistent with the fact that the two dimensional Burgers equation does not allow triple points [1, 7]. If the reflected wave behind the incident shock is continuous and $a > 1/\sqrt{2}$, it follows rigorously that the sonic line (3.3) must be embedded inside the reflected wave [7]. The numerical solution shown in Figure 4 indicates that the sonic line is embedded even when $a = 0.5 < 1/\sqrt{2}$.

The numerical solution is not sufficiently resolved to determine the detailed structure of the reflection directly. However, there are strong theo-

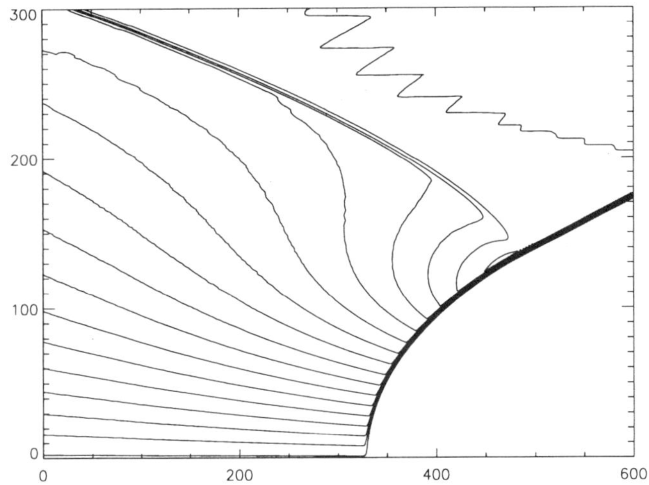

Figure 1: Contour plot of v after 80000 iterations

retical constraints on the possible structure of a continuous reflected wave. We show a schematic diagram of a proposed structure in Figure 5. To the best of our knowledge, this structure is consistent with all theoretical facts concerning (1.1).

The incident shock is associated with the plus family of characteristics in (3.2) below. The supersonic wave originates from the reflection of incoming minus characteristics off the sonic line. One part of the reflected plus wave steepens into the supersonic reflected shock, while the other part hits the incident shock and causes a supersonic bend. The plus characteristics are bent by the strong compression wave. The numerical plot of the plus characteristics shown in Figure 3 suggests that the characteristics bend back towards the incident shock near the reflected wavefront. The nonuniform wave in the hyperbolic region and the supersonic bend in the incident shock are both caused by characteristics reflected off the sonic line, so this structure does not violate the hyperbolic uniqueness argument given in [7]. The continuous nonuniform wave behind the incident shock is separated from the uniform state above it by a limiting plus characteristic. This limiting characteristic is tangent to the reflected shock at the point Z where its strength approaches zero. The supersonic bend in the incident shock lies between

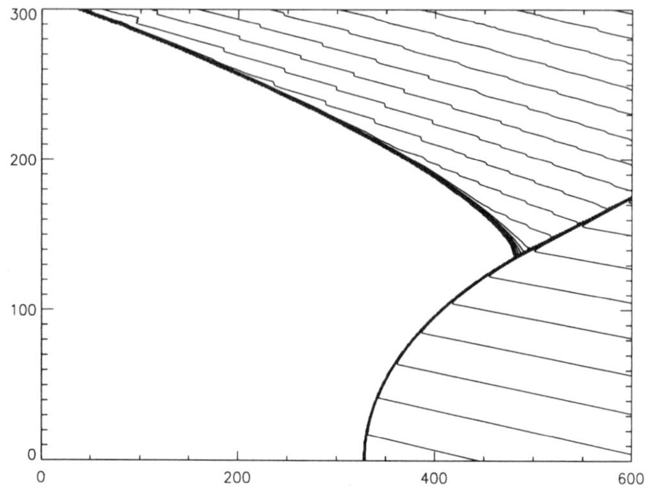

Figure 2: Minus characteristics

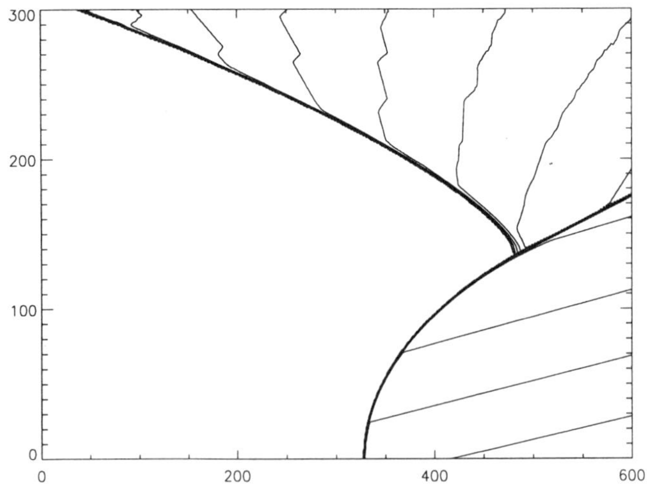

Figure 3: Plus characteristics

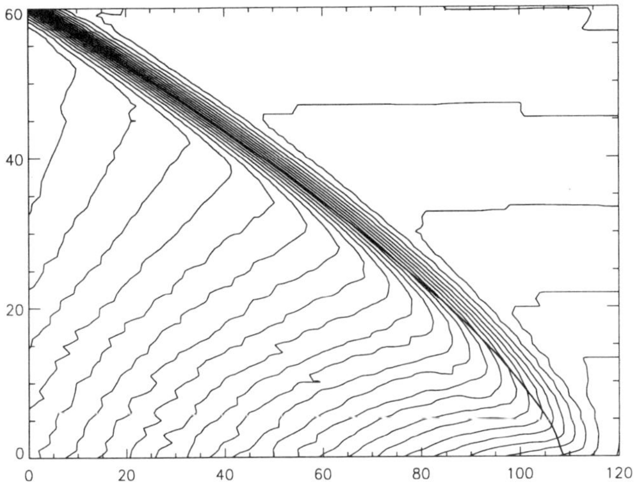

Figure 4: Blow up of v-contours and the sonic line

the sonic point S and the point C where the limiting characteristic hits the shock.

A count of the boundary conditions suggests that this problem is well-posed. We have one initial condition on all characteristics entering the nonuniform hyperbolic region, which is bounded by the sonic line, the reflected shock, the limiting characteristic, and the incident shock. For the elliptic region, we have one boundary condition on the sonic line and one on the incident and reflected shocks. Here, we count one boundary condition for each of the hyperbolic and elliptic regions on the sonic line, as in a standard Tricomi problem, and one on each shock, with the second shock condition determining the shock location.

A key point is that when the sonic line is embedded inside a nonuniform wave, the characteristics hit the sonic line at a nonzero angle. The problem therefore resembles a nonlinear Tricomi problem.

To show this last fact, we write (1.1) in terms of self-similar coordinates $\xi = x/t$ and $\eta = y/t$, which gives

$$\xi u_\xi + \eta u_\eta = u u_\xi + v_\eta, \qquad v_\xi - u_\eta = 0. \tag{3.1}$$

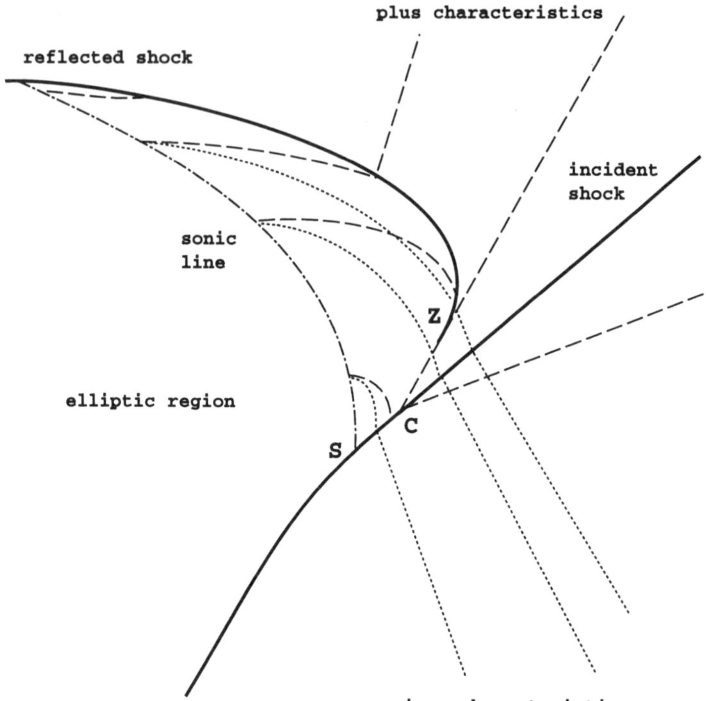

Figure 5: Proposed structure of a von Neumann reflection.

The slopes of the characteristics curves of (3.1) are

$$\left(\frac{d\xi}{d\eta}\right)_c = -\frac{1}{2}\eta \pm \sqrt{\xi + \frac{1}{4}\eta^2 - u}. \tag{3.2}$$

Equation (3.1) therefore changes type across the sonic line,

$$u(\xi, \eta) = \xi + \frac{1}{4}\eta^2. \tag{3.3}$$

Differentiating equation (3.3) for the sonic line with respect to η gives the sonic line slope,

$$\left(\frac{d\xi}{d\eta}\right)_s = \frac{-\frac{1}{2}\eta + u_\eta}{1 - u_\xi}.$$

On the sonic line, both characteristics have the same slope,

$$\left(\frac{d\xi}{d\eta}\right)_c = -\frac{1}{2}\eta.$$

The sonic line slope therefore differs from the characteristic slope, except at points where $u_\eta = \eta u_\xi/2$. The case in which the state ahead of the sonic line is constant is degenerate since the characteristics are always tangent to the sonic line.

Acknowledgments. The work of MB was partially supported by the AFOSR under grant number F49620-92-J-0054. The work of JKH was partially supported by the NSF under grant number DMS-9011548.

References

[1] M. Brio, and J. K. Hunter, Mach reflection for the two-dimensional Burgers equation, *Physica D* **60** (1992), 194-207.

[2] M. Brio, and J. K. Hunter, A von Neumann reflection for the 2-D Burgers equation, to appear in *Proc. Symp. Appl. Math.*

[3] P. Colella, and R. Henderson, The von Neumann paradox for the diffraction of weak shock waves, *J. Fluid Mech.* **213** (1990), 71-94.

[4] J. K. Hunter, Nonlinear geometrical optics, in *Multidimensional Hyperbolic Problems and Computations*, Vol. 29, IMA Volumes in Mathematics and its Applications, ed. A. J. Majda and J. Glimm, Springer-Verlag, New York (1991), 179-197.

[5] C. Morawetz, Potential theory for regular and Mach reflection of a shock at a wedge, to appear in *Comm. Pure Appl. Math.*

[6] E. Tabak, *Focusing of Weak Shock Waves and the von Neumann Paradox of Oblique Shock Reflection*, Ph.D. thesis, M.I.T., 1992.

[7] E. Tabak, and R. R. Rosales, Weak shock focusing and the von Neumann paradox of oblique shock reflection, *Phys. Fluids* **6** (1994), 1874-1892.

A kinetic formulation for chromatography

F. James[1], Y.-J. Peng[2], and B. Perthame[3]

1 Introduction

We present a kinetic formulation for the $N \times N$ chromatography system. The theory of kinetic formulations was introduced by P.L. Lions, B. Perthame and E. Tadmor [8]. It consists in representing a whole family of entropy inequalities by a single equation, using a supplementary variable. It can be used to derive rather easily some properties of the system based upon the manipulation of entropies, such as invariant regions or stability for weakly convergent initial data.

The new point here is that we obtain $N+1$ kinetic equations, which are necessary to get the N maximum *and* minimum principles on the Riemann Invariants, a very specific property of this kind of system. Also, $N+1$ kinetic equations are necessary to perform the compensated compactness argument for the stability property.

We shall give the details of the kinetic formulation for the chromatography system in Section 2. In Section 3, we show how to use it in order to recover the invariant regions and weak stability by the compensated compactness method. We refer to [7] for a more detailed presentation, and some references on related equations, such as the p-system and gas dynamics.

2 Kinetic formulation of the chromatography system.

We consider an $N \times N$ system which appears in chemical engineering and is called chromatography for Langmuir isotherms (see H.K. Rhee, R. Aris and N.R. Admundson [12] for the applications, variants and origins of this model). The unknowns are N functions $u_i(x,t)$, solutions of

$$\partial_t u_i + \partial_x \frac{k_i u_i}{D} = 0,\ t \geq 0,\ x \in \mathbb{R},\ 1 \leq i \leq N, \tag{2.1}$$

$$u_i(x,0) = u_i^\circ(x) \geq 0, \tag{2.2}$$

where the k_i are given numbers, $0 < k_1 < k_2 < \ldots < k_N$, and $D = 1 + u_1 + u_2 + \cdots + u_N$. In the following, we set $u(x,t) = (u_1(x,t), \ldots, u_N(x,t))$. For this system we shall exhibit $N+1$ one-parameter families of entropies, and use these entropies to write a kinetic formulation of system (2.1). This means that we can easily identify the convex entropies in this family. Knowledge of some algebraic properties of the system is required, and will be recalled in the first subsection. The special entropies we will use are presented in subsection 2.2, and the kinetic formulation in subsection 2.3.

[1]Centre de Mathématiques Appliquées, URA CNRS 756, Ecole Polytechnique, 91128 Palaiseau Cedex, FRANCE

[2](current address) CeReMaB, URA CNRS 226, Université Bordeaux I, 351 cours de la Libération, 33405 Talence Cedex, FRANCE

[3](current address) Laboratoire d'Analyse Numérique, URA CNRS 189, Université Paris 6, 4 place Jussieu, 75252 Paris Cedex 05, FRANCE

2.1 Some properties of the chromatography system

The system (2.1) has been widely studied (E. Canon and F. James [1],[2], H.K. Rhee and al [12], D. Serre [13]), and we recall some fundamental properties in the next proposition. We denote by A the Jacobian matrix $A(u) = \left(\frac{\partial}{\partial u_j}\left(\frac{k_i u_i}{D}\right)\right)_{1\le i,j\le N}$ of the system.

Proposition 2.1 *We consider the system (2.1) for* $u_i \geq 0$, $1 \leq i \leq N$.
(i) $A(u)$ *admits* N *eigenvalues* $0 < \lambda_1(u) \leq \lambda_2(u) \leq \cdots \leq \lambda_N(u)$;
(ii) $w_i = \lambda_i D$ *is an i-Riemann Invariant, i.e. for smooth solutions of (2.1), we have*

$$\partial_t w_i + \lambda_i \partial_x w_i = 0; \tag{2.3}$$

(iii) $$0 \leq w_1 \leq k_1 \leq w_2 \leq k_2 \leq \cdots \leq k_{N-1} \leq w_N \leq k_N;$$
and $u_i = 0$ *is equivalent to* $w_i = k_i$ *or* $w_{i+1} = k_i$;

(iv) $$D = \prod_{i=1}^{N} \frac{k_i}{w_i},$$

(v) $$u_i \prod_{j\neq i}\left(1 - \frac{k_i}{k_j}\right) = -\prod_{j=1}^{N}\left(1 - \frac{k_i}{w_j}\right);$$

(vi) $$\sigma_j(u) = \sum_{1\le i_1,\ldots,i_j\le N} \frac{1}{w_{i_1}\cdots w_{i_j}}, \quad \textit{for } 1 \leq j \leq N, \quad \sigma_0 = \prod_{i=1}^{N} k_i,$$
are $N+1$ *independent affine functions of* $(u_1, \ldots, u_N)$.

Of course these properties are very strong, and were already mentioned by D. Serre in [15, 16], as well as Lemma 2.3 below. D. Serre [14], [15], called (i), (ii) richness. In fact this system also belongs to the Temple class (B. Temple [18]), for which some existence and uniqueness results are known in the case of strict hyperbolicity (see D. Serre [15], A. Heibig [6]). We skip the proof of Proposition 2.1, which can be found in [7].

Remark 2.2 Let us note an important point (see [2] for further details). The properties (i) and (iii) allow a degeneracy of the system (two equal eigenvalues). This can only happen if $w_i = w_{i+1} = k_i$, in which case $\lambda_i = \lambda_{i+1}$, and $u_i = 0$. For this to occur, the initial data must satisfy $w_i^0(x) = w_{i+1}^0(x) = k_i$ for some $x \in \mathbb{R}$.

2.2 Some special entropies

In this subsection, we build some particular entropy-entropy flux pairs (E,F). This implies that additional conservation laws hold for smooth solutions of (2.1), namely

$$\partial_t E + \partial_x F = 0. \tag{2.4}$$

It is easier to look for functions depending upon the Riemann invariants $E = E(w_1, \ldots, w_N)$, $F = F(w_1, \ldots, w_N)$, rather than the conserved variables. Using Equation (2.3) for w_i, (2.4) then reduces to a system of $N(N-1)/2$ Euler-Darboux equations for E. Some straightforward computations then lead to the following Lemma, which is a first step toward the kinetic formulation.

Lemma 2.3 *For each* $\xi \in \mathbb{R}^+$ *the function* $E_0(\xi; w_1, ..., w_N) = \prod_{i=1}^{N}\left(1 - \frac{\xi}{w_i}\right)$ *is an entropy, affine with respect to* u. *The corresponding entropy flux is given by* $F_0(\xi; w_1, \ldots, w_N) = \frac{\xi E_0}{D}$.

A particular phenomenon occurs for this pair (E_0, F_0). Both functions vanish for $\xi = w_i$, so that the restriction $E_0 \mathbb{1}_{]w_i,w_{i+1}[}$ is also an entropy, with flux $F_0 \mathbb{1}_{]w_i,w_{i+1}[}$. Now taking the positive part of these $N+1$ families of nonlinear entropies preserves convexity, so that making the convention $w_0 = 0$, $w_{N+1} = +\infty$, we can state

Corollary 2.4 *For* $0 \le i \le N$, *the function*

$$\chi_i(\xi; w_1, \ldots, w_N) = \prod_{j=1}^{N} |1 - \frac{\xi}{w_j}|.\mathbb{1}_{\{\xi \in]w_i,w_{i+1}[\}} \tag{2.5}$$

is an entropy of the system (2.1), convex in u, *with the entropy flux*

$$\phi_i(\xi; w_1, \ldots, w_N) = \frac{\xi}{D}\chi_i. \tag{2.6}$$

This method of building entropies is common in the literature (see [3, 4, 5] or [9]). In particular, for general systems in the Temple class, D. Serre [15] builds another N parameter family of entropies using the positive part of an affine entropy. The relation with our entropies is still unclear to the authors.

2.3 Kinetic formulation

We are now interested in non-smooth solutions of the system (2.1). We shall call an "entropy solution" of (2.1) a weak solution satisfying

$$\partial_t E + \partial_x F \le 0 \quad \text{in } \mathcal{D}'([0, +\infty[\times\mathbb{R}) \tag{2.7}$$

for any convex entropy in the class defined by

$$E(u) = \int_{\mathbb{R}^+} g(\xi)\chi_i(\xi; u)d\xi, \tag{2.8}$$

for some $0 \le i \le N$, $g \in L^1(\mathbb{R}^+)$, $g \ge 0$. ¿From (2.7)-(2.8) we derive the kinetic formulation of the system (2.1).

Theorem 2.5 *Let* $u \in L^\infty(\mathbb{R}^+ ; L^1(\mathbb{R}))^N$, $u_i \ge 0$ *a.e. for* $1 \le i \le N$. *Then* u *is an entropy solution of (2.1), (2.2) if and only if* $\chi_i(\xi, u)$, $0 \le i \le N$, *satisfy (2.2) and (iii) below, and if there exist* $N+1$ *non-positive measures* $m_i(x, \xi, t)$ *such that*

$$\partial_t \chi_i(\xi, u(x,t)) + \partial_x \tfrac{\xi}{D}\chi_i(\xi, u(x,t)) = m_i \text{ in } \mathcal{D}'(\mathbb{R}_x \times \mathbb{R}^+_\xi \times \mathbb{R}^+_t). \tag{2.9}$$

Moreover, the measures m_i *in (2.9) satisfy:*
(i) *For some constant* $C_1(k_1, ..., k_N)$

$$0 \le -\int_{t=0}^{+\infty}\int_{x\in\mathbb{R}}\sum_{i=0}^{N} m_i(x,\xi,t)dxdt \le C_1(1+\xi^N)\|u^\circ\|_{L^1(\mathbb{R})^N}$$

(ii) $$\left|\frac{\partial}{\partial \xi}\int_{t=0}^{\infty}\int_{x\in\mathbb{R}}\sum_{i=0}^{N} m_i(x,\xi,t)dxdt\right| \leq C_1(1+\xi^{N-1})\|u^\circ\|_{L^1(\mathbb{R})^N}$$

(iii) $$m_i(x,k_i,t)=0 \quad , \quad 0\leq i\leq N.$$

This is very close to a transport equation because of the additional kinetic variable ξ, and explains the terminology "kinetic formulation". $\chi_i(\xi,u)$ looks like a Gibbs equilibrium in physical kinetics.

We do not give the detailed proof of this result here, but merely make a few remarks about it. We refer once again to [7] for the proof, in particular for the bounds (i) and (ii) on the measures m_i. First notice that, for a fixed i, χ_i, and thus m_i, has its support in ξ included in the interval $[k_i, k_{i+1}]$. Thus, the estimates (i) and (ii) give a uniform bound on m_i except for $i=N$ (then $k_{N+1}=+\infty$). A consequence of (i) and (ii) is that the total mass in x,t of m_i is continuous in ξ and thus (iii) is meaningful.

Now the equivalence between the two formulations is formally quite easy. If (2.7) holds for convex entropies given by (2.8), then (2.9) also holds. Indeed, it is enough to take $g(\xi)=\delta(\xi-\xi_0)$ (or a sequence of $g_n\geq 0$, which converges weakly to this g), and we know from corollary 2.4 that $\chi(\xi_0,u)$ is convex in u.

The converse is also clear. From (2.9) we recover (2.7), for any entropy of the form (2.8) with $g(\xi)\geq 0$ (except for $\xi=k_i$), which is a necessary and sufficient condition for E to be convex. Finally, we also recover the original system (2.1) up to a multiplicative constant, taking $\xi=k_i$ in the equation for χ_i and using Proposition 2.1 (v) and (iii).

The non-negativity of solutions $u_i(x,t)$ for $u_i^0\geq 0$, is in fact a consequence of the invariant regions that we present in Section 3.1.

Let us point out that the $N+1$ equations in (2.9) seem necessary to use this formulation in order to derive the set of all invariant regions and to perform the compensated compactness method in the "easy way", see Section 3.2.

The following result is an easy variant of Theorem 2.5, following from the finite speed of propagation. We give it here because it is a necessary tool to state a precise result on the invariant regions.

Theorem 2.6 *Let $u\in L^\infty(0,T;L^1_{loc}(\mathbb{R}))$, $u_i\geq 0$ a.e. for $1\leq i\leq N$. Then u is an entropy solution of (2.1) if and only if $\chi_i(\xi,u)$, $0\leq i\leq N$, satisfy (2.9) and (iii) of Theorem 2.5. Moreover, we have*

$$0\leq -\int_{[0,T]\times[-R,R]}\sum_{i=0}^{N} m_i(x,\xi,t)dxdt\leq C(R,T)(1+\xi^N)\|u\|_{L^\infty(0,T;L^1(-R,R))}.$$

The same estimate, replacing $\int\sum m_i$ by $\left|\frac{\partial}{\partial\xi}\int\sum m_i\right|$, and ξ^N by ξ^{N-1}, respectively, also holds.

3 Applications of the kinetic formulation.

In this section, we shall give the applications of the kinetic formulation (2.9) of the chromatography system (2.1), along the lines of [9]. We have two applications in mind. First, we recover the maximum and minimum principle on the Riemann Invariants, which are clearly related to the existence of $N+1$ kinetic equations for convex entropies. Second, we show how to handle the compensated compactness method on the χ_i.

3.1 Invariant regions.

Here we prove a result which is already known for the Riemann problem or for BV solutions (see [1], [13]):

Theorem 3.1 *Let u be an entropy solution of (2.1) satisfying the assumptions of Theorem 2.6. Then $k_{i-1} \leq \inf_{y\in\mathbb{R}} w_i(y,0) \leq w_i(x,t) \leq \sup_{y\in\mathbb{R}} w_i(y,0) \leq k_i$ almost everywhere.*

To prove this maximum principle on w_i, we use the equation on χ_{i-1} (therefore χ_0 is used here). We take $g(\xi) = \mathbb{I}_{\{\xi\geq \sup w_i(y,0)\}}$, so that $\int_{\mathbb{R}^+\times\mathbb{R}} g(\xi)\chi_{i-1}(\xi, u^0(x))d\xi dx = 0$. From $m_{i-1} \leq 0$, we deduce that $0 \leq \int_{\mathbb{R}^+\times\mathbb{R}} g(\xi)\chi_{i-1}(\xi, u(x,t))d\xi dx \leq 0$, and thus $w_i(x,t) \leq \sup_{y\in\mathbb{R}} w_i(y,0)$. The minimum principle follows similarly, using χ_i, $1 \leq i \leq N$.

¿From this theorem we easily obtain a maximum principle: for some constant $C > 0$, $C \inf_{y\in\mathbb{R}} u_j^0(y) \leq u_j(x,t) \leq \frac{1}{C} \sup_{y\in\mathbb{R}} u_j^0(y)$, for $1 \leq j \leq N$.

3.2 Compensated compactness

As soon as the initial data are uniformly bounded, the maximum principle allows us to consider a family $(u_{i,n}(x,t))_{n\geq 0}$ of uniformly bounded entropy solutions. In such a situation we may assume that as n tends to $+\infty$, $u_{i,n}(x,t) \rightharpoonup u_i(x,t)$ in $L^\infty((0,T)\times\mathbb{R})$ weak-$*$, for all $T \in (0,\infty)$ and each $1 \leq i \leq N$. The question, answered first by R.J. Di Perna [4], [5], for isentropic gas dynamics and strictly hyperbolic, genuinely non-linear systems, is to prove that this convergence is strong, which allows us to show that u_i is an entropy solution of (2.1).

The kinetic formulation allows the same kind of method as in G.Q. Chen [3] for isentropic gas dynamics, D. Serre [16] for electrophoresis and A. Heibig [6] for strictly hyperbolic systems in the Temple class, because we indeed have enough entropies. The $N \times N$ chromatography system can be treated in a manner similar to isentropic gas dynamics, for the full range $\gamma \geq 3$ and the p-system (see [10]). The case $1 < \gamma < 3$ is harder, because of both the degeneracy of the system close to the vacuum and the lack of entropies. Here some kind of a "vacuum" problem appears, corresponding to $\lambda_i = \lambda_{i+1} = k_i$, which implies $u_i = 0$. The lack of hyperbolicity is compensated here by the number of entropy families.

More precisely we have the following.

Theorem 3.2 *Let $u_{i,n}$ be a bounded, weakly convergent sequence of solutions. Then $(u_{i,n})_n$ converges pointwise to u_i, an entropy solution of (2.1).*

Let us denote by $d\nu_{x,t}(w)$, $w = (w_1, \ldots, w_N)$, or $d\nu$, the Young measure associated with the weak limit on the Riemann Invariants. We refer to L. Tartar [17], F. Murat [11] for the definition of Young measures and the original ideas about compensated compactness. We also denote by an overline the weak-$*$ limit.

Now, let us notice that adding the kinetic formulations (2.9) for $0 \leq i \leq N$, we obtain

$$\partial_t G_0(\xi, w) + \partial_x \frac{\xi}{D} G_0(\xi, w) = m := \sum_{i=0}^{N} m_i, \tag{3.1}$$

with $G_0 = |E_0|$, where E_0 was defined in Lemma 2.3. Next, applying the compensated compactness lemma to (3.1), for two values ξ and ξ', we obtain after a few computations

$$\frac{\partial}{\partial \xi}\frac{(\xi\overline{G_0(\xi)/D})}{\overline{G_0(\xi)}} = \frac{\overline{G_0(\xi)^2/D}}{\overline{G_0(\xi)}^2}. \tag{3.2}$$

Of course (3.2) has to be justified at points where $\overline{G_0(\xi)} = 0$, and this is rather technical. Let us merely mention that it follows from a careful study of the function $\overline{G}_0$ in the hyperbolic case, and $\overline{G}'_0$ for the non-strictly hyperbolic case, near the points where G_0 or G'_0 vanish.

Now (3.2) is not completely satisfactory because its right-hand side does not vanish when $d\nu$ is a Dirac mass. Therefore, we again apply compensated compactness to (3.1) and the equation $\partial_t D + \partial_x \frac{\alpha}{D} = 0$, where $\alpha = k_1 u_1 + \cdots + k_N u_N$. We combine this result with the left-hand side of (3.2) to get

$$-\partial_\xi \frac{\overline{G_0(\xi)\alpha/D}}{\overline{G_0(\xi)}} = \frac{\overline{G_0(\xi)^2/D}}{\overline{G_0(\xi)}^2} - \frac{1}{\overline{D}} \geq 0, \tag{3.3}$$

this last inequality being the Cauchy-Schwarz Inequality.

The inequality (3.3) is the keystone to proving that $d\nu$ is in fact a Dirac mass. Indeed, if in (3.3) the inequality is strict at some point, by studying the non-increasing function $\overline{G_0(\xi)\alpha/D}/\overline{G_0(\xi)}$, we find $\overline{\alpha} = \overline{D\alpha/D} > \overline{\alpha}$. Of course this means that the inequality in (3.3) is just an equality for all $\xi \geq 0$. Then, the equality case in the Cauchy-Schwarz inequality asserts the existence of a function $\lambda(\xi)$ such that, for all $\xi \geq 0$, $G_0(\xi, w) = \lambda(\xi)\, D(w)$ a.e. in the support of $d\nu(w)$. The proof that the support of $d\nu$ is a single point follows from Proposition 2.1 (iv).

To our knowledge, this is the first time that the compensated compactness method has been applied to an $N \times N$ system, $N > 2$, which is not strictly hyperbolic. The situation is however better than the gas dynamics case where the Young measure in (ρ, u) is not a Dirac mass when $\rho = 0$, that is at vacuum points.

To conclude, we state the following existence result:

Theorem 3.3 *There exists at least one entropy solution to (2.1) with initial data in* $L^\infty(\mathbb{R})$.

Such initial data can be approximated by a sequence of positive locally BV functions. For these initial data we can use the previous existence results, thus building a uniformly bounded sequence of entropy solutions. Theorem 3.2 allows us to conclude the result.

References

[1] E. Canon and F. James *Resolution of the Cauchy problem for several hyperbolic systems arising in chemical engineering.*, Ann. Inst. H. Poincaré, Analyse non linéaire, 9, 2, pp 219-238, (1992).

[2] E. Canon and F. James *Resolution of the Cauchy problem for non-strictly hyperbolic systems arising in chemical engineering*, Proceedings 3rd International conference on hyperbolic problems, Uppsala 1990, Engquist B, Gustafsson B. eds, Studentlitteratur, Chartwell Bratt, pp 212-225, (1991).

[3] G.Q. Chen *The theory of compensated compactness and the system of isentropic gas dynamics*, Preprint MCS-P154-0590, University of Chicago.

[4] R.J. Di Perna *Convergence of approximate solutions to conservation laws* Arch. Rat. Mec. and Anal., 82, 1, pp 27-70, (1983).

[5] R.J. Di Perna *Convergence of the viscosity method for isentropic gas dynamics* Comm. Math. Phys., 91, 1, pp 1-30, (1983).

[6] A. Heibig *Existence et unicité des solutions pour certains systèmes de lois de conservation*, C.R. Acad. Sc. Paris, t. 311, série I, pp. 861-866, To appear in Arch. Rat. Mech. Anal.

[7] F. James, B. Perthame and Y.-J. Peng *Kinetic formulation for chromatography and some other hyperbolic systems*, To appear in Jour. Maths Pures & Appl.

[8] P.L. Lions, B. Perthame, and E. Tadmor *Kinetic formulation of scalar conservation laws*, To appear in Journal of American Society.

[9] P.L. Lions, B. Perthame, and E. Tadmor *Kinetic formulation of the isentropic gas dynamics and p-system*, preprint.

[10] P.L. Lions, B. Perthame, and P.E. Souganidis, In preparation.

[11] F. Murat *Compacité par compensation*, Ann. Sc. Normale. Sup. Pisa, 5, pp 489-507, (1978).

[12] H.K. Rhee, R. Aris and N.R. Admundson *On the theory of multicomponent chromatography*, Philos. Trans. Roy. Soc. London, A267, pp 419-455, 1970.

[13] D. Serre *Solutions à variations bornées pour certains systèmes hyperboliques de lois de conservation*, J. Diff. Eq. 68, pp 137-168, (1987).

[14] D. Serre *Systèmes hyperboliques riches de lois de conservation*, Collège de France, Seminar X, Pitman Res. Notes Math. Series, Longmann (1993).

[15] D. Serre *Richness and the classification of quasilinear hyperbolic systems*, I.M.A. Vol. in Mathematics and its applications n. 29, J. Glimm and A. Majda eds, Springer-Verlag, Heidelberg.

[16] D. Serre *Oscillations non linéaires des systèmes hyperboliques : méthodes et résultats qualitatifs*, Ann. Inst. H. Poincaré, Analyse non linéaire, 8, 3-4 (1991), pp 351-417.

[17] L. Tartar *Compensated compactness and applications to partial differential equations*, In research notes in mathematics, nonlinear analysis and mechanics: Heriot-Watt, symposium., vol. 4, ed R.J. Knops, Pitman press (1979).

[18] B. Temple *Systems of conservation laws with invariant submanifolds*, Trans. A.M.S., vol. 280, pp 781-795, (1983).

The Relaxation Schemes

Shi Jin[1] and Zhouping Xin[2]

ABSTRACT. We present a class of numerical schemes (called the relaxation schemes) for systems of conservation laws in several space dimensions. The idea is to use a local relaxation approximation. We construct a linear hyperbolic system with a stiff lower order term that approximates the original system with a small dissipative correction. The new system can be solved by underresolved stable numerical discretizations without using either Riemann solvers spatially or a nonlinear system of algebraic equations solver temporally. Some numerical results for 2-D gas dynamics equations are presented.

In this paper we present a class of nonoscillatory numerical schemes for systems of conservation laws in several space dimensions. The basic idea is to use a local relaxation approximation. Consider systems of conservation laws in one space variable

$$\frac{\partial}{\partial t}\mathbf{u} + \frac{\partial}{\partial x}F(\mathbf{u}) = 0\ , \quad (x,t)\in \mathbb{R}^1\times\mathbb{R}\ , \quad \mathbf{u}\in\mathbb{R}^n\ , \tag{1}$$

where $F(\mathbf{u})\in\mathbb{R}^n$ is a smooth vector-valued function. We introduce the *relaxation system* corresponding to (1) as

$$\begin{aligned} &\frac{\partial}{\partial t}\mathbf{u} + \frac{\partial}{\partial x}\mathbf{v} = 0\ , && \mathbf{v}\in\mathbb{R}^n\ , \\ &\frac{\partial}{\partial t}\mathbf{v} + A\,\frac{\partial}{\partial x}\mathbf{u} = -\frac{1}{\epsilon}\left(\mathbf{v}-F(\mathbf{u})\right)\ , && \epsilon > 0\ , \end{aligned} \tag{2}$$

where $A = \operatorname{diag}\{a_1, a_2, \cdots, a_n\}$ is a positive diagonal matrix satisfying

$$A - F'(\mathbf{u})^2 \geq 0 \quad \text{for all} \quad \mathbf{u}\ . \tag{3}$$

How well does the relaxation system (2) approximate the original system of conservation laws (1)? In the same spirit as the Chapman-Enskog expansion, one can see that for small ϵ the first order behavior of the relaxation system (2) is governed by

$$\frac{\partial}{\partial t}\mathbf{u} + \frac{\partial}{\partial x}F(\mathbf{u}) = \epsilon\,\frac{\partial}{\partial x}\left(\left(A - F'(\mathbf{u})^2\right)\frac{\partial}{\partial x}\mathbf{u}\right), \tag{4}$$

[1]School of Mathematics, Georgia Institute of Technology, Atlanta, GA 30332, jin@math.gatech.edu

[2]Courant Institute of Mathematical Sciences, New York, NY 10012 xinz@cims.nyu.edu

Research of the first author was supported in part by NSF grant DMS 9404157 and AFOSR Grant F49620-92-J0098. Research of the second author was supported in part by a Sloan Foundation Fellowship, NSF Grants DMS 9304580 and DMS-93-03887, and DOE Grant DE-FG02-88ER-25053.

where $F'(\mathbf{u})$ is the Jacobian matrix of the flux function F. Thus (2) provides the desired *dissipative* approximation for (1) for small ϵ as long as (3) holds. Condition (3) can always be satisfied for $\mathbf{u}$ varying in a bounded domain by choosing sufficiently large A; however, due to the CFL constraints on numerical stability, a more accurate condition which ensures (3) is

$$\frac{\lambda^2}{a} < 1 \,. \tag{5}$$

Here $a = \min_{1\le i\le n} a_i$, and λ is the maximum of the absolute values of the eigenvalues of $F'(\mathbf{u})$ [4].

The 1-D Relaxation Scheme. One would expect that appropriate numerical discretizations to the relaxation system (2) yield accurate approximations to the original equation (1) when the relaxation rate ϵ is sufficiently small. The main advantage of numerically solving the relaxation system (2) over the original conservation laws lies in the special structures of the linear characteristic fields and the localized lower order term. The linear hyperbolic structure enables one to avoid the time consuming Riemann solvers, which are loaded by standard high resolution methods for nonlinear hyperbolic systems. In this respect, the relaxation scheme bears some spirit of the kinetic schemes. While there are numerous ways to discretize the relaxation system (2), we will only present a second order accurate scheme based on van Leer's MUSCL [7].

A Spatial Discretization of (2) in conservation form can be written as

$$\begin{aligned} &\frac{\partial}{\partial t}\mathbf{u}_j + \frac{1}{h_j}(\mathbf{v}_{j+1/2} - \mathbf{v}_{j-1/2}) = 0\,, \\ &\frac{\partial}{\partial t}\mathbf{v}_j + \frac{1}{h_j}A\,(\mathbf{u}_{j+1/2} - \mathbf{u}_{j-1/2}) = -\frac{1}{\epsilon}\,(\mathbf{v}_j - F(\mathbf{u}_j))\,. \end{aligned} \tag{6}$$

Here the point value quantities $\mathbf{u}_{j+1/2}$ and $\mathbf{v}_{j+1/2}$ will be defined by a second order upwind scheme specified below. Since the relaxation system (2) has two characteristic variables $\mathbf{v} \pm A^{\frac{1}{2}}\,\mathbf{u}$ that travel with the *frozen* characteristic speeds $\pm A^{\frac{1}{2}}$ respectively, it is natural to determine the point value quantities $\mathbf{u}_{j+1/2}$ and $\mathbf{v}_{j+1/2}$ in (6) by applying the upwind approximations to $\mathbf{v} \pm A^{\frac{1}{2}}\,\mathbf{u}$ componentwise to obtain

$$\begin{aligned} (v + \sqrt{a_p}\,u)_{j+1/2} &= (v + \sqrt{a_p}\,u)_j + \tfrac{1}{2}h_j\sigma_j^+\,, \\ (v - \sqrt{a_p}\,u)_{j+1/2} &= (v - \sqrt{a_p}\,u)_{j+1} - \tfrac{1}{2}h_{j+1}\sigma_{j+1}^-\,. \end{aligned} \tag{7}$$

Here u and v are the p-th ($1 \le p \le n$) components of $\mathbf{u}$ and $\mathbf{v}$ respectively, and $\sigma_j^\pm$ is the slope of $v \pm \sqrt{a_p}\,u$ on the jth cell which, in Sweby's notation [6], is given by

$$\begin{aligned} \sigma_j^\pm &= \frac{1}{h_j}\left(v_{j+1} \pm \sqrt{a_p}\,u_{j+1} - v_j \mp \sqrt{a_p}\,u_j\right)\,\phi(\theta_j^\pm)\,, \\ \theta_j^\pm &= \frac{v_j \pm \sqrt{a_p}\,u_j - v_{j-1} \mp \sqrt{a_p}\,u_{j-1}}{v_{j+1} \pm \sqrt{a_p}\,u_{j+1} - v_j \mp \sqrt{a_p}\,u_j}\,, \end{aligned} \tag{8}$$

with a slope-limiter function ϕ introduced by van Leer [7] as

$$\phi(\theta) = \frac{|\theta| + \theta}{1 + |\theta|} . \tag{9}$$

This completely determines the numerical fluxes in (6). We should emphasize here that the employment of MUSCL here is not essential, and other high resolution methods such as the ENO [2] or the PPM [8] may be applied as well.

The Time Discretization. We use the second order TVD Runge-Kutta splitting scheme introduced by Jin [3] for hyperbolic systems with stiff relaxations. This splitting scheme takes two implicit stiff source steps and two explicit convection steps alternatively. Denote

$$D_+ \mathbf{w}_j = \frac{1}{h_j}(\mathbf{w}_{j+1/2} - \mathbf{w}_{j-1/2}) . \tag{10}$$

Then applying the second order TVD Runge-Kutta splitting scheme to (2) gives

$$\begin{aligned}
\mathbf{u}^* &= \mathbf{u}^n , \\
\mathbf{v}^* &= \mathbf{v}^n + \frac{k}{\epsilon}(\mathbf{v}^* - F(\mathbf{u}^*)) ; \\
\mathbf{u}^{(1)} &= \mathbf{u}^* - k\, D_+ \mathbf{v}^* , \\
\mathbf{v}^{(1)} &= \mathbf{v}^* - k\, A\, D_+\, \mathbf{u}^* ; \\
\mathbf{u}^{**} &= \mathbf{u}^{(1)} , \\
\mathbf{v}^{**} &= \mathbf{v}^{(1)} - \frac{k}{\epsilon}(\mathbf{v}^{**} - F(\mathbf{u}^{**})) - 2\frac{k}{\epsilon}(\mathbf{v}^* - F(\mathbf{u}^*)) ; \\
\mathbf{u}^{(2)} &= \mathbf{u}^{**} - k\, D_+ \mathbf{v}^{**} , \\
\mathbf{v}^{(2)} &= \mathbf{v}^{**} - k\, A\, D_+ \mathbf{u}^{**} ; \\
\mathbf{u}^{n+1} &= \tfrac{1}{2}(\mathbf{u}^n + \mathbf{u}^{(2)}) , \\
\mathbf{v}^{n+1} &= \tfrac{1}{2}(\mathbf{v}^n + \mathbf{v}^{(2)}) .
\end{aligned} \tag{11}$$

Here without ambiguity we have omitted the subscript j.

This splitting scheme allows a CFL constraint solely dependent on the convection part (independent of ϵ). Although the source term is implicit, one can avoid numerically solving any systems of algebraic equations, since one can always update $\mathbf{u}$ from the first equation, applying it to $F(\mathbf{u})$, then solving the second equation for $\mathbf{v}$ exactly, since $\mathbf{v}$ is linear in the relaxation system.

The Relaxation System for Multi-Dimensional Systems of Conservation Laws. Consider the following m-dimensional $n \times n$ system

$$\frac{\partial}{\partial t}\mathbf{u} + \sum_{i=1}^{m} \frac{\partial}{\partial x_i} F_i(\mathbf{u}) = 0 , \qquad (x_i, t) \in \mathbb{R}^1 \times \mathbb{R}_+, \ \mathbf{u} \in \mathbb{R}^n , \tag{12}$$

with each $F_i\colon \mathbb{R}^n \to \mathbb{R}^n$ being a smooth function ($i = 1, 2, \ldots, m$). We will introduce the relaxation system corresponding to (12) as

$$(13)\qquad \begin{aligned} &\frac{\partial}{\partial t}\mathbf{u} + \sum_{i=1}^{m} \frac{\partial}{\partial x_i}\mathbf{v}_i = 0\,, && \mathbf{v}_i \in \mathbb{R}^n \\ &\frac{\partial}{\partial t}\mathbf{v}_i + A_i \frac{\partial}{\partial x_i}\mathbf{u} = -\frac{1}{\epsilon}\left(\mathbf{v}_i - F_i(\mathbf{u})\right)\,, && i = 1, 2, \ldots, m\,. \end{aligned}$$

As is the case for 1-D, one usually chooses A_i with distinct positive diagonal elements so the relaxation system (13) is non-degenerate. Here, for simplicity in our presentation, we will take that $A_i = a_i I\,,\quad a_i > 0\,,\quad i = 1, \ldots, m\,.$

In a similar way as for (4), one can show formally that the first order asymptotic behavior of (17) with a small relaxation rate is governed by

$$(14)\qquad \frac{\partial}{\partial t}\mathbf{u} + \sum_{i=1}^{m} \frac{\partial}{\partial x_i}F_i(\mathbf{u}) = \epsilon \sum_{i,j=1}^{m} \frac{\partial}{\partial x_i}\left(\left[\delta_{ij} A_i - F_i'(\mathbf{u})\, F_j'(\mathbf{u})\right] \frac{\partial}{\partial x_j}\mathbf{u} \right)$$

where δ_{ij} is the Kronecker delta. System (14) is dissipative if

$$(15)\qquad \delta_{ij} A_i - F_i'(\mathbf{u})\, F_j'(\mathbf{u}) \geq 0 \quad \text{on} \quad \mathbb{R}^{mn}\,.$$

Note that the left-hand side of (15) is an $(mn) \times (mn)$ matrix. The condition (15) is satisfied for $\mathbf{u}$ in a bounded domain by choosing a sufficiently large A. However, for the same reason as in the 1-D case, we want to choose the smallest A satisfying (15). Assume that the system (12) admits convex entropy extensions, and $\eta(\mathbf{u})$ is a convex entropy function. Let $\lambda_k^{(i)}$ and $\mu_k^{(i)}$ be the eigenvalues of the symmetric matrices $F_i'(\mathbf{u})\eta''(\mathbf{u})^{-1}$ and $\eta''(\mathbf{u})\, F_i(\mathbf{u})$, respectively, $i = 1, \ldots, m$, $k = 1, \ldots, n$. Setting

$$(16)\qquad \lambda^{(i)} = \max_{k,\ell}\{|\lambda_k^{(i)}|,\, |\mu_\ell^{(i)}|\}\,,$$

one can then obtain the following desirable dissipative condition [4].

Proposition 1. *The reduced system (14) is dissipative if*

$$(17)\qquad \frac{(\lambda^{(1)})^2}{a_1} + \frac{(\lambda^{(2)})^2}{a_2} + \cdots + \frac{(\lambda^{(m)})^2}{a_m} \leq 1\,.$$

Remarks:

(i) (17) generalizes the subcharacteristic condition of Liu [5] to higher dimensions.

(ii) For the scalar equation, $n = 1$, one may take $\eta(u) = \frac{1}{2}u^2$, then the dissipative condition becomes (17) with $\lambda^{(i)}$ replaced by $F_i'(bu)$.

(iii) If, instead of requiring that there exists a convex entropy extension, one assumes that (12) is a symmetric system, then (17) can be derived quite easily with $\lambda^{(i)}$'s being genuine eigenvalues of $F_i'(\mathbf{u})$.

The Higher Dimensional Relaxation Schemes. In the case of multidimensions, a similar MUSCL discretization can be applied to each space dimension. Consider the multidimensional relaxation system (13). Then the MUSCL discretization

(7) and (8) can be applied to the characteristic variables $\mathbf{u} \pm A_i \mathbf{v}_i$ in the i-th space dimension for $i = 1, 2, \cdots, m$. The second-order Runge-Kutta splitting scheme (11) applies here, with the only difference being that D_+ now becomes a multidimensional discrete derivative operator. For details see [4].

Since the formulation of the relaxation system is multidimensional, and the structure of the multidimensional relaxation system is similar to the 1-D system, the numerical implementations for higher dimensional problems based on the dimension-by-dimension discretizations are formally not much harder than for the 1-D problems. Again, all the multidimensional algorithms here are highly efficient and well vectorized. In fact, the only IF command in the entire algorithm appears in the slope limiter function (9).

Numerical Results.

Strong Mach Reflection. A planar shock is incident on an oblique surface, with the surface at a 30° angle to the direction of propagation of the shock. Our test problem involves a Mach 10 shock, ahead of it is the undisturbed air with a density of 1.4, a pressure of 1, and $\gamma = 1.4$. We use the boundary conditions described in [8]. In this problem we take $a_1 = b_1 = 98, a_2 = b_2 = 162, a_3 = b_3 = 242$, and $a_4 = b_4 = 338$. The density at $t = 0.2$ is plotted in Fig. 1 with $\Delta x = \Delta y = \frac{1}{120}$ and $k = 0.0002$. In each plot 30 equi-distributed contours are shown. There are some oscillations behind the shocks, but this is standard for any second order method in 2-D, since no method higher than first order can be TVD.

Weak Mach Reflections. The setup of the problems are the same as the Strong Mach reflection, but we test two sets of weak shock data. (1) $\gamma = 5/3$, Mach number $= 1.47$, angle$=34.60^{\mathrm{o}}$, and the 50 contours of density at $t = 1.5$ are plotted. (2) $\gamma = 5/3$, Mach number $= 1.0483$, angle$=10^{\mathrm{o}}$, and the 50 contours of density at $t = 1.8$ are plotted. This is the von Neumann reflection defined by Collela and Henderson [1]. In both problems we take $a_1 = b_1 = 1.21, a_2 = b_2 = 1.44, a_3 = b_3 = 1.69, a_4 = b_4 = 1.96$, $\Delta x = \Delta y = \frac{1}{120}$, and $\Delta t = 0.00125$. These results seem to agree with the results in [1].

Remark: All the numerical results presented here were obtained using the second order relaxation scheme with van Leer's slope limiter, uniform Cartesian grids, and $\epsilon = 10^{-8}$. No other techniques, such as artificial compression, subcell resolution, adaptive mesh and flattening were applied to sharpen the results. These techniques can be combined with the relaxing schemes and are under consideration.

Acknowledgments: The authors are grateful for the encouragement and support of Peter Lax, David Levermore, Tai-Ping Liu, Catheleen Morawetz, George Papanicolaou and Philip Roe, and many helpful discussions of Jian-Guo Liu.

References

1. P. Colella and L.F. Henderson, *The von Neumann paradox for the diffraction of weak shock waves,* J. Fluid Mech. **213** (1990), 71-94.
2. A. Harten, B. Engquist, S. Osher and S.R. Chakravarthy, *Uniformly High-Order Accurate Essentially Nonoscillatory Schemes, III,* J. Comp. Phys. **71** (1987), 231-303.

3. S. Jin, *Implicit numerical schemes for hyperbolic conservation laws with stiff relaxation terms,* submitted to J. Comp. Phys.
4. S. Jin and Zhouping Xin, *The relaxation schemes for systems of conservation laws in arbitrary space dimensions,* to appear in Comm. Pure Appl. Math.
5. T. P. Liu, *Hyperbolic conservation laws with relaxation,* Comm. Math. Phys. **108** (1987), 153-175.
6. P. R. Sweby, *High resolution schemes using flux limiters for hyperbolic conservation laws,* SIAM J. Num. Anal. **21** (1984), 995-1011.
7. B. van Leer, *Towards the ultimate conservative difference schemes V. A second-order sequal to Godunov's method,* J. Comp. Phys. **32** (1979), 101-136.
8. P. Woodward and P. Colella, *The numerical simulation of two-dimensional fluid flow with strong shocks,* J. Comp. Phys. **54** (1984), 115-173.

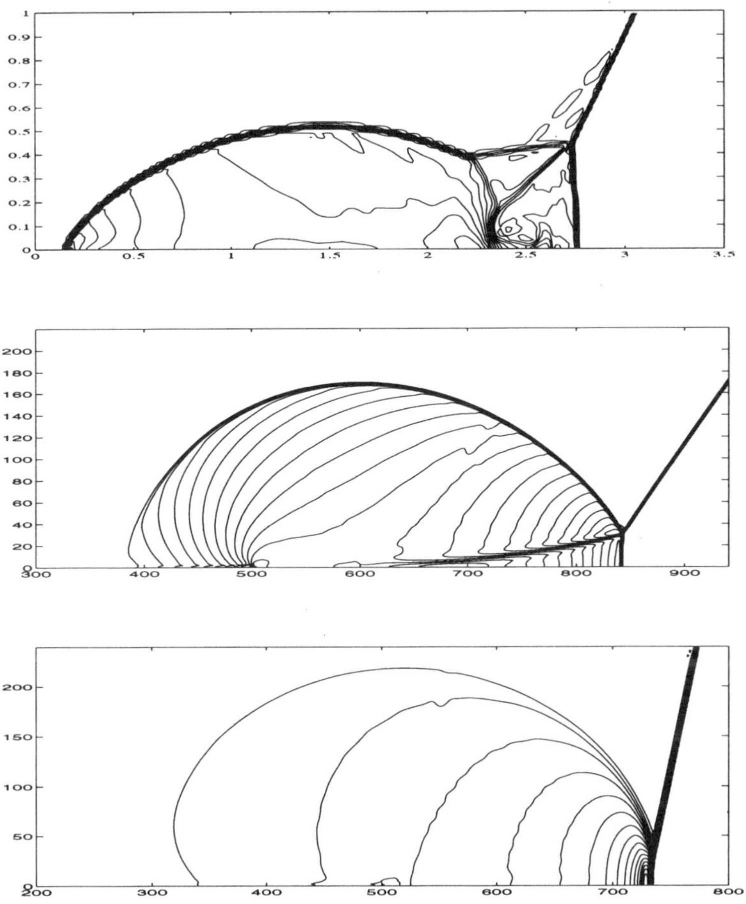

FIGURE 1. (a) The strong mach reflection at t=0.2. 30 equi-distributed density contours are shown. (b) The weak Mach reflection at t=1.5. 50 equi-distributed density contours are shown. (c) The von Neumann reflection at t=1.8. 50 equi-distributed density contours are shown.

Global Solutions to Hyperbolic Conservation Laws with Umbilic Degeneracy

Gui-Qiang Chen[1], Pui Tak Kan[2]

Abstract

We are concerned with systems of hyperbolic conservation laws with umbilic degeneracy. In this paper a compactness framework for approximate solutions to nonlinear hyperbolic systems with umbilic degeneracy is established by combining ideas in modern nonlinear analysis with classical methods, and by a detailed analysis of a highly singular Euler-Poisson-Darboux-type equation. Then this framework is successfully applied to prove the convergence of the viscosity method, and the existence of global entropy solutions for the Cauchy problem with large initial data for a canonical class of the quadratic flux systems.

1. Introduction

We are concerned with hyperbolic systems of conservation laws with umbilic degenerate points. A point in the state space $U_0 \in \mathbf{R}^n$ is called an umbilic degenerate point for a system of hyperbolic conservation laws

$$\partial_t U + \partial_x F(U) = 0, \qquad U \in \mathbf{R}^n, \tag{1.1}$$

if some wave speeds coincide at this point U_0, that is, at least two eigenvalues $\lambda_i(U)$ and $\lambda_j(U), i \neq j$, among the n real eigenvalues of the matrix $\nabla F(U)$ are such that $\lambda_i(U_0) = \lambda_j(U_0)$. Such umbilic degeneracy allows a degree of interaction, or nonlinear resonance, between distinct modes, and leads to high singularities, which is missing in the strictly hyperbolic case.

Nonlinear hyperbolic systems with such degeneracy occur in multiphase flows in porous media, elasticity, water wave problems, and maganetohydrodynamics. We refer the readers to [F], [G1], [G2], [KK], [SS1], [TT] and the references therein. Such umbilic points appear naturally in multidimensional systems of conservation laws (cf. [FR, J, La1]). In particular, Lax [La1] showed that in three space variables there must be degenerate points if the number of equations $n = 2(mod\ 4)$. More generally, the same result holds if $n = \pm 2, \pm 3, \pm 4(mod\ 8)$ (see [FR]). This means that, in general, plane wave solutions for such multi-dimensional systems are governed by one-dimensional hyperbolic systems with umbilic degeneracy.

Two kinds of degeneracy were classified. A degenerate point U_0 is a hyperbolic degeneracy if $\nabla F(U_0)$ is diagonalizable; it is parabolic otherwise.

In this paper, we summarize the results in the papers [CK1] and [CK2]. We focus on isolated umbilic points with hyperbolic degeneracy. Near such an isolated umbilic point, a normal form theory was provided in [SS1]. A family of quadratic flux systems was derived as the normal form systems that describe local qualitative wave structures near the umbilic point. The solution of the Riemann problem for such systems was constructed in [IMPT], [IT], [SS2], [SSMP]. Nonstandard (undercompressive and overcompressive) shocks were observed. Their sensitive dependence of admissibility on viscosity terms and

[1]Department of Mathematics, University of Chicago, Email: cheng@math.uchicago.edu

[2]Indiana University-Purdue University, Email: kan@math.iupui.edu

their asymptotic stability were considered in e.g. [IMP], [IMPP], [Liu], [LZ]. See [G1, G2] for review and references on applications of bifurcation theory and geometry to the analysis of Riemann solutions.

The global existence of weak solutions to the Cauchy problem for a special case of such quadratic flux systems was solved in [K] using the vanishing viscosity method. The method of compensated compactness was used and the main analysis involves detailed estimates and characterization of the singularities of solutions to the entropy equation which is of Euler-Poisson-Darboux type. For applications of the analysis in [K] to different systems, see [FS], [Rb]. See also [Lu] for a different method.

In this paper we establish an L^∞ compactness framework for sequences of approximate solutions to general nonstrictly hyperbolic systems with umbilic points. Under this compactness framework, approximate solution sequences, which are apriori bounded in L^∞ and which produce correct entropy dissipations, lead to the compactness of the corresponding Riemann invariant sequences.

The proof of this compactness framework theorem involves the analysis of the structure of the Young measures associated with the approximate solution sequence using regular entropy functions of (1.1). One of the principal difficulties is the general lack of enough classes of entropy functions that satisfy certain weak compactness conditions in the div-curl lemma of Tartar [Ta] and Murat [Mu]. This is due to possible singularities of entropy functions near the regions of nonstrict hyperbolicity. This difficulty is overcome by a detailed analysis of the singularities of the Riemann function of the entropy equation which is highly singular and of Generalized Euler-Poisson-Darboux type.

This compactness framework is successfully applied to prove the convergence of the vanishing viscosity method for the normal form quadratic flux systems in Section 3. As a consequence, we establish global existence of entropy solutions for such systems for large data in L^∞. Similar compactness and convergence results for the Lax-Friedrichs Scheme and the Godunov Scheme will be discussed in a forthcoming paper.

2. General Classes of Degenerate Systems and A Canonical Family of Quadratic Flux Systems

We consider a general 2×2 nonstrictly hyperbolic system (1.1) with an isolated umbilic point P in state space:

$$\lambda_1(P) = \lambda_2(P),$$

where λ_1, λ_2 are the eigenvalues of $\nabla F(U)$ in (1.1).

We list the basic assumptions on the structure of (1.1) for the compactness framework and global existence theorems that will be stated in section 3:

$(H1)$ There exist Riemann invariant coordinates (w_1, w_2) such that $w_1 \le 0 \le w_2$ and the umbilic point $P = (0,0)$;

$(H2)$ $$\frac{\lambda_{i w_j}}{\lambda_2 - \lambda_1} = \frac{\mathcal{A}_i(\frac{w_2}{w_1})}{w_2 - w_1} = \frac{\tilde{\mathcal{A}}_i(\theta)}{w_2 - w_1}, \quad \theta \in [\frac{\pi}{2}, \pi],$$
where $tan\theta = \dfrac{w_2}{w_1}$, $\tilde{\mathcal{A}}_i(\theta)$ is real analytic in $\theta \in (0, \dfrac{3\pi}{2})$, and $i \neq j,\ i,j = 1,2$.

An important example of systems satisfying the structural assumptions (H1) and (H2) is the canonical family of quadratic normal form systems:

$$\partial_t U + \partial_x(dC(U)) = 0,$$
$$U = (u, v)^T,$$
$$C(U) = \frac{1}{2}(\frac{1}{3}au^3 + bu^2v + uv^2),\tag{2.1}$$

where a and b are real parameters with $a \neq 1 + b^2$.

The eigenvalues of (2.1) are,

$$\lambda_i = \frac{1}{2}\left((a+1)u + bv + (-1)^i\sqrt{((a-1)u+bv)^2 + 4(bu+v)^2}\right), \quad i = 1, 2.\tag{2.2}$$

It is then immediate that as long as $a \neq 1 + b^2$, $\lambda_1 = \lambda_2 \iff (u, v) = (0, 0)$, so that $(0, 0)$ is the unique umbilic point for (2.1).

We remark that the characteristic fields of (2.2) are not genuinely nonlinear.

The quadratic systems (2.1) was derived in [SS1] as the normal form systems describing local wave structures near the umbilic point of the general system (1.1). There has been a lot of work on the admissibility and structures of elementary waves and their applications to the solution of the Riemann problem for these quadratic systems. Many new mathematical phenomena such as nonstandard shock waves and their nonlinear stability, sensitive dependence of admissibilities of waves on viscosity matrices, bifurcation of waves, and very complicated geometries of Hugoniot loci and rarefaction wave curves were found. We refer the readers to [CK1-CK2], [K], [G1], [G2], [Liu], [LZ], [IMP], [IT], [SS1], [SS2], [SSMP] and the references inside.

3. Compactness Framework for Approximate Solutions and Applications

3.1. Parabolic Approximation and Young Measures

We consider parabolic approximations to the general system (1.1) by adding artificial viscosities.

$$\partial_t U^\epsilon + \partial_x F(U^\epsilon) = \epsilon\partial_x(D\partial_x U^\epsilon), \qquad x \in \mathbf{R}, t > 0,$$
$$U^\epsilon(x, 0) = U_0(x),\tag{3.1}$$

where $D \geq 0$ is a non-negative matrix (viscosity matrix) and $\epsilon > 0$ measures the amount of artificial viscosities in (3.1). For simplicity, we shall take $D = I$ below.

We are concerned with the convergence of U^ϵ to a weak solution U of (1.1) as $\epsilon \to 0^+$. This problem can be reduced to the analysis of the structure of Young measures associated with the approximate solution sequence U^ϵ. Suppose that $||U^\epsilon||_{L^\infty} < constant$. (Such a uniform L^∞ apriori bound is typically established using the invariant region theory of [CCS].) Then there exists a subsequence (still labeled U^ϵ) and a family of Young measures $\nu_{x,t}(\lambda) \in \text{Prob.}(\mathbf{R}^2)$, such that, for any continuous function g,

$$w^* - lim_{\epsilon\to 0}\, g(U^\epsilon(x,t)) =< \nu_{x,t}(\lambda), g(\lambda) >\equiv \int_{\mathbf{R}^n} g(\lambda)d\nu_{x,t}(\lambda).$$

The problem of showing the convergence of U^ϵ to a weak entropy solution of (1.1) is equivalent to showing that $\nu_{x,t}$ is a Dirac Mass for *a.e.* (x,t). For the general theory of Young measures and compensated compactness, see [Ta], [Mu], [Di].

3.2 Entropy Functions and Entropy Dissipation Measures

The analysis of the structure of the Young measure $\nu_{x,t}$ and the convergence of U^ϵ uses the method of compensated compactness ([Ta], [Di]) and relies on detailed analyses of the dissipation of entropy functions associated with (1.1) (see [CK1-CK2], [K]). A pair of scalar functions $(\eta(u,v), q(u,v))$ is an entropy pair associated with (1.1) iff they satisfy the system of linear partial differential equations ([La2]):

$$(\eta_u, \eta_v)(\nabla F(U)) = (q_u, q_v)^T.$$

The loss of strict hyperbolicity for (1.1) means that the wave speeds of different wave modes can coincide. This allows a degree of interaction which leads to high singularities in the entropy functions and the wave curve geometries. In particular, the entropies satisfy a highly singular Generalized Euler-Poisson-Darboux equation.

Theorem 1 ([CK1-CK2]).

Consider the system (1.1) which satisfies the structural assumptions (H1) and (H2) (For instance, system (2.1)). Then we have

(1) η satisfies a second order singular hyperbolic equation of Generalized Euler-Poisson-Darboux type. In characteristic form, this equation takes the form:

$$\eta_{w_1 w_2} + \frac{A(w_1, w_2)}{w_2 - w_1}(\eta_{w_2} - \eta_{w_1}) = 0, \tag{3.2}$$

where w_1 and w_2 are Riemann invariants of (1.1), $w_2 - w_1 \geq 0$, and $A(w_1, w_2)$ is an L^∞ function which is multi-valued at $(w_1, w_2) = (0,0)$. In particular, $A(w_1, w_2) = \tilde{A}(\theta)$ where $tan\theta = \frac{w_2}{w_1}$ and $\tilde{A}$ is real analytic in $\theta \in (0, \frac{3\pi}{2})$.

(2) Given any integer m and any compact set K, there is a subspace $\mathcal{M}$ in the space of $L^2 \cap C^m$ Goursat data for the Goursat problem of (3.2), such that codim($\mathcal{M}$) $< \infty$, and if the Goursat data belongs to this subspace $\mathcal{M}$ then $\eta \in C^m$, and $\max_{0 \leq j+l \leq m} \sup_{(w_1,w_2) \in K} |\partial^j_{w_1} \partial^l_{w_2} \eta| \leq$ constant $|w_1^j w_2^l|$. (For details and more precise statements of the result, see [CK1-CK2], [K].)

(3) For any entropy pairs (η, q) constructed in (2), we have

$$\{\partial_t \eta(U^\epsilon) + \partial_x q(U^\epsilon)\}_{\epsilon > 0} \subset \text{ a compact set in } H^{-1}_{loc}(x,t),$$

where $\{U^\epsilon\}_{\epsilon>0}$ denotes a sequence of approximate solutions generated by the vanishing viscosity method (3.1).

The proof of Theorem 1 relies on the construction of the Riemann function to (3.2) and a very detailed analysis of its singular as well as regular behaviors to construct general classes of solutions (as described in (2) of Theorem 1) where the singularities are absent. (For the general definition of Riemann function, see e.g. [CH]) The weak compactness information on entropy dissipations ((3) in Theorem 1) is crucial in the proof of the convergence of the approximate solutions in Theorems 3 and 4. This information is used with the div-curl lemma in the compensated compactness theory to reduce the Young measure associated with the weak convergence of the approximate solution sequence to a Dirac mass. For details, see [CK1-CK2], [K].

3.3 Convergence of Approximate Solutions

Now we establish a structure framework theorem characterizing the structure of the Young measures, which are determined by the approximate solutions to a general non-strictly hyperbolic system (1.1) with an isolated umbilic point. Then we conclude a corresponding compactness framework theorem for the convergence of approximate solution sequences to the system (1.1).

The proof of these framework theorems rely on detailed studies of the structure of the Young measures associated with the approximate sequences. This is achieved by a delicate use of Serre's technique [Se] and regular entropy functions, constructed in Theorem 1, in the Tartar-Murat commutation equation ([Ta]) for Young measures associated with the approximate solution sequences.

Theorem 2. *Assume that (1.1) satisfies conditions (H1)-(H2). Suppose that a Young measure $\nu \in Prob.(\mathbf{R}^2)$ satisfies*

$$\text{supp } \nu(w_1, w_2) \subset [w_1^-, w_1^+] \times [w_2^-, w_2^+], \tag{3.3}$$

$$< \nu, \eta_1 q_2 - \eta_2 q_1 > = < \nu, \eta_1 >< \nu, q_2 > - < \nu, \eta_2 >< \nu, q_1 >, \tag{3.4}$$

for all C^2 Goursat entropy-entropy flux pairs $(\eta_i, q_i), i = 1, 2,$ as constructed in Theorem 1 for the systems (1.1) with an isolated umbilic point P. Then

$$\text{supp } \nu(w_1, w_2) \subset \{ \ (w_1, w_2) \mid \lambda_{iw_i} = 0, \ i = 1, 2 \ \}.$$

Combining Theorem 2 with the theory of Young measures, we have

Theorem 3 (Compactness Framework for Approximate Solutions). *Suppose that (1.1) satisfies (H1) and (H2) and that*

$$\lambda_{iw_i} \neq 0, \ i = 1, 2. \tag{3.5}$$

Suppose that $\{U^\epsilon(x,t)\}_{\epsilon>0}$ is a sequence of measurable functions satisfying

$$\|U^\epsilon\|_{L^\infty} \leq M < +\infty, \tag{3.6}$$

$$\eta(U^\epsilon)_t + q(U^\epsilon)_x \qquad \text{compact in } H^{-1}_{loc}, \tag{3.7}$$

for any C^2 Goursat entropy-entropy pair (η, q) as constructed in Theorem 1 for the systems (1.1) with an isolated umbilic point. Then

$$(w_1(U^\epsilon), w_2(U^\epsilon)) \longrightarrow (w_1(x,t), w_2(x,t)), \qquad a.e. \tag{3.8}$$

where $w_1(U)$ and $w_2(U)$ are the Riemann invariants.

Remark 1. If $U^\epsilon(x,t)$ are generated by the vanishing viscosity method then the condition (3.7) for $U^\epsilon(x,t)$ can be achieved by a standard argument (see e.g. [Di]).

Remark 2. The condition (3.5) for the system can be relaxed. For instance, the same result in Theorem 3 holds even if $\lambda_{iw_i} = 0$ along the line $w_i = 0$.

As a concrete application of the Compactness Framework Theorem, we have

Theorem 4 ([CK1-CK2], [K]).
Consider the quadratic normal form system

$$\begin{aligned} &\partial_t U + \partial_x(dC(U)) = 0, \\ &U = (u,v)^T, \\ &C(U) = \frac{1}{2}(\frac{1}{3}au^3 + bu^2v + uv^2), \\ &U(0,x) = U_0(x). \end{aligned} \tag{3.9}$$

Assume that $U_0 \in L^\infty \cap L^2$ and that $a > 1 + b^2$. Then approximate solutions generated by the vanishing viscosity method (i.e. (3.1)) applied to (3.9) converge to a global weak entropy solution for (3.9).

Acknowledgments.

Gui-Qiang Chen's research was supported in part by NSF grants DMS-9207080 and DMS-9201581, by ONR grant N00014-91-J-1384, and by an Alfred P. Sloan Foundation Fellowship. Pui Tak Kan's research was supported in part by NSF grants DMS-9100383 and DMS-9404341, and by the U.S. Army.

References

[CK1] Chen, G.-Q., and Kan, P.-T.: *Hyperbolic conservation laws with umbilic degeneracy I*, To appear in Arch. Rat. Mech. Anal.

[CK2] Chen, G.-Q., and Kan, P.-T.: *Hyperbolic conservation laws with umbilic degeneracy II*, In preparation

[CCS] Chueh, K.N., Conley, C.C., and Smoller, J.: *Positively invariant regions for systems of nonlinear diffusion equations*, Indiana Univ. Math. J. 26, 372–411 (1977).

[CH] Courant, R., and Hilbert, D.: *Methods of Mathematical Physics*, Vol. 2, Wiley-Interscience: New York, 1948.

[Di] DiPerna, R.: *Convergence of approximate solutions to conservation laws*, Arch. Rat. Mech. Anal. 82, 27–70 (1983).

[F] Freisthüler, H.: *Rotational degeneracy of hyperbolic systems of conservation laws*, Arch. Rat. Mech. Anal. 113, 39-64 (1990).

[FR] Friedlands, S., Robbin, J. W., and Sylvester, J: *On the crossing rule*, Comm. Pure Appl. Math. 37, 19-38 (1984).

[FS] Frid, H., and Santos, M. M.: *Nonstrictly hyperbolic systems of conservation laws of the conjugate type*, Preprint 1993.

[G1] Glimm, J.: *The continuous structure of discontinuities*, Lecture Notes in Physics 344, 177–186 (1986).

[G2] Glimm, J.: *Nonlinear waves: overview and problems*, In: Multidimensional hyperbolic problems and computations, ed. J. Glimm and A. Majda, IMA Vol. 29, Springer-Verlag: New York, 1991.

[IMP] Isaacson, E., Marchesin, D., and Plohr, B.: *Transitional shock waves*, Contemporary Mathematics 100, 125-145 (1988).

[IMPP] Isaacson, E., Marchesin, D., Palmeira, C., and Plohr, B.: *A global formalism for nonlinear waves in conservation laws*, Commun. Math. Phys. 146, 505-552 (1992).

[IMPT] Isaacson, E., Marchesin, D., Plohr, B., and Temple, B.: *The Riemann problem near a hyperbolic singularity: the classification of solutions of quadratic Riemann problem (I)*, SIAM J. Appl. Math. 48, 1-24 (1988).

[IT] Isaacson, E., and Temple, B.: *The classification of solutions of quadratic Riemann problem (II)-(III)*, SIAM J. Appl. Math. 48, 1287-1301, 1302-1318 (1988).

[J] John, F.: *Algebraic conditions for hyperbolicity of systems of partial differential equations*, Comm. Pure Appl. Math. 31, 89-106, 787-793 (1978).

[K] Kan, P.-T.: *Convergence of the viscosity method for a system of nonstrictly hyperbolic conservation laws*, Preprint 1992.

[KK] Keyfitz, B. L., and Kranzer, H. C.: *A system of nonstrictly hyperbolic conservation laws arising in elasticity theory*, Arch. Rat. Mech. Anal. 72, 219-241 (1980).

[La1] Lax, P.D.: *The multiplicity of eigenvalues*, Bull. AMS. 6, 213-214 (1982).

[La2] Lax, P.D.: *Shock waves and entropy*, in: Contributions to Nonlinear Functional Analysis, ed. E. A. Zarantonello, Academic Press, 603-634 (1971).

[Liu] Liu, T.-P.: *Geometric theory of shock waves*, In: Multidimensional hyperbolic problems and computations, ed. J. Glimm and A. Majda, IMA Vol. 29, Springer-Verlag: New York, 1991.

[LZ] Liu, T.-P., and Zumbrun, K.: *Nonlinear Stability of General Undercompressive Shocks*, Preprint.

[Lu] Lu, Y.-G.: *Convergence of the viscosity method for a nonstrictly hyperbolic system*, Acta Mathematica Sinica, 12, 1992.

[Mu] Murat, F.: *L'injection du cone positif de H^{-1} dans $W^{-1,q}$ est compacte pour tout $q < 2$*, J. Math. Pures Appl. 60, 309–322 (1981).

[Rb] Rubino, B.: *On the vanishing viscosity approximation to the Cauchy problem for a 2×2 system of conservation laws*, Preprint 1992.

[SS1] Schaeffer, D., and Shearer, M.: *The classification of 2×2 systems of nonstrictly hyperbolic conservation laws, with application to oil recovery*, Comm. Pure Appl. Math. 40, 141-178 (1987).

[SS2] Schaeffer, D., and Shearer, M.: *Riemann problems for nonstrictly hyperbolic 2×2 systems of conservation laws*, Trans. AMS. 304, 267-306 (1987).

[SSMP] Shearer, M., Schaeffer, D., Marchesin, D., and Paes-Leme, P. J.: *Solution of the Riemann problem for a prototype 2×2 system of nonstrictly hyperbolic conservation laws*, Arch. Rat. Mech. Anal. 97, 299-320 (1987).

[Se] Serre, D.: *La compacite par compensation pour les systemes hyperboliques nonlineaires de deux equation a une dimension d'space*, J. Math. Pures Appl., 65, 423-468 (1986).

[Ta] Tartar, L.: *Compensated compactness and applications to partial differential equations*, In: Research Notes in Mathematics, Nonlinear Analysis and Mechanics: Heriot-Watt Symposium, Vol. 4, ed. R. J. Knops, pp 136–212, Pitman Press, New York, 1979.

[TT] Tang, Z., and Ting, T.C.T.: *Wave Curves for the Riemann Problem of Plane Waves in Simple Isotropic Elastic Solids*, Int. J. Eng. Sciences, 25, 1343-1381 (1987).

Semi-Implicit High Resolution Schemes for Low Mach Number Flows[1]

R. Klein,[2] K. Lange,[2] W. Willems[2] and C.D. Munz[3]

Abstract

This paper highlights the general concept and some applications of a new class of semi-implicit, Low-Mach number extensions of Godunov type compressible flow solvers that we have proposed recently. For moderate and large Mach number the original explicit compressible flow solver is active without modification. For low Mach numbers, the new schemes employ an operator splitting technique that is motivated by a detailed asymptotic analysis of Euler solutions in the limit of vanishing Mach number, M. Advection of mass and momentum as well as long wave acoustics are discretized explicitly, while in solving the *energy equation*, the schemes use an implicit pressure correction formulation to guarantee both divergence-free flow in the zero Mach number limit, and appropriate representation of weakly nonlinear acoustic effects for small but nonzero Mach number. An analysis of this implicit pressure correction step as $M \to 0$ shows that the proposed splitting technique automatically reduces to a projection method for variable density incompressible flows as $M = 0$. A numerical example shows that the scheme successfully operates at CFL numbers of order $O(1/M)$ and that inertia-induced $O(M^2)$ pressure fluctuations, as well as weakly nonlinear acoustic effects, are captured reliably.

1 Introduction

One motivation for the present work comes from combustion applications. Reliable simulations of slow premixed flames, flame accelerations and the deflagration to detonation transistion require an "all Mach number capability" of the simulation code; laminar premixed combustion generates flows with Mach numbers $M_{\text{lam}} = O(10^{-4}...10^{-3})$, while a detonation wave attains lead shock Mach numbers $M_{\text{deto}} = 5...10$.

A considerable variety of numerical techniques for zero and/or low Mach numbers and for weakly compressible flows have been proposed in recent years. There are compressible extensions of incompressible flow solvers [3, 1], fully implicit (and computationally costly) versions of compressible flow codes [11], and semi-implicit methods. These are geared toward minimizing computational effort by treating only those parts of the flow equations implicitly that are responsible for the singular stiffness as $M \to 0$, [2, 5]. We have found that each of these schemes fails to satisfy at least one of the following list of requirements, motivated by the deflagration-to-detonation-transition phenomenon. The scheme should:

1. yield an acceptable numerical approximation for the equations of zero Mach number flow with large amplitude density variations, [7], as $M = 0$,

[1]This work has been supported by the DFG Schwerpunktprogramm "Flow Simulation on High Preformance Computers"

[2]Institut für Technische Mechanik, RWTH, Templergraben 64, 52056 Aachen, Germany, Telephone: ++49-241-803544, Fax: ++49-241-8888223, correspondence: r.klein@itm.rwth-aachen.de

[3]Kernforschungszentrum Karlsruhe, INR, PO-Box 3640, 76021 Karlsruhe, Germany, Telephone: ++49-7247-822449, Fax: ++49-7247-824874

2. provide an accurate representation of the long wave acoustics responsible for flow acceleration to higher and higher Mach numbers, and

3. reduce to a reasonable explicit full compressible Euler solver for characteristic flow Mach numbers of order $M = O(1)$ and larger.

The principal idea behind the present new class of schemes is to first perform an asymptotic analysis for small Mach numbers, and to use the insight so gained in designing a suitable numerical discretization, [6]. We emphasize that this procedure yields a scheme for the *full* Euler equations, not for solving some reduced asymptotic system. In this sense, the use of asymptotic analysis here is quite unusual. Rather than providing simplified limit equations which are then solved by highly specialized methods with restricted applicability, the asymptotics serve as a guide in designing numerical methods for the *full* equations which operate efficiently even in a certain singular limit regime.

The remainder of this paper is organized as follows. The rest of the introduction provides a summary of the asymptotic analysis in [6]. Section 2 explains the new concept for semi-implicit Godunov-type flow solvers and discusses the zero Mach number limit. Section 3 presents computational results for one space dimension.

1.1 Summary of the asymptotics

The flow acceleration from low to high Mach numbers is achieved in the deflagration-to-detonation process through the action of acoustic pulses that are generated by the unsteady premixed (turbulent) flame. The flow inherits the characteristic flame fluctuation time scale and the relevant asymptotic regime involves a single time scale but multiple length scales (a convective and an acoustic time scale). An appropriate asymptotic expansion for, say, density, velocity and pressure, $\underline{U} = (\rho, v, p)$, reads

$$\underline{U} = \underline{U}^{(0)}(x, Mx, t) + M\underline{U}^{(1)}(x, Mx, t) + M^2\underline{U}^{(2)}(x, Mx, t)\cdots, \qquad M = \frac{v_{\text{ref}}}{\sqrt{p_{\text{ref}}/\rho_{\text{ref}}}} \ll 1. \tag{1}$$

The pressure expansion deserves special attention. For the single time/multiple space scale regime, one finds that the leading order pressure is spatially homogeneous, i.e., $p^{(0)} = P_0(t)$, and that it plays the role of a thermodynamic variable. As we are interested in flows undergoing considerable changes of the characteristic flow Mach number, we must include acoustic pressure amplitudes large enough to produce leading order effects on the velocity field. It follows that the term $Mp^{(1)}$ in the pressure expansion may not be exluded as is usually done in asymptotic derivations of the incompressible limit [7]. The first order pressure varies spatially only on the acoustic length scale, i.e., $p^{(1)} = P^{(1)}(Mx, t)$, and it reflects the influence of long wave acoustics [6]. The second order pressure has small scale *and* acoustic scale spatial structure, i.e., $p^{(2)} = p^{(2)}(x, Mx, t)$, and it is responsible for guaranteeing the divergence-free constraint for incompressible flows for $M = 0$. Hence, the pressure expansion scheme reads

$$p = P_0(t) + MP^{(1)}(Mx, t) + M^2 p^{(2)}(x, Mx, t) + \cdots. \tag{2}$$

This ansatz, inserted in the compressible Euler equations (for an ideal gas with constant specific heats), yields separate equations for the short wave and long wave length dynamics. On the small convective length scale, the reference length l_{ref}, the leading order continuity, the second order momentum and the leading order energy conservation equations yield

$$\begin{array}{rcrcl} \rho_t^{(0)} & + & \nabla_x \cdot (\rho v)^{(0)} & = & 0\,, \\ (\rho v)_t^{(0)} & + & \nabla_x \cdot (\rho v \circ v)^{(0)} + \nabla_x p^{(2)} & = & -\nabla_\xi P^{(1)}\,, \\ & & \nabla_x \cdot v^{(0)} & = & -\frac{1}{\gamma P_0}\frac{dP_0}{dt}\,. \end{array} \tag{3}$$

These are the zero Mach number, inviscid, compressible flow equations, supplemented by a long wave acoustic pressure gradient in the momentum equation and a global compression source term in $(3)_3$, which is left-over from the *energy equation*. Depending on the overall system dimensions, only one of these supplementary terms can be present. The global compression term in $(3)_3$ is readily determined by the boundary conditions when the system dimensions L are of order $O(l_{\text{ref}})$, in which case no long wave acoustics can be accommodated and $\nabla_\xi P^{(1)} \equiv 0$. When the system *is* large enough to accomodate long wave acoustics, i.e., when $L = O(l_{\text{ref}}/M)$, then $dP_0/dt = \nabla_x \cdot v \equiv 0$, while $\nabla_\xi P^{(1)} \neq 0$. Leading order pressure changes then occur only due to accumulation of first order pressure wave effects on time scales $O(l_{\text{ref}}/M v_{\text{ref}})$, and the local flow divergence is $O(M)$ only.

The second order pressure gradient, $\nabla_x p^{(2)}$, adjusts to guarantee satisfaction of the small scale divergence constraint, $(3)_3$. Thus, the second order pressure in a low Mach number expansion attains the role of "the pressure" of the incompressible flow equations as $M = 0$, [7, 6].

The long wave dynamics are obtained from the x-scale-averages of $(3)_2$ and of the first order energy equation:

$$\left.\begin{array}{rcl} \overline{m}_t + \nabla_\xi P^{(1)} & = & 0 \\ P_t^{(1)} + \nabla_\xi \cdot (\overline{c}^2 \overline{m}) & = & \nabla_\xi \cdot (\overline{c}^2 \overline{\tilde{\rho}\tilde{v}}) \end{array}\right\} \quad \text{where} \quad \xi \equiv Mx \quad \text{and} \quad \left\{\begin{array}{rcl} \tilde{\rho} & = & \rho^{(0)} - \overline{\rho} \\ \tilde{v} & = & v^{(0)} - \overline{v} \end{array}\right. . \tag{4}$$

Here $\overline{\rho}, \overline{m}, \overline{v}$ denote the x-averaged leading order density, momentum and velocity, and $\overline{c}^2 = \gamma P_0/\overline{\rho}$ is the square of an average sound speed. This is the system of linearized acoustics except for the inhomogeneous term in the second equation. This term is generated by the x-scale fluctuation correlation of density and velocity, and it introduces a nontrivial coupling between the small-scale, quasi-incompressible flow and the superimposed long wave acoustics.

We now identify the singular and non-singular contributions to the fluxes of the conserved quantities as $M \to 0$. The convection terms for mass and momentum are obviously non-singular. The energy equation and the pressure gradient in the momentum equation are responsible for the degeneration of the Euler equations, which originally are hyperbolic. The divergence constraint in (3) leads to an *elliptic* Poisson-type equation for $p^{(2)}$, while the acoustic system (4) predicts infinitely fast signal propagation in the limit. This justifies a numerical operator splitting.

First, we have advection of mass and momentum and their changes due to flow divergence:

$$\left.\begin{array}{rcl} \rho_t + \nabla \cdot (v\,\rho) & = & 0 \\ m_t + \nabla \cdot (v \circ m) & = & 0 \end{array}\right\} \quad \text{System I} \tag{5}$$

Secondly, there is flow acceleration due to pressure forces and energy conservation,

$$\left.\begin{array}{rcl} m_t + \frac{1}{M^2} \nabla p & = & 0 \\ e_t + \nabla \cdot (H\,m) & = & 0 \end{array}\right\} \quad \text{System II}, \tag{6}$$

where $H = (e+p)/\rho$ and $p = (\gamma - 1)(e - M^2\, m^2/2\rho)$. The formal sum of (5) and (4) reproduces the compressible Euler equations in nondimensional form with a scaling that guarantees that all the nondimensional quantities ρ, v, p, e etc. remain of order $O(1)$ as $M = v_{\text{ref}}/\sqrt{p_{\text{ref}}/\rho_{\text{ref}}} \to 0$.

2 The multiple pressure variable method

In this section, we describe the concept of our new scheme for one-dimensional flows. In this case, the acoustic subsystem (4) decouples from the small scale quasi-incompressible flow and there is no feedback from the small scale flow to the long wave acoustics. In more than one dimension, there *is* a nontrivial coupling and the numerical techniques will have to be modified to account for this effect. This is work in progress and beyond the scope of this paper.

2.1 Discretization of System I

One of the requirements for our new discretization is that it operate reliably for low *and* high Mach number flows (including strong shocks). Thus, we wish to embed a full scale explicit higher order compressible flow solver in the scheme which becomes active at high Mach numbers. This is done in the solution step for System I from (5). We observe that the auxiliary System I*

$$\left.\begin{array}{lclcl} \rho_t & + & (v\,\rho)_x & = & 0 \\ m_t & + & (v\,m + p)_x & = & 0 \\ e_t & + & (v\,[e + \overline{\overline{p}}_{\mathrm{NL}} + M^2 p]) & = & 0 \end{array}\right\} \quad \text{System I}^* \tag{7}$$

with $\overline{\overline{p}}_{\mathrm{NL}}$ a fixed number, is very similar to the original Euler equations (note that the pressure gradient does not have a factor of $1/M^2$!). The system is hyperbolic, but all of its (nondimensional) signal speeds, $v - c_*, v, v + c_*$, with $c_*^2 = c^2 + (\gamma - 1)(\overline{\overline{p}}_{\mathrm{NL}} - (1 - M^2)p)/\rho$, are order $O(1)$ for all M. Hence, an explicit numerical solution will not suffer from CFL-restrictions as $M \to 0$.

For data in the asymptotic regime considered in the last section, the first two equations coincide with those of the split System I except for the $O(M^2)$ pressure gradient. (Recall that $\nabla p/M^2 = O(1)$ in the regime considered.) Thus, if we compute numerical fluxes of mass and momentum for this auxiliary system, equating $\overline{\overline{p}}_{\mathrm{NL}}$, e.g., with the average pressure in the computational domain, we have an $O(M^2)$ approximation to the fluxes of System I. As we are using a split step appoach, we may correct for this deviation when solving System II, (6), (see Section 2.2). Due to the similarity of System I* with the full compressible Euler equations, we obtain these fluxes by using a Godunov-type scheme for 1D gas dynamics, only slightly modified to adjust for the specific features of System I*. The energy fluxes in this step are simply discarded. Such a scheme, based on the Harten-Lax-van Leer/Einfeldt approach [4], has been proposed in [6], but any other higher order shock capturing method would do, too.

Other, more efficient discretizations of System I are certainly possible and even Godunov-type ideas have been introduced to this end in [1], in the framework of a projection method for incompressible flows. However, the key advantage of the present approach is that one may employ any existing fully compressible flow solver to handle System I, and any such solver can thus be extended to perform weakly compressible and even incompressible flow computations by (i) adapting the scheme to System I* for small M and (ii) adding suitable subroutines that deal with System II.

2.2 Discretization of System II

Notice that, due to the singular nature of the small Mach number limit, the first and second order pressures both affect the velocity field at leading order, albeit through completely different physical mechanisms. This observation led us to introduce a multiple pressure variable method, which operates with separate discrete fields for the actual pressure p and the second order pressure $p^{(2)}$ for small M. Also, suitable separate discretizations are used to account for the hyperbolicity of long wave acoustics and the ellipticity of the divergence constraint, both of which are hidden in the second split system.

In fact, we solve System II, i.e., (6), in a predictor-corrector fashion. The predictor step yields the dominant first order ($O(M)$) pressure changes through an *explicit* and non-dissipative discrete solution of the long wave acoustic system (4) using special large time step techniques. The final *implicit* discretization of (6) then computes the left-over small scale $O(M^2)$ pressure fluctuations. This correction step not only guarantees compliance with the small-scale divergence constraint as $M \to 0$, but it also produces the weakly nonlinear acoustic effects which, for small but nonzero Mach number, are neglected in the linear acoustic predictor.

One version of the acoustic predictor is described in detail in [6]. It involves (i) the determination of small scale averaged momenta $\overline{m}$ and approximations to the asymptotic first order pressure $P^{(1)}$ by suitable averaging and summation procedures, and (ii) the application of a large time step method to solve the linearized acoustic equations (4). (Notice that in one space dimension the coupling term on the r.h.s of $(6)_2$ is absent.) The results of this step are preliminary acoustic updates $\delta^t\overline{m}$ and $\delta^t e = M\delta^t P^{(1)}/(\gamma-1)$ of momentum and energy, and these estimates are employed subsequently in the final correction step to be described in the next paragraph. Klein [6] uses LeVeque's explicit large time step method [10] for the acoustic predictor and the results of this paper have been produced with this approach. We are currently implementing an implicit scheme which is more easily extendable to more than one space dimension. One may also use multi-grid techniques to take advantage of the space scale separation.

Given the linear acoustic predictions $\delta^t\overline{m}$ and $\delta^t e$, we proceed to solve for second order, $O(M^2)$, updates of total energy and pressure and for the final momentum changes. We describe here a version that is first order in time for simplicity of exposition.

$$\begin{aligned} m^{n+1} - (m^* + \delta^t\overline{m}) &= -\Delta t \;\; \tilde{\nabla}_1 \left[p^{(2),n+1} - p^*\right] , \\ M^2\left[\left(\tfrac{p^{(2)}}{\gamma-1} + \tfrac{m^2}{2\rho}\right)^{n+1} - e^{(2),*}\right] + \delta^t e &= -\Delta t \;\; \tilde{\nabla}_2 \left[H^{**}\, m^{n+1}\right] , \end{aligned} \tag{8}$$

where

$$H^{**} = (e+p)^{**}/\rho^* \qquad \text{and} \qquad (e+p)^{**} = (e+p)^* + \frac{\gamma}{\gamma-1}\delta^t e\,. \tag{9}$$

Here a single *-superscript denotes the data after the first split step, while a double star ** indicates data modified by the acoustic predictor. The symbols $\tilde{\nabla}_1, \tilde{\nabla}_2\cdot$ denote suitable discrete approximations of gradient and divergence. The term $\Delta t\,\tilde{\nabla}_1 p^*$ makes up for the $O(M^2)$ difference between System I and System I* as discussed in the previous subsection. Upon elimination of m^{n+1} from (8) one has a discrete Poisson-type equation for $p^{(2)}$ with a weak $O(M^2)$ nonlinearity due to the time dependence of the kinetic energy. This can be accounted for, e.g., by an outer iteration, (see [6], where a cell-centered finite volume approach for (8) is proposed).

2.3 The zero Mach number limit

Here we indicate the importance of the multiple pressure variable formulation and relate the present approach to projection methods for incompressible flows (see also [8]). We rewrite $(8)_1$ as

$$v^{n+1} = v^{**} - \frac{\Delta t}{\rho^*}\tilde{\nabla}_1 p^{(2),n+1} \quad \text{with} \quad v^{**} = v^* + \frac{1}{\rho^*}(\delta^t\overline{m} + \Delta t\,\tilde{\nabla}_1 p^*)\,, \tag{10}$$

substitute into $(8)_2$ and find an elliptic expression for the $O(M^2)$ pressure $p^{(2)}$:

$$\begin{aligned} &\tilde{\nabla}_2\cdot\left[\frac{(e+p)^{**}}{\rho^*}\,\tilde{\nabla}_1\, p^{(2),n+1}\right] - \frac{M^2}{(\gamma-1)\Delta t^2}p^{(2),n+1} = \\ &\frac{1}{\Delta t}\tilde{\nabla}_2\cdot\left[(e+p)^{**}\,v^{**}\right] + \frac{1}{(\gamma-1)\Delta t^2}\,\delta^t e - M^2\left[\frac{1}{(\gamma-1)\Delta t^2}\,p^{(2),n} - \frac{1}{\Delta t}\left(\frac{m^2}{2\rho}\right)_t\right]. \end{aligned} \tag{11}$$

In absence of $O(M)$-acoustics and global compression ($\delta^t\overline{m} = \delta^t e = 0$), the limit $M\to 0$, using $\tilde{\nabla}_2(e+p)^{**}\to 0$, yields

$$\tilde{\nabla}_2\cdot(\frac{\Delta t}{\rho^*}\tilde{\nabla}_1 p^{(2),n+1}) = \tilde{\nabla}_2\cdot v^{**}\,. \tag{12}$$

This is the pressure equation of the projection method for incompressible flows, [9, 1].

Consider now a projection method that is extended to the compressible equations without the introduction of multiple pressure variables. The discrete momentum equation then reads

$$v^{n+1} = v^* - \frac{\Delta t}{M^2 \rho^*} \tilde{\nabla}_1 p \,. \tag{13}$$

Insertion of this equation into the discretized energy equation yields

$$p^{n+1} - \frac{(\gamma-1)\Delta t^2}{M^2} \tilde{\nabla}_2 \cdot \left((e+p)^* \frac{1}{\rho^*} \tilde{\nabla}_1 p \right) = p^* - (\gamma-1)\Delta t \tilde{\nabla}_2 \cdot ((e+p)^* v^*) - \Delta t (\gamma-1)(\rho v^2)_t \,. \tag{14}$$

Multiplying by M^2 and letting $M \to 0$, we find at leading order that

$$\tilde{\nabla}_2 \cdot \left(\frac{1}{\rho^*} \tilde{\nabla}_1 p^{n+1} \right) = 0 \,. \tag{15}$$

Hence the $O(M^2)$ pressure fluctuations that are known to occur have to be represented numerically through subtle cancellations of large numbers and the performance of such a scheme will depend in an undesirable way on machine accuracy for small M. A pressure decomposition separating the second order pressure $p^{(2)}$ from the actual thermodynamic pressure p within the numerical framework seems necessary to guarantee the proper limiting behaviour as $M \to 0$.

3 Results

Here we describe the results for the repeated passage of a long wave acoustic pulse over a small-scale, but large amplitude density layering at a characteristic flow Mach number $M = 0.02$. For a periodic system $(-l_{\text{ref}}/M \le x < l_{\text{ref}}/M)$, resolved by 1020 grid cells, we start with a sinusoidal right running acoustic wave, such that the pressure at time zero satisfies $p(x,0) = 1.0 + 1.25M(1.0 + \cos(2\pi Mx))$. The velocity data are such that a clean right-running wave is obtained. The density profile, in addition to an $O(M)$ perturbation $M\tilde{\rho}_{\text{ac}}(Mx)$ due to the adiabatic compression from the acoustic wave, has a large amplitude fluctuation for $0 \le x \le 20$: $\rho(x,0) = 1.0 + 0.5\Phi(x)\sin(40\pi Mx) + M\tilde{\rho}_{\text{ac}}(x)$, with $\Phi(x) = 0.5(1.0 - \cos(5\pi Mx))$ for $0 \le Mx \le 0.4$ and zero otherwise.

Figures 1 and 3 show the pressure and density initial data in addition to the distributions at the later time $t = 5.071$. By this time, the pressure pulse has passed the density layering 2 1/2 times and a considerable distortion of the pressure profile due to weakly nonlinear acoustic wave steepening is visible. Figure 1 actually includes computational results obtained with the new semi-implicit algorithm (after 189 time steps, CFL $\approx 0.8/(2M)$), as well as results from a fully explicit computation (4086 time steps, CFL $= 0.8$) with only the original high resolution scheme being active. There is no difference in the pressure profiles within the graphical resolution.

A considerable difference does occur in the density distributions displayed in Figs. 3,4. The density fluctuation amplitude, which should be constant in time in this case, is damped roughly by a factor of two in the explicit computation (Fig. 4), while it is practically preserved in the semi-implicit run. The reason is the excessive dissipation in the explicit computation that occurs simply because of the large number of time steps needed to reach this stage.

Figure 2 displays the second order pressure as computed by the semi-implicit method, exhibiting fluctuations on the length scale of the density stratification. Mass elements are accelerated equally by the long wave acoustic pulse, according to the asymptotics, but due to their varying density they require different pressure gradients to this end. These $O(M^2) \sim 10^{-4}$ differential pressure gradients are fully resolved by the $p^{(2)}$-field in the semi-implicit code, while they have to be established through subtle differences of large numbers in the explicit computation.

References

[1] Bell J.B., Colella P., Glaz H.M. (1989), JCP, **85**, 257

[2] Patnaik G., Guirguis R.H., Boris J.P., Oran E.S. (1987), JCP, **71**, 1

[3] Chorin A.J. (1969), Math Comput., **23**, 341

[4] Einfeldt B. (1988), SIAM J. Num. Anal., **25**, 294

[5] Gustaffson B., Stoor H. (1991), SIAM J. Num. Anal., **28**, 1523

[6] Klein R. (1993), submitted to JCP, Dec. 1993

[7] Majda A., Sethian J. (1985), Comb. Sci & Technol., **42**, 185

[8] Munz C.D., Klein R. (1994), Intl. Conf. Num. Mod. in Cont. Mech., Prague, Aug. 1994

[9] Peyret R., Taylor T.D. (1983) Computational Methods for Fluid Flow, Springer, NY

[10] LeVeque R.J. (1985), SIAM J. Num. Anal., **25**, 294

[11] Yee, H. (1989), AIAA Jnl., **27**, 299

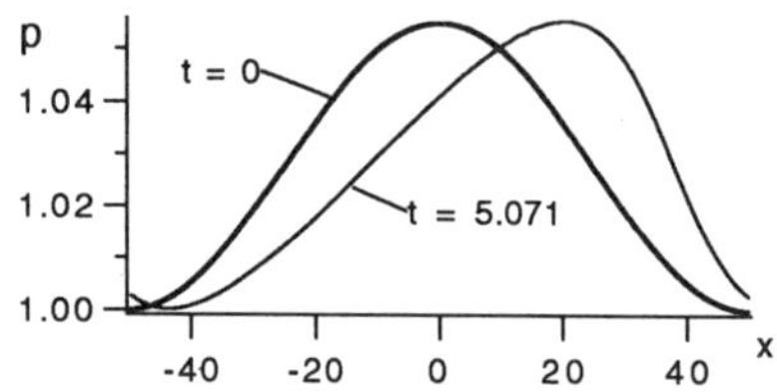

Figure 1: Total pressure distributions (explicit and semi-implicit) at times 0 and 5.071.

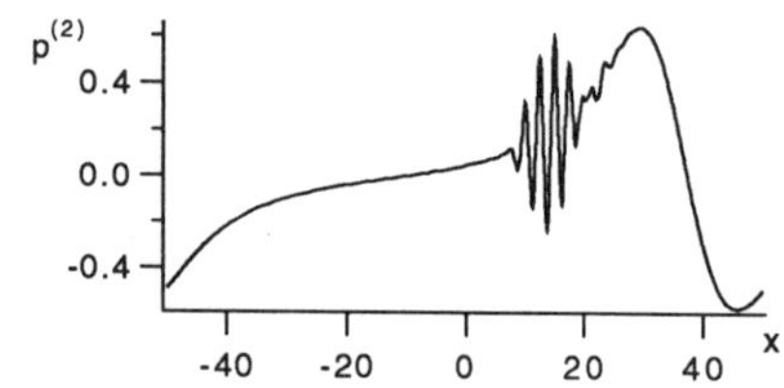

Figure 2: Second order pressure distribution at time 5.071 from the semi-implicit computation.

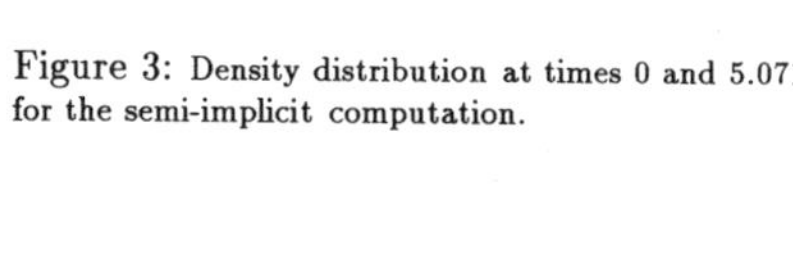

Figure 3: Density distribution at times 0 and 5.071 for the semi-implicit computation.

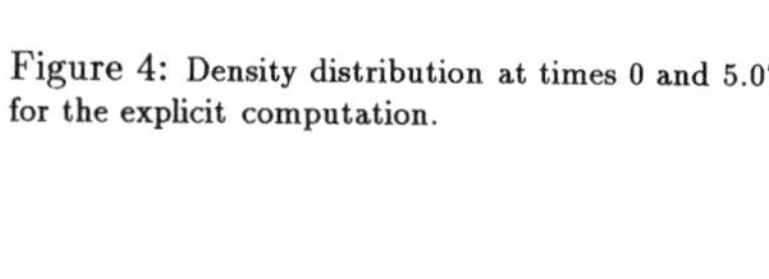

Figure 4: Density distribution at times 0 and 5.071 for the explicit computation.

Existence of Solutions to Resonant systems of Conservation Laws

Christian Klingenberg[1] and Yun-guang Lu[2]

Abstract

Using the method of compensated compactness global existence of weak solutions to resonant systems of two conservation laws (1.1), (1.2)) is shown. For system (1.2) entropy-entropy flux pairs of Lax type are constructed and estimates from singular perturbation theory of ODEs are used.

1 Introduction

In this paper we consider the Cauchy problem for an extended model of isothermal flow [9]:

$$\left.\begin{aligned} \rho_t + (\rho u)_x + h(x,t)\rho u &= 0 \\ (\rho u)_t + (\rho u^2 + P(\rho))_x + h(x,t)\rho u^2 &= 0 \end{aligned}\right\} \tag{1.1}$$

where ρ is the density, u the velocity, $p = p(\rho)$ is the pressure. For the special case $h(x,t) = \dfrac{a'(x)}{a(x)}$ the function a (which depends on x only) represents the crossectional area of a variable duct.

We also consider the related system

$$\left.\begin{aligned} (a\rho)_t + (a\rho u)_x &= 0 \\ u_t + (u^2 + \int_0^\rho \frac{p'(s)}{s} ds)_x &= 0 \end{aligned}\right\} \tag{1.2}$$

where $a = a(x)$. For $a(x) = 1$, system (1.2) was first derived by S. Earnshaw [3] for isentropic flow (see also [15], p. 168).

Systems (1.1) for $h(x,t) = \dfrac{a'(x)}{a(x)}$ and (1.2) are of interest because resonance occurs. This means there is a coincidence of wave speeds from different families of waves. To see this, we may augment either system by the equation

[1]Applied Mathematics, Würzburg University, Am Hubland, Würzburg 97074, Germany, also: Applied Mathematics, Heidelberg University, Heidelberg 69120, Germany

[2]This research was supported by the Humboldt Foundation. Address: Young Scientist Laboratory of Mathematical Physics, Wuhan Institute of Matehmatical Sciences, Academia Sinica, Wuhan, China. Current address: Applied Mathematics, Heidelberg University, Heidelberg 69120, Germany

$$a_t = 0. \tag{1.3}$$

Thus one wave of the augmented system has zero wave speed $\lambda_0 = 0$. One of the other waves may have coinciding wave speed in the transsonic regime:

$$\lambda_k(\tilde{U}) = \lambda_0 \quad with \quad \nabla\lambda_k \cdot R_k \mid_{\tilde{U}} \neq 0 \tag{1.4}$$

(here R_k is an eigenvector and $\tilde{U} = (\tilde{a}, \tilde{u}, \tilde{\rho}\tilde{u})$, thus we may have non-linear resonance.

T.P. Liu studied existence of solutions for near constant data to resonant systems of this type by using Glimm's random choice method [9], [10]. In [5] Glimm, Marshall and Plohr solved (1.1) numerically by using the random choice method. In [4] Embid, Goodman and Majda considered steady state solutions to varying channel flow. Recently in [6], Isaacson and Temple solved the Riemann problem for a general inhomogeneous system of this type. Our interest in studying this resonant problem is motivated by their papers.

Remark 1 *After finishing the work, the authors learned that Chen and Glimm studied global existence of solutions to compressible Euler equations through a varying channel using finite difference schemes as an approximation to the solution [1].*

Theorem 1 (existence of solutions to (1.2)) *Let*
1. $a(x) \geq c > 0$ *be bounded and nondecreasing (or nonincreasing) on* $\mathbf{R}$, *where* c *is a constant.*
2. $w_0(x) \leq M$, $z_0(x) \geq 0$ *(or* $w_0(x) \leq 0$, $z_0(x) \geq -M$*), where*

$$w = u + \int_0^\rho \frac{\sqrt{p'(s)}}{s} ds \qquad z = u - \int_0^\rho \frac{\sqrt{p'(s)}}{s} ds \tag{1.5}$$

are two Riemann invariants of the systems (1.1), (1.2). We use $w_0(x) = w|_{t=0}$, $z_0(x) = z|_{t=0}$.
3. $g(\rho) := \frac{\sqrt{p'(\rho)}}{\rho} \in C^2(0,\infty)$, $g \geq d > 0$, $g' \geq 0$ *for* $\rho \geq 0$, *where* d *is a positive constant.*

Then the Cauchy problem (1.2) with bounded initial data $(\rho, u)|_{t=0} = (\rho_0(x), u_0(x))$, $(\rho_0(x) \geq 0)$ *has a global weak solution in the sense of distributions.*

Remark 2 : *For the special choice of* $p(\rho) = \int^\rho s^2(\rho+\delta)^{\gamma-3} ds$ $(\gamma > 3, \delta > 0)$ *(see [11]), assumption 3 in theorem 1 is satisfied.*

Remark 3 : *The existence theorem 1 above includes the vacuum case* $\rho = 0$.

Remark 4 : *Here we give a proof only for theorem 1. The existence proof for 1.2 and another related system is given in [8]*

2 Proof of theorem 1

We rewrite system (1.2) in the form:

$$\left.\begin{aligned} \rho_t + (\rho u)_x &= -\frac{a'(x)}{a(x)}\rho u \\ (u_t + (\frac{u^2}{2} + \int_0^\rho s g^2(s) ds)_x &= 0 \end{aligned}\right\} \tag{2.1}$$

with g given in assumption 3 of theorem 1. First smoothing $a(x)$ by a mollifier, we get $a^\delta(x) \in C^\infty(R)$, $c \le a^\delta(x) \le M$, $0 \le \delta(a^\delta(x))' \le M$ (or $-M \le \delta(a^\delta(x))' \le 0$) by assumption 1 in theorem 1, and $a^\delta(x) \to a(x)$, a.e. in any compact set of R as $\delta \to 0$. Adding viscosity terms to the right hand side of the system (2.1) yields the following parabolic system

$$\left.\begin{aligned} \rho^\epsilon{}_t + (\rho^\epsilon u^\epsilon)_x &= -\frac{(a^\delta(x))'}{(a^\delta(x))'}\rho^\epsilon u^\epsilon + \epsilon\rho^\epsilon{}_{xx} \\ u^\epsilon{}_t + (\frac{(u^\epsilon)^2}{2} + f(\rho^\epsilon))_x &= \epsilon u^\epsilon{}_{xx} \end{aligned}\right\} \tag{2.2}$$

with the initial data

$$(\rho^\epsilon, u^\epsilon)_{t=0} = (\rho_0(x), u_0(x)) \tag{2.3}$$

where $f(\rho) = \int_0^\rho s g^2(s) ds$.

Lemma 1 *If the conditions in theorem 1 are satisfied and the solutions $(\rho^\epsilon, u^\epsilon)$ of the Cauchy problem (2.2) , (2.3) exist in $R \times [0,T]$, then $(\rho^\epsilon, u^\epsilon)$ satisfy the following estimates:*

$$w(\rho^\epsilon, u^\epsilon) \le M \quad , \quad z(\rho^\epsilon, u^\epsilon) \ge 0 \quad , \quad \rho^\epsilon \ge 0 \tag{2.4}$$

or

$$w(\rho^\epsilon, u^\epsilon) \le 0 \quad , \quad z(\rho^\epsilon, u^\epsilon) \ge -M \quad , \quad \rho^\epsilon \ge 0 \tag{2.5}$$

where w and z are the Riemann invariants.

Proof: If $(a^\delta)'(x) \ge 0$, we have from (2.2)

$$\begin{aligned} w_t + (u^\epsilon + \sqrt{p'(\rho^\varepsilon)})w_x &= \varepsilon w_{xx} - \varepsilon g'(\rho^\epsilon)(\rho^\epsilon)_x{}^2 - \frac{(a^\delta)'}{a^\delta}u^\epsilon\sqrt{p'(\rho^\varepsilon)} \\ &\le \epsilon w_{xx} - \frac{(a^\delta)'}{a^\delta}\sqrt{p'(\rho^\varepsilon)}z - \frac{(a^\delta)'}{a^\delta}\sqrt{p'(\rho^\varepsilon)}\int_0^{\rho^\varepsilon} g(s)ds \\ z_t + (u^\epsilon - \sqrt{p'(\rho^\epsilon)})z_x &= \epsilon z_{xx} + \epsilon g'(\rho^\epsilon)(\rho^\epsilon)_x{}^2 + \frac{(a^\delta)'}{a^\delta}u^\epsilon\sqrt{p'(\rho^\epsilon)} \\ &\ge \epsilon z_{xx} + \frac{(a^\delta)'}{a^\delta}\sqrt{p'(\rho^\varepsilon)}z + \frac{(a^\delta)'}{a^\delta}\sqrt{p'(\rho^\varepsilon)}\int_0^{\rho^\epsilon} g(s)ds \\ &\ge \epsilon z_{xx} + \frac{(a^\delta)'}{a^\delta}\sqrt{p'(\rho^\varepsilon)}z \end{aligned} \tag{2.6}$$

So the second equation in (2.6) and the maximum principle [12] give the estimate $z \geq 0$ first and then we have from the first equation in (2.6) that

$$w_t + (u^\epsilon + \sqrt{p'(\rho^\epsilon)})w_x \leq \epsilon w_{xx}, \tag{2.7}$$

so applying the maximum principle again to (2.7), we obtain $w \leq M$. $\rho^\epsilon \geq 0$ is trivial by the first equation in (2.2).

Similarily we can get the estimates (2.5) if $(a^\delta)'(x) \leq 0$, thus ending the proof of lemma 1.

Using the general contraction mapping principle and the estimate (2.4) or (2.5) we have

Lemma 2 *If the conditions in Theorem 1 are satisfied, then for any fixed $\epsilon > 0$, the Cauchy problem (2.2), (2.3) has a unique global solution $(\rho^\epsilon, u^\epsilon)$, satisfying*

$$0 \leq \int_0^{\rho^\epsilon} \sqrt{p'(s)}ds \leq u^\epsilon \leq M \tag{2.8}$$

or

$$-M \leq u^\epsilon \leq -\int_0^{\rho^\epsilon} \sqrt{p'(s)}ds \leq 0 \tag{2.9}$$

Remark 5 *Dropping the zero-th order term from the right hand side of (2.1) yields the system of conservation laws*

$$\left.\begin{aligned} \rho_t + (\rho u)_x &= 0 \\ (u_t + (u^2 + \int_0^\rho sg^2(s)ds)_x &= 0 \quad . \end{aligned}\right\} \tag{2.10}$$

Noticing the system (2.10) has a strictly convex entropy

$$\eta = \frac{1}{2}u^2 + \int_0^\rho \int_0^y g^2(s)dsdy$$

we deduce that

$\sqrt{\epsilon}\partial_x\rho^\epsilon$, $\sqrt{\epsilon}\partial_x u^\epsilon$ *are uniformly bounded in* $L^2_{loc}(R \times R^+)$.

¿From remark 5 and the boundedness of $(\rho^\epsilon, u^\epsilon)$, we have

Lemma 3 *For any C^2 entropy-entropy flux pair $(\eta(\rho, u), q(\rho, u))$ of the system (2.10)*

$\eta(\rho^\epsilon, u^\epsilon)_t + q(\rho^\epsilon, u^\epsilon)_x$ *compact in* $H^{-1}_{loc}(R \times R^+)$

with respect to the vicosity solutions $(\rho^\epsilon, u^\epsilon)$ of the Cauchy problem (2.2), (2.3).

Lemma 3 guarantees the commutativity relation for the representing measure to be true. We are going to construct the entropy-entropy flux pairs of Lax's type and give their desired estimates by using the theory of singular perturbations of ordinary differential equations [13].

We recall that a pair (η, q) of real-valued maps is an entropy-entropy flux pair of (2.10) if all smooth solutions of (2.10) satisfy

$$(u\eta_\rho + f'(\rho)\eta_u, \rho\eta_\rho + u\eta_u) =_\gamma (q_\rho, q_u) \quad . \tag{2.11}$$

Eliminating the q from (2.11), we have

$$\eta_{\rho\rho} = \frac{f'(\rho)}{\rho}\eta_{uu} \quad . \tag{2.12}$$

We choose special Lax entropies

$$\eta_k^1 = e^{kw(u,\rho)}(\alpha(\rho) + \frac{\beta(\rho,k)}{k}) \tag{2.13}$$

(here $k \in N$ and w is defined in (1.5)) and α and β are to be determined below. Now substitute η_k^1 into (2.12)

$$k[2g(\rho)\alpha' + g'(\rho)\alpha] + \alpha'' + sg(\rho)\beta' + g'(\rho)\beta + \frac{\beta''}{k} = 0. \tag{2.14}$$

Since this should hold for $k \in N$, we have

$$2g(\rho)\alpha' + g'(\rho)\alpha = 0 \tag{2.15}$$

$$\alpha'' + 2g(\rho)\beta' + g'(\rho)\beta + \frac{\beta''}{k} = 0 \tag{2.16}$$

then

$$\alpha = g^{-\frac{1}{2}}. \tag{2.17}$$

The existence of b and the uniformly bounded estimates of β and β' w.r.t. k can be obtained by (2.15), (2.16) and the following lemma [14], also referred to in [7], p. 114.

Lemma 4 *Let $Y(x) \in C^2[0,h]$ be solutions of the equation*

$$F(x, Y, Y') = 0 \tag{2.18}$$

and let functions $f(x,y,\zeta,\lambda)$, $F(x,Y,Y')$ be continuous on the regions $0 \le x \le h$, $|y - Y(x)| \le l(x)$, $|\zeta - Y'(x)| \le m(x)$ for some positive functions $l(x)$ and $m(x)$ and $\lambda_o > \lambda > 0$. In addition

$$\begin{aligned} |f(x,y,\zeta,\lambda) - F(x,y,\zeta)| &\le \epsilon \\ |F(x,y_2,\zeta) - F(x,y_1,\zeta)| &\le M|y_2 - y_1|, \\ \frac{F(x,y,\zeta_2) - F(x,y,\zeta_1)}{\zeta_2 - \zeta_1} &\ge L \end{aligned}$$

for some positive constants ϵ , M, L. If $y = y(x,\lambda)$ is a solution of the following ordinary differential equation of second order

$$\lambda y'' + f(x,y,y',\lambda) = 0 \tag{2.19}$$

with $y(0) = Y(0)$ and $y'(0)$ being arbitrary, then for sufficiently small $\lambda > 0$, $\epsilon > 0$ and $P = |y'(0) - Y'(0)|$, $y(x)$ exists for all $0 \le x \le h$ and satisfies

$$|y(x,\lambda) - Y(x)| < [\frac{\epsilon}{M} + \lambda(\frac{P}{L} + \frac{N}{M})]e^{\frac{Mx}{L}} \tag{2.20}$$

where $N = max|Y''(x)|$ for $0 \le x \le h$.

Using (2.11), we have

$$q_\rho = f'(\rho)\eta_u + \eta_\rho$$
$$q_u = \rho\eta_\rho + u\rho_u.$$

Then a progressing wave of the system (2.11) is provided by

$$\begin{aligned}
\eta_k^1 &= e^{kw}(\alpha(\rho) + \frac{\beta(\rho,k)}{k}) \\
q_k^1 &= (u + \sqrt{p'(\rho)})\eta_k^1 + e^{kw}(\frac{\rho\alpha' - \alpha}{k} + \frac{\rho\beta' - \beta}{k^2}).
\end{aligned}$$

In a similar way we can obtain the other entropy-entropy pairs of Lax type as follows:

$$\begin{aligned}
\eta_{-k}^2 &= e^{-kw}(\alpha(\rho) + \frac{\beta_1(\rho,k)}{k}) \\
q_{-k}^2 &= (u + \sqrt{p'(\rho)})\eta_{-k}^2 + e^{-kw}(\frac{\alpha - \rho\alpha'}{k} + \frac{\beta_1 - \rho\beta_1'}{k^2}) \\
\eta_k^3 &= e^{kz}(\alpha(\rho) + \frac{\beta_2(\rho,k)}{k}) \\
q_k^3 &= (u - \sqrt{p'(\rho)})\eta_k^3 + e^{kz}(\frac{\rho\alpha' - \alpha}{k} + \frac{\rho\beta_2' - \beta_2}{k^2}) \\
\eta_{-k}^4 &= e^{-kz}(\alpha(\rho) + \frac{\beta_3(\rho,k)}{k}) \\
q_{-k}^4 &= (u - \sqrt{p'(\rho)})\eta_{-k}^4 + e^{-kz}(\frac{\alpha - \rho\alpha'}{k} + \frac{\beta_3 - \rho\beta_3'}{k^2})
\end{aligned}$$

where β_i (i=1,2,3) satisfy respectively

$$\alpha'' - 2g(\rho)\beta_1' - g(\rho)\beta_1 + \frac{\beta_1''}{k} = 0 \tag{2.21}$$

$$\alpha'' - 2g(\rho)\beta_2' - g(\rho)\beta_2 + \frac{\beta_2''}{k} = 0 \tag{2.22}$$

$$\alpha'' + 2g(\rho)\beta_3' + g(\rho)\beta_3 + \frac{\beta_3''}{k} = 0 \tag{2.23}$$

Using lemma 4, we can get from (2.23) the existence of β_3 and its uniformly bounded estimate w.r.t. k. Differentiating (2.23) with respect to ρ, we can obtain the uniformly bounded estimate of β_3' w.r.t. k. By making the transformation $\rho_1 = \rho - M$ in the equations (2.22) and (2.21), where M is the upper bound of ρ, we also obtain existence of β_1, β_2, β_1' and β_2' by using lemma 4.

By assumption 3 of theorem 1

$$\alpha - \rho\alpha' = g^{-\frac{1}{2}} + \frac{1}{2}\rho g^{-\frac{3}{2}} g' > 0 \tag{2.24}$$

Thus system (1.2) is genuinly nonlinear. Thus we end the proof just as in section 4 of [11], where the framework of Diperna [2] gets applied to a non-strictly hyperbolic system.

References

[1] G.Q. Chen, J. Glimm, *Global solutions to the compressible Euler equations with geometric structure*, in preparation (1995)

[2] R.J. Diperna, *Convergence of approximate solutions to conservation laws*, Arch. Rat. Mech. Anal., vol. 82 (1983)

[3] S. Earnshaw, *On the mathematical theory of sound*, Phil. Trans., 150 (1858)

[4] P. Embid, J. Goodman, A. Majda, *Multiple steady steady states for 1-d transsonic flow*, SIAM J. Sci. Stat. Comp. 5, 1, (1984)

[5] J. Glimm, G. Marshall, B. Plohr, *A generalized Riemann Problem for quasi-one-dimensional gas flows*, Adv. Appl. Math., 5 (1984)

[6] E. Isaacson, B. Temple, *Nonlinear resonance in systems of conservation laws*, SIAM J. Appl. Math., 52 (1992)

[7] E. Kamke, *Differentialgleichungen, Lösungsmethoden und Lösungen: 1. Gewöhnliche Differentialgleichungen*, sixth ed., Akademische Verlagsanstalt, Leipzig (1959)

[8] C. Klingenberg, Y. Lu, *Existence of solutions to hyperbolic conservation laws with a source*, Heidelberg University preprint, SFB 359, (1994)

[9] T.P. Liu, *Quasilinear hyperbolic systems*, Comm. Math. Phys., 68, (1979)

[10] T.P. Liu, *Resonance for a quasilinear hyperbolic equation* , J. Math. Phys., 28, (1987)

[11] Y.G. Lu, *Convergence of the viscosity method for a nonstrictly hyperbolic conservation law*, Comm. Math. Phys., 150, (1992)

[12] Y.G. Lu, *Hölder continuous solutions of isentropic gas dynamics*, Proc. Royal Soc. Edinburgh, Vol 123A (1993)

[13] Y.G. Lu, *Convergence of the viscosity method for some nonlinear hyperbolic systems*, Proc. Royal Soc. Edinburgh, Vol 124A, (1994)

[14] M. Nagumo, Proc. Phys.-math. Soc. Japan, 3, 21 (1937)

[15] G.B. Whitham, *Linear and nonlinear waves*, John Wily and Sons (1973)

A TWO-DIMENSIONAL RIEMANN SOLVER FOR HYDRO-ELASTIC-PLASTIC MATERIALS

X. Lin[1] and J. Ballmann[1]

Abstract The paper presents an explicit second-order method to treat two-dimensional Riemann problems for high velocity impacts of hydro-elastic-plastic materials, which behave under strong stress conditions as compressible fluids in the volume change and obey the von Mises plastic yield law in the shape changes. The numerical scheme for the treatment of stress waves in such materials is divided into two steps consisting of the flux calculation with the proposed two-dimensional Riemann solver and the function updating. Finally, results of the successful application of the scheme to an impact problem are presented.

1. Introduction

Dynamic problems of hydro-elastic-plastic materials have been often considered by a non-conservative difference scheme in the HEMP code [1], e.g., for the modelling of shock waves and deformations in solids under very strong explosive loading conditions. The volume change of this material is non-linear, obeying the Mie-Grüneisen equation of state. When the pressure is several orders of magnitude higher than the yield stress, the shear components can be neglected, and the material behaves like a fluid. When the pressure decreases to the order of the yield stress, the deviatoric components of the stress tensor are again taken into account.

Godunov's difference method with a Riemann solver possesses a conservative property and promises therefore a more reasonable modelling of the shock waves in solids. One argument is that this method is successful in modelling the stress wave propagation in one-dimensional solids, see, e.g., [2,3]. For multi-dimensional problems, a widely used idea is the dimensional splitting technique. However, it was shown in [4] that a non-physical cutting trace will be produced if dimensional splitting technique is applied to a two-dimensional impact problem. Therefore, a two-dimensional Riemann problem should be considered in the numerical computation.

2. Basic equations for the shock wave

First we deal with the shock wave propagation under one-dimensional conditions in order to derive some Riemann invariants which will be applied to solve the two-dimensional Riemann problem later. Suppose $\mathcal{V} = 1/\rho$ and $\mathcal{E}$ to be the volume and internal energy per unit mass, respectively. The Mie-Grüneisen equation of state [5] can be written as

$$\mathcal{P} = \mathcal{P}_{\mathrm{H}}(\mathcal{V}) + \Gamma\rho_0[\mathcal{E} - \mathcal{E}_{\mathrm{H}}(\mathcal{V})], \tag{1}$$

where ρ_0 is the original mass density, and Γ the Mie-Grüneisen coefficient. The functions $\mathcal{P}_{\mathrm{H}}(\mathcal{V})$ and $\mathcal{E}_{\mathrm{H}}(\mathcal{V})$ are taken from a reference state, e.g., the Hugoniot state at a strong shock front. Consider a plane shock with speed $\mathcal{D}$. Applying the conservation laws of mass, momentum and energy as well as the shock speed relation

$$\mathcal{D} - u_0 = c_0 + \varsigma(u - u_0), \tag{2}$$

[1]Lehr- und Forschungsgebiet für Mechanik der RWTH Aachen, Templergraben 64, 52062 Aachen, Germany

where u is the particle velocity, $c_0 = \sqrt{K/\rho_0}$ is the volume wave speed, and ς is an experimental coefficient, the reference state can be derived as:

$$\mathcal{P}_{\rm H}(\mathcal{V}) = \rho_0 c_0^2 \frac{\eta}{(1-\varsigma\eta)^2}, \qquad \mathcal{E}_{\rm H}(\mathcal{V}) = \frac{c_0^2}{2}\frac{\eta^2}{(1-\varsigma\eta)^2}. \tag{3}$$

Here $\eta = 1 - \mathcal{V}/\mathcal{V}_0 = 1 - \rho_0/\rho$, and $\mathcal{P}_0 = 0$ is assumed for the initial state.

Consider an one-dimensional flow in fluid dynamics. Suppose as the initial conditions at $t = 0$ two different homogeneous states in $\bar{x} < 0$ and $\bar{x} \geq 0$. These states are denoted by subscripts 1 and 2, respectively. When $t > 0$, there will arise two families of centered simple waves, one family propagating to the left-hand side, and another to the right-hand side, see figure 1. Assuming $\bar{x}$ as a Lagrangean coordinate the contact interface will stay in $\bar{x} = 0$. Solving this Riemann problem we get the solution at both sides of the interface for $t > 0$.

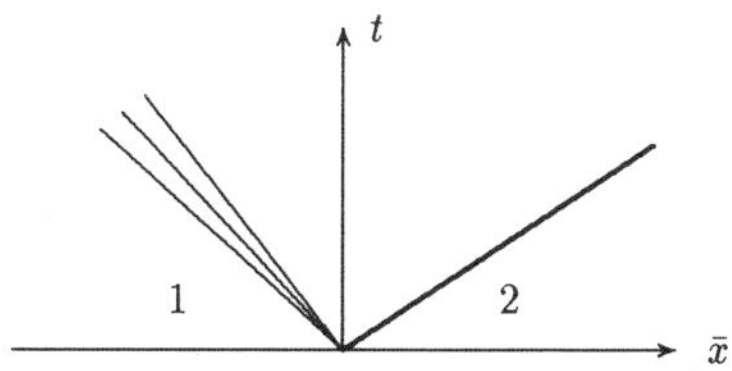

Figure 1: Riemann Problem in $(\bar{x}, t)$ plane.

If the wave is a continuous rarefaction wave, the differential equations of motion and continuity,

$$\rho\frac{\partial u}{\partial t} + \frac{\partial \mathcal{P}}{\partial \bar{x}} = 0, \qquad \frac{1}{\rho}\frac{\partial \rho}{\partial t} + \frac{\partial u}{\partial \bar{x}} = 0, \tag{4}$$

will be applied. To close the system, one needs a relation between $\mathcal{P}$ and ρ. Differentiating eq. (1) with respect to $\mathcal{V}$ and then eliminating $\mathrm{d}\mathcal{E}$ by the isentropic relation $\mathrm{d}\mathcal{E} = -\mathcal{P}\mathrm{d}\mathcal{V}$ one gets an ordinary differential equation

$$\frac{\mathrm{d}\mathcal{P}}{\mathrm{d}\mathcal{V}} + \Gamma\rho_0\mathcal{P} = \rho_0^2 c_0^2 \frac{(\Gamma-\varsigma)\eta - 1}{(1-\varsigma\eta)^3}. \tag{5}$$

Suppose $\mathcal{P}_1$ and $\mathcal{V}_1$ are the initial conditions. The solution of this differential equation will give the relation between $\mathcal{P}$ and $\mathcal{V}$

$$\mathcal{P} = \mathcal{P}_1 e^{\Gamma(\eta-\eta_1)} - \rho_0 c_0^2\, e^{\Gamma\eta} \int_{\eta_1}^{\eta} \frac{(\Gamma-\varsigma)\eta - 1}{(1-\varsigma\eta)^3} e^{-\Gamma\eta}\, \mathrm{d}\eta. \tag{6}$$

Denote the characteristic wave speed by $\mathcal{C}$, where $\mathcal{C} = \sqrt{\mathrm{d}\mathcal{P}/\mathrm{d}\rho}$, the characteristic relations of system (4) and (6) will give

$$u = u_1 - \int_{\mathcal{P}_1}^{\mathcal{P}} \frac{\mathrm{d}\mathcal{P}}{\rho\mathcal{C}}. \tag{7}$$

If the wave is a shock wave, the conservation law of mass, momentum and energy can be applied to obtain

$$u = u_1 - \sqrt{(\mathcal{P}-\mathcal{P}_1)(\mathcal{V}_1-\mathcal{V})}, \tag{8}$$

$$\mathcal{E} - \mathcal{E}_1 = \frac{1}{2}(\mathcal{P} + \mathcal{P}_1)(\mathcal{V}_1 - \mathcal{V}). \tag{9}$$

One should find a relation between $\mathcal{P}$ and $\mathcal{V}$ in eq. (8). To do this, substituting eq. (9) into eq. (1), we obtain

$$\mathcal{P} = \mathcal{P}_{\text{H}}(\mathcal{V}) + \Gamma\rho_0\Big[\mathcal{E}_1 + \frac{1}{2}(\mathcal{P} + \mathcal{P}_1)(\mathcal{V}_1 - \mathcal{V}) - \mathcal{E}_{\text{H}}(\mathcal{V})\Big]. \tag{10}$$

This is the so-called the Hugoniot equation for the shock wave. Making use of eq. (3), eq. (10) can be rewritten as

$$\Big[1 - \frac{\Gamma}{2}(\eta - \eta_1)\Big]\mathcal{P} = \rho c_0^2\frac{\eta(1 - \Gamma\eta/2)}{(1 - \varsigma\eta)^2} + \Gamma\rho_0\mathcal{E}_1 + \frac{\Gamma}{2}(\eta - \eta_1)\mathcal{P}_1, \tag{11}$$

which represents the relation between $\mathcal{P}$ and $\mathcal{V}$. Hence, the system of equations is closed to calculate the velocity u by eq. (8) for the known pressure $\mathcal{P}$.

Combining the two possibilities of a rarefaction wave and a shock wave, one can define the following function

$$\phi_k(\mathcal{P}) = \begin{cases} \displaystyle\int_{\mathcal{P}_k}^{\mathcal{P}} \frac{d\mathcal{P}}{\rho C}, & \mathcal{P} \le \mathcal{P}_k; \\ \sqrt{(\mathcal{P} - \mathcal{P}_k)(\mathcal{V}_k - \mathcal{V}(\mathcal{P}))}, & \mathcal{P} > \mathcal{P}_k, \end{cases} \tag{12}$$

with $k = 1$ corresponding to eqs. (7) and (8). Then the relation for the left ward running wave can be rewritten as

$$u = u_1 - \phi_1(\mathcal{P}). \tag{13}$$

The relation for the right ward running wave can be deduced by a similar procedure as

$$u = u_2 + \phi_2(\mathcal{P}). \tag{14}$$

Then the solution of the Riemann problem in the interface $\bar{x} = 0$ results in the pressure relation

$$\phi_2(\mathcal{P}) + \phi_1(\mathcal{P}) = u_1 - u_2. \tag{15}$$

The calculation of $\mathcal{P}$ needs generally an iterative procedure, in which $\mathcal{V}$ should be calculated from eq. (11) for the given $\mathcal{P}$. Eqs. (6) and (11) represent the functions $\mathcal{P} = \mathcal{P}(\mathcal{V})$ which are to be applied across the rarefaction wave and the shock wave, respectively. They can be inverted numerically.

A solid is different from a fluid in that the shear strength cannot be neglected. In this case there will exist two characteristic wave speeds, i.e., the longitudinal wave speed and the shear wave speed. Then the form of the Riemann solution eq. (15) must be changed. The function ϕ_k is to be replaced by

$$\Phi_k(\mathcal{P}) = \begin{cases} \displaystyle\int_{\mathcal{P}_k}^{\mathcal{P}} \frac{c_1}{\rho C^2} d\mathcal{P}, & \mathcal{P} \le \mathcal{P}_k; \\ \displaystyle\frac{D(\mathcal{P})}{c_0}\sqrt{(\mathcal{P} - \mathcal{P}_k)(\mathcal{V}_k - \mathcal{V}(\mathcal{P}))}, & \mathcal{P} > \mathcal{P}_k, \end{cases} \tag{16}$$

where c_0 is defined as in eq. (2); $D(\mathcal{P})$ is the Lagrangean shock wave speed modified by the shear effect and c_1 is the longitudinal wave speed,

$$D(\mathcal{P}) = \sqrt{\frac{1}{\rho_k^2}\frac{\mathcal{P} - \mathcal{P}_k}{\mathcal{V}_k - \mathcal{V}} + \frac{4}{3}c_2^2}, \qquad c_1 = \sqrt{\frac{K(\mathcal{P}) + 4\mu/3}{\rho}} = \sqrt{C^2 + \frac{4}{3}c_2^2}, \tag{17}$$

where μ is the shear modulus. The Riemann solution for the pressure $\mathcal{P}$ is then determined by

$$\Phi_2(\mathcal{P}) + \Phi_1(\mathcal{P}) = -(u_2 - u_1). \tag{18}$$

3. Flux calculation

Now we present the numerical scheme for hydro-elastic-plastic materials under plane strain conditions. Since the deformation of the solid will become large under a high velocity impact, at least locally, a fixed rectangular mesh is no longer suitable. Here an adaptive mesh based on material fixed coordinates (Lagrangean coordinates) is introduced for the numerical calculation.

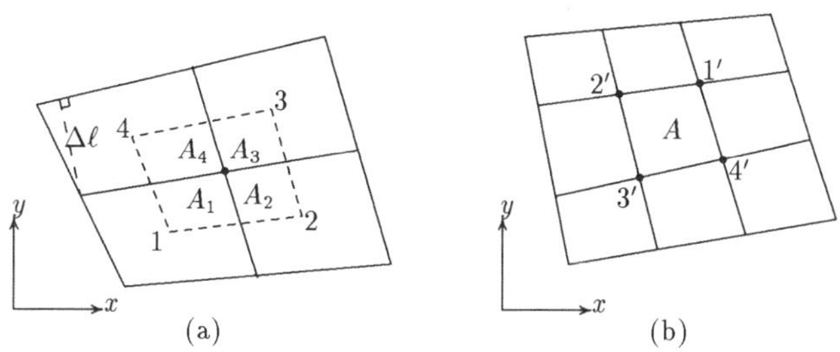

Figure 2: A sketch of an irregular mesh showing (a) the flux calculation at a grid point and (b) the function updating at a cell center.

This section discusses only the first step of the scheme, i.e., the flux calculation by an approximate two-dimensional Riemann solver. Let us consider a mesh with four cells as shown in figure 2(a). We assume the values in the cell centers at time level t^n to be given and want to calculate the flux at the central grid point at time level $t^{n+\frac{1}{2}}$. For such an irregular mesh, the space derivatives and their finite difference approximations should be represented by an integral form:

$$\begin{aligned} \frac{\partial \mathbf{f}}{\partial x} \approx \frac{1}{A}\oint_{\partial A} \mathbf{f} dy &= \frac{1}{A}\Big[(\mathbf{f}_3 - \mathbf{f}_1)(y_4 - y_2) - (\mathbf{f}_4 - \mathbf{f}_2)(y_3 - y_1)\Big], \\ \frac{\partial \mathbf{g}}{\partial y} \approx \frac{-1}{A}\oint_{\partial A} \mathbf{g} dx &= \frac{-1}{A}\Big[(\mathbf{g}_3 - \mathbf{g}_1)(x_4 - x_2) - (\mathbf{g}_4 - \mathbf{g}_2)(x_3 - x_1)\Big], \end{aligned} \tag{19}$$

where $A = A_1 + A_2 + A_3 + A_4$ is the quadrilateral area around the central grid point.

The flux components to be calculated in the grid point in question are the hydro-pressure $\mathcal{P}$, the three further stress components represented by the stress quantities $p - r$, $q - r$, τ, and two velocities u, v, where $p = \sigma_x$, $q = \sigma_y$, $r = \sigma_z$, $\tau = \tau_{xy}$. The differential equation for $\mathcal{P}$ is

$$\frac{1}{K(\mathcal{P})}\frac{\partial \mathcal{P}}{\partial t} = -\Big(\frac{\partial u}{\partial x} + \frac{\partial v}{\partial y}\Big). \tag{20}$$

The integral form for calculating the flux $\mathcal{P}$ is then

$$A_1\Phi_1(\mathcal{P}) + A_2\Phi_2(\mathcal{P}) + A_3\Phi_3(\mathcal{P}) + A_4\Phi_4(\mathcal{P}) = -\frac{\Delta \ell}{4}\oint_{\partial A} u dy - v dx, \tag{21}$$

where $\Delta\ell$ denotes the smallest distance in the whole mesh, which is introduced as the reference length for the numerical stability. Solving eq. (21), not only the flux $\mathcal{P}$ is obtained but also the maximum wave speed c_1 (or D for a shock) will be available for the choice of the time step Δt in order to satisfy the *CFL* stability condition.

The differential equations for the afore mentioned stress quantities are

$$\frac{1}{\mu(\kappa)}\frac{\partial}{\partial t}(p-r) = 2\frac{\partial u}{\partial x}, \qquad \frac{1}{\mu(\kappa)}\frac{\partial}{\partial t}(q-r) = 2\frac{\partial v}{\partial y}, \qquad \frac{1}{\mu(\kappa)}\frac{\partial \tau}{\partial t} = \frac{\partial v}{\partial x} + \frac{\partial u}{\partial y}, \tag{22}$$

where κ is the von Mises yield stress, $\mu(\kappa)$ is the shear modulus which is double valued depending on the elastic or plastic range. The iterative scheme for solving $p-r$, $q-r$ and τ can be developed in a similar manner, e.g., for the first term $p-r$,

$$\sum_{k=1}^{4} A_k \int_{p_k-r_k}^{p-r} \frac{c_1}{\mu(\kappa)} \mathrm{d}(p-r) \quad = \quad \frac{\Delta\ell}{2}\oint_{\partial A} u\mathrm{d}y. \tag{23}$$

Here c_1 (or in the shock case D) is set equal to the value obtained from the evalution of eq. (21).

Finally, the differential equations for u and v are

$$\rho\frac{\partial u}{\partial t} = \frac{\partial p}{\partial x} + \frac{\partial \tau}{\partial y}, \qquad\qquad \rho\frac{\partial v}{\partial t} = \frac{\partial \tau}{\partial x} + \frac{\partial q}{\partial y}. \tag{24}$$

Therefore, u and v can be calculated by (taking u as an example)

$$\sum_{k=1}^{4}(\rho c_1 A)_k(u-u_k) \quad = \quad \frac{\Delta\ell}{4}\oint_{\partial A} p\mathrm{d}y - \tau\mathrm{d}x, \tag{25}$$

The shock wave speed D or the largest longitudinal wave speed c_1 obtained in the solution of the Riemann problem for $\mathcal{P}$ will be used to determine the time step Δt for the function updating, $\Delta t = CFL \cdot \min(\Delta\ell/D)$, where min represents the minimum value of all cells in the domain. Since the velocity components u and v in a grid point are known after the flux calculation, the new position of the grid point at time level t^{n+1} can be obtained by

$$x^{n+1} = x^n + u\Delta t, \qquad\qquad y^{n+1} = y^n + v\Delta t. \tag{26}$$

If the Δt in eqs. (26) is replaced by $\Delta t/2$, one will get the grid point $(x^{n+\frac{1}{2}}, y^{n+\frac{1}{2}})$ at half time step which forms also a mesh. This mesh will be used for the updating of the unknown functions. It is most important to know that the *CFL* number cannot be set to the limit value of one, since the distance $\Delta\ell$ is taken from the mesh at time level t^n, while the mesh will move with the change of time.

4. The updating of the unknown functions

Suppose a cell to have four corner points $1'$ to $4'$ as shown in figure 2(b), and the flux components in the corner points are calculated already.

First, the cell's area at the time level t^{n+1} is calculated by the new coordinates of the corners:

$$A^{n+1} = \frac{1}{2}[(x_3^{n+1} - x_1^{n+1})(y_4^{n+1} - y_2^{n+1}) - (x_4^{n+1} - x_2^{n+1})(y_3^{n+1} - y_1^{n+1})]. \tag{27}$$

From the mass conservation law, the mass density of the cell can be obtained by

$$\rho^{n+1} = \rho^n A^n / A^{n+1}. \tag{28}$$

Second, the conservation of momentum,

$$\frac{\mathrm{d}}{\mathrm{d}t}\iint\limits_{A(t)} \rho u \, \mathrm{d}A = \oint\limits_{\partial A} p \, \mathrm{d}y - \tau \, \mathrm{d}x, \qquad \frac{\mathrm{d}}{\mathrm{d}t}\iint\limits_{A(t)} \rho v \, \mathrm{d}A = \oint\limits_{\partial A} \tau \, \mathrm{d}y - q \, \mathrm{d}x, \tag{29}$$

will be applied to calculate the velocity components u and v. The numerical schemes are written by (taking the component u as example)

$$\begin{aligned} (\rho A)^{n+1}u^{n+1} &= (\rho A u)^n + \frac{\Delta t}{2}[(p'_3 - p'_1)(y'_4 - y'_2) - (p'_4 - p'_2)(y'_3 - y'_1)] \\ &- \frac{\Delta t}{2}[(\tau'_3 - \tau'_1)(x'_4 - x'_2) - (\tau'_4 - \tau'_2)(x'_3 - x'_1)], \end{aligned} \tag{30}$$

where the quantities with a prime, like p'_3, y'_4 etc, are values in the corner points at the time level $t^{n+\frac{1}{2}}$.

Third, the deviatoric stress will be calculated. The strain components $\alpha = \varepsilon_x$, $\beta = \varepsilon_y$, $\gamma = 2\varepsilon_{xy}$ are obtained by

$$\frac{\mathrm{d}}{\mathrm{d}t}\iint\limits_{A(t)} \alpha \, \mathrm{d}A = \oint\limits_{\partial A} u \, \mathrm{d}y, \quad \frac{\mathrm{d}}{\mathrm{d}t}\iint\limits_{A(t)} \beta \, \mathrm{d}A = \oint\limits_{\partial A} -v \, \mathrm{d}x, \quad \frac{\mathrm{d}}{\mathrm{d}t}\iint\limits_{A(t)} \gamma \, \mathrm{d}A = \oint\limits_{\partial A} v \, \mathrm{d}y - u \, \mathrm{d}x. \tag{31}$$

The schemes are similar to eqs. (30) and are therefore not repeated here. With the strain increments the stress quantities $p - r$, $q - r$ and τ are then integrated in the stress space by

$$\frac{1}{\mu(\kappa)}\mathrm{d}(p - r) = 2\mathrm{d}\alpha, \qquad \frac{1}{\mu(\kappa)}\mathrm{d}(q - r) = 2\mathrm{d}\beta, \qquad \frac{1}{\mu(\kappa)}\mathrm{d}\tau = \mathrm{d}\gamma. \tag{32}$$

During the integration the stress path should be specified according to the plastic yield condition.

The energy conservation should be applied in order to calculate the hydro-pressure $\mathcal{P}$. Let E be the internal energy per unit volume stored in the material. Since the elastic-plastic effect is taken into account, E will include a part of strain energy due to the shape change. The energy conservation law is written in an integral form for the computation,

$$\frac{\mathrm{d}}{\mathrm{d}t}\iint\limits_{A(t)} \left[\frac{\rho}{2}(u^2 + v^2) + E\right] \mathrm{d}A = \oint\limits_{\partial A} (pu + \tau v)\, \mathrm{d}y - (\tau u + qv)\, \mathrm{d}x. \tag{33}$$

The energy increment ΔE can be obtained by eq. (33). Let

$$\mathrm{d}e = (p + \mathcal{P})\mathrm{d}\alpha + (q + \mathcal{P})\mathrm{d}\beta + \tau \mathrm{d}\gamma, \qquad \mathrm{d}\mathcal{E} = -\mathcal{P}\mathrm{d}\mathcal{V}, \tag{34}$$

where $\mathrm{d}e$ is the energy part due to the shape change and $\mathrm{d}\mathcal{E}$ is that due to the volume change. Then

$$\mathrm{d}E = \mathrm{d}e + \rho \mathrm{d}\mathcal{E}. \tag{35}$$

Since $\mathrm{d}e$ can be obtained together with the calculations of $p - r$, $q - r$ and τ, one will separate the increment $\mathrm{d}\mathcal{E}$ from the total internal energy increment $\mathrm{d}E$. Therefore, the hydro-pressure $\mathcal{P}$ is then calculated by the Mie-Grüneisen equation (1). So far the updating of functions for a cell is finished.

5. Example – Taylor's pressure bar

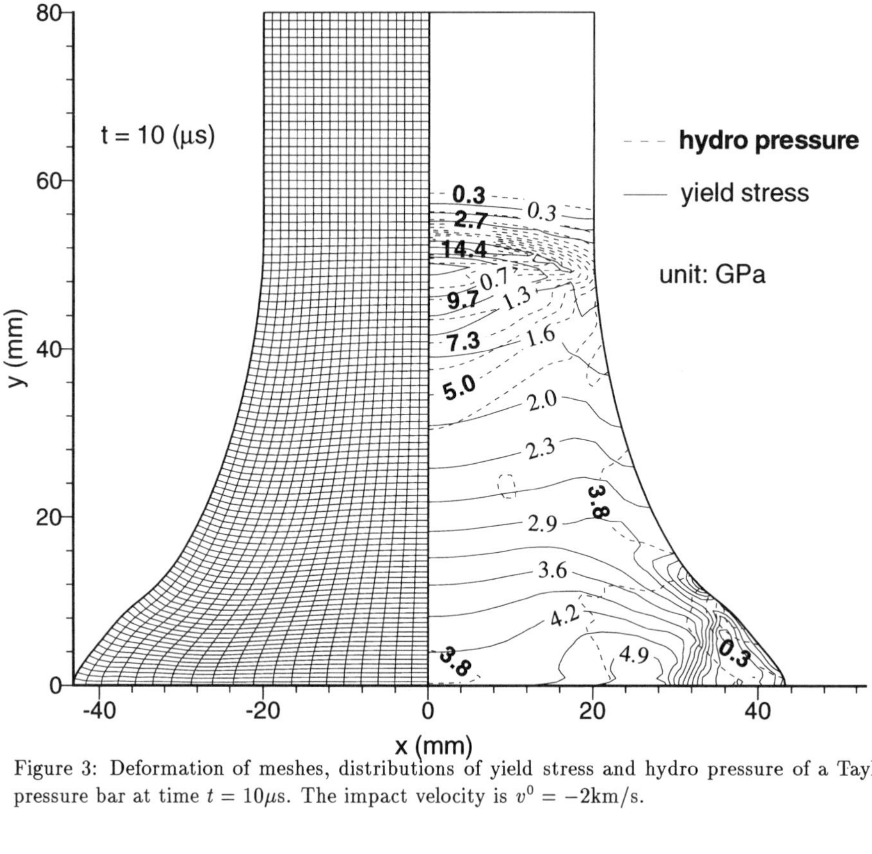

Figure 3: Deformation of meshes, distributions of yield stress and hydro pressure of a Taylor pressure bar at time $t = 10\mu s$. The impact velocity is $v^0 = -2 km/s$.

As a numerical example, the Taylor pressure bar is considered. The bar is 100 mm in length and 40 mm in width. The material is assumed to be aluminum with the parameters $\rho = 2.785$ gm/cm^3, $c_1 = 6.39$ km/s, $c_2 = 3.15$ km/s, $\kappa^0 = 1$ GPa, $\mu(\kappa)/\mu = 0.1$, $\Gamma = 2$, $\varsigma = 1.338$.

The bar flies with the initial velocity $v^0 = -2$ km/s in the negative y-direction and hits a rigid wall at $y = 0$. The numerical scheme is applied to model the deformation of the bar. Figure 3 shows the mesh in the time $t = 10\mu$s. The contour lines of the yield stress and the hydro pressure are plotted in the figure. It can be recognized that the yield stress has its highest values close to the rigid wall region where the material undergoes a large deformation. However, the hydro pressure has its highest values at the wave front running into the positive $y-$direction.

References

1. M.L. Wilkins, Calculation of elastic-plastic flow. In: B. Alder, S. Fernbach and M. Rotenberg (eds.): *Methods in Computational Physics*, Volume 3, Academic Press, New York and London (1964).

2. K. Nagayama, Solution of the high-pressure Riemann problem for solids including rigidity effects. *Journal of the Physical Society of Japan* **58**, 1631–1638 (1989).

3. X. Lin and J. Ballmann, A Riemann solver and a second-order Godunov's method for elastic-plastic wave propagation in solids, *International Journal of Impact Engineering* **13**, 463–478 (1993).

4. Lin X and Ballmann J, Numerical method for elastic waves in cracked solids, part 1: anti-plane shear problem, *Arch Appl Mech* **63**, 261–282 (1993).

5. M.A. Meyers and L.E. Murr (eds.), *Shock Waves and High-Strain-Rate Phenomena in Metals*, Appendix, Plenum Press, New York and London (1981).

On the Convergence of Vortex Methods for Weak Solutions to the 2-D Euler Equations

Jian-Guo Liu[1] Zhouping Xin[2]

ABSTRACT. In this note, we present a convergence theorem for the vortex blob methods to the classical weak solutions for the two-dimensional incompressible Euler equations with initial data satisfying the conditions that the vorticity is a finite Radon measure of distinguished sign and the kinetic energy is locally bounded. This includes the important example of vortex sheets. The proof assumes that the computational grid size h does not exceed the smoothing blob size ε, i.e., $h/\varepsilon \leq C$.

1. INTRODUCTION

Computational vortex methods approximate the vorticity by a finite sum of "blobs", or cores of prescribed shape, which are advected according to the corresponding desingularized velocity field. One application of these methods is in the numerical simulation of singular flows such as in the vortex sheet problems. In particular, Krasny applied such a method in his calculations of the evolution of vortex sheets even past the time when the sheet develops a singularity [12]. Even though the convergence of the vortex method is now well-documented for smooth solutions to the Euler equations (see [2,10,11] and the references therein) there is yet much to be done for the analysis of vortex methods for non-smooth flows. This is particularly so in the case of the vortex sheet problem, in which the vorticity is a measure concentrated on a curve or surface which is initially smooth. This defines a classical Hadamard ill-posed initial value problem. At later time, a singularity in the sheet may develop and the nature of solutions past the singularity formation is unknown [6,15]. It is thus important to analyze the structure of the approximate solutions generated by vortex methods.

The aim of this paper is to prove the convergence to a classical weak solution of the vortex methods for the 2-D incompressible Euler equations with a class of "rough" initial data, including the vortex sheet initial data in the limit as the grid size and regularization parameter tend to zero, provided that the initial vorticity has a distinguished sign. Here, vortex sheet initial data is defined to be the one for which the vorticity is a finite Radon measure and the corresponding velocity field has locally finite kinetic energy [9]. This is strongly motivated by the recent result of Delort on the existence of weak solutions to the 2-D incompressible Euler

[1]Department of Mathematics, Temple University, Philadelphia, PA 19122 jliu@math.temple.edu

[2]Courant Institute of Mathematical Sciences, New York, NY 10012 xinz@cims.nyu.edu

Research of the first author was supported in part by NSF grant DMS 9304580. Research of the second author was supported in part by a Sloan Foundation Fellowship, NSF Grants DMS 9304580 and DMS-93-03887, and DOE Grant DE-FG02-88ER-25053.

equation with vortex sheet initial data of positive vorticity [8], the work of DiPerna and Majda on the structure of weak solutions to the 2-D Euler equations [9], and the numerical demonstration of convergence by Krasny [12].

In the vorticity stream formulation, the incompressible 2-D Euler equations can be expressed in the form (see [9]):

$$\partial_t \omega + \nabla\cdot(\mathbf{u}\,\omega) = 0\,, \tag{1.1}$$

with initial data

$$\omega(\mathbf{x},0) = \omega_0(\mathbf{x})\,, \tag{1.2}$$

where the vorticity $\omega = \partial_{x_1} u_2 - \partial_{x_2} u_1$ is the curl of the velocity field $\mathbf{u}(\mathbf{x},t)$ and

$$\mathbf{u}(\mathbf{x},t) = (K*\omega)(\mathbf{x},t) \tag{1.3}$$

with the kernel K given by

$$K(\mathbf{x}) = \frac{1}{2\pi|\mathbf{x}|^2}(-x_2, x_1) \equiv \frac{\mathbf{x}^\perp}{2\pi|\mathbf{x}|^2}\,. \tag{1.4}$$

For simplicity in presentation, we will assume in the following that ω_0 has compact support. The vortex method, introduced by Chorin [7], consists of approximating the vorticity field at each time by a sum of approximate delta functions centered at particle positions $\mathbf{x}_j(t)$, which are advected according to the regularized velocity field. To describe this more precisely, we choose a vortex blob function ϕ with the following properties:

$$\phi(\mathbf{x}) = \phi(|\mathbf{x}|) > 0, \quad \int \phi(\mathbf{x})\,d\mathbf{x} = 1, \quad \int_{|\mathbf{x}|\le 1} \phi(\mathbf{x})\,d\mathbf{x} \ge \tfrac{1}{2}\,. \tag{1.5}$$

We shall assume also that ϕ is C^2 and decays at infinity at least as fast as $|\mathbf{x}|^{-3}$. Let ε be the smoothing parameter, and set $\phi_\varepsilon(\mathbf{x}) = \varepsilon^{-2}\phi(\mathbf{x}/\varepsilon)$. We cover the support of the initial vorticity by non-overlapping squares, R_j, with side length h and centered at $\alpha_j = jh$ with $j \in Z \times Z$. Denote by ξ_j the initial circulation strength in R_j, i.e.,

$$\xi_j = \int_{R_j} \omega_0(\mathbf{x})\,d\mathbf{x}\,. \tag{1.6}$$

The vortex method is

$$\frac{d}{dt}\mathbf{x}_j(t) = \sum_\ell \xi_\ell\, K_\varepsilon(\mathbf{x}_j(t) - \mathbf{x}_\ell(t))\,, \quad \mathbf{x}_j(0) = \alpha_j\,, \tag{1.7}$$

$$\mathbf{u}^\varepsilon(\mathbf{x},t) \equiv \sum_j \xi_j\, K_\varepsilon(\mathbf{x} - \mathbf{x}_j(t))\,, \tag{1.8}$$

$$\omega_\varepsilon(\mathbf{x},t) \equiv \sum_j \xi_j\, \delta(\mathbf{x} - \mathbf{x}_j(t))\,. \tag{1.9}$$

Here $\delta(\cdot)$ is the Dirac delta function, and $K_\varepsilon = K * \phi_\varepsilon$. It is easy to show that the finite system of ordinary differential equations (1.7) has a unique solution globally under the assumptions (1.5) on the blob function.

Now our main convergence theorem can be stated as follows.

Theorem 1.1. *Suppose that the* $\omega_0 \in \mathcal{M}(\boldsymbol{R}^2) \cap H^{-1}_{loc}(\boldsymbol{R}^2)$ *and* ω_0 *has a distinguished sign. Let* $\mathbf{u}^\varepsilon$ *be the approximate solution generated by the vortex method as described above. Then there is a subsequence of* $\mathbf{u}^\varepsilon$ *which converges strongly (weakly) to a function* $\mathbf{u} \in L^\infty([0,T), L^2_{loc}(\boldsymbol{R}^2))$ *in* $L^p_{loc}(\boldsymbol{R}^2 \times \boldsymbol{R}_+)$ *for* $1 \le p < 2$ $(L^\infty([0,T), L^2_{loc}(\boldsymbol{R}^2)))$, *provided* $h \le C\varepsilon$ *for any constant* $C > 0$. *Furthermore, the limit,* $\mathbf{u}(\mathbf{x},t)$, *is a classical weak solution to the 2-D Euler equations with the given initial data* $\mathbf{u}_0(\mathbf{x})$, *i.e.,*

(*i*) *for all test function* $\theta(\mathbf{x},t) \in C_0^\infty(\boldsymbol{R}^2 \times (0,T))$,

$$\iint \Big(\nabla^\perp \theta_t \cdot \mathbf{u} + (\nabla^\perp \otimes \nabla\theta) : (\mathbf{u} \otimes \mathbf{u}) \Big)\, d\mathbf{x}\, dt = 0\,; \tag{1.10}$$

(*ii*) *div* $\mathbf{u} = 0$ *in the sense of distribution;*

(*iii*) $\mathbf{u}(\mathbf{x},t) \in Lip\ ([0,T), H^{-m}_{loc}(\boldsymbol{R}^2))$ *for some* $m > 0$ *and* $\mathbf{u}(\mathbf{x},0) = \mathbf{u}_0(\mathbf{x})$. *where* $\nabla^\perp = (-\partial_{x_2}, \partial_{x_1})$, *and* $A : B$ *denote the trace of the matrix product.*

Remarks (i) The vortex methods associated with the class of blob functions satisfying (1.5) include, in particular, the one used by Krasny

$$K_\varepsilon(\mathbf{x}) = \frac{1}{2\pi} \frac{\mathbf{x}^\perp}{|\mathbf{x}|^2 + \varepsilon^2}\,,$$

with corresponding blob function $\phi(\mathbf{x}) = \frac{1}{\pi}\left(\frac{1}{1+|\mathbf{x}|^2}\right)^2$.

(ii) The main difficulty in analyzing the vortex method for singular flows such as vortex sheet is due to the weak consistency error. This in turn makes the stability analysis harder, since a simple energy estimate is not available for the velocity field $\mathbf{u}^\varepsilon$, in contrast to the continuous case [8,9,14]. The difficulty is overcome by employing *a priori* optimal estimates on the maximal function of the discrete vorticity (see section 2). We note that similar estimates on the vorticity maximal function have been used by Majda [14] in his recent simplified proof of Delort's theorem. It shall be also emphasized that in Theorem 1.1, there is no initialization procedure required (see (1.6)), in contrast to the previous studies [1,9].

(iii) It is remarkable that the only requirement on the ratio of the grid size h to the blob size ε is that h/ε is bounded. This is mainly due to the positivity of the initial vorticity.

(iv) In both theoretical and numerical analysis and engineering applications [3,6,12,15], the evolution of the vortex sheets is studied through the well-known Birkhoff-Rott integro-differential equation whose validity after the formation of singularity is not well understood. Our convergence to a classical weak solution for Krasny's desingularization procedure to the Birkhoff-Rott equation gives an appropriate interpretation of this integro-differential equation even past the formation of a singularity.

There is a large amount of literature on the convergence of vortex methods (See [1,2,4,5,9,10,11,13] and the references therein). The formulations of the problem and the analysis in this paper are closely related to those of DiPerna and Majda [9] in which they proved the convergence of the vortex method to a measure-valued solution of the Euler equations under the assumption that the initial energy is finite with a special initialization. For vortex sheets initial data, Beale obtained a convergence of vortex method with an initialization in the context of measure-valued solutions provided that the grid size h is exponentially smaller than the blob size ε [1]. Assuming the sheet is analytic and close to horizontal, Caflisch and Lowengrub [5] show that the vortex blob approximations to the sheet converge strongly for a time interval before the singularity formation.

The rest of the paper is organized as follows. In section 2, first we define the maximal function for the discrete vorticity field $\omega_\varepsilon(\mathbf{x},t)$ and derive *a priori* estimates on it. The optimal estimates play an essential role both in stability and weak consistency analysis. Then stability of the scheme is derived independent of consistency error. Finally we sketch the consistency analysis in section 3. With both stability and consistency at hand, the convergence can be obtained easily by employing Delort's idea [8].

2. Stability Analysis

Following DiPerna and Majda [9], one can define the maximal vorticity function for a discrete vorticity field $\omega_\varepsilon(x,t)$ as

$$M_r(\omega_\varepsilon) = \sup_{\substack{\mathbf{x}\in \mathbf{R}^2\\ 0<t\le T}} \sum_{|x_j - x|\le r} |\xi_j|\,, \quad \text{for } 0 < r \le \tfrac{1}{2}\,. \tag{2.1}$$

The crucial *a priori* estimate in our analysis is that

$$M_r(\omega_\varepsilon) \le C\left(\ln \tfrac{1}{r+\varepsilon}\right)^{-1/2} \tag{2.2}$$

with a positive constant C independent of ε. To derive (2.2), we set

$$G_\varepsilon(\mathbf{x}) = \phi_\varepsilon * \tfrac{1}{2\pi}\ln|\mathbf{x}|\,, \quad \psi^\varepsilon(\mathbf{x},t) = G_\varepsilon * \omega_\varepsilon\,. \tag{2.3}$$

It is evident that

$$K_\varepsilon(\mathbf{x}) = \nabla^\perp G_\varepsilon \quad \text{and} \quad \mathbf{u}^\varepsilon = \nabla^\perp \psi^\varepsilon\,. \tag{2.4}$$

We also define

$$I(t) = -\int_{\mathbf{R}^2} \psi^\varepsilon(\mathbf{x},t)\,\omega_\varepsilon(\mathbf{x},t)\,d\mathbf{x} \equiv \sum_{j,\ell} \xi_j\,\xi_\ell\,G_\varepsilon(\mathbf{x}_j - \mathbf{x}_\ell)\,. \tag{2.5}$$

Then it is easy to check that $I(t) \equiv I(0)$ for all $t > 0$. Furthermore, if $h \le C\varepsilon$ for any given positive constant C, then $I(0)$ is bounded independent of ε and h.

Consequently, we have

$$-\sum_{|\mathbf{x}_j-\mathbf{x}_\ell|\le 1/2} \xi_j\,\xi_\ell\, G_\varepsilon(\mathbf{x}_j-\mathbf{x}_\ell) = I(t) + \sum_{|\mathbf{x}_j-\mathbf{x}_\ell|>1/2} \xi_j\,\xi_\ell\, G_\varepsilon(\mathbf{x}_j-\mathbf{x}_\ell) \tag{2.6}$$
$$\le C + \sum_{j,\ell} \xi_j\,\xi_\ell\,(1+|\mathbf{x}_j|^2+|\mathbf{x}_\ell|^2) \le C_6 \,.$$

It follows that

$$\sum_{|\mathbf{x}_j-\mathbf{x}_\ell|\le 1/2} \xi_j\,\xi_\ell \ln \frac{1}{|\mathbf{x}_j-\mathbf{x}_\ell|+\varepsilon} \le C\,, \tag{2.7}$$

which implies the desired decay estimate (2.2) on the vorticity maximal function.

To derive the stability bound, we need the following potential estimates.

Lemma 2.1. *Assume $\theta(\mathbf{x}) \in C_0^\infty(\boldsymbol{R}^2)$. Let the matrix-valued function $\mathcal{K}^\varepsilon(\mathbf{y},\mathbf{z};\theta)$ be defined as*

$$\mathcal{K}^\varepsilon(\mathbf{y},\mathbf{z};\theta) = \int \theta(\mathbf{x})\, K_\varepsilon(\mathbf{x}-\mathbf{y}) \otimes K_\varepsilon(\mathbf{x}-\mathbf{z})\, d\mathbf{x}\,. \tag{2.8}$$

Then the following estimates holds for $|\mathbf{y}-\mathbf{z}| \le 1/2$,

$$|\mathcal{K}_{11}^\varepsilon(\mathbf{y},\mathbf{z};\theta)| + |\mathcal{K}_{22}^\varepsilon(\mathbf{y},\mathbf{z};\theta)| \le C \ln \frac{1}{|\mathbf{y}-\mathbf{z}|+\varepsilon}\,, \tag{2.9}$$

$$|\mathcal{K}_{12}^\varepsilon(\mathbf{y},\mathbf{z};\theta)| + |\mathcal{K}_{21}^\varepsilon(\mathbf{y},\mathbf{z};\theta)| \le C\,, \tag{2.10}$$

$$|\mathcal{K}_{11}^\varepsilon(\mathbf{y},\mathbf{z};\theta) - \mathcal{K}_{22}^\varepsilon(\mathbf{y},\mathbf{z};\theta)| \le C \tag{2.11}$$

where C depends only on θ.

The proof of lemma 2.1 can be find in [13]. It should be noted that the boundness on the inequality (2.10) in the case $\varepsilon = 0$ is one of the key observation of Delort [8].

With this preparation, we now turn to the main stability analysis. Choosing a nonnegative function $\chi \in C_0^\infty(\boldsymbol{R}^2)$ which is equal to one on the ball $\{\mathbf{x} : |\mathbf{x}| \le R\}$, we may bound the local kinetic energy as

$$\int_{|\mathbf{x}|\le R} |\mathbf{u}^\varepsilon(\mathbf{x},t)|^2\, d\mathbf{x} \le \int \chi(\mathbf{x})\, |\mathbf{u}^\varepsilon(\mathbf{x},t)|^2\, d\mathbf{x}$$
$$\le \Bigg(\sum_{|\mathbf{x}_j-\mathbf{x}_\ell|\le 1/2} + \sum_{|\mathbf{x}_j-\mathbf{x}_\ell|>1/2} \Bigg) \xi_j\,\xi_\ell \int \chi(\mathbf{x})\, K_\varepsilon(\mathbf{x}-\mathbf{x}_j)\cdot K_\varepsilon(\mathbf{x}-\mathbf{x}_\ell)\, d\mathbf{x}$$
$$\equiv I_1 + I_2\,.$$

For I_2, since there is no singularity in the kernel for $|\mathbf{x}_j - \mathbf{x}_\ell| \ge 1/2$, it is evident that $I_2 \le C$. As for I_1, we may use Lemma 2.2 to obtain

$$I_1 \le C \sum_{|\mathbf{x}_j-\mathbf{x}_\ell|\le 1/2} \xi_j\,\xi_\ell \ln \frac{1}{|\mathbf{x}_j-\mathbf{x}_\ell|+\varepsilon} \le C\,,$$

where the last step follows from (2.7). This proves the following proposition.

Proposition 2.2. *Let* $\mathbf{u}^\varepsilon(\mathbf{x},t)$ *be the approximate velocity field given in* (1.8). *Then for any given* $R > 0$, *there exists a positive constant* $C = C(R)$ *independent of* ε *and* h *such that*

$$\int_{|\mathbf{x}|\leq R} |\mathbf{u}^\varepsilon(\mathbf{x},t)|^2\, d\mathbf{x} \leq C\,. \tag{2.12}$$

3. Weak Consistency

We now check the weak consistency of the approximation generated by the vortex method with the 2-D Euler equations as blob size ε tends to zero, we note first that by definition, $(\mathbf{u}^\varepsilon, \omega_\varepsilon)$ satisfies the following equation

$$\partial_t\, \omega_\varepsilon + \nabla\cdot(\mathbf{u}^\varepsilon \omega_\varepsilon) = 0 \tag{3.1}$$

in the sense of distribution. Defining

$$\omega^\varepsilon(\mathbf{x},t) \equiv \sum_j \xi_j\, \phi_\varepsilon(\mathbf{x}-\mathbf{x}_j(t))\,, \quad \mathbf{u}_\varepsilon(\mathbf{x},t) \equiv \sum_j \xi_j\, K(\mathbf{x}-\mathbf{x}_j(t))\,, \tag{3.2}$$

we can rewrite (3.1) as

$$\iint (\nabla\theta_t\cdot\mathbf{u}^\varepsilon + \nabla^\perp \otimes \nabla\theta : \mathbf{u}^\varepsilon \otimes \mathbf{u}^\varepsilon)\; d\mathbf{x}\, dt = \iint \nabla\theta_t\cdot(\mathbf{u}^\varepsilon - \mathbf{u}_\varepsilon)\, d\mathbf{x}\, dt + E_\theta \tag{3.3}$$

for any given test function $\theta \in C_0^\infty(\boldsymbol{R}^2 \times \boldsymbol{R}_+)$, where

$$E_\theta = \iint \nabla\theta\cdot\mathbf{u}^\varepsilon(\omega^\varepsilon - \omega_\varepsilon)\, d\mathbf{x}\, dt\,. \tag{3.4}$$

The first term on the right hand side of (3.3) can be estimated directly as

$$\int_{|\mathbf{x}|\leq R} |\mathbf{u}^\varepsilon - \mathbf{u}_\varepsilon|\, d\mathbf{x} \leq C(R)\,\varepsilon \ln \tfrac{1}{\varepsilon} \tag{3.5}$$

by noting that

$$|K_\varepsilon(\mathbf{x}) - K(\mathbf{x})| \leq C\frac{\varepsilon}{|\mathbf{x}|+\varepsilon}\,\frac{1}{|\mathbf{x}|}\,.$$

The main consistency error estimate is to show

$$E_\theta \to 0\,, \quad \text{as } \varepsilon \to 0^+\,. \tag{3.6}$$

This can be obtained by utilizing the symmetry properties of the kernel and the vortex blob functions, and by using decay estimate (2.2) for the vorticity maximal function. Detail can be found in [13]. We can summarize the consistency estimate as the following proposition.

Proposition 3.1. *The vortex blob approximation* (1.7)-(1.11) *is weakly consistent with the 2-D Euler equations in the sense that for any* $\theta \in C_0^2(\boldsymbol{R}^2 \times \boldsymbol{R}_+)$,

$$\iint (\nabla\theta_t\cdot\mathbf{u}^\varepsilon + \nabla^\perp \otimes \nabla\theta : \mathbf{u}^\varepsilon \otimes \mathbf{u}^\varepsilon)\; d\mathbf{x}\, dt \to 0\,, \quad \textit{as } \varepsilon \to 0^+\,. \tag{3.7}$$

The convergence theorem now follows easily as a consequence of Proposition 2.2, Proposition 3.1, Lemma 2.1, and a priori estimate (2.2).

Acknowledgments: This work was done when the authors are visiting IAS. The hospitality of the faculty and staff at IAS is greatly appreciated. The authors wish to thank Professors Weinan E and R. Krasny for many useful discussions and their suggestions. The authors would like also to thank Professors J. Goodman and P. Lax for their interest in this work.

References

1. J.T. Beale, *The approximation of weak solutions to the 2-D Euler equations by vortex elements*, Multidimensional Hyperbolic Problems and Computations, Edited by J. Glimm and A. Majda, IMA (1991).
2. J.T. Beale and A. Majda, *Vortex methods. II: higher order accuracy in two and three dimensions*, Math. Comp. **39** (1982), 29–52.
3. G. Birkhoff, *Helmholtz and Taylor instability*, Proc. Symp. Appl. Math. **13** (1962), 55–76.
4. Y. Brenier and G.H. Cottet, *Convergence of particle methods with random rezoning for the 2-D Euler and Navier-Stokes equations*, preprint.
5. R. Caflisch and J. Lowengrub, *Convergence of the vortex method for vortex sheets*, SIAM J. Numer Anal. **26** (1989), 1060–1080.
6. R. Caflisch and O. Orellana, *Singularity formulation and ill-posedness of vortex sheets*, SIAM J. Math. Anal. **20** (1988), 293–307.
7. A.J. Chorin, *Numerical study of slightly viscous flow*, J. Fluid Mech. **57** (1973), 785–796.
8. J.-M. Delort, *Existence de nappes de tourbillon en dimension deux*, J. Amer. Math. Soc. **4** (1991), 553-586.
9. R. DiPerna and A. Majda, *Concentrations in regularizations for 2-D incompressible flow*, Comm. Pure Appl. Math. **40** (1987), 301–345.
10. J. Goodman, T.Y. Hou and J. Lowengrub, *Convergence of the point vortex method for 2-D Euler equations*, Comm. Pure Appl. Math. **43** (1990), 415–430.
11. O.H. Hald, *Convergence of vortex methods for Euler's equations, II*, SIAM J. Numer. Anal. **16** (1979), 726–755.
12. R. Krasny, *Computing vortex sheet motion*, Proc. International Congress of Math, Kyoto, (1990), 1573–1583.
13. J.-G. Liu and Z. Xin, *Convergence of vortex methods for weak solutions to the 2-D Euler equations with vortex sheets data*, Submitted to Comm. Pure Appl. Math., (1994).
14. A. Majda, *Remarks on weak solutions for vortex sheets with a distinguished sign*, Indiana Univ. Math. J. **42** (1993), 921–939.
15. D.W. Moore, *The spontaneous appearance of a singularity in the shape of an evolving vortex sheet*, Proc. Roy. Soc. Lond. A **365** (1979), 105–119.

Recent topics on the compressible Euler equation

Tetu Makino[1], Kiyoshi Mizohata[2], Seiji Ukai[3]

1. Introduction

In this paper we shall describe recent topics on the compressible Euler equation and the relativistic Euler equation for the multidimensional case. The compressible Euler equation for an isentropic gas in $\mathbf{R}^n$ is given by

$$(1) \qquad \begin{cases} \rho_t + \sum_{j=1}^{n} \frac{\partial}{\partial x_j}(\rho u_j) = 0 \, , \\ (\rho u_i)_t + \sum_{j=1}^{n} \frac{\partial}{\partial x_j}(\rho u_i u_j + \delta_{ij} p) = \rho f_i \, , \quad (i = 1, 2, \cdots, n) \end{cases}$$

with the equation of state

$$(2) \qquad p = a^2 \rho^\gamma \, ,$$

where the density ρ, velocity $\mathbf{u} ={}^t (u_1, u_2, \cdots, u_n)$ and pressure p are functions of $x \in \mathbf{R}^n$ and $t \geq 0$, while f is a given external force and $a > 0$ and $\gamma \geq 1$ are given constants.

This is a typical example of conservation laws and has attracted many mathematicians. In particular, for the one dimensional case (n=1), the Cauchy problem for (1.1) with (1.2) has been studied extensively.

On the other hand, little is known for the case $n \geq 2$. No global solutions are known to exist, but only local classical solutions, in the full generality.

In 1992, we have presented global weak solutions first for the case $n \geq 2$. We have done this, however, only for the case of spherically symmetry with $\gamma = 1$ in the domain outside a unit ball.

However, we could not content ourselves since our class of initial data does not contain stationary solutions. In 1993 we have shown that if we use a non-uniform mesh chosen carefully, Glimm's method still works and gives global weak solutions for the initial data in the class containing the stationary solutions. We shall explain these results and related topics in section 2. It was observed in [13] that these stationary solutions are stable if initial data are sufficiently close to these solutions.

For the case $\gamma > 1$, our method can't be applied. T. Makino and S. Takeno have proved, by using compensated compactness method, the existence of temporally local weak solution for (1.1) with spherical symmetry in [2]. Recently we have known that Glimm and Chen have succeeded to prove the existence of global weak solution for (1.1) with spherical symmetry. For viscous barotropic gas M. Okada and T. Makino [10] have proved the existence of global solutions.

[1]Department of Liberal Arts, Osaka Sangyo University
[2]Department of Mathematical and Computing Sciences, Tokyo Institute of Technology
[3]Department of Mathematical and Computing Sciences, Tokyo Institute of Technology

Next, let us consider the relativistic Euler equation describing a motion of perfect fluid in special relativity. The relativistic Euler equation for an isentropic gas in the Minkowski space-time is given by

$$
(3)\qquad
\begin{aligned}
&\frac{\partial}{\partial t}\left(\frac{\rho c^2+p}{c^2}\frac{1}{1-\frac{u^2}{c^2}}-\frac{p}{c^2}\right)+\sum_{i=1}^{3}\frac{\partial}{\partial x_i}\left(\frac{\rho c^2+p}{c^2}\frac{u_i}{1-\frac{u^2}{c^2}}\right)=0\,,\\
&\frac{\partial}{\partial t}\left(\frac{\rho c^2+p}{c^2}\frac{v_i}{1-\frac{u^2}{c^2}}\right)+\sum_{j=1}^{3}\frac{\partial}{\partial x_j}\left(\frac{\rho c^2+p}{c^2}\frac{u_iu_j}{1-\frac{u^2}{c^2}}+\delta_{ij}p\right)=0,\ (i=1,2,3),
\end{aligned}
$$

with the equation of state

$$
(4)\qquad p=\sigma^2\rho\,,
$$

where the density ρ, velocity $\mathbf{u}={}^t(u_1,u_2,u_3)$ and pressure p are functions of $x\in\mathbf{R}^3$ and $t\geq 0$, while c is the speed of the light and σ is the speed of the sound which are constants. According to the relativistic theory, σ never exceeds c. Especially, the case $\sigma^2=\frac{c^2}{3}$ is important in the context of the physics. If $c\to\infty$, (1.3) reduces to the classical compressible Euler equation (1.1).

In 1993, Smoller and Temple [11] have constructed uniformly bounded weak solutions for 1 dimensional case by using Glimm's method. After constructing approximate solutions, they have showed that the variation of $\log\rho$ is monotone decreasing.

In 1994, T.Makino and S.Ukai [3] have proved, by using Lax's theorem, the existence of the local classical solutions for (1.3). Recently we have constructed global weak solutions for (1.3) with spherical symmetry. we shall explain this result in section 3.

2. Global weak solution for the compressible Euler equation

In this section we shall present our recent results on the compressible Euler equation with spherical symmetry. We look for solutions of the form

$$
(5)\qquad \rho=\rho(t,|x|),\ \vec{u}=\frac{x}{|x|}\cdot u(t,|x|).
$$

Then, denoting $r=|x|$, (1.1) with $f=0$ becomes

$$
(6)\qquad
\begin{aligned}
&\rho_t+\frac{1}{r^{n-1}}(r^{n-1}\rho u)_r=0\,,\\
&\rho(u_t+uu_r)+p_r=0\,.
\end{aligned}
$$

This equation has a singularity at r=0. To avoid the difficulty caused by this singularity, we simply deal with the boundary value problem for (2.2) in the domain $1\leq r<\infty$ (the exterior of a sphere) with the boundary condition $u(t,1)=0$, which is identical, under the assumption (2.1), to the usual boundary condition $\vec{n}\cdot\vec{u}=0$ for (1.1) where $\vec{n}$ is the unit normal to the boundary.

Put $\tilde{\rho}=r^{n-1}\rho$. Then we get from (2.2)

$$\tilde{\rho}_t + (\tilde{\rho} u)_r = 0,$$
$$u_t + u u_r + \frac{a^2\gamma\tilde{\rho}_r}{\tilde{\rho}^{2-\gamma} r^{(n-1)(\gamma-1)}} = \frac{a^2\gamma(n-1)\tilde{\rho}^{\gamma-1}}{r^n \cdot r^{(n-1)(\gamma-2)}}. \tag{7}$$

Introduce the Lagrangian mass coordinates

$$\tau = t, \quad \xi = \int_1^r \tilde{\rho}(t,r)\, dr. \tag{8}$$

Then $\xi > 0$ as long as $\tilde{\rho} > 0$ for $r > 1$, and (2.3) is reformulated as

$$\tilde{\rho}_\tau + \tilde{\rho}^2 u_\xi = 0,$$
$$u_\tau + \frac{a^2\gamma\tilde{\rho}_\xi}{\tilde{\rho}^{1-\gamma} r^{2\gamma-2}} = \frac{a^2\gamma(n-1)\tilde{\rho}^{\gamma-1}}{r^n \cdot r^{(n-1)(\gamma-2)}}. \tag{9}$$

Put $v = 1/\tilde{\rho}$ and note that the inverse transformation to (2.4) is given by

$$t = \tau, \; r = 1 + \int_0^\xi v(t,\zeta) d\zeta. \tag{10}$$

Then after changing τ to t and ξ to x, (2.5) is written as

$$v_t - u_x = 0,$$
$$u_t + \left(\frac{a^2}{v^\gamma}\right)_x \cdot \frac{1}{r^{(n-1)(\gamma-1)}} = \frac{a^2\gamma(n-1)v^{1-\gamma}}{r^n \cdot r^{(n-1)(\gamma-2)}}, \tag{11}$$

where r is now defined by $r = 1 + \int_0^x v(t,\zeta)d\zeta$. Now we restrict ourselves to the case $\gamma = 1$. Then (2.7) becomes

$$v_t - u_x = 0,$$
$$u_t + \left(\frac{a^2}{v}\right)_x = \frac{K}{1 + \int_0^x v(t,\zeta)d\zeta}. \tag{12}$$

where $K = a^2(n-1)$. Let us consider the initial boundary value problem for (2.8) in $t \geq 0$, $x \geq 0$ with the following boundary and initial conditions.

$$u(0,x) = u_0(x), \quad v(0,x) = v_0(x), \quad for\ x > 0, \tag{13}$$

$$u(t,0) = 0, \quad for\ t > 0. \tag{14}$$

Theorem 2.1. *Suppose that $u_0(x)$ and $v_0(x)$ are of bounded variation, and that $v_0(x) \geq \delta_0 > 0$ for all $x > 0$ with some positive constant δ_0. Then (2.8), (2.9) and (2.10) have a global weak solution.*

This theorem can be proved by following Nishida's argument [9] based on Glimm's method [1]. For the detail, see [4]. Note that the principal part of (2.8) coincides exactly with the one-dimensional compressible Euler equation, but this coincidence does not occur for $\gamma > 1$.

Remark 1. Note that it is not obvious that this result also implies the existence of global weak solutions of (1.1). If solutions in Lagrangian coordinate are smooth functions, we can show that $\mathbf{u}$ and ρ deduced from these solutions satisfy (1.1) by using the chain rule. But if solutions are weak solutions, we must be more careful. In [8] K. Mizohata has proved that weak solutions in Lagrangian coordinate are weak solutions in Eulerian coordinate at least they are spherically symmetric and that vice versa. Instead of using the chain rule, we use the fact that the Lagrangian transformation is a bi-Lipschitz homeomorphism to prove that $(\mathbf{u}, \rho)$ is also weak solution of (1.1). This is the main idea of Wagner [4]. He has showed the equivalence for the Cauchy problem in one space dimension. In [8], K. Mizohata has given the detailed proof of this equivalence for the more general case.
Remark 2. K. Mizohata has extended this result to the case in which the gravitational force $f = -\rho M/r^2$ appears in the right-hand side of the second equation of (2.2). For the detail, see [7].

It is clear that the equation (2.2) admits the stationary solutions $\rho = \bar{\rho} = constant > 0$, $u = 0$. The corresponding stationary solutions for (2.8) are

$$v = \frac{1}{\bar{\rho}\left(1 + \frac{n}{\bar{\rho}}x\right)^{1-\frac{1}{n}}} \, , \quad u = 0 \, . \tag{15}$$

For these stationary solutions, we have $\inf v = 0$, i.e., v is not bounded away from zero. Now, if we attempt to enlarge the class of the initial data of Theorem 2.1 so that it includes these stationary solutions, we readily encounter a difficulty in construction of the approximate solutions used in Glimm's method, [1], that is, the mesh lengths Δx and Δt should be chosen so that $\Delta x/\Delta t > a/\inf v$ in view of the Courant-Friedrichs-Lewy condition, whereas this could not be possible as long as the mesh is supposed to be uniform. In [5] we have showed that if we use a non-uniform mesh chosen carefully, Glimm's method still works and gives global weak solutions for the initial data in the class containing the stationary solutions (2.11).

More precisely, we deal with the initial data u^0 and v^0 satisfying the following conditions:

$$\begin{gathered} v_0 \, , \; |u_0| \, , \; T.V.u_0 \, , \; T.V.v_0 \;\leq\; C_0 \, , \\ \frac{\delta_0}{(1+x)^{1-\epsilon}} \;\leq\; v_0(x) \, , \end{gathered} \tag{16}$$

where C_0, δ_0 and ϵ are positive constants independent of x and $0 < \epsilon \leq 1$. By using a non-uniform mesh method, we have succeeded to enlarge the class of the initial data so that it includes these stationary solutions. For the detail, see [5].

Theorem 2.2. *If u_0 and v_0 satisfy (2.12), then the initial-boundary value problem (2.8), (2.9) and (2.10) admit a global weak solution (u, v).*

3. Global weak solution for the relativistic Euler equation

In this section we shall present our result on the relativistic Euler equation with spherical symmetry. Similarly we look for solutions of the form

$$\rho = \rho(t, |x|),\ \vec{u} = \frac{x}{|x|} \cdot u(t, |x|). \tag{17}$$

where $r = \sqrt{x_1^2 + x_2^2 + x_3^2}$. Then (1.3) becomes

$$\begin{aligned} &\frac{\partial}{\partial t}\left(\frac{\rho c^2 + p}{c^2}\frac{1}{1-\frac{u^2}{c^2}} - \frac{p}{c^2}\right) + \frac{\partial}{\partial r}\left(\frac{\rho c^2 + p}{c^2}\frac{u}{1-\frac{u^2}{c^2}}\right) + \frac{\rho c^2 + p}{c^2}\frac{1}{1-\frac{u^2}{c^2}}\frac{2u}{r} = 0\,, \\ &\frac{\partial}{\partial t}\left(\frac{\rho c^2 + p}{c^2}\frac{u}{1-\frac{u^2}{c^2}}\right) + \frac{\partial}{\partial r}\left(\frac{\rho c^2 + p}{c^2}\frac{u^2}{1-\frac{u^2}{c^2}} + p\right) + \frac{\rho c^2 + p}{c^2}\frac{u}{1-\frac{u^2}{c^2}}\frac{2u^2}{r} = 0\,. \end{aligned} \tag{18}$$

We again simply deal with the boundary value problem for (3.2) in the domain $1 \leq r < \infty$ (outside a unit ball) with the boundary value condition $u(t,1) = 0$ to avoid the singularity caused at the origin. Introduce Lagrangian coordinate

$$x = \int_1^r \rho r^2 J dr\,, \tag{19}$$

where

$$J = 1 + \frac{(c^2+\sigma^2)u^2}{c^2(c^2-u^2)}\,. \tag{20}$$

Putting

$$W = \frac{1}{Jq},\ V = \frac{\frac{c^2+\sigma^2}{c^2-\sigma^2}}{J}u = \frac{(c^2+\sigma^2)c^2}{c^4+\sigma^2u^2}u\,,\ q = r^2\rho\,. \tag{21}$$

Then (3.2) becomes,

$$\begin{aligned} W_t \ - \ V_x \ &= \ 0\,, \\ V_t \ + \ \left(\frac{\sigma^2}{J^2W}\right)_x \ &= \ \frac{2\sigma^2}{Jr}\,, \end{aligned} \tag{22}$$

where r is now defined by $r = 1 + \int_0^x W dx$. Let us consider the initial boundary value problem for (3.6) in $t \geq 0,\ x \geq 0$ with the following initial and boundary conditions.

$$q(0,x) \ = \ q_0(x),\ \ u(0,x) \ = \ u_0(x),\ \ x > 0\,, \tag{23}$$

$$v(t,0) = 0\,. \tag{24}$$

Our main result is as follows.

Theorem 3.1. *Suppose that* $\log q_0(x)$ *and* $\log \frac{c+u_0(x)}{c-u_0(x)}$ *are of bounded variation. Then there exists a global weak solution for (3.6), (3.7) and (3.8) satisfying*

$$|u| < c\,,\ \ q > 0\,. \tag{25}$$

We shall briefly explain the outline of proof. First, consider Riemann problem for the homogeneous equation corresponding to (3.6) which is given by

$$
\begin{aligned}
W_t \ - \ V_x &= 0\,, \\
V_t + \left(\frac{\sigma^2}{J^2 W}\right)_x &= 0\,,
\end{aligned}
\tag{26}
$$

Then we can show that all shock curves for (3.10) have the same figure in the plane of Riemann invariants. Fortunately, their figure is similar to the figure of shock curves of the classical case. This fact was also discovered by Smoller and Temple [11] for the one dimensional case of (1.3). We construct approximate solutions via the Glimm's difference scheme. More precisely, we construct our approximate solutions of the form

$$\{\textit{solution of Riemann problem for (3.10)}\} + \{\textit{inhomogeneous term}\} \times t\,.$$

Next we estimate the total variation of the approximate solutions. We do it by analyzing the waves in the plane of Riemann invariants z_1 and z_2 where they are given by

$$
\begin{aligned}
z_1 &= \log q + \frac{c^2+\sigma^2}{2\sigma c}\log\frac{c+u}{c-u}\,, \\
z_2 &= -\log q + \frac{c^2+\sigma^2}{2\sigma c}\log\frac{c+u}{c-u}\,.
\end{aligned}
\tag{27}
$$

Using the geometry of shock waves, we estimate the variation of $\log q$. This is the main idea of Smoller and Temple [9]. But in our case, we must be more careful since there is an inhomogeneous term in (3.6). To obtain our desired uniform estimates, we use the transformation

$$u \ = \ c\tanh w \ = \ c\,\frac{e^w - e^{-w}}{e^w + e^{-w}}, \tag{28}$$

Instead of estimating u itself, we estimate w. Fortunately, it follows that

$$\log\frac{c+u}{c-u} \ = \ 2w\,. \tag{29}$$

The transformation (3.12) plays a crucial role in our paper. Using (3.12) and (3.13), we can obtain uniform estimates of the approximate solutions and thus we can construct global weak solution. For the detail, see [6].

4. Concluding remarks

Conservation laws are important nonlinear PDEs since many problems in science have conserved quantities. Especially, the Euler equation of gas dynamics is a typical example of it and many mathematician have studied it. Consequently, much of interesting theory of conservation laws have been discovered by studying the Euler equation.

But although there are many results for the one dimensional case, there are few results for the multi-dimensional case. In this case it is indeed difficult to obtain desired uniform estimates for a system of conservation laws. If we look for the solutions with spherical symmetry, the problems become a one-dimensional problem and thus we can use one-dimensional methods and theory. But in this case we encounter a singularity at the origin. Unfortunately we do not yet succeed to deal with this singularity. To avoid this singularity, we considered the problems outside a unit ball. We thus obtained, by using one-dimensional methods and our new methods, several results for the multi-dimensional case.

We hope that by considering the problems with spherical symmetry, we can find solution to the problem for the general multi-dimensional case.

References

[1] J. Glimm, *Solutions in the large for nonlinear hyperbolic systems of equations,* Comm. Pure Appl. Math, **18**, (1965), 697-715.

[2] T. Makino and S. Takeno, *Initial Boundary Value Problem for the Symmetric Motion of Isentropic Gas,* preprint.

[3] T. Makino and S. Ukai, *Local Smooth Solutions of the Relativistic Euler Equation,* preprint.

[4] T. Makino, K. Mizohata and S. Ukai, *The Global Weak Solutions of the Compressible Euler Equation with spherical Symmetry,* Japan J. Indust. Appl. Math., 9, (1992), 431-449.

[5] T. Makino, K. Mizohata and S. Ukai, *The Global Weak Solutions of the Compressible Euler Equation with spherical Symmetry II,* to appear in Japan J. Indust. Appl. Math.

[6] T. Makino, K. Mizohata and S. Ukai, *Global Solutions to the Relativistic Euler Equation with spherical Symmetry,* preprint.

[7] K. Mizohata, *Global weak solutions for the equation of isothermal gas around a star,* to appear in J. Math Kyoto Univ.

[8] K. Mizohata, *Equivalence of Eulerian and Lagrangian weak solutions of the compressible Euler equation with spherical symmetry,* Kodai Mathematical Journal., 17, (1994), 69-81.

[9] T. Nishida, *Global solutions for an initial boundary value problem of a quasilinear hyperbolic system,* Proc. Japan Acad, **44**, (1968), 642-646.

[10] M. Okada and T. Makino, *Free Boundary Problem for the Equation of Spherically Symmetric Motion of Viscous Gas,* to appear in Japan J. Indust. Appl. Math.

[11] J. Smoller and B. Temple, *Global solutions of the relativistic Euler equations* Comm. Math. Phys., 156 (1993), 67-99.

[12] D. H. Wagner, *Equivalence of the Euler and Lagrangian Equations of Gas Dynamics for Weak Solutions,* Journal of Differential Equations., 68, (1987), 118-136.

[13] Tong Yang, *A fuctional integral approach to shock wave solutions of the Euler equations with spherical symmetry,* preprint.

The Effect of a Non-Zero Shock Width on Wave Propagation in Multi-Dimensions

Ralph Menikoff[1], Klaus S. Lackner[2], Bruce G. Bukiet[3]

ABSTRACT. It is known that a numerical shock width in one dimension can lead to a localized entropy error when waves interact. In multi-dimensions the wave width, either numerical or physical, can also affect the propagation of the wave, and thus have a global effect on fluid flow. A well known physical example is the diameter effect for detonation waves. This effect is due to reaction zone dynamics; the competition between a source term for the release of chemical energy and a geometric source term due to front curvature. When the reaction zone is underresolved, the effective reaction zone dynamics leads to mesh dependent numerical results. The mesh dependence is due to an artificial numerical curvature effect. An explanation of this effect is given based on the conservation laws. Because of the non-zero wave width, the standard Hugoniot jump conditions must be modified. Correction terms are proportional to the wave width. It follows when the width is proportional to the cell size, as occurs in shock capturing schemes, that the curvature effect is mesh dependent. A similar modification of the jump conditions can be expected whenever there are multiple length scales determining the dissipation which gives rise to the wave width. As a consequence, the wave curve depends on the local front curvature and the divergence of the velocity field in the tangent plane in addition to the usual state variables. In computations, the length scales affecting the wave curve may either be physical or artificial in nature. Artificial length scales lead to numerical errors which may not vanish under mesh refinement.

1. INTRODUCTION

Shock capturing algorithms are widely used to obtain numerical solutions of hyperbolic partial differential equations (PDEs). In these algorithms a shock wave is not truly discontinuous but rather a very rapid change in state over a small number of cells. There are two reasons for the numerical shock width: (i) to control the truncation errors for numerical stability, and (ii) to add the necessary dissipation in order to select out the physical shock waves. The heuristic motivation for shock capturing algorithms is that the state behind a shock wave is determined by the Hugoniot jump conditions and giving a shock a narrow profile causes only a local error that doesn't affect the global solution.

Though shock capturing algorithms are generally effective, the artificial shock width can lead to non-trivial errors which manifest themselves as incorrect values for the entropy. Only for a steady-state shock wave is the entropy increase determined solely by the Hugoniot jump conditions and independent of the form of

[1]Theoretical Division, MS-B214, Los Alamos National Laboratory, Los Alamos, NM 87545 rtm@lanl.gov

[2]Theoretical Division, MS-B216, Los Alamos National Laboratory, Los Alamos, NM 87545 ksl@lanl.gov

[3]Department of Mathematics, Center for Applied Mathematics and Statistics, New Jersey Institute of Technology, Newark, NJ 07102, bukiet@shock.njit.edu

This work was supported by the U. S. Department of Energy.

the numerical dissipation. A transient, when a shock wave is formed or when a shock profile changes abruptly, gives rise to a localized entropy error. This has been analyzed in one-dimension for shock wave interactions [3]. The same analysis explains many commonly occurring and widely ignored numerical errors such as excess wall heating, shock start-up error, the error when a shock impinges on a material interface, and the error when a shock passes through an abrupt change in cell size.

Here we show that for a curved shock the wave width gives rise to correction terms in the Hugoniot jump conditions. As a consequence, there is a curvature effect on propagating shock waves. This is a generalization of the well known diameter effect for detonation waves. The diameter effect has been observed experimentally and in numerical simulations. We will briefly describe the diameter effect in Section 2. Latter sections will use detonation waves as an illustrative example. A qualitatively similar curvature effect occurs for partially and fully dispersed shock waves resulting from relaxation effects described by source terms in hyperbolic PDEs.

Our analysis of a curved shock with non-zero wave width is based on a quasi-steady 1-D approximation. The approximation is developed in Section 3. Assuming the dimensionless product of wave width and front curvature is small, we show that there is a distinguished frame of reference in which the dominant multi-dimensional effects are well approximated by steady nozzle flow, *i.e.*, a flow in a duct with variable cross sectional area. The geometric source term for the nozzle is determined by the front curvature and the divergence of the velocity in the plane tangent to the front.

The quasi-steady 1-D approximation is used to derive modified jump conditions. In Section 4 we show that the leading order correction is proportional to the dimensionless product of wave width and front curvature. As discussed in Section 5, the same analysis applies equally well to conservative difference schemes. Consequently, the numerical shock width can give rise to an artificially large and mesh dependent curvature effect. This is a generic problem with shock capturing algorithms.

2. Diameter Effect for Detonation Waves

An explosive may be modeled by the reactive fluid flow equations [2]. In 1-D, this leads to the Zel'dovich-von Neumann-Doering (ZND) model for a detonation wave. In the ZND model a detonation wave consists of a lead shock followed by a thin reaction zone. It can be viewed as a partially dispersed wave in which chemical equilibrium is established in the reaction zone.

An important property of a detonation is that there exists a self-sustaining detonation wave known as a Chapman-Jouguet (CJ) detonation; the shock heating initiates the chemical reaction and the energy release drives the shock wave. The state behind a CJ detonation wave is sonic relative to the front, and hence the wave is acoustically decoupled from the flow behind. Moreover, a CJ detonation has the minimum wave speed of any steady detonation wave. The CJ detonation speed can be determined from the equation of state of the detonation products and the Hugoniot jump conditions. It is independent of the reaction rate law.

Rate stick experiments are used to determine the CJ detonation speed. In a rate stick experiment, a cylinder of explosive is detonated at one end. After a distance of run of a few diameters a steady-state is approached and the wave speed

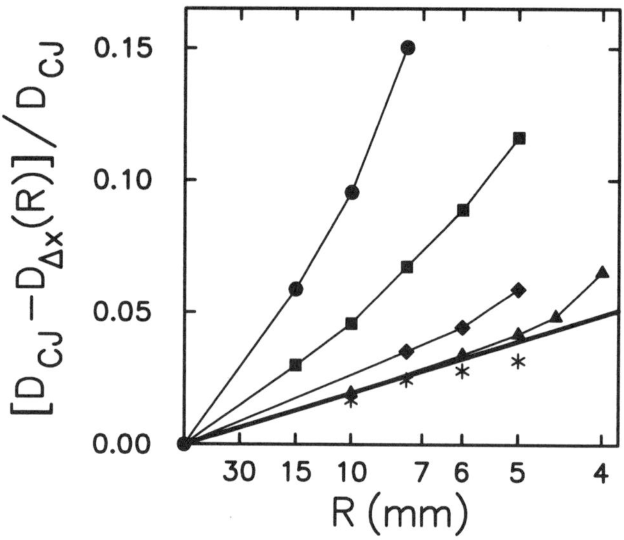

FIGURE 1. Mesh dependence of curvature effect; numerical data from Table I of Ref. [1]. Thick solid line is fit to experimental data. Circle, square, diamond and triangle are calculations with cell size of $\Delta x = 0.5, 0.25, 0.125$ and 0.075 mm respectively. The symbol $*$ corresponds to the extrapolation of the last two points to $\Delta x = 0$ assuming second order convergence.

is measured. It is found that the detonation speed decreases as the diameter of the explosive cylinder is decreased. This is known as the diameter effect. The diameter effect is related to rarefactions from the side which lead to a curved detonation front. An infinite diameter would correspond to a 1-D planar CJ detonation wave. This gives rise to the question; Why is the detonation speed of a curved wave less than the minimum allowed value for the planar case?

Furthermore, in numerical experiments the magnitude of the diameter effect is mesh dependent. A systematic numerical study varying the cell size was carried out by Donguy & Legrand [1]. The normalized velocity deficit, showing the diameter effect as a function of cell size is plotted in Fig. 1. We note that the numerical effect is qualitatively the same as the physical effect.

With the typical cell size used for a calculation, the curvature effect is quantitatively inaccurate and leads to a significant error when the cylinder diameter is small. As the cell size is decreased and the reaction zone is resolved, rather than captured, the artificial numerical effect decreases compared to the real physical effect. Moreover, as the number of cells in the reaction zone increases, the width of the reaction zone decreases until the value intrinsic to the reaction rate model is reached. The subsequent analysis shows that the artificially large numerical width is the dominant cause of the numerical curvature effect.

Our analysis shows that the curvature effect is not limited to detonation waves. However, the effect stands out in detonation waves. Because of the minimum wave speed, it can be demonstrated with the measurement of a single variable. In addition, the effect is insensitive to boundary conditions because acoustic waves cannot catch up with and influence a self-sustaining detonation. Most importantly, the effect has been measured experimentally and cannot be accurately computed with shock capturing algorithms.

3. Quasi-Steady 1-D Approximation

Our analysis is based on a long wavelength, low frequency approximation. We assume that (i) the reaction zone width w is much smaller than the principal radii of curvature of the front, and that (ii) the wave speed normal to the front is slowly varying in both position and time. A consequence of these assumptions is that the derivatives of the flow variables tangent to the front are much smaller than those normal to the front. Locally, the tangential flow can be neglected relative to the normal flow. However, the normal directions are not parallel to each other when the front is curved. This effect cannot be neglected, but to leading order it can be accounted for with geometric source terms in a quasi one-dimensional model.

For a small areal patch on the detonation front, the streak lines form a flow tube. Let A be the cross sectional area of the flow tube, and x the coordinate along a stream line. Locally, the reaction zone is described by the reactive flow equations in a duct of variable cross sectional area

$$(\rho A)_t + (\rho A u)_x = 0 \,, \tag{3.1}$$

$$(\rho A u)_t + \left(\rho A\left[u^2 + PV\right]\right)_x = P A_x \,, \tag{3.2}$$

$$(\rho A \mathcal{E})_t + \left(\rho A u\left[\mathcal{E} + PV\right]\right)_x = -P A_t \,, \tag{3.3}$$

$$(\rho A \lambda)_t + (\rho A u \lambda)_x = \rho A \mathcal{R} \,, \tag{3.4}$$

where ρ, u, E and λ are the fluid density, particle velocity along a stream line, specific internal energy (thermal + chemical), and mass fraction of the reaction products ($\lambda = 0$ unburnt, $\lambda = 1$ completely burnt), $V = 1/\rho$ is the specific volume, $\mathcal{E} = \frac{1}{2}u^2 + E$ is the total specific energy, $P(V, E, \lambda)$ is the pressure, and $\mathcal{R}(V, E, \lambda)$ is the specific reaction rate.

It is important to note that the flow tube is not Galilean invariant. There are two special frames in which the equations simplify because to leading order $\partial_t A = 0$: (i) the co-moving frame, and (ii) the normal frame. In these frames the only geometric term that enters the equations is the local divergence of the streak lines, $\kappa = \lim_{A\to 0}(\partial_x A/A)$. The co-moving frame is not a good choice because $\kappa(x)$ is not *a priori* known.

The angle between a streamline and the local normal to the front depends on the ratio of the tangential and normal velocity components. In our approximation, the tangential component is constant but the normal component varies dramatically through the reaction zone. In 2-D, the shape of the stream tube depends on the reaction zone dynamics unless one chooses the normal frame in which the tangential velocity vanishes in first order. In 3-D there is no frame in which the tangential velocity vanishes over a local region. However, we have shown in Ref. [4] that there is a unique frame in which the residual tangential velocity affects only the shape but not the cross sectional area $A(x)$. Since only the area enters into the quasi 1-D model, this is sufficient for our analysis to apply in 3-D as well.

In the normal frame the velocity ahead of the front is on average normal to the front and has the magnitude u_{ahead} determined by the relation

$$\kappa_f u_{ahead} = \nabla_T \cdot \vec{u} \ , \tag{3.5}$$

where κ_f is the local front curvature and $\nabla_T \cdot \vec{u}$ is the velocity divergence in the tangent plane. Moreover, $\kappa = \kappa_f$ is constant to $O(\kappa_f w)$. In the typical case with solid high explosives, the material ahead of the front is standing still. This happens to correspond to the normal frame.

The time and length scales in the reaction zone are much smaller than the bulk flow. Consequently, we can assume the reaction zone profile varies adiabatically, *i.e.*, rapidly equilibrates to slow changes of the front curvature. To leading order the PDEs reduce to ODEs for the quasi-steady reaction zone profile

$$[\rho(D-u)]_x = \rho u \kappa_f \ , \tag{3.6}$$

$$\left[\rho(D-u)^2 + P\right]_x = \rho \cdot (D-u)\, u \kappa_f \ , \tag{3.7}$$

$$\left[E + PV + \tfrac{1}{2}(D-u)^2\right]_x = 0 \ , \tag{3.8}$$

$$(u-D)\lambda_x = \mathcal{R} \ , \tag{3.9}$$

where D is the detonation wave speed and x is the distance relative to the front. For these ODEs, κ_f and D are to be treated as parameters determined by the bulk flow.

4. Modified Jump Conditions

The Hugoniot jump conditions are an expression of the conservation laws across a shock wave. When the wave has a width, the jump conditions are obtained by integrating the quasi-steady 1-D equations. To leading order this gives the modified jump conditions

$$\Delta[\rho(D-u)] = \kappa_f w \,\langle \rho u \rangle \ , \tag{4.1}$$

$$\Delta\left[\left(\rho(D-u)\right)^2 V + P\right] = \kappa_f w \,\langle \rho(D-u)u \rangle \ , \tag{4.2}$$

$$\Delta\left[E + PV + \tfrac{1}{2}(D-u)^2\right] = 0 \ , \tag{4.3}$$

where $\Delta[f] = f(x_0) - f(x_1)$ is the change of variable f across the detonation wave, $w = x_0 - x_1$ is the reaction zone width, and

$$\langle f \rangle = \frac{1}{w}\int_{x_1}^{x_0} dx\, f \tag{4.4}$$

is the average value of f in the reaction zone. The averages result from the fact that the source terms are not perfect derivatives.

In contrast to the standard jump conditions, additional information about the reaction zone dynamics is needed for the modified jump conditions. In particular,

$\lim_{w \to 0}\langle f \rangle$ does not vanish, nor can it be expressed in terms of the values of f at the end points. Thus, the conservation relations applied to a wave with non-zero width are no longer algebraic conditions.

The modified jump conditions reduce to the standard Hugoniot jump conditions when either the front is planar $\kappa_f = 0$ or the width of the wave is zero $w = 0$. In effect the correction terms express the dynamics of the reaction zone; the competition between the source term for energy release due to the chemical reaction and the geometric source term due to front curvature. We note that the correction terms are not Galilean invariant. The modified jump conditions are valid *only* in the normal frame described in the previous section.

Even though the short length scale of the reaction zone may not be of interest, the reaction zone dynamics does affect the propagation of a detonation wave and hence the flow on the hydrodynamic length scale of interest. Instead of computing a quasi-steady reaction zone, its dynamics can be accounted for in the wave curve; *i.e.*, locus of final states connected to a given initial state. In contrast to a discontinuous shock, the detonation wave curve depends on front curvature and the tangential velocity divergence ahead of the wave as well as the initial state. The curvature effect is a material property described by the wave curve. The diameter effect is merely a special application of the curvature effect to a rate stick experiment.

5. Conclusion

To capture shock waves correctly, it is important that numerical algorithms are written in conservation form. Moreover, when the wave profile is not resolved, the assumptions of the quasi-steady 1-D approximation are applicable. Since the modified jump conditions are derived from the conservation laws, they also apply to finite difference algorithms. Hence, there is an artificial curvature effect from the numerical shock width.

The averages in the modified jump conditions are slowly varying functions of three parameters; D, κ_f and $\nabla_T \cdot \vec{u}$. The factor $w\kappa_f$ has a large effect on the magnitude of the correction terms. In shock capturing algorithms, the wave width is several cells and hence proportional to the cell size. Consequently, the numerical curvature effect is mesh dependent.

The mesh dependence of the numerical curvature effect can be reduced with adaptive mesh algorithms. This has the effect of partially resolving the reaction zone and is computationally expensive. Since an adaptive algorithm must place a fine mesh around the detonation wave, in effect, it is a crude tracking algorithm. Front tracking is a more natural algorithm for dealing with the curvature effect. It provides a clean separation between the long length scales of the non-reactive flow and the short length scales required for the reaction zone dynamics. The analytic information needed to propagate a detonation wave is contained in the wave curve. By empirically calibrating the average quantities in the modified jump conditions, the wave curve can be determined without a detailed knowledge of the complicated chemical reactions which are impractical to obtain experimentally.

Our analysis is based on the conservation equations and the quasi-steady approximation. It is not limited to detonation waves. The curvature effect occurs in general whenever a wave has a width. Therefore, partially dispersed and fully dispersed shock waves due to relaxation phenomena are also subject to a curvature effect.

A narrow wave front may have a local region of high curvature. This can occur when waves interact to form multi-dimensional wave patterns. In this case the 1-D assumption in the quasi-steady model breaks down. Consequently, the region where wave profiles overlap may be fully multi-dimensional. This undoubtedly will affect the threshold for the bifurcation of wave patterns, such as the transition between regular and Mach reflection. Thus, we hypothesize that the physical dissipative mechanisms that determine shock profiles can in certain instances affect the global behavior of fluid flow.

References

1. P. Donguy and N. Legrand, *Numerical simulations of non-ideal detonations of a heterogeneous explosive with the two dimensional Eulerian code C.E.E.*, Proceedings of the Seventh International Symposium on Detonation, 1981, pp. 695–702.
2. W. Fickett and W. C. Davis, *Detonation*, Univ. of California Press, Berkeley, 1979.
3. R. Menikoff, *Errors when shock waves interact due to numerical shock width*, SIAM J. Sci. Comp. **15** (1994), 1227–1242.
4. R. Menikoff, K. S. Lackner, and B. G. Bukiet, *Modeling flows with curved detonation waves*, preprint LA-UR-94-481, Los Alamos National Lab., 1994, (submitted to Combustion Flame).

Viscoplastic Relaxation in Granular Flow

E. Bruce Pitman[1,2]

Abstract

The equations governing the elastic-plastic deformation of granular materials are prone to instability at large strain. This instability signals the onset of strain localization; that is, deformation becomes confined to a narrow band within the sample. Phenomenological constitutive models do not contain a regularization mechanism, and additional, often ad-hoc, modeling of the shearband is required to follow subsequent deformation. We present a viscoplastic model of granular materials for anti-plane shearing, define an appropriate scale for the regularization, and derive energy bounds for the solution. We also present computational results from one-dimensional calculations.

1 Introduction

Models of the elastic-plastic deformation of granular materials often constitute hyperbolic systems of equations. At large values of the strain, these model equations typically lose "evolutionarity", and become ill-posed [13, 14, 15, 16, 18]. Higher-order strain gradients, couple-stresses, and rate effects have been proposed as mechanisms for regularizing the governing equations [6, 7, 17, 19]. In this paper we present a viscoplastic model similar to the model in [6]. Our purpose is to (i) define a structure for general viscoplastic relaxation models of granular flow; (ii) establish simple bounds for the solution of the equations; (iii) set-up a framework for computations of granular deformation. We mention that the regularization here relies on a small relaxation time; in contrast, many models of metals rely on a large exponent; see, e.g., [8].

The underlying elastic-plastic granular flow model in which we are interested is the anti-plane shearing of a linearly-elastic-plastic material, with a von Mises yield surface and strain-hardening, using a non-associative flow rule [15]. In this model, non-associativity is crucial in the loss of hyperbolicity and the formation of shearbands, reflecting the special role of non-associative flow rules for granular materials. This loss of hyperbolicity means that important physics is working at lengthscales which have not been included in the modeling. In numerical simulations, shear localization may lead to misleading, mesh dependent results. In order to continue the integration procedure once a shearband has formed, some form of regularization must be included in the governing equations, either implicitly or explicitly [3, 10]; see also [4]. The viscoplastic model presented here provides this regularization, and gives "structure" to the shearband. The viscoplastic system is always well-posed, and numerical intergration of the equations may proceed into the post-localization regime.

The mathematics of hyperbolic systems with relaxation, and convergence as the relaxation time approaches zero, has been studied in [1]. However, convergence for general models of granular deformation does not follow directly from these results; in particular, the underlying elastic-plastic system of equations are not conservation laws. At the same time, finite difference schemes based on the Godunov method for computing the numerical solution of hyperbolic systems with stiff

---

[1]Research supported by the National Science Foundation under Grant DMS-9201062, which includes additional funds from the Air Force Office of Scientific Research.

[2]Department of Mathematics, State University of New York, Buffalo, New York 14214-3093, Phone: (716) 829-2144, Fax: (716) 829-2299, E-mail: pitman@galileo.math.buffalo.edu

relaxation have been the subject of recent attention [5, 9]. These separate trends in mechanics, mathematics and computing all have a common focus, and it is appropriate to develop a unified perspective of the problem of relaxation effects in hyperbolic systems.

2 Model Equations

In this section we present model equations governing the deformation of a sample of granular material.

The fundamental balance law of mass and momentum may be written

$$d_t\rho \ + \ \rho\partial_j u_j \ = \ 0 \tag{1}$$

$$\rho d_t u_i \ + \ \partial_j T_{ij} \ = \ 0 \ . \tag{2}$$

Here ρ is the density of the granular sample, u_j is the velocity in the j-direction, $j = 1, 2, 3, T$ is the (symmetric) stress tensor, d_t denotes the convective time derivative, and summation convention is employed. These equations are four in number, and there are ten dependent variables. Constitutive relations are needed to describe the granular material and to close the governing system.

2.1 The Visco-Plastic Model

We now describe a viscoplastic constitutive model which closes the system of equations. In the viscoplastic model, stresses may lie inside, on, or outside of a loading surface given by $\Phi(T, \gamma) = 0$, where γ is a hardening parameter (taken here to be the accumulated inelastic strain) whose evolution is described below. Inside the loading surface ($\Phi < 0$) deformation is elastic; outside ($\Phi > 0$) the inelastic stresses relax in a characteristic time ϵ.

Define the strain-rate $V_{ij} = \frac{1}{2}(\partial_j u_i + \partial_i u_j)$, and decompose this strain-rate into elastic and inelastic parts: $V = V^{(e)} + V^{(i)}$. For granular materials, the elastic deformation is often assumed to be governed by linear elasticity, i.e., $CV^{(e)} = \nabla_t T$, where ∇_t is the time derivative taken in an objective frame of reference (e.g., the Jaumann derivative). The inelastic strain-rate is governed by a (possibly non-associative) flow rule, i.e. the direction of inelastic deformation is given by $\frac{d\Psi}{dT}$, where Ψ is the "flow potential". If $\Psi = \Phi$, the direction of inelastic deformation is normal to the loading surface and flow is called associative. Thus the general structure of the stress rate equation is:

$$\nabla_t T - CV \ = \ \frac{-1}{\epsilon} < \Phi > \widetilde{\frac{d\Psi}{dT}} \ . \tag{3}$$

Here the tilde denotes a unit vector, and the brackets $< \zeta >= \max(\zeta, 0)$.

Now specialize to one dimensional deformations describing the anti-plane shearing of a sample of material with space dependence on only one variable, say $x_1 = x$. Let v be the out-of-plane velocity, say the velocity in the x_3 direction, and assume all space derivatives in the x_2 and x_3 directions are zero. Then the continuity equation (1) is satisfied with a constant density. There is but one non-trivial momentum equation. Moreover, the convective derivative reduces to the simple partial time derivative, as does the Jaumann derivative. The stress tensor contains only two independent entries, $T_{13} = \sigma$ and $T_{23} = \tau$; write $T = (\sigma, \tau)^{\mathbf{t}}$. The strain-rate may be written as $(\partial_x v, 0)^{\mathbf{t}}$. For the constitutive model, we shall use $\Phi = |T| - H(\gamma) = 0$, where $|T|$ is the Euclidean norm of the stress. We assume the direction of inelastic flow makes a fixed angle α with the normal $\frac{d\Phi}{dT}$. The governing equations may then be written

$$\rho\partial_t v - \partial_x \sigma \;=\; 0 \tag{4}$$

$$\partial_t \sigma - c\partial_x v \;=\; \frac{-1}{\epsilon} < \sqrt{\sigma^2+\tau^2} - H(\gamma) > \frac{\sigma\cos(\alpha)+\tau\sin(\alpha)}{\sqrt{\sigma^2+\tau^2}} \tag{5}$$

$$\partial_t \tau \;=\; \frac{-1}{\epsilon} < \sqrt{\sigma^2+\tau^2} - H(\gamma) > \frac{-\sigma\sin(\alpha)+\tau\cos(\alpha)}{\sqrt{\sigma^2+\tau^2}} \tag{6}$$

Here c is the shear modulus. We also need to describe the evolution of strain. Using the strain-rate decomposition together with the stress rate evolution, one finds $\partial_t\gamma \;=\; \frac{1}{\epsilon} < \sqrt{\sigma^2+\tau^2} - H(\gamma) >$.

2.2 Scaling Issues

A characteristic time ϵ enters the viscoplastic system; this is the timescale on which stresses relax back to the loading surface. In experiments on granular materials, no such timescale appears to enter the constitutive framework. However a characteristic lengthscale does enter when shearbands are present. Shearbands are typically about 10 grain diameters thick [7, 12]. We translate this lengthscale into a timescale by dividing by the elastic wavespeed. To understand the relative scales, it is helpful to nondimensionalize the governing viscoplastic equations.

Denote nondimensional quantities by an overbar so that, for example, $V\bar{v} = v$ is the velocity, where V is a typical velocity scale in the problem. Nondimensionalize the velocity v by V, stresses σ and τ by Σ, length x by L, and time t by T. In a typical biaxial constitutive test, the velocity is set at the boundary, and is on the order of 0.2 cm./hr. (i.e. about two grain widths per hour), $L = 10$ cm., $\Sigma = 1$ atm., $\rho = 1.2\,\mathrm{g}/cm^3$, and the shear modulus $c = 50$ atm. The characteristic time is then defined through the equation $\frac{cVT}{\Sigma L} = 1$, so $T = 1$ hr. The nondimensional elastic wavespeed is ${c_e}^2 = \frac{\Sigma T}{\rho L V} \approx 5.5 \times 10^{12}$. Also, the relaxation time ϵ is taken to be the time it takes to travel across a typical shearband (10 particles or about 1 cm.) at the (dimensional) elastic wavespeed $\sqrt{c/\rho}$. Thus the non-dimensional ϵ is about 4×10^{-8}.

The nondimensional equations (dropping the overbars) are

$$\partial_t v - {c_e}^2\partial_x \sigma \;=\; 0 \tag{7}$$

$$\partial_t \sigma - \partial_x v \;=\; \frac{-1}{\epsilon} < \sqrt{\sigma^2+\tau^2} - H(\gamma) > \frac{\sigma\cos(\alpha)+\tau\sin(\alpha)}{\sqrt{\sigma^2+\tau^2}} \tag{8}$$

$$\partial_t \tau \;=\; \frac{-1}{\epsilon} < \sqrt{\sigma^2+\tau^2} - H(\gamma) > \frac{-\sigma\sin(\alpha)+\tau\cos(\alpha)}{\sqrt{\sigma^2+\tau^2}} \tag{9}$$

The strain-rate equation remains $\partial_t\gamma \;=\; \frac{1}{\epsilon} < \sqrt{\sigma^2+\tau^2} - H(\gamma) >$. This system of equations is semi-linear and always hyperbolic, with eigenvalues $\mu_0 = 0$ and $\mu_\pm = \pm\mu = \pm c_e$.

3 Energy Bounds

In this section we establish a simple energy estimate for the viscoplastic system. Consider a sample of unit length and, for simplicity, assume periodic boundary conditions. Define the energy

$$e(t) \;=\; \frac{1}{2}\int_0^1 (\frac{v^2}{{c_e}^2} + \sigma^2 + \tau^2) \tag{10}$$

Differentiating in t and substituting from the equations, we have

$$\dot{e}(t) \;=\; \int_0^1 (\,\partial_x(\sigma v) \;-\; \frac{1}{\epsilon} < \sqrt{\sigma^2+\tau^2} - H(\gamma) > (\sqrt{\sigma^2+\tau^2}\,\cos(\alpha))\,)\,dx \tag{11}$$

Thus, modulo work done at the boundaries ($\int_0^1 \partial_x(\sigma v)\,dx$, which we assume to be zero), $\dot{e}(t) \leq 0$, and the velocity and stresses are bounded in $\mathcal{L}^2$.

Now integrate Eqn. (11) in t. After rearranging, we have

$$\frac{\epsilon}{\cos(\alpha)}\, e(0) \geq \int_0^t \int_0^1 \; < \sqrt{\sigma^2+\tau^2} - H(\gamma) > \sqrt{\sigma^2+\tau^2}\; dx\, dt \tag{12}$$

Eqn. (12) says that, as the relaxation time approaches zero, the stresses approach the underlying elastic-plastic yield condition.

4 Sample Calculation

To understand the regularizing effect of relaxation, we performed numerical simulations on the viscoplastic system. A Strang splitting separated the wave propagation from the stiff source terms. We used an unsplit, explicit second-order Godunov method [2] for the (linear) hyperbolic portion, and an explicit trapezodial rule for the source terms. The stability constraint based on the hyperbolic part is $\Delta t \leq \Delta x / c_e$; for the ODE solver, $\Delta t \leq K\epsilon$, where K bounds the nonlinear relaxation terms. On a grid with $\Delta x = 0.01$, the hyperbolic step is $\frac{1}{2} \times 10^{-8}$ and the source step is about 4×10^{-8}. Thus explicit integration of the source term imposes no further timestep constraint on the system.

Needless to say, such small timesteps make simulations costly. We are currently studying an implicit solver for the viscoplastic system. We mention that, in a simplier system studied in [11], an idea of Jin [5] allowed one to avoid solving nonlinear equations; however his idea does not carry over to the system presented in this paper. To reduce computer time, for the runs in this paper we artificially set $c_e = 1$ and $\epsilon = 0.001$.

In Figure 1, we show the computed solution at time $t = 0.5$. For this calculation, the initial data is $v = -0.75$ for $-0.15 < x < -0.05$ and $v = 0.75$ for $0.05 < x < 0.15$, $T^{13} = 0.75$ for $-0.15 < x < -0.05$ and $0.05 < x < 0.15$. The loading function is $H(\gamma) = 0.5 * (1 + \sqrt{1-(\gamma-1)^2})$ We see the propagating elastic waves and the localizing strain; the second plot compares the final value of the loading function and the maximum of the norm of the stress, illustrating convergence of the stress to the elastic-plastic yield surface.

Acknowledgements This research is supported by the National Science Foundation under Grant DMS-9201062, which includes funds from the Air Force Office of Scientific Research. Thanks to the Pittsburgh Supercomputing Center for time on the C90 used to perform the calculations reported here.

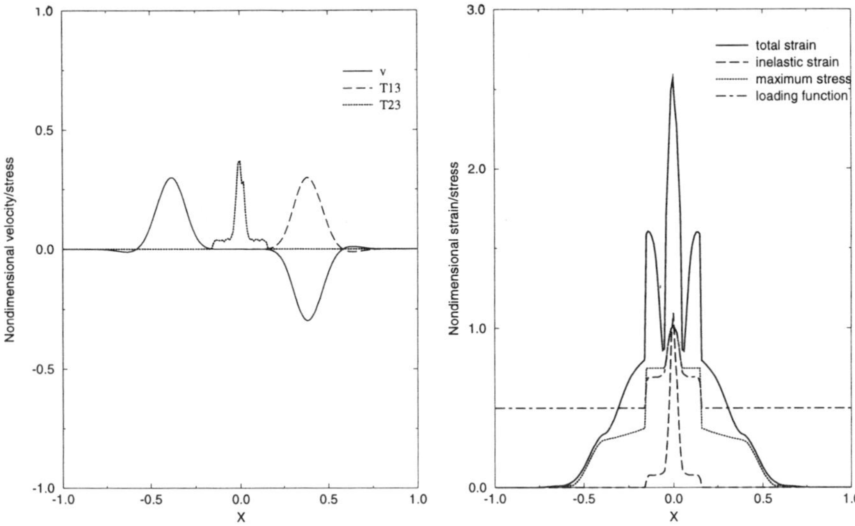

Figure 1: Velocity and stresses at times $t = 0.5$, showing wave propagation and the inelastic stress (left). The total and inelastic strain, and the maximum stress and loading function, at $t = 0.5$ (right).

References

[1] G.-Q CHEN, T.-P. LIU AND C. D LEVERMORE, *Convergence of hyperbolic conservation laws with relaxation*, Commun. Pure Appl. Math., to appear

[2] P. COLELLA, *Multidimensional upwind methods for hyperbolic conservation laws*, J. Comput. Phys., 87 (1990), pp. 171–200.

[3] X. GARAIZAR AND D. G. SCHAEFFER, *Numerical computations for shearband in an antiplane shear model*, J. Mech. Phys. Solids, to appear.

[4] X. GARAIZAR, these proceedings.

[5] S. JIN AND Z. XIN, *The relaxing schemes for systems of conservation laws in arbitrary space dimensions*, Commun. Pure Appl. Math., to appear

[6] B. LORET AND J. PREVOST, *Dynamic strain localization in elasto-(visco)-plastic solids, Part 1. General formulation and one-dimensional examples*, Comput. Meth. Appl. Mech. Eng., 83 (1990), pp. 247–273.

[7] H. MUHLHAUS AND I. VARDOULAKIS, *The thickness of shearbands in granular material*, Geotech., **37** (1987), pp. 271-283.

[8] A. NEEDLEMAN, *Computational micromechanics*, in *Theoretical and Applied Mechanics*, P. Germain, M. Piau and D. Caillerie (eds.) Elsevier, New York, (1989)

[9] R. PEMBER, *Numerical methods for hyperbolic conservation laws with stiff relaxation II: high order Godunov methods* , SIAM J. Sci. Comput. 14 (1993), pp. 824-859.

[10] E. B. PITMAN, *A Godunov method for localization in elasto-plastic granular flow*, Int. J. Analy. Num. Methods in Geomech., 17 (1993), pp. 385–400.

[11] E. B. PITMAN, *Viscoplastic relaxation: convergence and localization*, preprint

[12] K. H. ROSCOE, *The Tenth Rankine Lecture*, Geotech., **20** 1970, pp. 129-170.

[13] J. RUDNICKI AND J. R. RICE, *Conditions for localization of deformation of pressure sensitive solids*, J. Mech, Phys. Solids, **23** (1975), pp. 371-394.

[14] I. SANDLER AND D. RUBIN, *The Consequences of non-associated plasticity in dynamic problems*, in *Constitutive Laws for Engineering Materials: Theory and Application*, C. Desai et. al. (eds.), Elsevier, New York (1987)

[15] D. G. SCHAEFFER, *A mathematical model for localization in granular flow*, Proc. R. Soc. London A, **436** (1992),pp. 217-250.

[16] D. G. SCHAEFFER, *Instability and illposedness in the deformation of granular materials*, Int. J. Analy. Num. Meth. Geomech., **14**, (1990), p. 253-278

[17] L. J. SLUYS, *Wave propagation, localization and dispersion in softening solids*, Doctoral Thesis, Dept. Civil Engineering, Delft University of Technology, 1992

[18] I. VARDOULAKIS, *Rigid granular plasticity model and bifurcation in the triaxial test*, Acta Mech., **49** (1983), pp. 55-79.

[19] I. VARDOULAKIS AND E. C. AIFANTIS, *Gradient dependent dilatancy and it implications in shearbanding and liquefaction*, Ing. Arch.. **59**, 1989, pp. 197-208.

Dense Oscillations for the Euler Equations II

J.-L. Joly[1], G. Metivier[2], J. Rauch[3]

The fact that the resonant interaction of a finite number of oscillatory wave trains might lead to the creation of an infinity of waves was suggested by Hunter, Majda, and Rosales in §6.2 of [1]. To avoid this possibility they systematically assumed suitable finiteness hypotheses on their phase functions. Joly and Rauch [2] constructed a system of semilinear wave equations in spacetime of dimension 1+2 where waves propagating in a set of directions dense in the unit circle were created. Their intent was to show that nonlinear geometric optics in higher dimensions would not be possible without a finiteness hypothesis. Exactly this unlikely goal was achieved in [3, 4]. Schochet [6] then gave an alternate proof in a special case sufficiently general for the examples of this note. Our paper [5] proved that such dense oscillations can be generated by the 1+2 dimensional inviscid compressible Euler equations. In this note we recall and refine that result, answering one of two questions posed at the end of that paper.

The isentropic Euler equations are a 3×3 system of quasilinear equations of the form

$$\partial_t u + \sum_{1\le j\le 2} A_j(u)\,\frac{\partial u}{\partial x_j} = 0,$$

with coefficient matrices $A_j(u)$ defined from

$$(\partial_t + v_1\partial_1 + v_2\partial_2)v_1 + (p'(\rho)/\rho)\partial_1\rho = 0 \tag{1}$$

$$(\partial_t + v_1\partial_1 + v_2\partial_2)v_2 + (p'(\rho)/\rho)\partial_2\rho = 0 \tag{2}$$

$$(\partial_t + v_1\partial_1 + v_2\partial_2)\rho + \rho(\partial_1 v_1 + \partial_2 v_2) = 0 \tag{3}$$

The same phenomena occur in the nonisentropic case. The present setting is chosen for simplicity. The state vector is denoted $u = (v_1, v_2, \rho)$. Our solutions will be close to a background state $\underline{u} = (0, 0, \underline{\rho})$ which has the fluid at rest and the density constant. In [5] we constructed a family of smooth solutions

$$u^\epsilon = (0,0,\underline{\rho}) + \epsilon\, U(t, t/\epsilon, x_1/\epsilon, x_2/\epsilon) + O(\epsilon^2) \in C^\infty([0,t_1]\times \mathbf{R}^2)\,. \tag{4}$$

[1]CEREMAB URA 226 CNRS, Université de Bordeaux I, 33405 Talence, FRANCE
[2]IRMAR URA 305 CNRS, Université de Rennes I, 35042 Rennes, FRANCE
[3]Department of Mathematics, University of Michigan, Ann Arbor MI 48109, U.S.A.

The smooth profile $U(t,T,X)$ is a periodic function of T, X and the solutions u^ϵ are periodic in x. The error is measured in the L^∞ norm with a loss of a factor $1/\epsilon$ for each derivative taken, that is

$$\left\|\partial^\beta\left(u^\epsilon - ((0,0,\underline{\rho}) + \epsilon U(t,t/\epsilon,x_1/\epsilon,x_2/\epsilon))\right)\right\|_{L^\infty([0,t_1]\times\mathbf{R}^2)} = O(\epsilon^{2-|\beta|}). \quad (5)$$

More general oscillatory solutions would have profiles $U(t,x,T,X)$ depending also on x. For the solutions in (4), the oscillations at time t are essentially the same at all positions x. For this reason we refer to them as homogeneous oscillations in analogy with homogeneous turbulence.

The theory of Nonlinear Geometric Optics [3, 4, 6] provides the following recipe. Given smooth periodic initial data $U(0,0,X)$ there is a nonlinear evolution equation which is locally solvable determining $U(t,T,X)$ for $0 \le t \le t_1$ for some $t_1 > 0$. Then for ϵ small, the solution of the Euler equation with initial data

$$u(0,x) = \underline{u} + \epsilon\, U(0,0,x/\epsilon)\,, \quad (6)$$

is smooth on $[0,t_1]\times\mathbf{R}^2$ and the relation (5) holds.

Denote by $L(\partial_T,\partial_X)$ the linearization of the Euler equations at the background state $\underline{u}$. Then L is a constant coefficient strictly hyperbolic partial differential operator. For any $\gamma = (\gamma_0,\gamma_1,\gamma_2) \in \mathbf{R}^3$, $L(\gamma)$ denotes the symbol of L. The characteristic variety of L is given by the equation

$$\det\, L(\gamma) = \gamma_0(\gamma_0^2/c^2 - \gamma_1^2 - \gamma_2^2) = 0\,, \quad c = \sqrt{p'(\underline{\rho})}\,. \quad (7)$$

Here c is the speed of sound and the characteristic variety is the union of the horizontal plane $\gamma_0 = 0$ and the speed c light cone.

The first of the equations determining U is that, as a function of the fast variables T, X, U is a solution of the linearized equation

$$L(\partial_T,\partial_X)\, U(t,T,X) = 0\,.$$

Thus standard linearization yields the approximate solution $\underline{u}+\epsilon U(0,t/\epsilon,x/\epsilon)$, which is accurate only for times t small compared to ϵ. Expanding U in a Fourier series yields

$$U(t) = \sum_\gamma U_\gamma(t)\; e^{i\gamma\cdot(T,X)}\,. \quad (8)$$

The sum is over γ from the characteristic variety and the corresponding amplitude U_γ belongs to the kernel of $L(\gamma)$. The solutions (4) have as principal contributions the terms

$$\epsilon\, U_\gamma(t)\; e^{i\gamma\cdot(T,X)/\epsilon}$$

which are plane waves of amplitude and wavelength of order ϵ. When $\gamma_0 = 0$, the corresponding waves are stationary and are called entropy or vorticity waves. Otherwise, they are acoustic waves which move with speed c in the direction of the unit vector $(\gamma_1, \gamma_2)/|\gamma_1, \gamma_2|$.

Denote by $\mathbf{E}_\gamma$ the spectral projection of $\mathbf{R}^3$ onto kernel $L(\gamma)$. Then $\mathbf{E}_\gamma = |r_\gamma >< l_\gamma|$ for normalized right and left eigenvectors r_γ and l_γ. Introduce the operator $\mathbf{E}$ on trigonometric series by

$$\mathbf{E}(\sum_\gamma V_\gamma(t)\, e^{i\gamma.(T,X)}) = \sum_\gamma \mathbf{E}_\gamma V_\gamma(t)\, e^{i\gamma.(T,X)}\,. \tag{9}$$

The equations determining the profile U from its initial data $U(0,0,X)$ are then

$$\mathbf{E}U = U\,, \quad \text{and} \quad \mathbf{E}\Big(U_t + \sum_j (\mathbf{A}_j\, U)\frac{\partial U}{\partial X_j}\,) = 0\,, \tag{10}$$

$$\mathbf{A}_j = D_u A_j(\underline{u})\,.$$

The main result of [3] studies the behavior of the solution of (10) with initial data so that

$$U(0,T,X) = r_\alpha\, f(\alpha.(T,X)) \; + r_\beta\, f(\beta.(T,X)) \; + \; r_\nu\, f(\nu.(T,X)) \tag{11}$$

$$\alpha = (c,1,0)\,, \quad \beta = (0,1,0)\,, \quad \text{and} \quad \nu = (0,3,4)\,,$$

consists of three plane waves. The function f is given by

$$f(\theta) = \sum_{n\in\mathbf{Z}} e^{-|n|}\, e^{in\theta}\,. \tag{12}$$

Definiton 1 *Denote by Λ the lattice of integer linear combinations of the vectors α, β, and, ν.*

The characteristic covectors in Λ are precisely the wavenumbers one expects to find in the spectrum of U generated by nonlinear effects. In [5] we analyse the initial value problem (10)-(11). There is a unique local solution $U \in C^\infty([0,t_1]\times\mathbf{R}^3)$ which is $2\pi/c, 2\pi, 2\pi$ periodic in T, X_1, X_2. The spectrum of $U(t)$ is contained in the characteristic points of Λ. Concerning the set of amplitudes which are ignited we prove the following.

Theorem 1 *The set of characteristic points $\gamma = (\tau,\xi)$ in $\Lambda\cap\{\tau = c|\xi|\}$ have directions $\xi/|\xi|$ dense in the unit circle. In addition, for any γ in this* *set which is not proportional to any of the five vectors* $(7,-24)$, $(\pm 3,4)$, $(1,0)$, $(0,1)$ *one has*

$$\frac{d^2U_\gamma(0)}{dt^2} \neq 0\,, \quad \text{and,} \quad \frac{d^jU_\gamma(0)}{dt^j} = 0 \quad \text{for} \quad j = 0,1\,. \tag{13}$$

This result asserts that waves traveling in directions dense in the unit sphere are generated by the nonlinear interaction of three waves present initially. The above result does not asssert that the $U_{\tau,\xi}(t)$ are all nonzero for a particular t. In this note we provide a simple proof that they are all nonzero except for at most a countable number of t. We leave unanswered the question of whether for some $0 < t_2 < t_1$, they are all nonzero for $0 < t < t_2$.

Theorem 2 *For the solution* $U(t,T,X)$ *of the last theorem, each of the functions* $U_\gamma(t)$ *is real analytic on* $[0,t_1]$. *In particular, for the* τ,ξ *for which (13) is proved, the* $U_{\tau,\xi}(t)$ *vanishes for at most a finite set of* $t \in [0,t_1]$.

Proof. We indicate two different arguments showing that $U(t,T,X)$ is real analytic. The first is an application of the abstract Cauchy-Kowalewski theorem to the initial value problem (11)-(12). Toward this end introduce a scale of Banach spaces.

Definiton 2 $\mathbf{B}_s$ *is the Hilbert space of triply periodic solutions of the linearized equation,*

$$F(T,X) = \sum_{\lambda \in \Lambda} F_\lambda e^{i\lambda.(T,X)}, \quad \mathbf{E}_\lambda F_\lambda = F_\lambda \tag{14}$$

such that

$$\sum ||F_\lambda||_{\mathbf{C}^3}\, e^{s|\lambda|} < \infty\,. \tag{15}$$

The abstract treatment of the Cauchy-Kowalewski Theorem (see [9] and the references therein) applies virtually without change in the present setting. It shows that if $U(0)$ belongs to $\mathbf{B}_\sigma$ for some $\sigma > 0$, then there is a $t_3 > 0$ and a unique solution U of (10), (11) on $t < t_3$ such that for all $\underline{t} < t_3$ U is continuous on $[0,\underline{t}[$ with values in $B_s(\underline{t})$ with $s(t)$ of the form $\sigma - \lambda t$ for some constant λ. Using the equation to express time derivatives in terms of spatial derivatives it follows that U is a real analytic function of t,T,X for $0 \le t < t_3$. Therefore all the Fourier coefficients $U_\lambda(t)$ are real analytic on $[0,t_3[$.

The advantage of this method is that it is quick and easy. The disadvantage is that it applies on a possibly shorter interval $[0, t_3]$. For hyperbolic Cauchy problems the analogous gap was filled by the result of Mizohata [7] in the linear case and by Alinhac and Metivier [8] in the nonlinear case. They show that for real analytic initial data, solutions remained real analytic as long as they are C^1. Their proof extends to the present context without essential modification. The main difference is the presence of the factor $\mathbf{E}$ in the equation for the profile U. However, $\mathbf{E}$ commutes with derivatives and is bounded on $H^2(\mathbf{T}^3)$ which is all that is needed to control it. For the proof one estimates the rate of growth of the derivatives of U as the order of differentiation increases. One proves by induction on $|\alpha| > 1$ that there are constants M, λ, σ such that for $t < t_1$

$$||\partial_{T,X}^{\alpha} U(t, ., .)||_{H^2(\mathbf{T}^3)} \leq M \sigma^{1-|\alpha|} \frac{(|\alpha| - 1)!}{|\alpha|^2} e^{(|\alpha|-1)\lambda t} . \tag{16}$$

To prove (16), differentiate the equation for U to find

$$\partial_t \partial^{\alpha} U + \mathbf{E}(\sum_j (\mathbf{A}_j U) \partial_{X_j} \partial^{\alpha} U) = \mathbf{E} G_{\alpha} , \tag{17}$$

$$G_{\alpha} = \sum_{0<\beta\leq\alpha} \binom{\alpha}{\beta} \mathbf{A}_j \partial^{\beta} U \, \partial_{X_j} \partial^{\alpha-\beta} U . \tag{18}$$

The basic energy estimate for the profile equation then shows that

$$||\partial^{\alpha} U(t)||_{H^2(\mathbf{T}^3)} \leq ||\partial^{\alpha} U(0)||_{H^2(\mathbf{T}^3)} + C \int_0^t ||G_{\alpha}(t')||_{H^2(\mathbf{T}^3)} \, dt' \tag{19}$$

with a constant C that depends on the C^1 norm of U. For the details of the proof of (16) using (18) and (19) see [8].

This shows that U is real analytic on $[0, t_1] \times \mathbf{R}^3$ which implies the analyticity of the Fourier coefficients $U_{\gamma}(t)$. ◊

Acknowledgements: The authors were partially supported by the North Atlantic Treaty Organization, the Office of Naval Research, and the National Science Foundation under grants CRG 890904, N 014 92 J 1245, and DMS 90 03256 respectively.

References

[1] J. Hunter, A. Majda, and R. Rosales, Resonantly interacting weakly nonlinear hyperbolic waves II: Several space variables, Stud. Appl. Math. 75(1986), 187-226.

[2] J.-L. Joly, and J. Rauch, Nonlinear resonance can create dense Oscillations, in *Microlocal Analysis and Nonlinear Waves* eds. M. Beals, R. Melrose, and J. Rauch, IMA Volumes in Mathematics and its Applications, vol. 30, Springer Verlag.

[3] J.-L. Joly, G. Metivier and J. Rauch, Remarques sur l'optique géométrique non linéaire multidimensionelle, in *Séminaire Equations aux Dérivées Partielles de l'Ecole Polytechnique*, (1990-1991), éxpose 1.

[4] J.-L. Joly, G. Metivier and J. Rauch, Coherent and focusing multidimensional nonlinear geometric optics, Ann. Sci. de l'Ecole Normale Superieur, to appear.

[5] J.-L. Joly, G. Metivier and J. Rauch, Dense Oscillations for the 2-d compressible Euler equations, in *Nonlinear Partial Differential Equations and their Applications, College de France Seminar 1992-1993* eds. J.-L. Lions and H. Brezis, North-Holland, to appear.

[6] S. Schochet, Fast singular limits of hyperbolic PDEs, Journal of Differential Eqns.

[7] S. Mizohata, Analyticity of solutions of hyperbolic systems with analytic coefficients, Comm. Pure Appl. Math. 13(1961), 547-559.

[8] S. Alinhac and G. Metivier, Propagation de l'analyticité des solutions de systèmes hyperboliques non-linéaires, Invent. Math 75(1984), 189-204.

[9] S. Baouendi, and C. Gaoulaouic, Remarks on the abstract form of the nonlinear Cauchy-Kovalevsky theorem, Comm. P. D. E., 2(1977), 1151-1162.

A SYSTEM OF CONSERVATION LAWS WITH A RELAXATION TERM

H. JOACHIM SCHROLL[1], ASLAK TVEITO[2], AND RAGNAR WINTHER[2]

ABSTRACT. We consider a 2×2 system of conservation laws including a stiff relaxation term. Well-posedness of the system, the rate of convergence to equilibrium, and the rate of convergence for a finite difference scheme is discussed. Also a numerical example is presented.

1. INTRODUCTION

The purpose of this paper is to review some recent results concerning the following system of conservation laws

$$
\begin{aligned}
(u+v)_t + f(u)_x &= 0 \\
\delta v_t &= A(u) - v.
\end{aligned} \tag{1}
$$

This system has been studied in the papers [4, 5], and proofs of the results presented here can be found in those papers. In (1), u and v are the unknowns, $\delta > 0$ is referred to as the relaxation time, and $f = f(u)$ and $A = A(u)$ are given functions. This system arises in chromatography, cf. [6].

For the system (1), we prove the existence of an entropy solution of bounded variation for any $\delta > 0$, and it is interesting to note that the variation bound is independent of δ. Thus, we may pass to the limit of zero-relaxation time, and prove that the limit is an entropy solution of the following scalar conservation law,

$$
(w + A(w))_t + f(w)_x = 0. \tag{2}
$$

Moreover, we show the rate of convergence towards equilibrium, measured in L^1, is 1/3. We also study a finite difference scheme applied to the system (1), and prove that the numerical solutions converge towards the entropy solution with the rate

[1]Numerische Mathematik, RWTH–Aachen, Templergraben 55, D-52056 Aachen. E-mail: schroll@igpm.rwth-aachen.de. H. J. Schroll has been supported by The Norwegian Research Council (NFR), program no. 100284/431.

[2]Department of Informatics, P. O. Box 1080 Blindern, University of Oslo, Norway. E-mail: aslak@ifi.uio.no and ragnar@ifi.uio.no. A. Tveito and R. Winther have been supported by The Norwegian Research Council(NFR), program no. 100284/431 and STP.29643.

1/2, again measured in L^1, as the mesh size goes to zero. Since the order 1/2 is known to be optimal for scalar conservation laws, it is also optimal for this system.

Systems of conservation with relaxation terms appear in a series of important applications. A nice introduction is provided in Whithams book [6], where such models arising in e.g. chromatography, traffic modelling, water waves and gas dynamics are discussed. For more recent results we refer to Chen, Levermore and Liu [1] and references given therein.

The rest of this paper is organized as follows; in the next section we review the results concerning the well-posedness of the system (1), in Section 3 the rate of convergence for the finite difference scheme is discussed, in Section 4 the rate of convergence towards the solution of the scalar equation is analyzed, and finally in Section 5 we present some numerical computations.

2. Well-posedness

In this section we will review the existence and uniqueness theory of the system (1) developed in [5], and we start by giving the precise assumptions on the model.

We assume that $f = f(u)$ is a smooth function satisfying $f(0) = 0$, and

$$f'(u) \geq 0 \quad \text{for } u \in [0,1].$$

The function $A = A(u)$ will be assumed to satisfy:

$$\begin{aligned} &A(0) = 0, A(1) = 1, \\ &A'(u) \geq 0, \\ &|A''(u)| \leq \alpha < 1, \end{aligned} \tag{3}$$

for all relevant values of u. We shall consider solutions of (1) in the state-space

$$\mathcal{S} = [0,1] \times [0,1]$$

and solutions of (2) in $[0,1]$.

The initial conditions (u^0, v^0) for the system (1) are supposed to satisfy:

$$\begin{aligned} &\text{i) } (u^0(x), v^0(x)) \in \mathcal{S} \quad \forall x \in \mathcal{R}, \\ &\text{ii) } TV(u^0) + TV(v^0) \leq M, \\ &\text{iii) } \|p^0\|_1 = \|A(u^0) - v^0\|_1 \leq M\delta, \\ &\text{iv) } u^0(\pm\infty), v^0(\pm\infty) = 0. \end{aligned} \tag{4}$$

Here M denotes a finite constant not depending on δ. Note that the variable $p = A(u) - v$ measures the deviation from equilibrium, and that we assume the initial conditions are almost in equilibrium.

For the precise definition of the various norms etc. used in the present paper, we refer to [5]. The entropy solution of the system is defined as follows.

Definition 2.1. *Let (u^0, v^0), satisfying (4), be the given initial data. Then (u, v) is called a entropy solution of the system (1) if the following requirements are satisfied:*

(1) $(u, v) \in \mathcal{S} \quad \forall (x,t) \in \mathcal{R} \times \mathcal{R}_0^+$,

(2) $TV(u(\cdot,t)), TV(v(\cdot,t)) \leq M \quad \forall t \in \mathcal{R}_0^+$,
(3) $\|u(\cdot,t) - u(\cdot,\tau)\|_1 + \|v(\cdot,t) - v(\cdot,\tau)\|_1 \leq M|t-\tau| \quad \forall t,\tau \in \mathcal{R}_0^+$,
(4) *For any* $(k,q) \in \mathcal{S}$ *and any* $\varphi,\psi \in \mathcal{D}_+(T)$, *where* $T > 0$ *is arbitrary,*

$$\begin{aligned}
&\int_0^T \int_{\mathcal{R}} [|u-k|\varphi_t + |f(u)-f(k)|\varphi_x + |v-q|\psi_t] dx\, dt \\
&+\int_{\mathcal{R}} [|u^0-k|\varphi(x,0) + |v^0-q|\psi(x,0)] dx \\
&-\int_{\mathcal{R}} [|u(x,T)-k|\varphi(x,T) + |v(x,T)-q|\psi(x,T)] dx \\
&\geq \frac{1}{\delta}\int_0^T \int_{\mathcal{R}} [\sigma(u-k)\varphi - \sigma(v-q)\psi](A(u)-v) dx\, dt,
\end{aligned} \tag{5}$$

where σ *denotes the sign function.*

In this definition, M denotes a finite constant not depending on δ, and $\mathcal{D}_+(T)$ denotes the set of all nonnegative C^∞-functions with compact support in $\mathcal{R} \times [0,T]$.

The following result concerning well-posedness of (1) is proved in [5].

Theorem 2.1. *Let* (u^0, v^0) *be initial data satisfying (4). Then there is a unique solution* (u,v) *of (1) satisfying the requirements of Definition 2.1. The unique solution also satisfies the following bound,*

$$\|p(\cdot,t)\|_1 = \|A(u)-v\|_1 \leq M\delta \quad \textit{for} \quad t \geq 0,$$

where M *is a finite constant not depending on* δ.

Furthermore, if $(\bar{u}^0, \bar{v}^0)$ *is another pair of initial data satisfying (4), there is a unique entropy solution* $(\bar{u},\bar{v})$ *such that*

$$\|u(\cdot,t) - \bar{u}(\cdot,t)\|_1 + \|v(\cdot,t) - \bar{v}(\cdot,t)\|_1 \leq \|u^0 - \bar{u}^0\|_1 + \|v^0 - \bar{v}^0\|_1 \quad \textit{for} \quad t \geq 0. \qquad \square$$

This strong regularity result enables us both to prove an error estimate for a finite difference scheme and to derive a rate of convergence towards equilibrium.

3. A Finite Difference Scheme

One way to prove the existence part of Theorem 2.1 is to study the following semi-implicit finite difference scheme

$$\begin{aligned}
&\frac{(u_j^{n+1} + v_j^{n+1}) - (u_j^n + v_j^n)}{\Delta t} + \frac{f(u_j^n) - f(u_{j-1}^n)}{\Delta x} = 0, \\
&\frac{\delta(v_j^{n+1} - v_j^n)}{\Delta t} = A(u_j^{n+1}) - v_j^{n+1}.
\end{aligned} \tag{6}$$

The discrete values (u_j^n, v_j^n) computed from (6) are considered as approximations of (u,v) in the cell $I_j := [(j-1/2)\Delta x, (j+1/2)\Delta x)$, at time level $t_n = n\Delta t$. The initial data for the scheme is taken as cell averages of u^0 and v^0

$$u_j^0 = \frac{1}{\Delta x}\int_{I_j} u^0(x)\, dx \quad \text{and} \quad v_j^0 = \frac{1}{\Delta x}\int_{I_j} v^0(x)\, dx. \tag{7}$$

Due to the explicit treatment of the convective term one has to impose a CFL-condition

$$\mu f'(u) \le 1 \text{ for all } u \in [0,1], \text{ where } \mu = \Delta t/\Delta x. \tag{8}$$

It should be noted that the step size is not related to the relaxation time δ.

In [5] the following a-priori bounds for solutions of the scheme are established.

Lemma 3.1. *Suppose that the initial data (u^0, v^0) satisfy (4), and that the grid-parameters Δt and Δx satisfy the CFL-condition (8). Then, for any $\delta > 0$, there is a finite constant M independent of δ, Δt and Δx such that:*

$$\begin{aligned}
&I. \quad (u_j^n, v_j^n) \in \mathcal{S} \quad \text{for all } (j,n) \in \mathcal{Z} \times \mathcal{Z}^+\\
&II. \quad TV(u^n) + TV(v^n) \le TV(u^0) + TV(v^0)\\
&III. \quad \|p^n\|_1 \le M\delta\\
&IV. \quad \|u^n - u^m\|_1 + \|v^n - v^m\|_1 \le M|n-m|\Delta t\\
&V. \quad \|u^n - \bar{u}^n\|_1 + \|v^n - \bar{v}^n\|_1 \le \|u^0 - \bar{u}^0\|_1 + \|v^0 - \bar{v}^0\|_1,
\end{aligned}$$

where $(\bar{u}^n, \bar{v}^n)$ is a discrete solution based on initial data $(\bar{u}^0, \bar{v}^0)$ satisfying (4). □

Here the total variation and norm for grid functions is defined by

$$TV(u) = \sum_{j \in \mathcal{Z}} |u_j - u_{j-1}|, \text{ and } \|u\|_1 = \Delta x \sum_{j \in \mathcal{Z}} |u_j|.$$

Furthermore, $p_j^n = A(u_j^n) - v_j^n$ is the discrete deviation from equilibrium.

By a proper application of Helly's theorem it follows from I-IV of Lemma 3.1 that a subsequence of the family of approximate solutions generated by (6) converges to a pair of functions $(u,v) \in (L^1_{loc}(\mathcal{R} \times \mathcal{R}_0^+))^2$ as the mesh size tends to zero. Using a discrete entropy inequality (see [5]), it can be seen that the limit (u,v) is a entropy solution of (1). Having established the existence of the entropy solution, we are able to derive an error estimate for solutions of the scheme:

Theorem 3.1. *Let (u^0, v^0) be initial data satisfying (4). Define (u_j^0, v_j^0) as cell averages by (7) and consider grid parameters satisfying the CFL-condition (8). Let (u,v) be the entropy solution of (1) and (u^N, v^N) a piecewise constant representation of the data $\{u_j^N, v_j^N\}$ generated by the scheme (6).*

Then for any fixed $T = N\Delta t > 0$ there is a finite constant M independent of Δt, Δx and δ such that

$$\|u(\cdot,T) - u^N\|_1 + \|v(\cdot,T) - v^N\|_1 \le M\sqrt{\Delta t}. \tag{9}$$ □

A proof of this theorem can be found in [4]. Here we will only sketch the main steps of the proof.

Sketch of proof: In order to compare the discrete data $\{u_j^n, v_j^n\}$ generated by the scheme to the entropy solution (u,v) of (1) we define a comparison function $(\bar{u}_\Delta, \bar{v}_\Delta)$ on $\mathcal{R} \times \mathcal{R}_0^+$ which extends the data in the sense that

$$(\bar{u}_\Delta, \bar{v}_\Delta)(x, t_n^+) = \lim_{t \to t_n} (\bar{u}_\Delta, \bar{v}_\Delta)(x,t) = (u_j^n, v_j^n) \text{ for } x \in I_j.$$

This comparison function is defined iteratively. First, we initialize

$$\begin{aligned} \bar{u}_\Delta(x,0) &= \frac{1}{\Delta x}\int_{I_j} u^0(z)\,dz, \\ \bar{v}_\Delta(x,0) &= \frac{1}{\Delta x}\int_{I_j} v^0(z)\,dz \text{ for } x \in I_j. \end{aligned}$$

Then, we iterate the following three steps for $n \in \mathcal{Z}_0^+$.

1) In (t_n, t_{n+1}), $(\bar{u}_\Delta, \bar{v}_\Delta)$ is the solution of

$$\begin{aligned} (\bar{u}_\Delta)_t + f(\bar{u}_\Delta)_x &= 0, \\ (\bar{v}_\Delta)_t &= 0 \end{aligned} \tag{10}$$

with initial data $(\bar{u}_\Delta, \bar{v}_\Delta)(x, t_n^+)$.

2) At t_{n+1} we take cell averages

$$\begin{aligned} &\bar{u}_\Delta(x, t_{n+1}) = \frac{1}{\Delta x}\int_{I_j} \bar{u}_\Delta(z, t_{n+1}^-)dz, \text{ for } x \in I_j \\ &\bar{v}_\Delta(x, t_{n+1}) = \bar{v}_\Delta(x, t_{n+1}^-), \end{aligned} \tag{11}$$

where $\bar{u}_\Delta(\cdot, t_n^-) := \lim\limits_{\substack{t\to t_n \\ t<t_n}} \bar{u}_\Delta(\cdot, t)$.

3) The initial data for the next iteration is defined by solving a backward Euler step

$$\begin{aligned} \bar{u}_\Delta(x, t_{n+1}^+) &= \bar{u}_\Delta(x, t_{n+1}) - \frac{\Delta t}{\delta} p(\bar{u}_\Delta(x, t_{n+1}^+), \bar{v}_\Delta(x, t_{n+1}^+)), \\ \bar{v}_\Delta(x, t_{n+1}^+) &= \bar{v}_\Delta(x, t_{n+1}) + \frac{\Delta t}{\delta} p(\bar{u}_\Delta(x, t_{n+1}^+), \bar{v}_\Delta(x, t_{n+1}^+)). \end{aligned} \tag{12}$$

From the monotonicity property of p it follows that $(\bar{u}_\Delta(x, t_{n+1}^+), \bar{v}_\Delta(x, t_{n+1}^+))$ are well-defined by (12).

For this pair of functions $(\bar{u}_\Delta, \bar{v}_\Delta)$ we derive an Kruzkov type inequality which can be compared to inequality (5). Similar as in Lucier's paper [3], the error estimate is a result of this comparison. For details see [4]. □

4. Convergence towards equilibrium

This section is devoted to the question how well the solution of the scalar equilibrium equation (2) approximates the δ–dependent solution of the relaxed system (1). The result is:

Theorem 4.1. *Let (u^0, v^0) be a pair of initial data satisfying (4), and let $w^0 = u^0$. Then, for any $T > 0$, there is a finite constant M such that*

$$\|u(\cdot,t) - w(\cdot,t)\|_1 \le M\delta^{1/3} \quad \textit{for all } \; 0 \le t \le T.$$

Here (u, v) and w are solutions of (1) and (2) respectively. □

There are two independent arguments to obtain this result. The first one was established in [5] and is based on a parabolic regularization of the system (1):

$$u_t^\epsilon + f(u^\epsilon)_x = \frac{1}{\delta}(v^\epsilon - A(u^\epsilon)) + \epsilon u_{xx}^\epsilon,$$
$$v_t^\epsilon = \frac{1}{\delta}(A(u^\epsilon) - v^\epsilon) + \epsilon v_{xx}^\epsilon.$$

In [5], Lemma 6.1 it is proved that

$$\|u^\epsilon(\cdot,t) - u(\cdot,t)\|_1 + \|v^\epsilon(\cdot,t) - v(\cdot,t)\|_1 \le M\epsilon^{1/2},$$

where (u^ϵ, v^ϵ) solves the regularized problem and (u,v) solves (1). Then we introduce a regularized equilibrium model

$$(13) \qquad (w^\epsilon + A(w^\epsilon))_t + f(w^\epsilon)_x = \epsilon(w^\epsilon + A(w^\epsilon))_{xx}.$$

For this scalar equation, it is well known, cf. Kuznetsov [2], that

$$\|w - w^\epsilon\|_1 \le M\epsilon^{1/2},$$

where w solves (2) and w^ϵ solves (13). Furthermore, Lemma 6.2 of [5] states that

$$\|u^\epsilon - w^\epsilon\|_1 \le M\delta/\epsilon.$$

Finally, choosing $\epsilon = \delta^{2/3}$, the result follows from

$$\|u - w\|_1 \le \|u - u^\epsilon\|_1 + \|u^\epsilon - w^\epsilon\|_1 + \|w^\epsilon - w\|_1 \le M(\epsilon^{1/2} + \delta/\epsilon + \epsilon^{1/2}).$$

The second alternative argument is stated in [4] and uses the error estimate for the finite difference scheme above. We introduce the discretized equilibrium equation

$$(14) \qquad \frac{w_j^{n+1} + A(w_j^{n+1}) - (w_j^n + A(w_j^n))}{\Delta t} + \frac{f(w_j^n) - f(w_{j-1}^n)}{\Delta x} = 0$$

and set $w(x,0) = u(x,0)$. Using the error estimate for the scheme (6) and the scalar Godunov scheme (14), we get

$$(15) \quad \begin{aligned} \|u(\cdot,T) - w(\cdot,T)\|_1 &\le \|u(\cdot,T) - u^N\|_1 + \|u^N - w^N\|_1 + \|w^N - w(\cdot,T)\|_1 \\ &= O(\Delta t^{1/2}) + \|u^N - w^N\|_1. \end{aligned}$$

The discrete deviation $\|u^N - w^N\|_1$ can be estimated by comparing the two schemes. ¿From (6) we get

$$u_j^{n+1} + A(u_j^{n+1}) = u_j^n + A(u_j^n) - \mu(f(u_j^n) - f(u_{j-1}^n)) + p(u_j^{n+1}, v_j^{n+1}) - p(u_j^n, v_j^n)$$

and similarly, from (14), we have

$$w_j^{n+1} + A(w_j^{n+1}) = w_j^n + A(w_j^n) - \mu(f(w_j^n) - f(w_{j-1}^n)).$$

Hence, by using the CFL-condition (8), we obtain

$$\begin{aligned} & \Delta x \sum_j |u_j^{n+1} + A(u_j^{n+1}) - (w_j^{n+1} + A(w_j^{n+1}))| \\ \leq\ & \Delta x \sum_j |u_j^n + A(u_j^n) - (w_j^n + A(w_j^n))| \\ +\ & \Delta x \sum_j (|p(u_j^{n+1}, v_j^{n+1})| + |p(u_j^n, v_j^n)|). \end{aligned}$$

Since $\|p(u^n, v^n)\|_1 \leq M\delta$ for all $n \in \mathcal{Z}_0^+$, cf. III of Lemma 3.1, we get

$$\Delta x \sum_j |u_j^N + A(u_j^N) - (w_j^N + A(w_j^N))| \leq \tilde{M} N \delta = \tilde{M} T \delta / \Delta t,$$

where the constant $\tilde{M}$ is independent of Δt and δ. Using the monotonicity properties of A, we obtain

$$\|u^N - w^N\|_1 \leq \tilde{M} T \delta / \Delta t. \tag{16}$$

Now, the desired result follows from (15) and (16) choosing $\Delta t = \delta^{2/3}$.

5. A NUMERICAL EXAMPLE

In this section will present a numerical example of convergence towards equilibrium. Further experiments are discussed in [4] where it is observed, numerically, that the rate of convergence towards equilibrium is about 1/2 for one particular example. This shows that the convergence towards equilibrium is slower than a straightforward analysis of smooth solutions would indicate, but we have not been able to find numerical evidence for the rate of 1/3. We therefore consider the problem of determining the optimal rate of convergence to be open.

In the present example, we consider the functions

$$A(u) = \frac{2u}{u+1}, \quad \text{and} \quad f(u) = \frac{1}{2}u^2,$$

and the initial data,

$$u(x,0) = w(x,0) = \begin{cases} 1, & x \leq 0 \\ 0, & x > 0, \end{cases}$$

and $v(x,0) = A(u(x,0))$.

Figure 1 shows the exact solution of the equilibrium model (2) and the numerical solution of the non-equilibrium model (1) generated by the finite difference scheme (6) for some values of the relaxation time δ. In the present computations, we used the parameters

$$\Delta x = 0.0005 \quad \Delta t = 0.00025 \quad \text{and } T = 1.$$

Note that the non-equilibrium solutions converge towards the single shock of the scalar equation. For this example, the rate of convergence is estimated to be about 1, cf. the numerical section of [4].

References

1. G. Q. Chen, C. D. Levermore, and T. P. Liu, *Hyperbolic conservation laws with stiff relaxation terms and entropy*, submitted to Comm. on Pure and App. Math. (1992).
2. N. N. Kuznetsov, *Accuracy of some approximate methods for computing the weak solutions of a first order quasi linear equation*, USSR. Comp. Math. and Math. Phys. **16** (1976), 105–119.
3. B. J. Lucier, *Error bounds for the methods of Glimm, Godunov and LeVeque*, SIAM J. Numer. Anal. **22** (1985), 1074–1081.
4. H. J. Schroll, A. Tveito, and R. Winther, *An error bound for a finite difference scheme applied to a stiff system of conservation laws*, Preprint, Dep. of Informatics, Univ. of Oslo, Norway, 1994.
5. A. Tveito and R. Winther, *On the rate of convergence to equilibrium for a system of conservation laws including a relaxation term*, Preprint, Dep. of Informatics, Univ. of Oslo, Norway, 1994.
6. G. B. Whitham, *Linear and nonlinear waves*, John Wiley & Sons, New York, 1973.

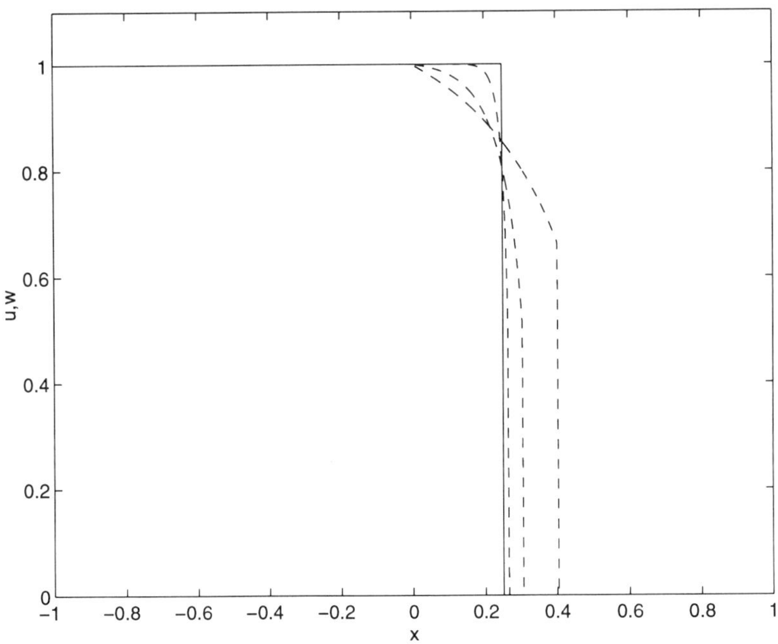

FIGURE 1. *The figure shows the solution of the equilibrium model (solid lines) and numerical solutions of the non-equilibrium model (dotted lines) for* $\delta = 1, 1/4, 1/16$, *at the time* $t = 1$.

FULLY NONLINEAR HYPERBOLIC SYSTEMS RELATED TO HYPOPLASTICITY

by

Michael Shearer[1]

1 Introduction

In the theory of hypoplasticity, the equations of motion for two-dimensional deformations take the form

$$\begin{array}{llll} (a) & \partial_t v_i & = \partial_k \tau_{ik}, & i = 1,2 \\ (b) & \partial_t \tau_{ij} & = G_{ij}(\tau, V), & 1 \leq i,j \leq 2. \end{array} \tag{1.1}$$

In these equations, the independent variables are time t and space $x = (x_1, x_2)$; the dependent variables are velocity $v = (v_1, v_2)$ and stress $\tau = (\tau_{ij})$. (We use the notation $\partial_t = \partial/\partial t, \partial_k = \partial/\partial x_k$.) The tensor $V = (V_{ij})$ is the strain rate tensor, defined by

$$V_{ij} = -\frac{1}{2}\left(\partial_i v_j + \partial_j v_i\right). \tag{1.2}$$

Equations (1.1a) are conservation of momentum. (We have normalized the density, which we take to be constant, to be 1.) Equations (1.1b) are the constitutive law. In hypoplasticity, the constitutive law is taken to be *rate-independent*, meaning the changes in the time scale do not affect the equations of motion. Rate independence implies that $G = (G_{ij})$ is homogeneous of degree one in strain rate:

$$G_{ij}(\tau, \alpha V) = \alpha G_{ij}(\tau, V), \quad \alpha > 0. \tag{1.3}$$

Since the stress tensor (τ_{ij}) is symmetric, $\tau_{ij} = \tau_{ji}$, equations (1.1) imply G_{ij} is also symmetric:

$$G_{ij}(\tau, V) = G_{ji}(\tau, V). \tag{1.4}$$

Thus, system (1.1) reduces to five equations in five unknowns $v_1, v_2, \tau_{11}, \tau_{22}, \tau_{12} = \tau_{21}$.

In this paper, we consider plane wave solutions of system (1.1), under the simplification that $G(\tau, V)$ is independent of τ. Without loss of generality, we may assume the direction of propagation

[1]Department of Mathematics, North Carolina State University, Raleigh, NC 27695. Research supported by National Science Foundation grant DMS 9201115, which includes funds from AFOSR, and by Army Research Office grants DAAL03-91-G-0122, DAAH04-94-G-0043.

of the plane waves is parallel to the x_1-axis. Relabeling x_1 as x, we have the 4×4 system

$$\begin{aligned} \partial_t v &= \partial_x T \\ \partial_t T &= F(\partial_x v), \end{aligned} \tag{1.5}$$

where $v = (v_1, v_2), T = (\tau_{11}, \tau_{12})$, and $F(v_x) = (G_{11}, G_{12})(\tau, V)$. The fifth equation

$$\partial_t \tau_{22} = F_{22}(\partial_x v) \equiv G_{22}(\tau, V) \tag{1.6}$$

decouples from those of (1.5). From (1.3), we see that F is homogeneous of degree one:

$$F(\alpha w) = \alpha F(w) \text{ for all } \alpha > 0. \tag{1.7}$$

The Riemann problem for (1.5), (1.6) is the initial value problem with piecewise constant initial data, having a single jump:

$$\begin{aligned} v(x,0) &= \begin{cases} v_L & \text{if } x < 0 \\ v_R & \text{if } x > 0 \end{cases} && (a) \\ T(x,0) &= \begin{cases} T_L & \text{if } x < 0 \\ T_R & \text{if } x > 0 \end{cases} && (b) \\ \tau_{22}(x,0) &= \begin{cases} \tau_{22}^L & \text{if } x < 0 \\ \tau_{22}^R & \text{if } x > 0 \end{cases} && (c) \end{aligned} \tag{1.8}$$

The solutions we seek are piecewise constant functions of x/t, with jump discontinuities (shocks) satisfying jump conditions similar to those of the 2×2 systems studied in [4, 5]. Indeed, our analysis of the Riemann problem (1.5), (1.6), (1.8) in many ways resembles our earlier analysis of the 2×2 system

$$v_t + F(v_x) = 0. \tag{1.9}$$

In Section 2, we give a geometric construction that reduces the Riemann problem to a problem of solving equations of the form

$$g(v_R - v_L, w) = T_R - T_L, \tag{1.10}$$

where $g : \mathbb{R}^2 \times \mathbb{R}^2 \to \mathbb{R}^2$ is piecewise linear in the second argument, and $g(0, \cdot)$ is odd. A simple degree theory argument then yields the existence of a solution for all data (1.8).

In Section 3, we demonstrate that, under the conditions that naturally yield existence of solutions (and which also yield uniqueness in the 2×2 case), the solution of the Riemann problem is not unique in general. Specifically, we exhibit F for which $g(0, \cdot)$ maps an entire quadrant onto a line.

This paper is a preliminary report on joint work with David Schaeffer [6].

2 Existence of solutions

As in [4, 5], we say a shock wave

$$(v,T)(x,t) = \begin{cases} (v_-,T_-) \text{ if } x < st \\ (v_+,T_+) \text{ if } x > st \end{cases} \tag{2.1}$$

is a weak solution of (1.5) if the jump $(w,R) = (v_+ - v_-, T_+ - T_-)$ is *radial* with *nonlinear wave speed* s, meaning

$$\begin{aligned} -sw &= R \\ -sR &= F(w). \end{aligned} \tag{2.2}$$

From (2.2), we derive

$$s^2 w = F(w), \tag{2.3}$$

motivating the following definition, analogous to the 2×2 case: w is a *radial direction* for F, with *nonlinear eigenvalue* λ if

$$F(w) = \lambda w. \tag{2.4}$$

The radial points for (1.5) are of the form $\begin{pmatrix} v \\ -sv \end{pmatrix}$, where v is a radial direction for F, with nonlinear eigenvalue $\lambda = s^2$.

In order to guarantee real wave speeds s, we impose the following condition on F.

(H) $F(v).v > 0$ for all $v \neq 0$ in $\mathbb{R}^2$.

Remark: Regarding G as a function only of $(\partial_i v_j)$ in (1.1), the corresponding condition is that plane waves (including smooth plane waves) have real wave speeds, which is implied by the condition $\xi \otimes w.G(\xi \otimes w) > 0$ for all $\xi \in \mathbb{R}^2, w \in \mathbb{R}^2, \xi \neq 0, w \neq 0$.

It is convenient to introduce polar coordinates (ρ, ϕ) for V. Let

$$\hat{r}(\phi) = (\cos\phi, \sin\phi)^T, \quad \hat{\theta}(\phi) = (-\sin\phi, \cos\phi)^T.$$

be unit vectors in the radial and angular directions on the unit circle. We identify radial and angular components of F as follows:

$$R(\phi) = \langle \hat{r}(\phi), F(\hat{r}(\phi)) \rangle \qquad \Theta(\phi) = \langle \hat{\theta}(\phi), F(\hat{r}(\phi)) \rangle \tag{2.5}$$

where $\langle \cdot, \cdot \rangle$ denotes the standard inner product.

Recall from [5] that V is a *radial point* if it is a radial direction and $|V| = 1$. Observe that $V = \hat{r}(\phi)$ is a radial point for F if and only if $\Theta(\phi) = 0$; the corresponding nonlinear eigenvalue is given by $\lambda = R(\phi)$. We say a radial point corresponding to angle ϕ_0 is *nondegenerate* if $\Theta'(\phi_0) \neq 0$, where prime indicates differentiation with respect to ϕ. Since Θ is a smooth 2π-periodic function, we have the following result.

If all the radial points for F are nondegenerate, then F has an even number of radial points.

If F has $2n$ nondegenerate radial points $V_1, \dots, V_{2n}$, we call the radial points *confined* if they all belong to some closed half plane $\{V : \langle u, V\rangle \geq 0\}$ where $u \in \mathbb{R}^2$ is nonzero. If no such half plane exists, we call the radial points *unconfined.* Note that there must be at least four radial points for them to be unconfined. We say that two radial points are *adjacent* if there is no radial point between them on the unit circle.

To solve the Riemann problem for (1.1), we impose the following condition:

(C1) There are exactly $2n$ radial points $V_1, V_2, \dots, V_{2n}$ for F, all nondegenerate, they are unconfined, and adjacent radial points have different nonlinear eigenvalues.

For $k = 1, \dots, 2n$, let λ_k be the nonlinear eigenvalue associated with V_k. The corresponding radial points for equation (1.5) (i.e., satisfying (2.2)) are $(V_k, \pm s_k V_k)$, where $s_k^2 = \lambda_k, s_k > 0$. These radial points have nonlinear wave speeds $\mp s_k$, respectively.

Our construction of solutions of the Riemann problem is based on the following piecewise linear, homogeneous continuous mapping $L : \mathbb{R}^2 \to \mathbb{R}^2$. By condition (C1), any $V \in \mathbb{R}^2$ may be uniquely expressed in the form

$$V = \alpha V_i + \beta V_j, \tag{2.6}$$

where V_i, V_j are adjacent radial points for F, $\alpha \geq 0, \beta \geq 0$. (Note that if V is a multiple of V_i, there is an inconsequential ambiguity in the choice of V_j. Similarly, if $V = 0$, the choice of both i and j is ambiguous, but irrelevant.) Define

$$L(V) = \alpha s_i V_i + \beta s_j V_j. \tag{2.7}$$

Then any $(v_L, T_L) \in \mathbb{R}^2 \times \mathbb{R}^2$ may be joined to each point $(v_L + V, T_L + L(V)), V \in \mathbb{R}^2$ by a pair of shock waves with negative speeds $-s_i, -s_j$. More precisely, if $s_i < s_j$, the function

$$(v,T)(x,t) = \begin{cases} (v_L, T_L) & \text{if } x < -s_j t \\ (v_L + \beta V_j, T_L + \beta s_j V_j) & \text{if } -s_j t < x < -s_i t \\ (v_L + V, T_L + L(V)) & \text{if } -s_i t < x \end{cases} \tag{2.8}$$

is a weak solution of (1.5).

Similarly, any point $(v_o, T_o) \in \mathbb{R}^2 \times \mathbb{R}^2$ may be joined to each point $(v_o + W, T_o - L(W)), W \in \mathbb{R}^2$ by a pair of shock waves with positive speeds s_i, s_j.

Now consider the Riemann problem. First, we observe that it is enough to solve the Riemann problem (1.5), (1.8a,b). The reason is that, for each shock wave with speed $s \neq 0$ that is a solution of (1.5), (1.6), the jump $\tau_{22}^+ - \tau_{22}^-$ in τ_{22} is given in terms of the jump in v, by (1.6), viz.,

$$-s(\tau_{22}^+ - \tau_{22}^-) = F_{22}(v_+ - v_-). \tag{2.9}$$

Thus, for $s \neq 0$, the jump in τ_{22} is determined by the corresponding shock wave solution of (1.5). For $s = 0$, we require v to be continuous (i.e., $v_- = v_+$), which implies T is continuous, but τ_{22} may have an arbitrary jump, according to (2.9), with $s = 0$ and $v_- = v_+$. Suppose we have a solution $(v,t)(x/t)$ of (1.5), (1.8a,b). We then construct $\tau_{22}(x/t)$ to be piecewise constant according to the above rule across shock waves for (1.5). Working from τ_{22}^L to the right through waves with negative speed, we arrive at the value τ_{22}^-, and working from τ_{22}^R to the left through waves with positive speed, we arrive at a value τ_{22}^+. We complete the definition of $\tau_{22}(x/t)$ by the jump discontinuity at $x = 0$ from τ_{22}^- to τ_{22}^+. The resulting function $(v, T, \tau_{22})(x/t)$ solves the Riemann problem (1.5), (1.6), (1.8).

To solve (1.5), (1.8a,b), we wish to connect (v_L, T_L) to a point (v_0, T_0) using shock waves with negative speed, and the same point (v_0, T_0) to (V_R, T_R) using waves with positive speed. Thus

$$\begin{pmatrix} v_0 \\ T_0 \end{pmatrix} = \begin{pmatrix} v_L \\ T_L \end{pmatrix} + \begin{pmatrix} V \\ L(V) \end{pmatrix}$$

for some $V \in \mathbb{R}^2$, and

$$\begin{pmatrix} v_R \\ T_R \end{pmatrix} = \begin{pmatrix} v_0 \\ T_0 \end{pmatrix} + \begin{pmatrix} W \\ -L(W) \end{pmatrix}$$

for some $W \in \mathbb{R}^2$. Eliminating (v_0, T_0), we have

$$\begin{aligned} V + W &= v_R - v_L \\ L(V) - L(W) &= T_R - T_L. \end{aligned} \tag{2.10}$$

Let $X = v_R - v_L, Y = T_R - T_L$. Then $W = X - V$, and we have

$$L(V) - L(X - V) = Y \tag{2.11}$$

Theorem 2.1 *Suppose F satisfies (H) and (C1). Then for every $(X, Y) \in \mathbb{R}^2 \times \mathbb{R}^2$, there is $V \in \mathbb{R}^2$ satisfying (2.11).*

Proof: Let $g(X, V)$ denote the left hand side of (2.11). Then $V \mapsto g(0, V) = L(V) - L(-V)$ is an odd continuous mapping H. Suppose for a contradiction that $H(V) = 0$ for some $V \neq 0$. Then

$$L(V) = L(-V) \tag{2.12}$$

But each wedge $W_{ij} = \{\alpha V_i + \beta V_j,\ \alpha \geq 0,\ \beta \geq 0\}$ between adjacent radial points V_i, V_j is invariant under L. Suppose $V \in W_{ij}$. Then $L(V) \in W_{ij}$. But by (C1), the radial points are unconfined. Consequently, $-V$ lies in a different wedge $W_{k\ell}$ and therefore $L(-V) \in W_{k\ell}$ also. This contradicts (2.12) and we conclude that $V = 0$ is the only zero of the odd map H.

Now let $B \subset \mathbb{R}^2$ be a large ball centered at $V = 0$. Since H is odd, and $H(V) = 0$ if and only if $V = 0$, the degree, denoted $\deg(H, 0, B)$, of H (with respect to the target point $0 \in \mathbb{R}^2$, and domain B) is an odd integer. By continuity of $g(X, \cdot)$ with respect to X, and the continuity of degree with respect to its arguments, there is $\epsilon > 0$ such that $\deg(g(p, \cdot), q, B)$ is odd for all $p \in \mathbb{R}^2, q \in \mathbb{R}^2$ satisfying $|p| < \epsilon, |q| < \epsilon$. Therefore, there is at least one solution $V_{p,q} \in B$ of the equation

$$g(p, V) = q \tag{2.13}$$

whenever $|p| < \epsilon, |q| < \epsilon$.

Now consider arbitrary $(X, Y) \in \mathbb{R}^2 \times \mathbb{R}^2$. Choose $\gamma > 0$ so that $\gamma|X| < \epsilon, \gamma|Y| < \epsilon$, and define $p = \gamma X, q = \gamma Y$. Let $V = V_{p,q}/\gamma$. Then $V_{p,q} = \gamma V$, and using the homogeneity of L, we see that (2.11) is satisfied. ■

Theorem 2.1 leads immediately to the following result.

Theorem 2.2 *The Riemann problem for system (1.5), (1.6) has at least one solution consisting of constants separated by shock waves.*

Remark. In paper [6], we show that the solutions constructed above are approximated by scale invariant solutions of the equations with the appropriate form of Dafermos regularization [1]. The analysis, which we omit here, is similar to the 2×2 case, but with some interesting additional technicalities.

3 An example with multiple solutions

Finally, we provide an example that demonstrates nonuniqueness of solutions, and gives some insight into what is required to recover uniqueness. Note that the map H:

$$H(V) = L(V) - L(-V)$$

is linear in the wedges between the lines

$$\{\alpha V_i : \alpha \in \mathbb{R}\}, \qquad i = 1, 2, \ldots, 2n.$$

Thus, H is defined by its action on the set of points $\{V_i, i = 1, \ldots, 2n\} \cup \{-V_i : i = 1, \ldots, 2n\}$. Somewhat more precisely, consider the map $h(V) = \frac{H(V)}{|H(V)|}$. Then h maps the unit circle onto itself.

Like H, h is defined by the locations of the points $h(V_i)$ and $h(-V_i)$, $i = 1, \ldots, 2n$. Now consider the V_i's and $-V_i$'s in order, say $W_1, \ldots, W_{4n}$ around the unit circle. If $h(W_1), \ldots, h(W_{4n})$ has the same order, then h, and hence H, is bijective. Such a property would provide uniqueness of solutions of the Riemann problem in the case $v_R = v_L$. However, it is not clear how this ordering property would affect uniqueness of solutions when $v_L \neq v_R$.

What is now clear is that to construct an example for which the Riemann problem has more than one solution, we need only construct an h that destroys the order of the W_k's. In fact, we shall do something more degenerate, which is to map two adjacent W_k's onto the same point. Here is the example. Note that H depends only on the choice of $V_1, \ldots, V_{2n}$ and $\lambda_1, \ldots, \lambda_{2n}$. We choose $n = 2$ and

$$V_1 = \begin{pmatrix} 1 \\ 0 \end{pmatrix}, \lambda_1 = 17/4;\ V_2 = \begin{pmatrix} 0 \\ 1 \end{pmatrix}, \lambda_2 = \frac{1}{2};\ V_3 = \begin{pmatrix} -1 \\ 1 \end{pmatrix}, \lambda_3 = 1;\ V_4 = \begin{pmatrix} 1 \\ -2 \end{pmatrix}, \lambda_4 = \frac{1}{8}.$$

In this example, (C1) is satisfied, and H maps the first quadrant onto the ray $\{c \begin{pmatrix} 7 \\ -2 \end{pmatrix} : c \geq 0\}$.

References

[1] C.M. Dafermos, Solution of a Riemann problem for a class of hyperbolic systems of conservation laws by the viscosity method. *Arch. Rat. Mech. Anal.*, **52** (1973), 1-9.

[2] D. Kolymbas, An outline of hypoplasticity. Arch. Appl. Mech. **61** (1991), 143-151.

[3] P.D. Lax, Hyperbolic systems of conservation laws II. *Comm. Pure Appl. Math.*, **10** (1957), 537-566.

[4] M. Shearer and D.G. Schaeffer, A Class of Fully Nonlinear 2×2 Systems of Partial Differential Equations. Submitted to Comm. PDE.

[5] M. Shearer and D.G. Schaeffer, Riemann problems for 2×2 systems of fully nonlinear equations related to hypoplasticity. Submitted to Comm. PDE.

[6] M. Shearer and D.G. Schaeffer, Riemann problems for 5×5 systems of fully nonlinear equations related to hypoplasticity. Preprint.

A genuinely multidimensional upwind scheme for the compressible Euler equations

David Sidilkover*†‡

Abstract

We present a new approach towards the construction of a genuinely multidimensional high-resolution scheme for computing steady-state solutions of the Euler equations of gas dynamics. The unique advantage of this approach is that Gauss-Seidel relaxation is stable when applied directly to high-resolution discrete equations, thus allowing us to construct a very simple and efficient multigrid steady-state solver. This is the only high-resolution scheme known to us that has this property. The two-dimensional scheme is presented in detail. It is formulated on triangular (possibly unstructured) meshes and can be interpreted as a genuinely two-dimensional extension of the Roe scheme. The quality of the solutions obtained using this scheme and the performance of the multigrid algorithm are illustrated by numerical experiments.

1 Introduction

The main motivation for constructing the truly multidimensional scheme in this work is the improvement of multigrid efficiency. It was observed in [10] that pointwise Gauss-Seidel relaxation is unstable when applied directly to high-resolution dimensionally split schemes even in the simple case of a two-dimensional scalar advection equation. Therefore, the steady-state Euler solver constructed in [11] relies on the defect correction technique, which is not a fully efficient way to utilize multigrid methods. Another possibility for avoiding this difficulty is the use of a multigrid algorithm that employs a well-known multi-stage Runge-Kutta relaxation technique which was introduced in [3].

However, any further improvement of multigrid efficiency requires us to address this problem directly: it is necessary to develop a new high-resolution (at least at the steady-state) discrete scheme, such that Gauss-Seidel relaxation is stable when applied directly to the resulting discrete equations. This was the main motivation for constructing the genuinely two-dimensional advection scheme in the control volume context in [6],[8]. The novel feature of this scheme was two-dimensional limiter, i.e., a limiter function that relies on the ratio of two finite differences in *different* directions.

Another class of genuinely two-dimensional advection schemes has also been developed — the so-called "fluctuation-splitting" schemes (see [2]).

A strong relationship between fluctuation-splitting and control volume advection schemes was established in [9]. As a result, a fluctuation-splitting scheme utilizing two-dimensional limiters was constructed. The action of the limiter function can be given a purely algebraic interpretation.

Much of the research effort in the last several years concentrated on finding a way to apply multidimensional advection schemes to discretize hyperbolic systems of equations, in particular

*Courant Institute of Mathematical Sciences, New York University, New York, NY 10012

†Current address: Institute for Computer Applications in Science and Engineering, Mail Stop 132C, NASA Langley Research Center, Hampton, VA 23681. sidilkov@icase.edu

‡This research was supported by DOE grant DE-FG02-92ER25139

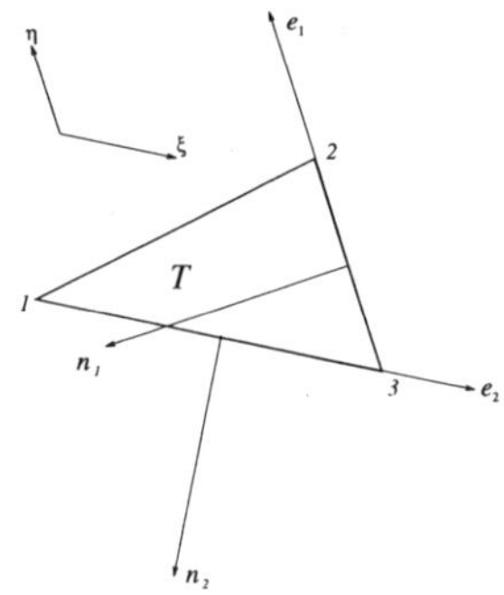

Figure 1: Triangle.

the Euler system of gas dynamics. One of the major directions was the so-called wave modelling (see [4],[1]). This approach concentrated on finding a way to represent (locally) the physics of two-dimensional flow by a finite number of simple waves, each having an associated advection equation.

The approach presented in this work is concerned not with applying an advection scheme to discretize a hyperbolic system of equations in two dimensions, but rather with applying the same strategy that was used when constructing a scalar advection scheme. The "algebraization" of the advection scheme formulation appeared to be crucial for this purpose. The resulting scheme for the Euler system is capable of producing high-resolution steady-state solutions. Its unique feature is that Gauss-Seidel relaxation is stable when applied directly to high resolution discrete equations.

The paper is organized as follows: in §2 we give a brief description of the fluctuation-splitting approach for scalar advection equations in two dimensions. Then we construct a truly two-dimensional scheme for the Euler systems (in §3). Numerical experiments are presented in §4, followed by a summary in §5.

2 Advection scheme in two dimensions

Consider an unstructured triangular grid covering our domain and triangle T belonging to this grid (see Fig. 1). Denote by h_i the length of the ith face of this triangle and by $\vec{e}_i$ a unit vector along the ith face in the clockwise direction. Also let $\vec{n}_i$ be a unit inward normal to the ith face, and let S_T be area of the triangle T.

Consider a linear constant coefficient advection equation

$$u_t + \vec{\lambda} \cdot \vec{\nabla} u = 0. \tag{2.1}$$

Choose a non-orthogonal coordinate system ξ, η aligned with the vectors e_1, e_2 (see Fig. 1) and introduce the quantities

$$\alpha = \frac{\vec{\lambda} \cdot \vec{n}_2}{\theta} \quad \text{and} \quad \beta = \frac{\vec{\lambda} \cdot \vec{n}_1}{\theta} \tag{2.2}$$

where

$$\theta = \vec{e}_1 \cdot \vec{n}_2 \equiv \vec{e}_2 \cdot \vec{n}_1. \tag{2.3}$$

Equation (2.1) can then be rewritten in the coordinates (ξ, η) as

$$u_t + \alpha u_\xi + \beta u_\eta = 0. \tag{2.4}$$

The fluctuation of the Eq. (2.1) on triangle T (Fig. 1) is defined as

$$R = R^\xi + R^\eta, \tag{2.5}$$

where

$$R^\xi = -S_T[\alpha(u_3 - u_1)/h_1] \quad \text{and} \quad R^\eta = -S_T[\beta(u_2 - u_3)/h_2] \tag{2.6}$$

The action of the fluctuation-splitting scheme can be interpreted as a two-stage process (see [2]):

- compute the fluctuation on each element and distribute (split) it among the vertices of this element;
- update the solution at each vertex due to the accumulated portions of fluctuations.

The fluctuation distribution formulae

$$\begin{aligned} S_1 u_1^{n+1} &= S_1 u_1^n + \tfrac{\tau}{2}(R^\xi - \bar{R}^\xi), \\ S_2 u_2^{n+1} &= S_2 u_2^n + \tfrac{\tau}{2}(R^\eta + \bar{R}^\eta), \\ S_3 u_3^{n+1} &= S_3 u_3^n + \tfrac{\tau}{2}[(R^\xi + \bar{R}^\xi) + (R^\eta - \bar{R}^\eta)], \end{aligned} \tag{2.7}$$

with the artificial viscosity terms defined by

$$\bar{R}^\xi = \text{sign}(\alpha) R^\xi \quad \text{and} \quad \bar{R}^\eta = \text{sign}(\beta) R^\eta, \tag{2.8}$$

result in a positive type upwind scheme.

Define

$$R^{\xi^*} = R^\xi + R^\eta \Psi(Q) \quad \text{and} \quad R^{\eta^*} = R^\eta + R^\xi \frac{\Psi(Q)}{Q}, \tag{2.9}$$

where Ψ is a limiter function and $Q = -R^\xi / R^\eta$. Consider the scheme obtained by substituting R^{ξ^*}, R^{η^*} instead of R^ξ, R^η in (2.7),(2.8). It is easy to verify that the resulting scheme is positive if

$$0 \le \Psi(Q) \le 1, \quad 0 \le \frac{\Psi(Q)}{Q} \le 1 \tag{2.10}$$

and linearity preserving if

$$\Psi(1) = 1. \tag{2.11}$$

Note also that

$$R = R^{\xi^*} + R^{\eta^*} = R^\xi + R^\eta, \tag{2.12}$$

so that the constructed scheme is conservative.

3 Genuinely two-dimensional scheme for the Euler system

Euler equations of gas dynamics in two dimensions can be written as

$$\boldsymbol{u}_t + \boldsymbol{F}(\boldsymbol{u})_x + \boldsymbol{G}(\boldsymbol{u})_y = \boldsymbol{0}, \tag{3.1}$$

where:

$$\boldsymbol{u} = \begin{pmatrix} \rho \\ \rho u \\ \rho v \\ e \end{pmatrix}; \quad \boldsymbol{F}(\boldsymbol{u}) = \begin{pmatrix} \rho u \\ \rho u^2 + p \\ \rho uv \\ \rho u H \end{pmatrix}; \quad \boldsymbol{G}(\boldsymbol{u}) = \begin{pmatrix} \rho v \\ \rho uv \\ \rho v^2 + p \\ \rho v H \end{pmatrix} \tag{3.2}$$

The enthalpy H is defined by

$$H = \frac{e+p}{\rho} = \frac{c^2}{\gamma - 1} + \frac{u^2+v^2}{2}, \tag{3.3}$$

and the speed of sound c and the pressure p are

$$c = \sqrt{\frac{\gamma p}{\rho}} \quad \text{and} \quad p = (\gamma-1)(e - \rho\frac{u^2+v^2}{2}) \tag{3.4}$$

The fluctuation of system (3.1) over the triangle T (see Fig. 1) is

$$\boldsymbol{R} = \int\!\!\int \boldsymbol{u}_t = -\int\!\!\int (\boldsymbol{F}_x + \boldsymbol{G}_y)\, dx\, dy = -S_T\left[\widehat{\boldsymbol{F}}_x + \widehat{\boldsymbol{G}}_y\right] \tag{3.5}$$

where $\widehat{\boldsymbol{F}}_x, \widehat{\boldsymbol{G}}_y$ are some averaged values of the flux derivatives over the triangle T.

Introduce the non-orthogonal coordinate system (ξ, η) aligning with two of the faces of triangle T (or the unit vectors $\vec{e}_1, \vec{e}_2$) as illustrated on Fig. 1. Define the quantities

$$\mathcal{U} = \vec{U}\cdot\vec{e}_1, \quad \mathcal{V} = \vec{U}\cdot\vec{e}_2 \tag{3.6}$$

and

$$\alpha = \frac{\vec{U}\cdot\vec{n}_2}{\theta}, \quad \beta = \frac{\vec{U}\cdot\vec{n}_1}{\theta}. \tag{3.7}$$

The non-conservative *auxiliary* variable formulation of the Euler system in these new non-orthogonal coordinates (ξ, η) can be written as follows

$$\begin{aligned} &s_t + \alpha s_\xi + \beta s_\eta = 0 \\ &\rho\mathcal{U}_t + \rho\alpha\mathcal{U}_\xi + \rho\beta\mathcal{U}_\eta + p_\xi = 0 \\ &\rho\mathcal{V}_t + \rho\alpha\mathcal{V}_\xi + \rho\beta\mathcal{V}_\eta + p_\eta = 0 \\ &p_t + \alpha p_\xi + \beta p_\eta + \rho c^2(\alpha_\xi + \beta_\eta) = 0 \end{aligned} \tag{3.8}$$

where $ds = d\rho - dp/c^2$. In order to compute the fluctuation of the system (3.8)

$$\boldsymbol{r} = \boldsymbol{r}^\xi + \boldsymbol{r}^\eta \tag{3.9}$$

on a general triangle T, we follow the two-dimensional linearization procedure presented in [5] and assume the quantity varying linearly on the triangle is the "parameter vector" $\boldsymbol{m} = \sqrt{\rho}(1, u, v, H)^T$. Then

$$\boldsymbol{r}^\xi = -S_T\left(\tilde{\alpha}\widehat{s_\xi},\ \tilde{\alpha}\widehat{\rho\mathcal{U}_\xi} + \widehat{p_\xi},\ \tilde{\alpha}\widehat{\rho\mathcal{V}_\xi},\ \tilde{\alpha}\widehat{p_\xi} + \tilde{c}^2\widehat{\rho\alpha_\xi}\right)^T \tag{3.10}$$

where the averaged values $\tilde{\alpha}, \widehat{s}_\xi, \widehat{p}_\xi, \widehat{\rho\mathcal{U}_\xi}, \widehat{\rho\mathcal{V}_\xi}, \widehat{\rho\alpha_\xi}$ are computed according to [5]. $\boldsymbol{r}^\eta$ can be evaluated in a similar way.

It can be verified easily that the fluctuations satisfy the relation

$$\boldsymbol{R} = C_a^\theta \cdot \boldsymbol{r}, \tag{3.11}$$

where

$$C_a^\theta = \begin{pmatrix} 1 & 0 & 0 & 1/\tilde{c}^2 \\ \tilde{u} & n_2^y/\theta & n_1^y/\theta & \tilde{u}/\tilde{c}^2 \\ \tilde{v} & n_2^x/\theta & n_1^x/\theta & \tilde{v}/\tilde{c}^2 \\ \tilde{U}^2/2 & \tilde{\alpha} & \tilde{\beta} & 1/(\gamma-1)+\tilde{U}^2/(2\tilde{c}^2) \end{pmatrix} \tag{3.12}$$

The fluctuation distribution formulae for the conservative variables update is then

$$\begin{array}{lll} S_1\boldsymbol{u}_1^{n+1} & = S_1\boldsymbol{u}_1^n & +\frac{\tau}{2}C_a^\theta(\boldsymbol{r}^\xi - \bar{\boldsymbol{r}}^\xi) \\ S_2\boldsymbol{u}_2^{n+1} & = S_2\boldsymbol{u}_2^n & +\frac{\tau}{2}C_a^\theta(\boldsymbol{r}^\eta + \bar{\boldsymbol{r}}^\eta) \\ S_3\boldsymbol{u}_3^{n+1} & = S_3\boldsymbol{u}_3^n & +\frac{\tau}{2}C_a^\theta[(\boldsymbol{r}^\xi + \bar{\boldsymbol{r}}^\xi) + (\boldsymbol{r}^\eta - \bar{\boldsymbol{r}}^\eta)] \end{array} \tag{3.13}$$

The relationship between artificial viscosity and fluctuations is given by

$$\bar{\boldsymbol{r}}^\xi = M_\xi \boldsymbol{r}^\xi \quad \text{and} \quad \bar{\boldsymbol{r}}^\eta = M_\eta \boldsymbol{r}^\eta \tag{3.14}$$

where:

$$M_\xi = \begin{cases} M_\xi^{sub}, & \text{if } |\tilde{\alpha}| \le \tilde{c} \\ M_\xi^{sup}, & \text{if } |\tilde{\alpha}| > \tilde{c}, \end{cases} \quad ; \quad M_\eta = \begin{cases} M_\eta^{sub}, & \text{if } |\tilde{\beta}| \le \tilde{c} \\ M_\eta^{sup}, & \text{if } |\tilde{\beta}| > \tilde{c}, \end{cases} \quad ; \tag{3.15}$$

$$M_\xi^{sub} = \begin{pmatrix} \text{sign}(\tilde{\alpha}) & 0 & 0 & 0 \\ 0 & 0 & 0 & 1/\tilde{c} \\ 0 & 0 & \text{sign}(\tilde{\alpha}) & 0 \\ 0 & \tilde{c} & 0 & 0 \end{pmatrix}; \quad M_\eta^{sub} = \begin{pmatrix} \text{sign}(\tilde{\beta}) & 0 & 0 & 0 \\ 0 & \text{sign}(\tilde{\beta}) & 0 & 0 \\ 0 & 0 & 0 & 1/\tilde{c} \\ 0 & 0 & \tilde{c} & 0 \end{pmatrix}; \tag{3.16}$$

$$M_\xi^{sup} = \text{sign}(\tilde{\alpha})I; \quad M_\eta^{sup} = \text{sign}(\tilde{\beta})I. \tag{3.17}$$

A linearity preserving genuinely multidimensional scheme can be constructed by introducing the vectors $\boldsymbol{r}^{\xi^*}, \boldsymbol{r}^{\eta^*}$ having elements defined by

$$r_i^{\xi^*} = r_i^\xi + \Psi(q_i) r_i^\eta, \quad r_i^{\eta^*} = r_i^\eta + \frac{\Psi(q_i)}{q_i} r_i^\xi \tag{3.18}$$

for $i = 1, 2, 3, 4$ and substituting them instead of $\boldsymbol{r}^\xi, \boldsymbol{r}^\eta$ respectively in (3.13),(3.14). Here Ψ is a limiter function and $q_i = -r_i^\xi / r_i^\eta$.

4 Numerical experiments

The purpose of the numerical experiments reported in this section is to verify the robustness of the scheme and the quality of our numerical solutions. Some preliminary experiments illustrating the performance of the multigrid algorithm using this scheme are presented as well.

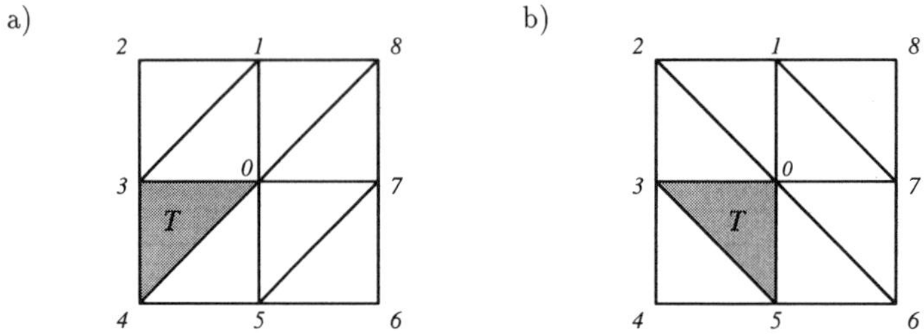

Figure 2: Examples of the possible triangulations.

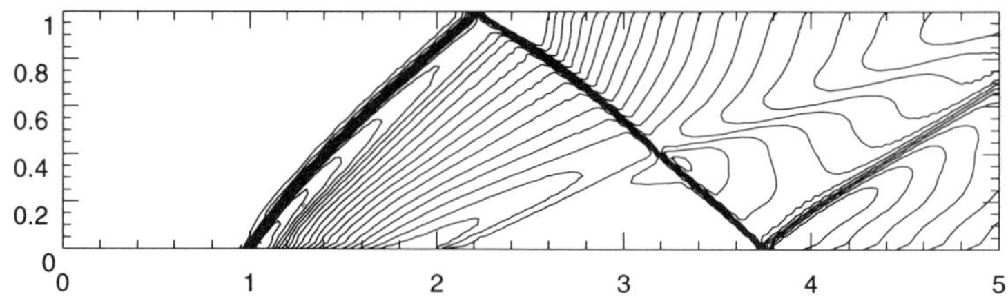

Figure 3: Supersonic flow in a channel over a circular bump: grid 200 × 40 pts., scheme from §4.

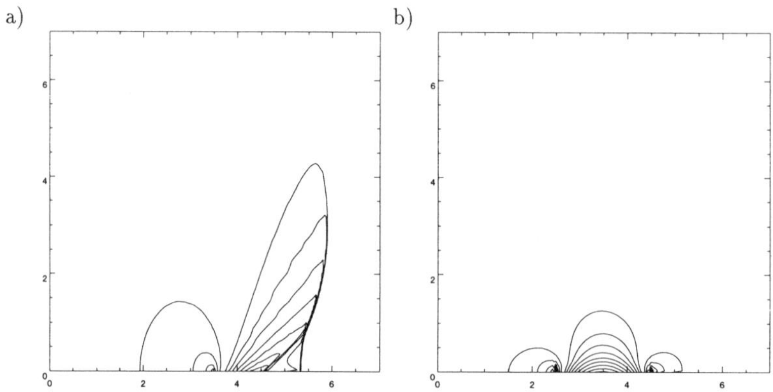

Figure 4: a) transonic flow (free stream Mach= .9 and b) low speed flow (Mach= .1) over a wall with a circular bump.

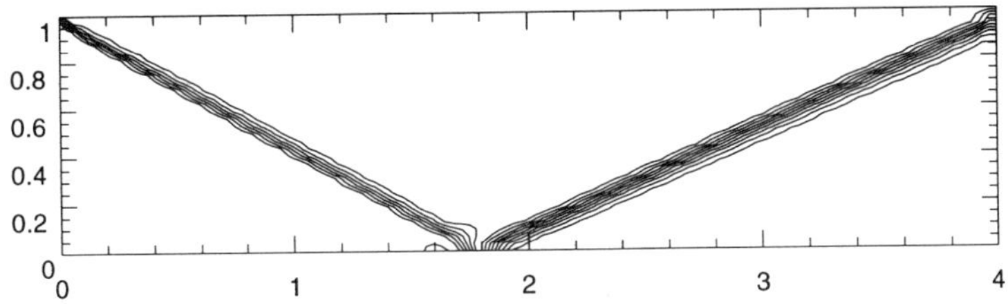

Figure 5: Performance of the multigrid algorithm, grid 129×33 pts.: solution obtained by $2FMG-W(2,1)$ algorithm.

4.1 Supersonic flow in a channel with a bump

The test case considered here is a supersonic (Mach=2.9) flow in a channel with a circular bump. The bump is located at the lower wall of the channel between $1 \leq x \leq 2$, and its surface is a circular arc of angle $\pi/3$ and diameter 1. Note that the actual shape of the domain is a rectangle. The influence of the bump on the flow is imposed through the boundary conditions: the velocity component normal to the surface of the bump at a certain location is reflected.

The experiment uses a grid of size 200×40 points. The density contour plots of the steady-state solution are presented in Fig. 3. The triangulation of the domain is as illustrated in Fig. 2(b). The derivatives are approximated along the diagonal and parallel to the x direction faces of the traingles. The limiter used in all the numerical experiments is *minmod*. Even though the grid is structured, the most essential feature of the unstructured grid formulation of the scheme — use of the non-orthogonal coordinate system — is present.

The purpose of this experiment is to test the performance of such a scheme and to illustrate the effect of the scheme and triangulation on the resolution of discontinuities. It can be seen in Fig. 3 that the shock reflected from the upper wall is resolved better than the stronger one that is incident to the upper wall. This is because for this scheme/grid combination, discontinuities whose direction is close to the grid diagonal (upper-left to lower-right) will be resolved very well (in addition to those which are close to the horizontal or vertical directions).

4.2 Low speed and transonic flow over a circular bump

The test problem considered here is a transonic flow (free-stream Mach= .9) over a flat wall with a bump (Fig. 4(a)). The surface of the bump is a circular arc with angle $\pi/2$ and diameter 1, and its location is between $3.5 \leq x \leq 4.5$. Again, in order to keep the experiments simple at this stage of the work, the bump is treated the same way as in the previous experiments. The grid is 200×200 points and the scheme is the same as in the first two experiments.

The shock with "fish-tail" shape can be clearly observed in Fig. 4(a).

We present also a numerical experiment for a low Mach number (=.1) flow over a flat wall with a circular (angle $\pi/2$ and diameter 2) bump. Here as well as in the previous case the presence of the bump is imitated through the appropriate boundary conditions. The grid is 200×200 points. The density contours of the steady-state solution are presented in Fig. 4(b).

4.3 Multigrid algorithm

To illustrate the performance of the multigrid algorithm we consider the well-known test problem of a shock reflecting from a flat wall. The multigrid algorithm involves five grids (levels): the finest consists of 129×33 points, the coarsest is 9×3 points.

The multigrid algorithm is based on the same scheme that was used in the previous two experiments. Lexicographic Gauss-Seidel relaxation is employed as a smoother. The restriction and prolongation procedures are the standard Full Weighting of the residuals and bilinear correction interpolation. The numerical solution obtained by the $2FMG - W(2,1)$ algorithm is presented in Fig. 5.

The rate of convergence observed in this test problem as well as in other simple experiments for a variety of flow regimes is very close to .75.

5 Summary

A new genuinely two-dimensional high-resolution (at the steady-state) scheme for the compressible Euler equations was presented. It is formulated on triangular (unstructured) grids with remarkable simplicity. The main advantage of this scheme is that Gauss-Seidel relaxation can be applied directly to the resulting discrete equations. This allows us to construct a simple and efficient multigrid steady-state solver.

A remarkable property of the constructed scheme is its very compact stencil: it involves only the immediate neighbors of the point of interest.

The description of the multigrid algorithm as well as more details concerning the multidimensional Euler scheme can be found in [7].

References

[1] H. Deconinck, Ch. Hirsch, and J. Peuteman, *Characteristic decomposition methods for for the multidimensional euler equations*, Lecture Notes in Physics 264, Springer, 1986, pp. 216–221.

[2] H. Deconinck, R. Struijs, and P. L. Roe, *Fluctuation splitting for multidimensional convection problem: an alternative to finite volume and finite element methods*, VKI Lecture Series 1990-3 on Computational Fluid Dynamics, Von Karman Institute, Brussels, Belgium, March 1991.

[3] A. Jameson, W. Schmidt, and E. Turkel, *Numerical simulation of the euler equations by finite volume method using runge-kutta time-stepping schemes*, AIAA Paper 81-1259, AIAA 5th Computational Fluid Dynamics Conference., 1981.

[4] P. L. Roe, *Discrete models for the numerical analysis of time-dependent multidimensional gas dynamics*, J. Comp. Phys. **63** (1986), 458–476.

[5] P. L. Roe, R. Struijs, and H. Deconinck, *A conservative linearization of the multidimensional euler equations*, To appear in J. Comp. Phys.

[6] D. Sidilkover, *Numerical solution to steady-state problems with discontinuities*, Ph.D. thesis, The Weizmann institute of Science, Rehovot, Israel, 1989.

[7] ———, *A genuinely multidimensional upwind scheme and efficient multigrid solver for the compressible Euler equations*, Report No. 94-84, ICASE, 1994, Submitted for publication.

[8] D. Sidilkover and A. Brandt, *Multigrid solution to steady-state 2d conservation laws*, SIAM J. Numer. Anal. **30** (1993), 249–274.

[9] D. Sidilkover and P. L. Roe, *Unification of some advection schemes in two dimensions*, Report No. 95-10, ICASE, 1995, Submitted for publication.

[10] S. Spekreijse, *Multigrid solution of monotone second-order discretization of hyperbolic conservation laws*, Math. Comp. **49** (1987), 135–155.

[11] ———, *Multigrid solution of the steady euler equations*, Ph.D. thesis, CWI, Amsterdam, 1987.

Flame front tracking via in-cell reconstruction

V. Smiljanovski and R. Klein*

Abstract

We present a new flame-front tracking algorithm for compressible, high-speed, premixed combustion that is compatible with any modern higher order shock-capturing method. The scheme relies on time updates of cell averages of conserved quantities only, i.e., the separate storage of pre- and post-flame states in grid cells cut by the flame is unnecessary. The flame front is represented as a level set of a scalar field, $G(\vec{x}, t)$, which makes topological changes of the flame front accessible in a straightforward fashion. We discuss 1-D and 2-D applications including a model for deflagration-to-detonation transition and pressure wave flame interactions in a 2-D rectangular closed box.

1 Introduction

One intriguing and as yet poorly understood phenomenon in combustion is the deflagration-to-detonation transition. This phenomenon is of major importance, e.g., in nuclear reactor safety, explosion hazards in industrial plants, coal mines etc., and it is thought to be active also under knocking conditions in spark ignition engines. A complete numerical description of this phenomenon requires (1) reliable and accurate representation of large amplitude pressure waves including shocks, (2) a chemical kinetic model for the combustible mixture in front of the flame, (3) an accurate model for turbulence in compressible flows, (4) a suitable numerical representation of a turbulent premixed combustion front.

In this paper we propose a solution to problem (4), the numerical representation of a high-speed deflagration in compressible flow. We consider a regime where deflagrative combustion occurs over length scales that are too small to be resolved numerically, so that we can consider the combustion front as a reactive gasdynamic discontinuity. Since, in the case of a deflagration wave, the Lax-Entropy-Condition is violated [1], the naive use of standard shock-capturing will produce spurious results, and flame tracking becomes necessary. We present a new flame-front tracking technique for multi-D flows that can be combined with any modern high resolution capturing scheme for the (reactive) Euler equations and has the following key features:

- the scheme guarantees conservation of mass, momentum and energy
- time updates are computed for cell-averages of conserved quantities only
- it relies on a level set representation of the front geometry
- it is compatible with directional operator splitting
- with minor modifications the scheme can be transformed into a shock-tracking algorithm.

*Institut für Technische Mechanik, RWTH-Aachen, Templergraben 64, 52056 Aachen, Germany. v.smiljan-ovski@itm.rwth-aachen.de

Importantly, the time update of all conserved quantities and of the species mass densities is cast into the form of a standard solver for the reactive Euler equations. Changes of cell-averages of conserved quantities are obtained from time averaged fluxes across the cell boundaries and the species mass densities are updated by combining (numerical) flux integrals and suitable source terms. These sources are equivalent to the species consumption/production due to flame burning relative to the flow within the cell during the time step considered. As a consequence, any conservative compressible flow solver can be extended to perform accurate high-speed flame-front tracking by addition of an algorithm for updating a level set function and by a routine that computes suitable numerical fluxes and source terms for the flame cells.

This paper is organized as follows. In Section 2, we briefly report on a very robust and accurate flame tracking method for 1-D that we have developed along the lines of LeVeque's "Shock tracking via wave propagation." We use this scheme as a reference for testing our new, much simpler, algorithm, the 1-D version of which we discuss in Section 3. Section 4 shows an example of deflagration-to-detonation transition (DDT) in a 1-D model problem and we use this quite dramatic example to compare the methods from Sections 2 and 3. This comparison comes out favourably and we proceed in Section 5 to describe the extension of our new flame tracking scheme to two space dimensions. An example computation of a Mach-0.05-flame in a closed rectangular domain exhibits pressure wave-flame interactions and the formation of a tulip flame.

2 Flame tracking via wave propagation in 1-D

Here we summarize a very robust flame tracking technique that borrows the principal concept from LeVeque. A detailed description is available in [2].

Consider a 1-D compressible flow, in which a single flame travels to the right. The computation of the compressible burnt and unburnt gas flows away from the flame are performed by means of a Roe-type shock-capturing algorithm on an equally spaced grid.

The "flame-cell," i.e., the cell currently containing the flame, is devided into a burnt and an unburnt part. As in LeVeque [2], we store the separate state vectors for the subcells left and right of the tracked discontinuity. At each interface between regular numerical grid cells, an approximate Roe Riemann problem is solved. At the tracked discontinuity, this is the same procedure as in LeVeque [2]. For the flame, we have to solve an appropriate approximate flame Riemann problem (FRP) as defined, e.g., by Chorin et al. [3], so that an accurate computation of the flame-flowfield-interaction is guaranteed. The FRP includes a change of the chemical species concentration across the flame; hence the species balance equations receive a source term. In addition to the flame, shocks, and contacts, there are continous expansion fans. A fan is conservatively treated as two discontinuities, i.e., the slowest and fastest characteristic of the fan, separated by a constant state. Hence, the wave propagation can be applied to the waves of an exact Riemann problem solution. The solution of the Riemann problems yields a series of distinct linearized discontinuity waves. The tracked discontinuity moves during the time step to its new position where an additional cell is inserted.

The actual computational grid now consists of a regular underlying grid plus two interfaces at the old and new front positions. Next the flow conditions for each computational cell of this irregular grid are updated by linearly superimposing the wave discontinuity contributions of all the interface Riemann problems; the increment in state space of each wave is added to the local state at the old time level at every point that the wave passes during the time step. The linear superposition makes it possible to choose a time step satisfying the CFL condition for the underlying regular grid but violates the criterion for the cut-cells in the vicinity of the tracked front. After the wave superposition step, the interface at the old flame location is erased and the emerging enlarged cell receives the appropriate average concluding the time update for a single step.

3 Flame tracking via in-cell reconstruction in 1-D

The scheme relies on reconstructing the pre- and post-flame states, $\vec{U}_u$ and $\vec{U}_b$, in cut cells using the integral conservation for mass, momentum, and energy

$$\vec{U} = \alpha\vec{U}_b + (1-\alpha)\vec{U}_u, \tag{1}$$

and the flame Hugoniot jump conditions for mass, momentum, and energy

$$\vec{F}_b - \vec{F}_u = D_F(\vec{U}_b - \vec{U}_u). \tag{2}$$

Here α denotes the burnt volume fraction of the flame cell, which is calculated from a known flame position, $\vec{U}$ is the state vector of the average cell quantity, $\vec{F}_b$ and $\vec{F}_u$ the burnt and unburnt fluxes, and D_F is the flame propagation speed. The jump conditions yield the burnt state as a function of the unburnt state and the relative burning velocity $s = D_F - u$ (u denotes the particle velocity). The system of equations is completed by the prescription of a burning law for s, which is evaluated in the unburnt state, and a fixed zero mass fraction ($Y_b = 0$) in the burnt state and is then solved iteratively.

The flame is the only wave, and its state increment will be distributed across neighbouring cells. With this approach, the new scheme can be written in conservation form with an additional source term:

$$\vec{U}_t + \nabla\vec{F} = \vec{Q}. \tag{3}$$

Any conservative compressible flow solver can be extended in this way. Using the reconstructed burnt and unburnt states we solve the Roe Riemann problems at the cell interfaces. The solution yields approximate numerical fluxes for the conservative update for $\vec{U}$. When the flame crosses a cell boundary an appropriate modification of the fluxes using the jump across the flame is applied yielding a correctly time averaged numerical flux.

Additionally we account for the flame propagation relative to the flow particles which leads to an additional source term, $\vec{Q}$, in the species balance equations. The species jump across the flame, $(\rho Y)_u - (\rho Y)_b$, is transported a distance $s \cdot \Delta t$. The jump has to be distributed on those cells touched by the flame during a timestep, Δt. Finally the new flame position is stored.

4 DDT in a 1-D model problem

In a tube, closed at the left end, filled with a quiescent explosive mixture ($p_0 = 2$ bar, $T_0 = 600$ K, $\rho_0 = 1.11$ kg/m^3), we ignite a planar (stoichiometric methane-air, $c_p/c_v = 1.28$, $p_b \approx 2$ bar, $T_b - T_u \approx 2038$ K) flame at time $t = 0$ at the left boundary. To avoid the complications associated with a turbulence closure for compressible high-speed flows, we prescibe an effective flame burning rate, which is determined by $S_T = A(t) \cdot S_L$, where the laminar burning speed relation $S_L = S_L(p_u, T_u, Y_u)$ mimics the behaviour of a hydrocarbon fuel according to [4]. The area increase factor $A(t)$, given by

$$\frac{A(t) - A_{min}}{A_{max} - A_{min}} = 1 - (1 + t/\tau) \cdot \exp(-t/\tau), \tag{4}$$

represents the net effect of turbulent wrinkling of the flame front. The flame begins as a laminar plane flame ($A_{min} = 1$). It then accelerates due to turbulent wrinkling ($A_{max} = 60$) in an acceleration period ($\tau = 0.44ms$), thereby radiating pressure waves which, in turn, precondition the explosive gas outfront. The end gas in front of the flame undergoes an autoignition process, which we model by a one-step Arrhenius reaction ($dY/dt = -YBexp(-E/RT)$) with a chosen non-dimensional activation energy, $E/RT_0 \approx 14$. It is solved in a chemistry operator splitting step with an adjusted rate coefficient $B = 5 \cdot 10^5/s$.

Figures 1 and 2 show a series of pressure and fuel mass fraction profiles at successive times computed with the wave propagation scheme. From the start, the combined discontinuity of pressure and fuel mass fraction represents the current flame location. During the early stages of the process, all the reactant consumption takes place in the flame while the cold gas in front of it has not yet ignited. The accelerating flame radiates a pressure pulse, which preheats the gas. At later times, the fuel mass fraction begins to diminish in the immediate vicinity of the flame front due to autoignition; associated with this process is the build-up of an increasing pressure pulse. In the sequel, a mutual amplification of the autoignition reactions and this steepening pressure pulse lead to a very sharp pressure reaction spike, which then decays towards a Chapman-Jonguet detonation. We have computed this same example with both the wave propagation and in-cell reconstruction schemes for tracking the flame front. In Figure 3, we compare the time history of the maximum pressure in the domain obtained with the two methods. The graph shows the average maximum pressure over time and the relative error.

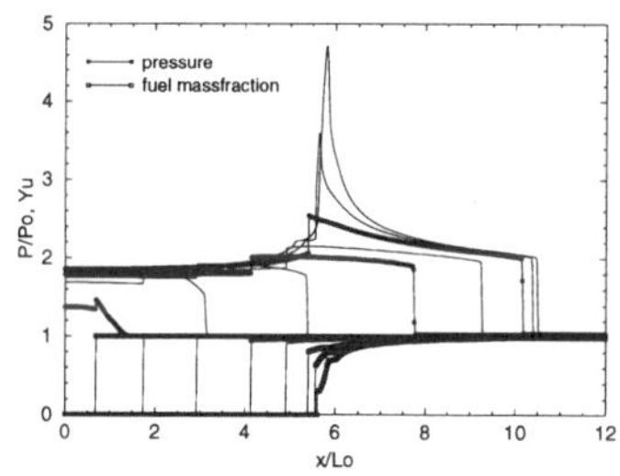

Figure 1: Pressure and unburnt mass fraction vs. space at different early times

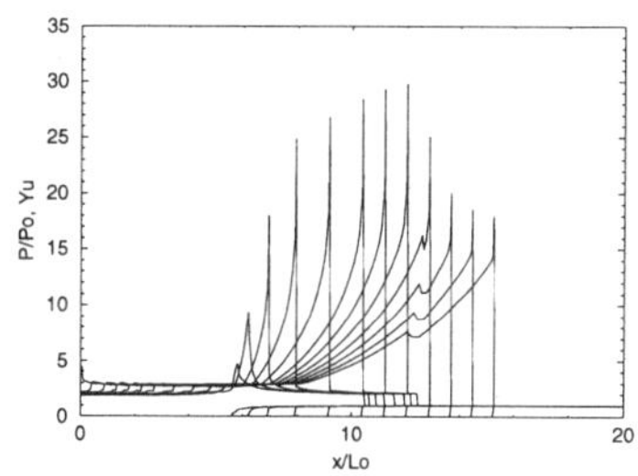

Figure 2: Pressure and unburnt mass fraction vs. space at different later times

5 In-cell reconstruction in 2-D

As mentioned in Section 3, we need the position of the flame to calculate the burnt volume fraction α necessary for the reconstruction step. To determine the front position we employ the level set representation (G-equation concept). The flame front is propagated as a particular level set, G_0, of a scalar field $G(\vec{x}, t)$. Relative to the flow, which moves with speed $\vec{v}$, the flame moves along its normal vector, $\vec{n}$, with speed s

$$G_t + (\vec{v} + s\vec{n})\nabla G = 0. \tag{5}$$

Hence, the reconstruction equations change in the way that the flame Hugoniot jump conditions (2) are solved normal to the flame, which implies a constant tangential velocity. The integral conservation equations (1) remain unchanged.

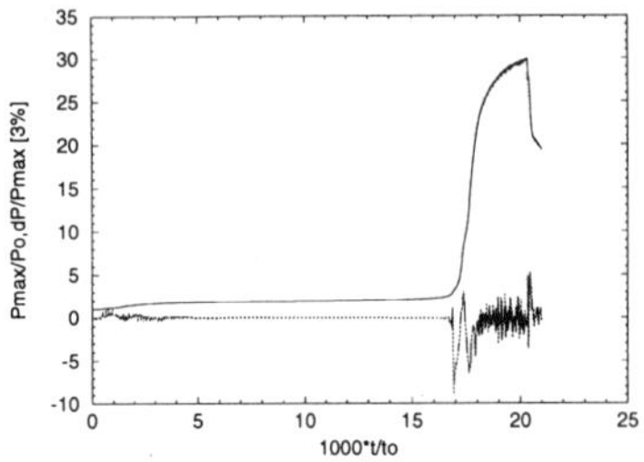

Figure 3: average maximum pressure and the relative error (multiplied by 300)

The maximum error is about 3% in the detonation regime. The coupling of the clipping of peaks in shock-capturing and chemistry results in oscillations about the pressure spike. Hence, it is not a consequence of the new scheme. In the more interesting acceleration period at the start the error is less than 0.4%.

The function G_0 is found simply by interpolation. In a flame cell, the flame is represented as piecewise linear, and the calculation of the (un)burnt volume fraction and the (un)burnt area fractions of the 4 cell interfaces as well as the normal vector, $\vec{n}$, is straightforward.

As mentioned in the introduction, we use directional operator splitting. Hence, the G-equation is also solved alternatingly with the flame propagation speed, $\vec{v} + s\vec{n}$, projected on the respective direction, yielding the new position in the next time step with new (un)burnt volume and area fractions.

In general, the appropriate cell interface, at which numerical fluxes or wave jumps have to be computed, can be intersected by a flame, determining a burnt area fraction $\beta(t)$ of the cell interface. Now neighbouring (un)burnt states are used to solve the respective (un)burnt 1-D Roe Riemann problems. The fluxes are calculated with a second order flux limiter scheme as long as a flame is not crossed, i.e., an (un)burnt state on one side of the interface has no corresponding neighbour on the other side. In that case the scheme reduces to first order. The total numerical flux is an area weighted average of unburnt and burnt fluxes. To obtain an appropriate time averaged flux one has to use the time averaged area fraction, $\bar{\beta}$, as in the 1-D case when the flame crosses the cell boundary.

In analogy, the source term in the species balance equations results from burnt volume differences between the old and the new flame position. In general the left and right neighbours can be flame cells too, and will transport their flame jumps into the appropriate cells depending on their respective flame speeds. With the corresponding numerical fluxes, and the source terms, the update is done.

To complete this section, the numerical treatment of the G-equation shall be summarized. It is composed of two steps. First the scalar field is transported by an explicit first order upwind scheme only in the vicinity of G_0, i.e., for a single flame the neighbouring cells, which are not flame cells, will be assigned the same flame speed as the flame cell itself. In the second step the remaining cells, which were not treated previously, will be computed according to an algorithm developed in [5]. For these cells the equation

$$|\nabla G| = 1 \tag{6}$$

is solved iteratively leaving the others unchanged. This way the G-value receives a physical meaning, i.e., the normal distance from the flame front. This guarantees that the level sets are always spaced equally and the actual dynamics of the G-field results only from the flame itself.

6 Tulip flame formation due to pressure wave-flame interactions

We consider a fast flame in a closed rectangular box (160 mm $\times$ 40 mm). The initial flame Mach number is $M_0 = 0.05$, which means a nondimensional turbulent flame area of $A = S_T/S_L = 50 =$ const. Initially the bulk of the cold gas ($p_0 = 1$ bar, $T_0 = 298$ K, $\rho_0 = 1.15$ kg/m^3) is at rest but we insert a half-cylindrical flame ($R_F = 6.5$ mm, $c_p/c_v = 1.4$, $T_b - T_u \approx 1902$ K) adjacent to the left wall of the box which is one of the shorter boundaries of the box. The initial data also include a weak half-cylindrical precursor shock wave ($R_S = 16$ mm), which connects the quiescent bulk gas with the expansion flow in front of the flame fulfilling zero normal velocities at the walls. We use the same laminar burning law as in Section 4 but include no autoignition model.

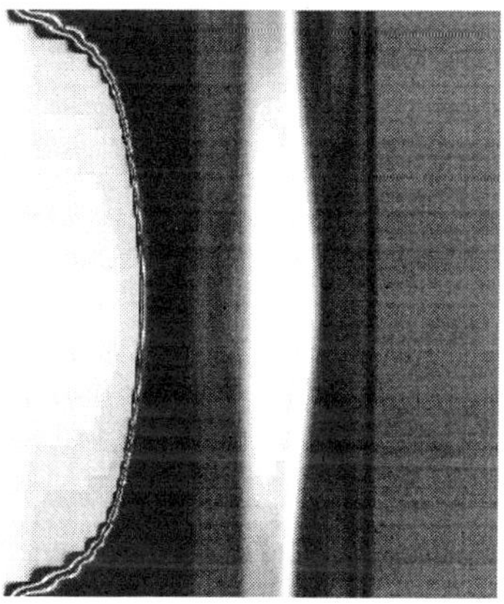

Figure 4: density at time t=0.263ms

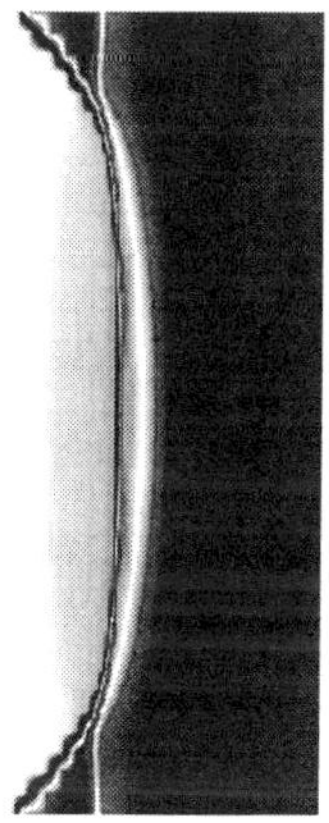

Figure 5: density at time t=0.507ms

We display grayscale graphs of density distributions at several instants in time. We plot averaged quantities only, even for flame cells. Hence, the flame may not appear as a sharp discontinuity on these graphs. Notice, however, that all the information needed to reconstruct the sharp interface, is available in the computational data.

As the precursor shock expands, it is reflected first off the side walls, and then propagates inwards to interact with the flame. The flame responds by forming a half-elliptical shape, which is then stabilized for a considerable time and travels down the tube in a quasi-steady fashion, (see Figure 4). Meanwhile, the precursor shock has reached the end of the box, has been reflected from it and now interacts again with the flame (see Figure 5). The flame tip is deformed and the pressure wave is partly transmitted into the burnt gases and partly reflected back into the unburnt. Figure 6 shows the further deformation of the flame due to inertial forces and due to the vorticity that has been generated during the pressure wave-flame collision. At a later time, the transmitted part of the pressure wave has returned from the left hand end of the tube and again passes the

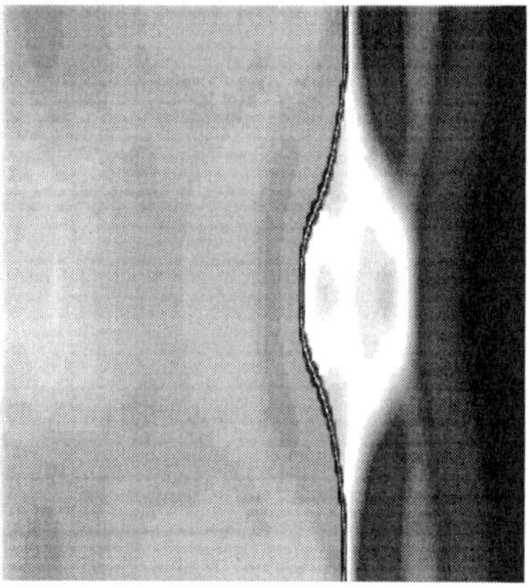

Figure 6: density at time t=0.585ms

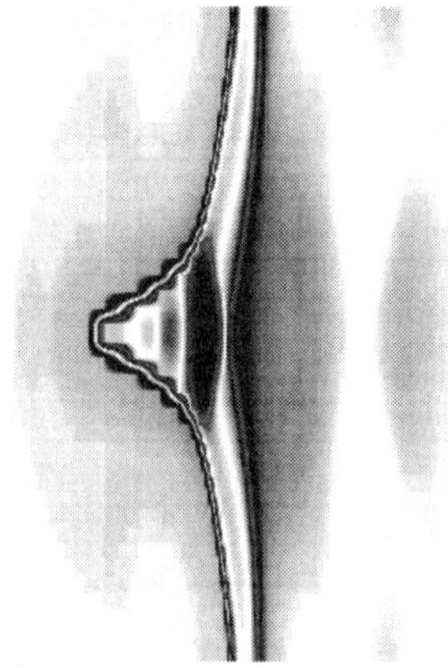

Figure 7: density at time t=0.719ms

flame. Figure 7 shows the situation just after this second collision; it is seen how the pressure wave structure bares the imprint of the flame geometry. Figures 8, 9 show how the narrow, sharp flame spike that is seen in Figure 7 quickly burns out, and how the flame shape becomes more and more curved back and forth as the distance from the end wall of the tube diminishes and the frequency of the flame-pressure wave interaction increases.

This example demonstrates that the flame tracking algorithm via in-cell reconstruction operates very robustly even, under extreme conditions of shock wave-flame interactions. Future work will, however, have to include detailed tests of the accuracy of the scheme, e.g., by reproducing the stability features of a plane flame.

7 Conclusions and future work

We have proposed here a new flame-front tracking algorithm that is based on cell-by-cell conservation, but circumvents the difficulties associated with seperate computations for the subcells generated by the cutting of the front through a regular grid cell. The scheme relies on reconstructing the pre- and post-flame states in cut cells using the integral conservation and flame Hugoniot conditions. The flame geometry is represented as the level set of a scalar function to allow an efficient computation of complex changes of the flame topology. We include nonzero heat release, so that the nontrivial flame-flow coupling is readily accounted for. The scheme extends to multiple dimensions by means of directional operator splitting, if one carefully takes into account that the flame is, in general, not aligned with any of the numerical grid lines. For future work, we plan to include autoignition chemistry in the pre-flame cold gas in order to simulate deflagration-to-detonation transition phenomena in 2-D. Also, we are currently working on extending the present tracking

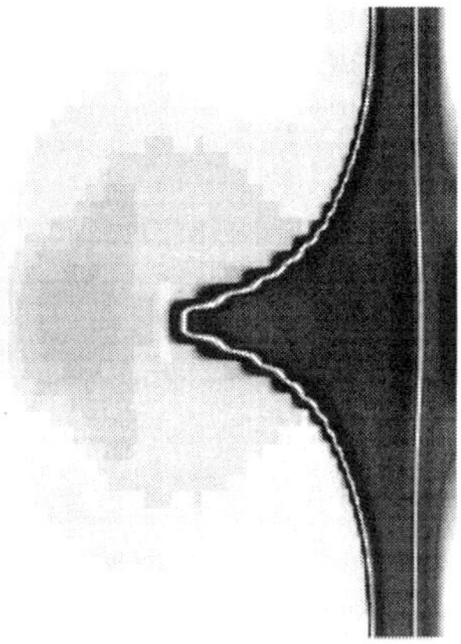

Figure 8: density at time t=0.841ms

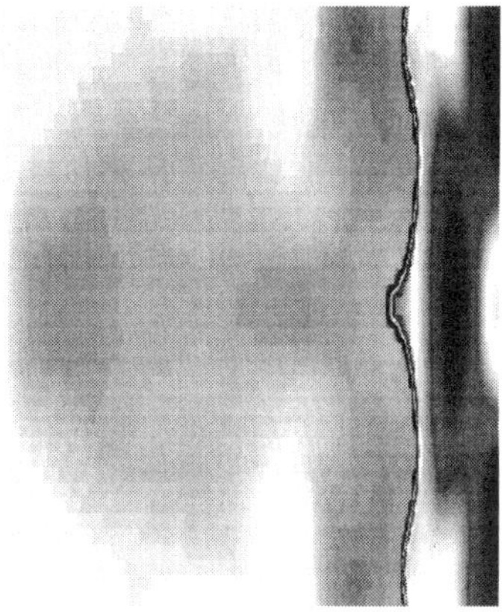

Figure 9: density at time t=0.979ms

approach to shock waves, contact discontinuities and to the representation of curved (moving) boundaries on Cartesian grids.

References

[1] R. Courant and K.O. Friedrichs, "Supersonic Flows and Shock Waves," Interscience Publishers, New York, 1948.

[2] R.J. LeVeque, "Shock tracking with the large time step method," in Proc. 7th Intl. Conf. Comput. Meth. in Appl. Sci. Eng., R. Glowinski and J.-L. Lions, eds., Versailles, 1985.

[3] Z.-H. Teng, A.J. Chorin, and T.-P. Liu, "Riemann Problems for Reacting Gas, with Application to Transition," SIAM J. Appl. Math, vol. 42, 1982.

[4] J. Göttgens, F. Mauss, and N. Peters, "Analytic Approximations of Burning Velocities and Flame thicknesses of Lean Hydrogen, Methane, Ethylene, Ethane, Acetelene, and Propane Flames," Twenty-Fourth Symposium (International) on Combustion, The Combustion Institute, 1992, pp. 129–135.

[5] M. Sussman, P. Smereka, and S. Osher, "A Level Set Approach for Computing Solutions to Incompressible Two-Phase Flow," J. Comput. Phys., vol. 114, 1994, pp. 146–159.

A Combined Adaptive Structured/Unstructured Technique for Essentially Unsteady Shock–Obstacle Interactions at High Reynolds Numbers

Mira Beryozkina[1], Dmitri Ofengeim[1], Maya Syshchikova[1], Eugene Timofeev[1]
Andrei Fursenko[2], Peter Voinovich[2]

Abstract

The aim of this work is to develop and to verify a numerical method to compute essentially unsteady viscous/inviscid interactions at high Reynolds numbers. The full Navier–Stokes equations underlie the mathematical model. The computational domain is covered by a hybrid structured/unstructured grid. In the course of computation the unstructured grid is adapted dynamically to the solution. The inviscid (convective) terms are approximated by a non–oscillatory Godunov–type scheme while central differences are applied to viscous terms. The implicit approximation is employed only on the structured grid for terms with second derivatives along the normal direction to the body surface. The initially plane shock wave diffraction over a circular cylinder is suggested as a verification problem. The comparative analysis of the numerical results and the experimental data obtained by the first (M. Beryozkina) and the third (M. Syshchikova) authors demonstrates the capability of the method.

Introduction

The investigation of shock wave interactions with obstacles (diffraction, reflection etc.) is of fundamental and practical interest. For high Reynolds numbers the Eulerian (inviscid, non–heat–conducting) gas model is widely used providing reasonable results. Because the total number of mesh points is typically determined by the desirable resolution at discontinuities, dynamic adaptation of the grid to the solution and high–accurate methods might save considerably computer memory and CPU time consumption. The respective numerical technique based on adaptive unstructured triangular grids and a high–resolution Godunov–type scheme has been developed previously by the authors. The numerical results [1]–[4] have demonstrated high accuracy, efficiency and reliability of the method for a wide range of transient shocked problems. However, for separated flows it fails to predict properly the location of separation points, the gasdynamic parameters on the body surface, and the surrounding wave pattern.

The present study is aimed as an extension of the above approach in such a way as to take into account viscous forces, primarily near solid surfaces. An attempt has been made to combine the locally adaptive unstructured solver with a Navier–Stokes method on body–fitted partially structured grids.

Due to the lack of analytic solutions for unsteady viscous problems with shock waves, the verification of the proposed technique involves difficulties. The unsteady gas flow which accompanies shock wave diffraction over a circular cylinder may be considered as a test problem because a variety of experimental data are available ([5]).

[1]Ioffe Physical–Technical Institute, Russian Academy of Sciences, Polytekhnicheskaya 26, St.Petersburg 194021, Russia
[2]Advanced Technology Center, P.O. Box 160, St.Petersburg 198103, Russia

1 Numerical Model

1.1 Governing Equations

The numerical model is based on the full Navier–Stokes equations written in an integral form with respective boundary and initial conditions. For two dimensions the governing equations are:

$$\int_{\mathrm{Vol}} U \mathrm{d}V \Big|_{t_0}^{t} + \int_{t_0}^{t} \mathrm{d}t \int_{\sigma} \left(F_x^{c+v} n_x \mathrm{d}s + F_y^{c+v} n_y \mathrm{d}s \right) = 0,$$

$$U = \begin{pmatrix} \rho \\ \rho u_x \\ \rho u_y \\ \rho e \end{pmatrix}, \quad F_x^c = \begin{pmatrix} \rho u_x \\ \rho u_x^2 + p \\ \rho u_x u_y \\ (\rho e + p) u_x \end{pmatrix}, \quad F_y^c = \begin{pmatrix} \rho u_y \\ \rho u_x u_y \\ \rho u_y^2 + p \\ (\rho e + p) u_y \end{pmatrix},$$

$$F_x^v = \begin{pmatrix} 0 \\ \tau_{xx} \\ \tau_{yx} \\ \tau_{xx} u_x + \tau_{xy} u_y + \lambda \frac{\partial T}{\partial x} \end{pmatrix}, \quad F_y^v = \begin{pmatrix} 0 \\ \tau_{yx} \\ \tau_{yy} \\ \tau_{yx} u_x + \tau_{yy} u_y + \lambda \frac{\partial T}{\partial y} \end{pmatrix},$$

where $\vec{u} = (u_x, u_y)$ – velocity vector, p – pressure, ρ – density, T – temperature, $e = \epsilon + 0.5u^2$ – specific total energy, $\tau_{xx}, \tau_{yx}, \tau_{yy}$ – shear stress terms, $\lambda = \mu C_p / Pr$ – heat conductivity coefficient (C_p – specific heat at constant pressure, Pr – Prandtl number). The viscosity coefficient μ is calculated using Sutherland's law. Assuming perfect gases, the specific internal energy is given by $\epsilon = (1/(\gamma - 1))p/\rho$, where γ – polytropic index. Vol represents a gas volume bounded by the closed surface σ with the outward normal $\vec{n} = (n_x, n_y)$.

1.2 Discretization of the Computational Domain and Grid Adaptation

To approximate adequately viscous terms near solid walls at high Reynolds numbers and to construct an effective numerical procedure, a thin layer of non–uniform structured orthogonal grid is constructed along the body surface. The thickness of the layer δ can not be defined rigorously. Two heuristic rules have been used: 1) $\delta \geq 10/\sqrt{Re}$, where Re is the Reynolds number. This means that δ should be at least ten times greater than the expected width of the boundary layer; 2) δ should be chosen so that the size of control volumes near the structured/unstructured grid interface would be nearly the same.

The employment of a structured grid layer may be partially justified by the following argument. Fine grids near the body surface are needed to resolve boundary layers at high Reynolds numbers. This, in turn, leads to implementation of implicit schemes. On the other hand, for unsteady problems, the solver is called at each time step and the time step value is restricted by the Courant–Friedrichs–Levy condition. Therefore, the implicit solver should be very fast. There are many efficient methods to solve a set of linear equations produced by discretization of the governing equations on structured grids in contrast to the unstructured grid case. Furthermore, unstructured grids for the problems under consideration typically contain very elongated triangles near the body and almost regular triangles for the outer flow. This considerably complicates grid generation and adaptation procedures.

The rest of the computational domain is covered by an unstructured triangular grid. In the course of computation the unstructured grid is adapted to the solution using a hierarchical refinement/unrefinement procedure outlined in [2]–[4] and based on the ideas from reference [6]. The structured grid remains fixed. An example of the combined grid is given in Figure 1.

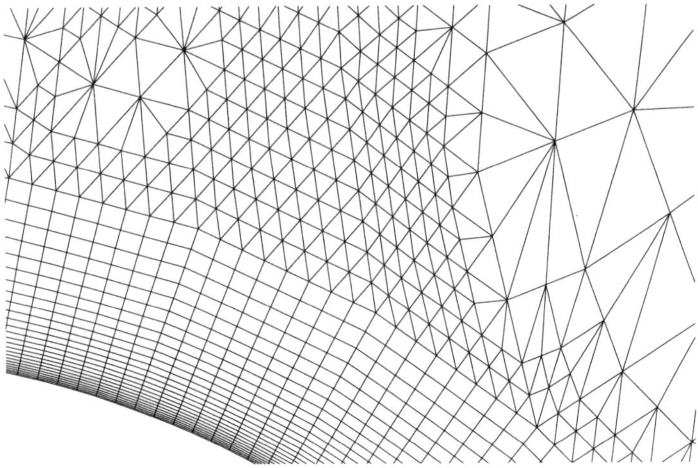

Figure 1: Detail of hybrid structured/unstructured grid.

1.3 Finite–Volume Schemes for Time Step Operator

Nonoverlapping control volumes are established for both the above grids. The numerical methods are formulated in terms of fluxes through the control volume interfaces. An unsplit explicit technique is employed on the unstructured grid for the time step operator $L(2\Delta t) = L^{c+v}(\Delta t)L^{c+v}(\Delta t)$ (the superscripts denote: c – convective operator, v – viscous operator). The inviscid (convective) terms are approximated by the high–order Godunov–type scheme ([2]–[4], [7]), while central differences are applied to the viscous terms.

The time step operator for the structured grid is split according to physical processes. To compensate partially the severe limitation of the time step, the inviscid operator is split, in its turn, in space:

$$L(2\Delta t) = L_1^c(\Delta t)(L_2^c(\frac{\Delta t}{m}))^m L^v(\Delta t)\times$$

$$\times L^v(\Delta t)(L_2^c(\frac{\Delta t}{m}))^m L_1^c(\Delta t) \ ,$$

where L_1^c – inviscid operator for the tangential direction to the body surface; L_2^c – inviscid operator for the normal direction to the body surface. The respective one–dimensional operators are approximated by an 1D version of the above Godunov–type scheme. A predictor–corrector procedure allows us to enhance the temporal accuracy for all operators.

The viscous operator $L^v(\Delta t)$ is unsplit and contains central differences written implicitly in time:

- Predictor

$$U_{i,j}^{n+1/2} = U_{i,j}^n - \frac{\Delta t}{\mathrm{Vol}_{i,j}} \sum F_o^n;$$

- Corrector

$$U_{i,j}^{n+1} = U_{i,j}^n - 0.5\frac{\Delta t}{\mathrm{Vol}_{i,j}}(\sum F_o^n + \sum F_o^{n+1/2})$$

$$-0.5\frac{\Delta t}{\text{Vol}_{i,j}}(\sum F_m^n + \sum F_m^{n+1}),$$

where i, j – spatial indices, n – time step counter, F_m – flux including "main" viscous terms, F_o – flux including the remaining viscous terms.

The implicit approximation is employed only for terms with second derivatives along the normal direction to the body surface (the above "main" terms). The resulting set of linear difference equations with a block tridiagonal matrix of coefficients is solved by the Thomas algorithm. The computational procedure for corrector step of $L^v(\Delta t)$ operator includes the following steps:

1. linearization: $\mu = \mu(T^n)$;
2. solving of coupled linear equations for u_x^{n+1} and u_y^{n+1};
3. solving of linear equation for T^{n+1} (terms of $u_x \partial u_x / \partial x$ type are calculated using u_x^{n+1} and u_y^{n+1} (see step 2);
4. iterations repeating steps 1–3 are required because the viscosity coefficient μ depends nonlinearly on the temperature T.

2 Numerical Results

For lack of unsteady viscous problems with shock waves having analytical solution or reliable computational data, the verification of the proposed technique involves difficulties. The following set of testing problems has been suggested:

A. Steady problems:

1. Boundary layer on a flat plate;
2. Shock wave – boundary layer interaction;
3. Supersonic flows over bodies (circular cylinder, NACA 0012 profile etc.);

B. Unsteady problems:

4. Initially plane shock wave propagation through a curved channel;
5. Shock wave diffraction over a circular cylinder.

The proposed explicit technique for essentially unsteady flows seems to be less efficient for steady problems as compared to conventional implicit procedures. Therefore, the stationary flows are computed solely to compare with the analytical solution (problem A1), numerical and experimental results by different authors (problems A2 and A3) and to demonstrate the accuracy of the method. The unsteady flows B4, B5 may be considered as test problems because a variety of experimental data are available ([2],[5]). Some results for problem B5 are presented below.

2.1 Shock Wave Diffraction Over a Circular Cylinder

The above approach has been applied to the numerical simulation of initially plane shock wave ($M_s = 3.0$) diffraction over a circular cylinder in CO_2 ($\gamma = 1.289$). The Reynolds number based on the cylinder radius r is 80,000. The temperature of the cylinder's surface is taken to be constant and equal to $293K$.

Some results are given in Figure 2 and in Figure 3 in comparison with data obtained from shock tube experiments. Figure 2 contains experimental shadowgraphs and computed instant

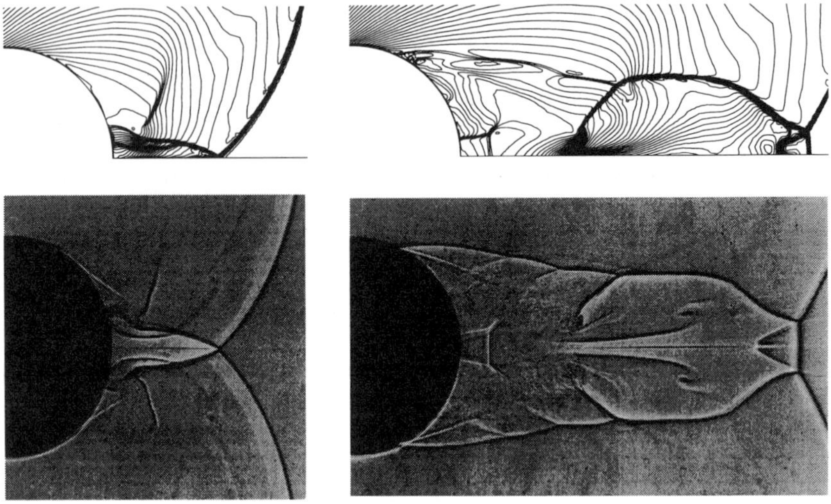

Figure 2: Diffraction of a shock wave over a cylinder: flow development behind the body. Computed density contours are placed at the top of the figure while experimental shadowgraphs - at the bottom. The left images correspond to the time moment $t_r = 3.7$, the right ones - to $t_r = 5.8$.

density contours for time moments $t_r = 3.7$ and $t_r = 5.8$ where time t_r is measured from the beginning of shock wave/cylinder interaction and the unit of measurement is the time it takes for the initial (undisturbed) shock to travel a distance r. The shadowgraphs and the contours display flowfields behind the cylinder. Due to total symmetry of the flow the computations have been carried out for half of the cylinder. Figure 3 presents experimental and numerical time history of the pressure P at two points on the surface of cylinder: $\varphi = 120°$ and $\varphi = 140°$. The angle φ is counted from the forward stagnation point. Pressure P is nondimensionalized by the pressure of gas at rest.

The diffraction process for $M_s = 3.0$ is characterized by strong viscous/inviscid interaction. A rearward–facing stagnation shock wave is generated in the course of diffraction (see left images in Figure 2). The shock interacts with the boundary layer and forces flow separation resulting in a slip surface, which rolls up into a vortex, and a new stagnation shock is formed under the slip surface near the wall. The flow pattern becomes more complicated once the diffracted shocks meet behind the cylinder, reflect and interact with gasdynamic structures arising in an early stage of the process . Additional gasdynamic discontinuities arise near the rear end of the body and on the reflected shocks (right images in Figure 2).

In spite of the flow complexity the comparison between experimental data and numerical results shows a good quality of computations which describe correctly the above–mentioned wave pattern, the location of separation point (Figure 2) as well as pressure values on the surface of cylinder (Figure 3).

The flow under consideration was previously investigated by the authors using an Eulerian — inviscid, non-heat-conducting — gas model (see [3], [8]). The results are in rather poor agreement with the experimentally observed wave pattern behind the body including the location of the separation point. This is obviously related to the extensive viscous/inviscid

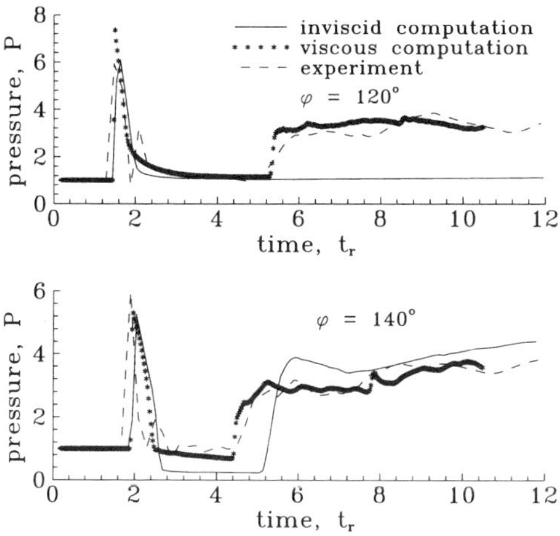

Figure 3: Time history of the pressure at two points ($\varphi = 120°$ and $\varphi = 140°$) on the surface of cylinder.

interactions mentioned above. The distinctions between flow structures computed by the Euler and Navier–Stokes approaches, in particular, influence the pressure distribution on the wall of the cylinder. The solid line in Figure 3 corresponds to inviscid computations. One can clearly see the Euler model fails to properly predict the time moment when the reflected shock reaches the point in question (moreover according to inviscid computations the reflected shock does not reach angle $\varphi = 120°$ at all) as well as pressure values before and behind the reflected shock front.

Nevertheless, the Eulerian model is not as useless as it may appear at first glance. The careful analysis of experimental results in combination with computations based on the two physical model — Eulerian and Navier–Stokes ones — provides a powerful tool to get a clearer insight into the gas dynamic phenomena under investigation. It becomes possible to study influence of viscous/inviscid interactions on the initiation and development of shock wave configurations as well as on the gasdynamic parameters on the body surface and in the near wake. The respective investigations are in progress now.

3 Concluding Remarks

In the present paper the first numerical results based on a new technique for essentially unsteady viscous/inviscid interactions are described. The results are of great promise and stimulate future development, verification and application of the numerical method as well as detailed investigation of viscous/inviscid interactions by means of a correlation between experimental data and computations based on Euler and Navier-Stokes models.

Acknowledgements

The encouraging discussions with Prof. Yu.Golovachov and valuable help of Dr. D.Sharov are kindly acknowledged.

This work is partially supported by the Russian Foundation of Fundamental Investigations (grants Nr. 93-02-14797 and Nr. 94-01-01456), Friedrich Alexander University Erlangen–Nurnberg (contract Nr. FAU4001-01IR303) and the International Science Foundation (personal short–term research grants).

References

[1] A.A. Fursenko, D.M. Sharov, E.V. Timofeev, P.A. Voinovich: *Numerical simulation of shock waves interactions with channel bends and gas non–uniformities*, Comp. & Fluids, Vol.21 (1992), 377-396.

[2] A.A. Fursenko, N.P. Mende, K. Oshima, D.M. Sharov, E.V. Timofeev, P.A. Voinovich: *Numerical simulation of propagation of shock waves through channel bends*, Comp. Fluid Dyn. J., Vol.2 (1993), 1-36.

[3] A.A. Fursenko, D.M. Sharov, E.V. Timofeev, P.A. Voinovich: *High–resolution schemes and unstructured grids in transient shocked flow simulation*, Lecture Notes in Phys., Vol.414 (1993), 250-254.

[4] A.A. Fursenko: *Unstructured Euler solvers and underlaying algorithms for strongly unsteady shocked flows*, in H. Daiguji (Ed.) Proc. of 5th Int. Symp. on CFD, JSCFD 1993, Vol.1, 227-234.

[5] M.P. Syshchikova, M.K. Beryozkina, I.V. Smirnov: *The effects upon bodies in the zone of interference of two explosions*, in Collection of Papers "Unsteady gas flows with shock waves", Ioffe Physical–Technical Institute USSR Academy of Sciences 1990, 7–29 (in Russian).

[6] R. Löhner: *The efficient simulation of strongly unsteady flows by the finite element method*, AIAA Paper 1987, 87–0555.

[7] A.V. Rodionov: *Improvement of the approximation order in the Godunov scheme*, Zh. Vych. Mat. Mat. Fiz., Vol.27 (1987), 1853–1860 (in Russian).

[8] D.Kh. Ofengeim, M.P. Syshchikova, M.K. Berezkina, D.M. Sharov, P.A. Voinovich: *Some features of the transient relaxation to steady-state pressure on the surface of a cylinder acted on by a shock wave*, Pis'ma Zh. Tekh. Fiz., Vol.19 (1993), 91–94.

CONSERVATION LAWS, COORDINATE TRANSFORMATIONS, AND DIFFERENTIAL FORMS

DAVID H. WAGNER[1]

ABSTRACT. It is shown that differential forms and exterior algebra can be very useful in studying weak solutions of hyperbolic systems of conservation laws. Particular examples are given in the area of solid mechanics.

1. INTRODUCTION

There are two different points of view that are used to describe the laws of physics for a continuum. The referential, or Lagrangian, point of view uses a coordinate system which describes points in (material)×(time), whereas the spatial, or Eulerian point of view uses a coordinate system that describes locations in (physical space)×(time). The mapping $(a,t) \to x$ of the material to physical space is called a *motion*. Typically, the motion is an unknown quantity which is determined by the solution of an initial value problem for a system of partial differential equations.

An interesting class of motions consists of those which exhibit shock waves. Mathematically, a shock wave occurs as the result of focussing characteristics for a nonlinear hyperbolic system, and is represented as a discontinuity in the unknowns, as functions of space (or material) and time. When the velocity of the motion $(\frac{\partial x}{\partial t})$ is discontinuous, then the motion will at best be only Lipschitz continuous, or perhaps even discontinuous, as may occur in a shearing motion.

The presence of discontinuities mandates the use of weak solutions to the relevant partial differential equations. The calculus for such solutions is limited. For example, there is no product rule for the weak derivative of a product of L^∞ functions. Thus a question arises, whether the two different coordinate systems are equivalent for such weak solutions. For the case when the map $(a,t) \to (x,t)$ is a bi-Lipschitz homeomorphism, this equivalence has been proved for one space dimension in [7]. The generalization of this result to several space dimensions is straightforward, and is given in [2]. In addition, Dafermos generalized the equivalence to fields in L^1.

In this paper, we will demonstrate some interesting consequences that come from the techniques used in [7, 2]. In particular we will see that the natural language for divergence form conservation laws is the language of differential forms. This language gives an essentially coordinate free description of continuum physics. In addition, this description enables us to make a clear distinction between conservation laws which are physical, and coordinate free, such as laws of conservation of mass, momentum and energy, and those which are mathematical constructs that are not coordinate free, such as equations of continuity.

One result that is particularly interesting is a product rule for weak exterior derivatives. In particular such a product rule holds when α and β are closed forms, i.e., $d\alpha = 0$ and $d\beta = 0$. This

Department of Mathematics, University of Houston, Houston, TX 77204-3476. e-mail: dwagner@uh.edu.
Research supported by the Texas Coordinating Board for Higher Education, Advanced Research Program, and by the Institute for Mathematics and its Applications with funds provided by the National Science Foundation.

simple fact enables us to resolve a mystery regarding redundant conservation laws for deformation gradients.

We begin by briefly reviewing the technique that underlies these results. Let t be the time variable, let $\mathbf{a}$ be the referential, or material coordinate, and let $\mathbf{x}$ be the spatial coordinate, so that $\mathbf{a}$ is constant on particle paths. Then $\mathbf{x}(\mathbf{a},t)$ is the motion of the material. Let

$$\mathbf{v} = \frac{\partial \mathbf{x}}{\partial t} \tag{1.1}$$

be the velocity of the motion, and let

$$F^i_{\ \alpha} = \frac{\partial x^i}{\partial a^\alpha}. \tag{1.2}$$

The matrix $\mathbf{F}$ is called the deformation gradient of the flow. By equality of mixed partials, which must hold in the sense of distributions, for any Schwarz distribution, we have:

$$\frac{\partial F^i_{\ \alpha}}{\partial t} = \frac{\partial v^i}{\partial a^\alpha} \tag{1.3}$$

We call this a continuity equation; its purpose is to permit us to forget about the motion as an unknown. Note that (1.3) is not complete in expressing equality of mixed partials. It is shown in [4] however, that (1.3) may be used as an evolution equation for the equality of mixed spatial partial derivatives, as follows. If we differentiate (1.3) with respect to a_β, permute α and β, and subtract, we obtain:

$$\frac{\partial}{\partial t}\left(\left(F^i_{\ \alpha}\right)_{a^\beta} - \left(F^i_{\ \beta}\right)_{a^\alpha}\right) = 0 \tag{1.4}$$

Thus, if F^i_α is a gradient in the entire body at $t=0$, then it is also a gradient for all $t>0$.

Suppose we have a balance law in material coordinates:

$$\frac{\partial D}{\partial t} + \frac{\partial G_\alpha}{\partial a^\alpha} = H, \tag{1.5}$$

where the repetition of the index α indicates summation from 1 to n, where n is the dimension of the body.

The weak formulation of (1.5) is:

$$\int_0^\infty \int_{\mathbb{R}^n} \phi_t D + \phi_{a^\alpha} G_\alpha + \phi H \; d\mathbf{a}dt = 0. \tag{1.6}$$

This must hold for every smooth test function ϕ with bounded support.

By changing variables from $\mathbf{a}$ to $\mathbf{x}$, we obtain:

$$\int_0^\infty \int_{\mathbb{R}^3} \left(\left(\phi_t + \phi_{x^i} v^i\right) D + \phi_{x^i} F^i_{\ \alpha}\, G_\alpha + \phi H\right) |\mathbf{F}^{-1}| d\mathbf{x}dt = 0. \tag{1.7}$$

Here, and henceforth, for any $n \times n$ matrix $\mathbf{A}$ we use $|\mathbf{A}|$ to denote the determinant of $\mathbf{A}$.

We regroup:

$$\int_0^\infty \int_{\mathbb{R}^3} \phi_t \left(D|\mathbf{F}^{-1}|\right) + \phi_{x^i} \left(\left(v^i D + F^i_{\ \alpha}\, G_\alpha\right) |\mathbf{F}^{-1}|\right) + \phi H |\mathbf{F}^{-1}| d\mathbf{x}dt = 0. \tag{1.8}$$

This looks like the weak form of

$$\frac{\partial}{\partial t}\left(D|\mathbf{F}^{-1}|\right) + \frac{\partial}{\partial x^i}\left(\left(v^i D + F^i_{\ \alpha}\, G_\alpha\right) |\mathbf{F}^{-1}|\right) = H|\mathbf{F}^{-1}|. \tag{1.9}$$

Formally this gives a simple way to transform a balance, or conservation, law from material coordinates to spatial coordinates. The formula for the transformation is the same as the one obtained by performing a change of variables on an integral balance law.

To make this rigorous at the level of distribution theory, we need to state hypotheses under which:

- The change of variables for integrals is valid.
- The Lagrangian test functions $\phi(\mathbf{a},t)$ are mapped into 'enough' Eulerian test functions $\psi(\mathbf{x},t) = \phi(\mathbf{a}(\mathbf{x},t),t)$ so that (1.8) is true for all smooth test functions with compact support.
- The integral in (1.8) makes sense, at least in terms of evaluation of a distribution on a test function ϕ.

A sufficient condition for the above result is that the coordinate transformation $(\mathbf{a},t) \to (\mathbf{x},t)$ is a bi-Lipschitz homeomorphism, and that the fields $(D,F,H) \in L^1_{loc}$ for the material coordinate system. The coordinate transformation is a bi-Lipschitz homeomorphism if there are positive constants C and K such that

$$C|\mathbf{a}_1 - \mathbf{a}_2| < |\mathbf{x}(\mathbf{a}_1,\ t) - \mathbf{x}(\mathbf{a}_2,\ t)| < K|\mathbf{a}_1 - \mathbf{a}_2|. \tag{1.10}$$

for every t, $\mathbf{a}_1$, and $\mathbf{a}_2$.

Thus the Lagrangian equations for the conservation of mass, momentum, and energy:

$$\begin{aligned} &(\rho_{ref})_t = 0 \\ &\rho_{ref} v^i_t - \left(F^i{}_\alpha S^{\alpha\beta}\right)_{a^\beta} = f^i \\ &\rho_{ref}\left(|\mathbf{v}|^2/2 + e\right)_t - \left(v_i F^i{}_\alpha S^{\alpha\beta}\right)_{a^\beta} = -Q^\beta{}_{a^\beta} + f^i v^i \end{aligned} \tag{1.11}$$

are transformed into the Eulerian equations, where $\rho = \rho_{ref}|\mathbf{F}^{-1}|$

$$\begin{aligned} &\rho_t + (\rho v_j)_{x^j} = 0 \\ &(\rho v^i)_t + (\rho v^i v^j)_{x^j} - \left(F^i{}_\alpha S^{\alpha\beta} F^j{}_\beta |\mathbf{F}^{-1}|\right)_{x^j} = f^i|\mathbf{F}^{-1}| \\ &\left(\rho\left(|\mathbf{v}|^2/2 + e\right)\right)_t + \left(\rho\left(|\mathbf{v}|^2/2 + e\right)v^j - \left(v^i F^i{}_\alpha S^{\alpha\beta} F^j{}_\beta |\mathbf{F}^{-1}|\right)\right)_{x^j} + v^i f^i |\mathbf{F}^{-1}| \\ &\qquad = -\left(Q^\beta F^j{}_\beta |\mathbf{F}^{-1}|\right)_{x^j} \end{aligned} \tag{1.12}$$

In addition one can obtain other useful 'conservation laws.' For example, the fact that

$$\begin{aligned} \frac{\partial 1}{\partial t} &= 0, \\ \frac{\partial 1}{\partial a^\alpha} &= 0 \end{aligned} \tag{1.13}$$

transforms into equations that seem to have significance:

$$(|\mathbf{F}^{-1}|)_t + (v^i|\mathbf{F}^{-1}|)_{x^i} = 0, \tag{1.14}$$

$$(F^i{}_\alpha|\mathbf{F}^{-1}|)_{x^i} = 0. \tag{1.15}$$

When ρ_{ref} is constant, (1.14) describes the conservation of mass. The second equation (1.15) is the Piola identity [1]. In addition, the equality of mixed partials,

$$\frac{\partial F^i{}_\alpha}{\partial t} = \frac{\partial v^i}{\partial a^\alpha} \tag{1.16}$$

transforms into a new continuity equation:

$$((F^i{}_\alpha|\mathbf{F}^{-1}|)_t + \left(\left(F^i{}_\alpha v^j - F^j{}_\alpha v^i\right)|\mathbf{F}^{-1}|\right)_{x^j} = 0. \tag{1.17}$$

In some physical problems (elasticity, plasticity) where the stress depends on the components of $\mathbf{F}$, or on the cofactor matrix, $\mathbf{F}|\mathbf{F}|^{-1}$, we need an equation for $\mathbf{F}$, such as (1.17), to close the system; it is a constitutive law. In other areas (gas dynamics) where the stress (pressure) depends on $|\mathbf{F}|$

and not on the components of $\mathbf{F}$, (1.17) is not needed; it is simply a mathematical law regarding the configuration (coordinate transformation) of the gas.

As we remarked concerning (1.3), the equation (1.17) may be understood as an evolution equation. It implies

$$\frac{\partial}{\partial t}\left(\sum_i (F^i{}_\alpha|\mathbf{F}^{-1}|)_{,x^i}\right) = 0, \tag{1.18}$$

so that if the Piola identity holds at $t = 0$, then it must hold for all $t > 0$. Curiously, however, (1.17) does not appear to directly imply an evolution law for the equality of mixed partials in any coordinate system.

An alternative to (1.17, 1.18) was proposed in [6]. This alternative is the equality of mixed partials for $\mathbf{a}(\mathbf{x},\ t)$:

$$(F^{-1\alpha}{}_i)_t = -\,(F^{-1\alpha}{}_j v_j)_{x^i} \tag{1.19}$$

This equation must hold in the weak sense because the inverse map $\mathbf{a}(\mathbf{x},\ t)$ is a well defined Schwarz distribution. This equation may also be treated as an evolution equation, obtaining:

$$\frac{\partial}{\partial t}\left((F^{-1\alpha}{}_i)_{x^j} - (F^{-1\alpha}{}_j)_{x^i}\right) = 0. \tag{1.20}$$

Thus if $\mathbf{F}^{-1}$ is a gradient at $t = 0$, (1.19) implies that it remains a gradient for $t > 0$. $\mathbf{F}^{-1}$ will be a gradient if the initial conditions

$$(F^{-1\alpha}{}_i)_{x^j}(x,0) = (F^{-1\alpha}{}_j)_{x^i}(x,0) \tag{1.21}$$

hold.

These two continuity equations (1.17, 1.19) do not appear to be consistent; normally the only transformation one can make on a conservation law, that preserves the jump (Rankine-Hugoniot) conditions is a linear transformation. However, as we have shown, both equations must be true in the weak sense. Do these equations supplement each other, as "extra conservation laws" in the sense of Lax [3], or are they interchangeable? We show, in fact, that they are consistent, although they do not appear to be interchangeable. The Eulerian continuity equation (1.19) implies (1.17); however (1.17) does not appear to imply (1.19). One proof, which we sketch briefly, relies on the coordinate transformation. Equation (1.19), plus suitable initial conditions, implies the existence of $\mathbf{a}(\mathbf{x}, t)$. We then invert this map to obtain $\mathbf{x}(\mathbf{a}, t)$, which in turn implies equation (1.17).

A second proof, which relies on wedge products of differential forms, is more direct. In the author's experience this is the first known example of conservation laws which are not linearly related and still consistent at the level of weak solutions.

2. Second Proof

We begin this discussion by clarifying a mysterious result that was mentioned in Section 1, namely, the equivalence of the trivial equation $\frac{\partial 1}{\partial a^\alpha} = 0$ with the Piola law, (1.15). In the language of differential forms, $\frac{\partial 1}{\partial a^\alpha} = 0$ has the expression $d((-1)^\alpha dt \wedge da_{\hat\alpha}) = 0$, in which d is the exterior derivative, and $da_{\hat\alpha}$ denotes the alternating product $da^1 \wedge \cdots \wedge da^n$ with the factor da^α omitted. If we formally transform this differential form into (x, t) coordinates, we obtain:

$$(-1)^\alpha dt \wedge da_{\hat\alpha} = \sum_{j=1}^n F^j_\alpha|\mathbf{F}|^{-1}(-1)^j dt \wedge dx_{\hat j}. \tag{2.1}$$

Equation (2.1) gives two different coordinate expressions for the same differential form. In the material coordinate system, this form is trivially closed. The Piola law expresses the fact that the form remains closed in the physical space-time coordinate system—which is a well known fact for smooth coordinate transformations. Yet, as we saw in Section 1, this must also be true for bi-Lipschitz coordinate transformations.

The foregoing calculation suggests that the wedge product of weakly exact forms is weakly exact. This is particularly interesting, because normally one cannot multiply a weak derivative by a non-smooth function. Yet we shall see that this is indeed the case; the exterior derivative of $\alpha \wedge \beta$ is $d\alpha \wedge \beta \pm \alpha \wedge d\beta$, as long as α, β, $d\alpha$ and $d\beta$ are in suitable L^p_{loc} spaces. Such theorems are also true for partial derivatives of functions, but are less useful for the study of nonlinear hyperbolic conservation laws, because a function with, for example, partial derivatives in L^∞ will be Lipschitz continuous. The exterior derivative does not have this property; indeed, in three space dimensions there are differential forms of degree 1 and 2, with distributional coefficients of any order, whose exterior derivative is zero.

We begin to make these remarks more precise with a few definitions. In particular, we need a definition of weak derivatives for differential forms on a Lipschitz manifold.

We denote by $Sh(p,n)$ the set of injective order preserving maps of $\{1,\dots,p\}$ into $\{1,\dots,n\}$, and for $\sigma \in Sh(p,n)$ we denote by dx_σ the p-form $dx_{\sigma(1)} \wedge \dots \wedge dx_{\sigma(p)}$.

Definition 1. *Let $\omega \in \bigwedge_p(U)$, $U \subset \mathbb{R}^n$. Let $x_1,\dots,x_n$ be coordinates on $\mathbb{R}^n$, and suppose that $\omega = \sum_{\sigma \in Sh(p,n)} \omega_\sigma(x) dx_\sigma$.*

(1) *Let $1 \le q \le \infty$. We say that $\omega \in L^q_{loc}\left(U, \bigwedge_p\right)$ if each coefficient ω_σ is in $L^q(K)$ for each compact set $K \subset U$.*

(2) *If $\omega \in L^q_{loc}\left(U, \bigwedge_p\right)$, for some $q \ge 1$, we say that $d\omega$ is the unique linear functional T on the space $Lip_0\left(U, \bigwedge_{(n-p-1)}\right)$ defined by $T(\beta) = \int_U (-1)^{p-1} \omega \wedge d\beta$, for every Lipschitz continuous $(n-p-1)$-form β, with compact support. Note that $d\beta$ will be defined almost everywhere, by Rademacher's theorem, and that $d\beta \in L^\infty_{loc}$.*

The next theorem shows that these definitions are invariant under bi-Lipschitz coordinate transformations.

Theorem 2.1. *Let ω be a locally integrable p-form on $U \subset \mathbb{R}^n$. Let $T : V \to U \subset \mathbb{R}^n$ be a bi-Lipschitz homeomorphism. Then*

(1) *$T^*\omega$ is a locally integrable p-form on V.*

(2) *$d(T^*\omega) = T^* d\omega$.*

Proof. (1) If $\omega = \sum_{\sigma \in Sh(p,n)} \omega_\sigma(x) dx_\sigma$, then

$$T^*\omega = \sum_{\sigma \in Sh(p,n)} \omega_\sigma(T(y)) \left(\sum_{j=1}^n \frac{\partial T_{\sigma(1)}}{\partial y_j} dy_j\right) \wedge \dots \wedge \left(\sum_{j=1}^n \frac{\partial T_{\sigma(p)}}{\partial y_j} dy_j\right) \tag{2.2}$$

Let K be a compact subset of V. Then $T(K)$ is a compact subset of U, and ω is integrable on $T(K)$. Then for any $\gamma \in Sh(p,n)$, the coefficient of $T^*\omega$ corresponding to dy_γ is a linear combination of terms of the form

$$\omega_\sigma(T(y)) \prod_{k=1}^p \frac{\partial T_{\sigma(k)}}{\partial y_{\gamma(k)}} \tag{2.3}$$

Since T is bi-Lipschitz, the factors $\frac{\partial T_{\sigma(k)}}{\partial y_{\gamma(k)}}$ are in L_∞, as is the Jacobian $|DT|^{-1}$ of T^{-1} Thus

$$\int_K |\omega_\sigma(T(y)) \prod_{k=1}^p \frac{\partial T_{\sigma(k)}}{\partial y_{\gamma(k)}}| dy = \int_{T(K)} |\omega_\sigma(x) \prod_{k=1}^p \frac{\partial T_{\sigma(k)}}{\partial y_{\gamma(k)}}| |DT|^{-1} dx. \tag{2.4}$$

Thus the coefficients of dy_γ are integrable on K.

(2) Let β be a Lipschitz continuous differential form of degree $n-p-1$, and of compact support. Then $T^{-1*}\beta$ is a Lipschitz continuous differential form on U, and $d(T^{-1*}\beta) = T^{-1*}d\beta$. Thus

$$\int_V (-1)^{p-1} T^*\omega \wedge d\beta = \int_U (-1)^{p-1}\omega \wedge d(T^{-1*}\beta) \tag{2.5}$$

$$= \int_U d\omega \wedge T^{-1*}\beta \tag{2.6}$$

$$= \int_V T^*(d\omega) \wedge \beta \quad \square \tag{2.7}$$

Next we define smoothing for differential forms, in a way that respects the exterior derivative.

Definition 2. *Let ω be a locally integrable p-form on $U \subset \mathbb{R}^n$. Extend ω as a p-form on $\mathbb{R}^n$ by defining $\omega(x) = 0$ if $x \notin U$. Let $\omega(x) = \sum_{\sigma \in Sh(p,n)} \omega_\sigma(x) dx_\sigma$. Let j be a smooth function with compact support in $\mathbb{R}^n$ such that $\int_{\mathbb{R}^n} j(x)dx = 1$. Then for each $\sigma \in Sh(p,n)$ and each $\epsilon > 0$, define*

$$\omega_{\sigma\epsilon}(x) = \int_{\mathbb{R}^n} \omega_\sigma(x-y) j\left(\frac{y}{\epsilon}\right) \epsilon^{-n} dy. \tag{2.8}$$

Finally, let $\omega_\epsilon = \sum_{\sigma \in Sh(p,n)} \omega_{\sigma\epsilon} dx_\sigma$.

We note that ω_ϵ is smooth in the x-coordinate system, and that locally Lipschitz coordinate transformations will transform ω_ϵ as a locally Lipschitz form. If $\omega \in L^p_{loc}$, for $1 \le p < \infty$, then ω_ϵ converges to ω in L^p_{loc} as $\epsilon \to 0$. Also, one may readily verify that $d(\omega_\epsilon) = (d\omega)_\epsilon$. Note that although our definition of smoothing is coordinate dependent, commutation of smoothing with d is coordinate-free.

Our main theorem is the following.

Theorem 2.2. *Let $\alpha \in L^p_{loc}$ and $\beta \in L^{p'}_{loc}$ be differential forms of degree q and q', respectively, where $\frac{1}{p} + \frac{1}{p'} = 1$, and further suppose that $d\alpha \in L^p$ and $d\beta \in L^{p'}$. Then $d(\alpha \wedge \beta) = d\alpha \wedge \beta + (-1)^q \alpha \wedge d\beta$, and is locally integrable.*

Proof. We approximate α and β by smooth forms α_ϵ and β_ϵ, as in Definition 2. Then $d(\alpha_\epsilon \wedge \beta_\epsilon) = d\alpha_\epsilon \wedge \beta_\epsilon + (-1)^q \alpha_\epsilon \wedge d\beta_\epsilon$. As $\epsilon \to 0$, we have for each compact set $K \subset \mathbb{R}^n$, using the bilinearity of the wedge product together with the triangle inequality,

$$\|d\alpha_\epsilon \wedge \beta_\epsilon - d\alpha \wedge \beta\|_{1,K} \le \|d\alpha_\epsilon - d\alpha\|_{p,K}\|\beta_\epsilon\|_{p',K} + \|d\alpha\|_{p,K}\|\beta_\epsilon - \beta\|_{p',K} \to 0, \tag{2.9}$$

$$\|\alpha_\epsilon \wedge d\beta_\epsilon - \alpha \wedge d\beta\|_{1,K} \to 0, \tag{2.10}$$

and we conclude by the completeness of $L^1(K)$ that $d(\alpha \wedge \beta) \in L^1(K)$, and

$$d(\alpha \wedge \beta) = d\alpha \wedge \beta + (-1)^q \alpha \wedge d\beta. \quad \square \tag{2.11}$$

Theorem 2.2 is not new (see, for example, [5], eq. 11). However the applications which we now present for this theorem are new and interesting. First, Theorem 2.2 implies that our formal calculation for the Piola identity is valid whenever, for example, the differential forms $da_\alpha, \alpha = 1, \ldots, n$ are in L_∞, and are closed, that is, equations (1.19, 1.21) hold. Thus equations (1.19, 1.21) imply the Piola identity, even for weak solutions.

Similar calculations may be performed with ease and simplicity. For example, we noted that the conservation of mass, (1.14) was implied by the Lagrangian equation $\frac{\partial 1}{\partial t} = 0$. In terms of differential forms, this says that

$$da_1 \wedge \ldots \wedge da_n = |\mathbf{F}^{-1}| dx_1 \wedge \ldots \wedge dx_n + v^i |\mathbf{F}^{-1}| (-1)^i dt \wedge d\hat{x_i}. \tag{2.12}$$

Thus, even for weak solutions, (1.14) is implied by (1.19, 1.21).

The cofactor equation, (1.17), was obtained by transforming (1.3) as a divergence form conservation law; that is, we transformed the underlying 1-form, $dx_i = v^i dt + \sum_{\alpha=1}^{3} F^i_\alpha da_\alpha$, as a form of degree 3. This was accomplished by forming the product $dx_i \wedge da_{\hat{\alpha}}$ and then expressing the result in Eulerian (x) coordinates. Thus, if we take the Eulerian expression which says that $dx_i \wedge da_{\hat{\alpha}}$ is closed, we obtain (1.17).

Of course, the natural way to express (1.3) in Eulerian coordinates is with the trivial equation $d(dx_i) = 0$. Similarly, it is natural to transform (1.19) as $d(da_\alpha) = 0$. The cofactor equations, while they do not appear to be natural continuity equations, may be useful for other purposes.

From these observations, it is apparent that a continuity equation belongs to a particular coordinate system, whereas physical conservation laws, such as conservation of mass, momentum and energy, are coordinate free. However the conservation of mass becomes trivial in a Lagrangian coordinate system. It may be somewhat distressing to think of conservation of mass as trivial, but in Lagrangian coordinates (the reference configuration), conservation of mass is built in to the coordinate system. Instead one must be concerned about the conservation of volume, as expressed by the density of volume with respect to mass.

Other conservation laws may be derived using these techiniques. For example, we can show that the characteristic polynomial of $\mathbf{F}$ is conserved, as follows. Let $\omega_i = dx_i - \lambda\delta^i_\alpha da_\alpha$; ω_i is the differential of $y_i = x_i - \lambda a_\alpha$. We can think of y_i as a new configuration of the material. The corresponding deformation gradient is $\mathbf{F} - \lambda\mathbf{I}$, and the corresponding velocity is $\mathbf{v}$. Then

$$d\omega_1 \wedge d\omega_2 \wedge d\omega_3 = |\mathbf{F} - \lambda\mathbf{I}| da_1 \wedge da_2 \wedge da_3 - \sum_{\alpha=1}^{3} \left((\mathbf{F} - \lambda\mathbf{I})^{-1} \mathbf{v} \right)_\alpha dt \wedge da_{\hat{\alpha}} (-1)^\alpha \tag{2.13}$$

From this one can derive conservation laws for each of the coefficients of the characteristic polynomial of $\mathbf{F}$.

References

1. Phillippe G. Ciarlet, *Mathematical elasticity, Volume I: Three-dimensional elasticity*, Studies in Mathematics and its Applications, vol. 20, North-Holland, Amsterdam, 1988.
2. Constantine M. Dafermos, *Equivalence of referential and spatial field equations in continuum physics*, Notes on Numerical Fluid Mechanics, Vieweg Verlag, to appear, Proceedings of the Fourth International Conference on Hyperbolic Problems: Theory, Numerical Methods, and applications, Taormina, Italy, April, 1992.
3. Peter Lax, *Shock waves and entropy*, Contributions to Nonlinear Functional Analysis (New York) (E. Zarantonello, ed.), Academic Press, 1971, pp. 603–634.
4. Bradley J. Plohr and David H. Sharp, *A conservative Eulerian formulation for the equations of elastic flow*, Advances in Applied Mathematics **9** (1988), 481–499.
5. Y. Meyer R. Coifman, P.L.Lions and S. Semmes, *Compensated compactness and Hardy spaces*, J. Math. Pures Appl. **72** (1993), 247–286.
6. J.A. Trangenstein and P. Colella, *A higher-order Godunov method for modeling finite deformation in elastic-plastic solids*, Comm. Pure Appl. Math. **XLIV** (1991), 41–100.
7. David H. Wagner, *Equivalence of the Euler and Lagrangian equations of gas dynamics for weak solutions*, Jour. Diff Eq. **68** (1987), 118–136.

3D SIMULATIONS OF COLLIDING HYPERSONIC RADIATIVE FLOWS IN ASTROPHYSICS

R. WALDER[1]

ABSTRACT. I present first 3D results of colliding stellar winds in binary star systems and some results on the structure and stability of radiative shock waves. Finally, I report on some numerical difficulties and failures in the treatment of such flows with modern high-resolution integrators.

1. INTRODUCTION

Colliding, hypersonic, radiative flows are a common flow type in Astrophysics. It can be found in *ejecta of supernovae, in stellar atmospheres and winds, in accretion flows onto massive compact objects like black holes or neutron stars, and in jets of active galactic nuclei.* Therefore, a proper understanding of this flow-type is of highest interest. However, the special character of the flow makes the numerical treatment demanding. Radiation affects the flow on time-scales which may differ by orders of magnitude from typical flow time-scales and relevant structures within the flow often have much smaller spatial extent than other characteristic lengths.

The intention of this paper is to present in a first part a few typical astrophysical flows which demonstrate the demanding computational needs and the methods (modern finite volume integrators and adaptive mesh refinement techniques) with which such flows can be successfully computed. In a second part, I will show how complex flow computations can give feedbacks to numerical analysis by exposing weak points of state of the art numerical methods which can lead to numerical failures in the computation.

2. COMPUTING ASTROPHYSICAL FLOWS: A DEMANDING NUMERICAL TASK

2.1. Equations and Numerical Methods. A simple mathematical model of radiative flows are the Euler equations with source terms and a polytropic equation of state. The sources consider external forces like gravitation as well as cooling and heating of the gas by radiation. In this simplest case, cooling can be described by $\dot{E} \propto -\rho^2 \cdot \Lambda^2(T)$, where $\Lambda(T)$ is a complex function of the temperature T. (Details can be taken from the paper by D.Folini and myself in this volume or from [7].)

[1]Institut für Astronomie, ETH-Zentrum, 8092 Zürich, Switzerland. e-mail: walder@astro.phys.ethz.ch

For our simulations we developed a code based on Cartesian grids. The adaptive mesh refinement algorithm (AMR) of Berger (see e.g. [1]) is implemented, boundaries are treated along ideas of Berger and LeVeque [2], we use the multidimensional MUSCL-integrator developed by Colella [3]. Details can be found in [7].

Generally, this code and the implemented methods are extremely successful and we obtained a variety of new results, including those presented in the following 3D computations of the dynamical behavior of flows in double star systems and stability investigations of radiative shock waves.

2.2. 3D Simulation of Colliding Winds in Double Star Systems. Colliding winds are a common feature in the sky since approximately 60 percent of all stars live in binary star systems and most of the stars lose mass by stellar winds. The astrophysical point of interest in the simulation shown in Fig. 1 is how the mat-

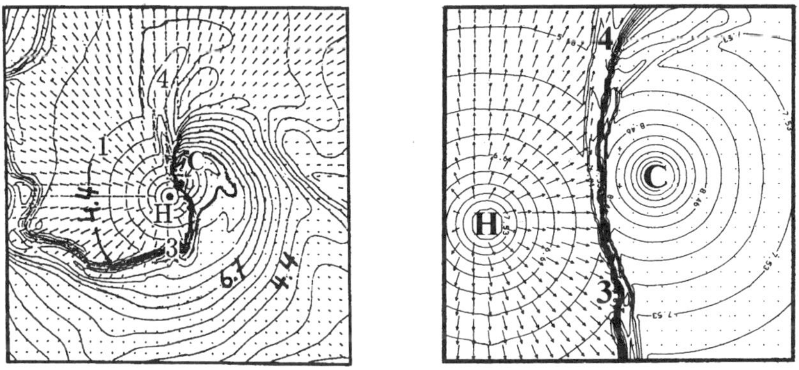

FIGURE 1. Colliding winds in a double star system. Shown are logarithmic scaled density contours and the projected velocity field in the orbital plane. **Left:** Large scale structure of the system. **Right:** Detailed structure of the inner part having a size which is 1/10 smaller than the computational domain.

ter originating from the two stars (H and C in the graphs) is distributed in the neighborhood of the system and what physical conditions (temperature, pressure, velocity) can be found there. These conditions determine the observable spectrum, the only information we have of such systems. The prediction of the emitted spectrum on the base of the correct simulation and the subsequent comparison with the measured spectrum is the only promising way to test our model idea.

However, the simulation of such a system is very demanding. As can be seen in the left graph of Fig. 1, the large scale structure of the system reaches out to a size of many diameters of the stellar orbits. On the other hand, significant small scale structures (see right graph in Fig. 1) are orders of magnitude smaller than this diameter. Both scales have to be correctly computed to predict e.g. the emitted X-ray spectrum originating in the small cooling layers downstreams of the shock wave (No. 3 in Fig. 1) and from the jet (No. 4). A further complication is that the system is never stationary since the stars move on elliptical orbits and their positions are steadily changing.

As Fig. 1 demonstrates, our code could essentially handle the posed tasks. The use of AMR has allowed to spatially resolve more than 3 orders of magnitude (laterally) and to capture all the relevant scales.

2.3. An Oscillatory Instability Related with Radiative Shock Waves. Radiative shock waves are subject to a variety of different instabilities. Most of them are hardly completely understood, especially in more than one space dimension. One of them, usually cited as *thin shell instability*, is shown in its dynamical development in Fig. 2. The graphs are showing the onset of a highly supersonic wind

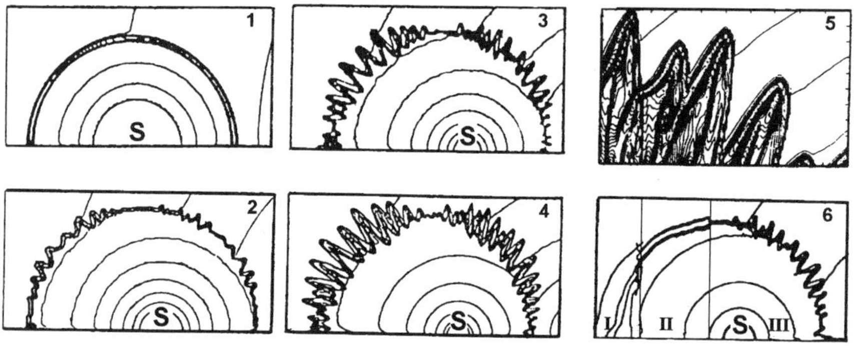

FIGURE 2. **Graphs 1–4:** Dynamical development of an oscillatory instability of the cold gas layer located downstream of a radiative shock. **Graph 5:** Detailed density structure in the unstable cold layer. **Graph 6:** The growth of the instability depends on the mesh size, which is finest in region III and coarsest in region I.

from a star (S in the graphs) into a very dense nebula, present in the neighborhood of the star. The wind acts like a snow-plow, collecting the dense material from the nebula. The cooling is so strong (due to $\dot{E} \propto -\rho^2$) that even very strongly shocked matter cools nearly instantaneously. The nebular material is then compressed into a very thin sheet which is highly unstable and shows a lot of internal structure (see graph 5). It should be emphasized that, again, the radiative emission from the thin sheet can nearly completely dominate the observable spectrum from the entire object. Therefore, it is absolutely necessary to correctly compute the internal structure.

Again, our code was able to fulfill the necessary tasks. In this computation, we resolved small scale structures within the cold shell which are more than four orders of magnitude smaller than the dynamical structure. The density jump at the inner rim of the thin shell is more than five orders of magnitude.

A remark on the investigated instability: In the investigated case, the instability develops spontaneously from small disturbances accidentally introduced by numerical truncation errors. Independently from its origin, the instability then develops a rather regular pattern which seems to grow steadily. Computations on coarser grids do not show the instability as demonstrated in graph 6 of Fig. 2. There, the grid in region II is coarsened by a factor of four (in one coordinate direction) compared to the one in region III; the grid in region I again by a factor four compared to the one in region II. Moreover, not only the stability property has changed, also the size of the computed cold shell in regions I and II is much too big.

The mesh-dependent stability properties can be understood when we look at a simplified analytical model, which shows that modes with smaller wavenumbers are growing faster (see e.g. Dgani et al. [4]). Therefore, up to a certain dynamical flow time, only modes with small wavenumbers can grow, which cannot be resolved by a grid coarser than the wavelengths of the growing modes.

Another important instability in this context, a thermal instability active in cooling layers downstream of shocks, the interested reader may find in the contribution to this Proceedings by D.Folini and myself.

3. A Numerical Instability associated with Stagnation Lines

3.1. An introductory remark. In Sect. 2 we have shown the successful use of advanced numerical methods in simulating complex astrophysical systems. But complex simulations can also expose weak points of numerical methods. In our simulations we have encountered several different numerical artifacts. Important for our applications is the numerical instability associated with a stagnation line which will be discussed in this section.

Other artifact are, for example, grid alignment effects which can be observed in the left graph of Fig. 3: The contour lines of the spherically symmetric undisturbed wind are not circles. Rather they have 'knots' and 'hollows'. This phenomenon is due to the fact that, up to now, no real multidimensional integrator for the Euler equation is known and the integration is performed by solving one-dimensional Riemann problems (for details see LeVeque and Walder [5]).

The following point is of importance: Even advanced methods can produce failures, as also shown by several other authors (see e.g. Quirk [6]). The numerical instability analyzed below resembles in its manifestation a physical one, known as 'thin shell instability'. Only a thorough analysis of possible physical driving mechanisms together with the exact knowledge of the numerical method can prevent numerical failures from being mistaken for real physical phenomena. In this context, I would like to emphasize the usefulness of a steady interaction between complex numerical simulations and numerical analysis.

3.2. The numerical instability. Two cases of this instability, an asymmetric and symmetric one, are shown in Fig. 3. To be independent from further complications, the cylindrically symmetric Euler equations are solved without any source terms.

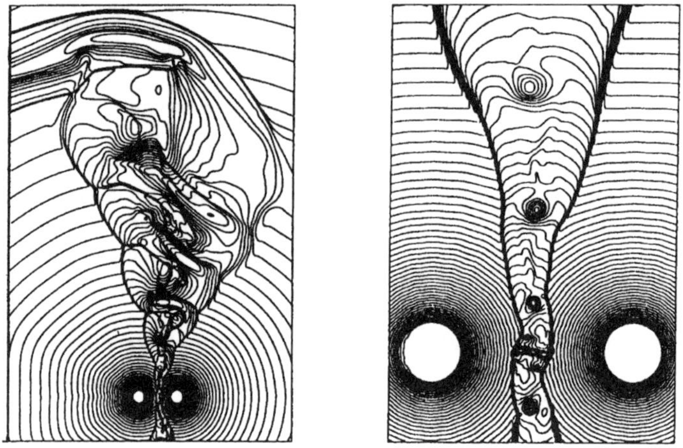

FIGURE 3. The asymmetric (left) and the symmetric (right) case of the numerical instability associated with stagnation lines.

The following *settings* are investigated. Two mass shedding stars with radii $R_i = 0.05, i = 1,2$ are located at positions (0.45,0.1) and (0.55,0.1) in the computational domain which is the unity squared. The following boundary conditions at the stellar surfaces are used: *asymmetric situation* (left graph in Fig. 3): $\rho_1 = 0.1633$, $\rho_2 = 6.123$, $v_1 = 4.44$, $v_2 = 1.77$, $M_1 = 4.44$ and $M_2 = 1.775$; *symmetric situation* (right graph in Fig. 3): $\rho_1 = \rho_2 = 1$, $v_1 = v_2 = 8.888$, $M_1 = M_2 = 8.888$.

The numerical treatment of the *boundaries* is similar to the one described by Berger and LeVeque [2]; details can be found in [7]. The computations were performed on a uniform 400x400 Cartesian mesh. *Initially*, the computational domain is filled with the undisturbed wind from the right star; at time $t = t_0$, the wind from the left star starts to blow and drives the interaction region towards the stationary position. It should be emphasized that the same instability also appears when other initial conditions are used.

How the instability is actually generated is illustrated in Fig. 4 where two graphs at the subsequent times 0.0246 (left side) and 0.0293 (right side) are shown. Visible on the graph are the right shock and the slip plane. The stable solution would show a stagnation point where the slip plane crosses the axis of symmetry. In this failed computation, the stagnation point is completely pushed to the right. This is the late consequence that the code does not correctly transport the material away from the stagnation line, leaving too much matter in some cells along the stagnation line (in fact, only two rows of cells are affected). Clearly visible are unphysical discontinuous rarefactions above and below these two cells. The failure focuses the matter along the stagnation line, leading to too much momentum along the line. The focussed flow acts as a rammer and pushes the slip plane forward. In the left graph, the slip plane has just reached the right shock. In the right

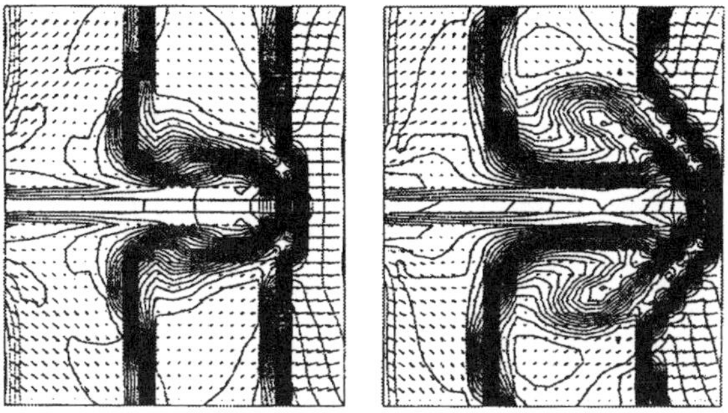

FIGURE 4. The stagnation line and its direct environment shown in density contours and the velocity field. The graphs illustrate the forming of the vortices at two subsequent times.

graph, the focussed flow has also pushed the shock to the right. The two deflected waves are then generating two vortices on each side of the axis of symmetry. The vortices detach from the central region. Through the detachment of the vortices, the discontinuous rarefaction is destroyed temporarily and the shock and partially also the slip plane can relax to their equilibrium position. After a while, the whole cycle starts again, leading to a second pair of vortices.

Fig. 3 impressively demonstrates that **the numerical instability prevails over the simulations** and, therefore, the simulation becomes worthless. From this central region, where the vortices and the disturbances in the shocks are generated, they are transported outwards by the flow as illustrated in the two graphs of Fig. 5. The vortices are spread out by their internal behavior and by the general expansion of the flow and are distributed in the whole collision zone of the two stellar winds.

Also in a **complete symmetric setting** the instability develops (see Fig. 3 right side), where, in addition, the symmetry with respect to the axis connecting the two stars is broken.

3.3. Final remarks. There is no physical reason for this instability, especially not in the complete symmetric case. On the other hand, there are clear indications for a numerical origin of the instability. It disappears when using coarser meshes, it strengthens when using finer meshes. When finer meshes are used, the enhancement of mass and momentum along the stagnation line is still present in two cells only, showing that we are dealing with a discretization and not with a physical effect. Also, the unphysical discontinuous rarefaction wave becomes stronger. Finally, the artificial addition of viscosity leads to a substantial damping of the instability.

These observations can be understood as follows: The flow is very subsonic in the direction normal to the axis of symmetry and the pressure gradient which drives the matter away from the central region is very small. Therefore, the flow normal to the stagnation line is very subsonic and its time-scale is very stiff against the time-scale of the flow parallel to the stagnation line. Finite volume methods based

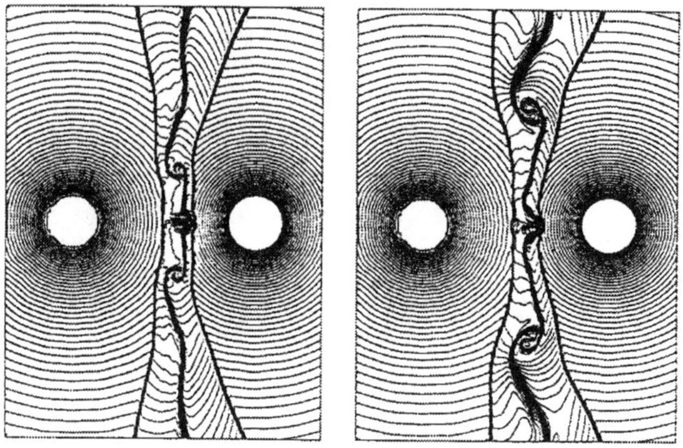

FIGURE 5. Density contours of the central flow region. The origin of the instability lies in the generation of vortices in the immediate neighborhood of the stagnation line. The vortices are then transported outward, determining finally the flow pattern completely.

on Riemann solvers are known for the bad capturing of very low Mach number flows, since, in this case, the equations are becoming more and more elliptic-like.

We believe that this numerical failure has some relation to the carbuncle phenomenon well known in the literature (described well e.g. by Quirk [6]). Also this author mentioned that adding viscosity often can suppress this phenomenon.

REFERENCES

1. M.J. Berger, *Adaptive mesh refinement for hyperbolic equations*, Lectures in applied Mathematics **22** (1985), 31–40.
2. M.J. Berger and R.J. LeVeque, *Stable boundary conditions for cartesian grid calculations*, Computing Systems in Engineering **1** (1990), no. Nos–4, 305–311.
3. P. Colella, *Multidimensional upwind methods for hyperbolic conservation laws*, Journal of Computional Physics **87** (1990), 171.
4. R. Dgani, R. Walder, and H. Nussbaumer, *Stability analysis of colliding winds in double star systems*, Astronomy and Astrophysics **267** (1993), 155–160.
5. R.J. LeVeque and R. Walder, *Grid alignment effects and rotated methods for computing complex flows in astrophysics*, Proceedings of the Ninth GAMM-conference on numerical methods in fluid mechanics, Notes on numerical fluid mechanics (J.B. Vos, A. Rizzi, and I.L. Ryhming, eds.), no. 35, 1992.
6. J.J. Quirk, *A contribution to the great Riemann solver debate*, Tech. Report 92–64, ICASE, NASA Langley Research Center, 1992.
7. R. Walder, *Some aspects of the computational dynamics of colliding flows in astrophysical nebulae*, Ph.D. thesis, ETH Zürich, 1993.

I Didn't Know Jiffy Lube® Did All That!

The Preventative Maintenance of your vehicle is important to us.

NEW

Ask About These Additional Services:

- [ ] **Signature Service Gold**
  Our complete Preventative Maintenance package includes: Signature Service Oil Change; Tire Rotation Service; Prestone® Express Antifreeze/Coolant Replacement; Aquapel® Glass Treatment; Windshield Wiper Replacement; Transmission Fluid Change. (please ask our store manager for details).

- [x] **Super Service**
  A compliment to our oil change that's a benefit to your vehicle. The Super Service includes our Signature Service oil change plus an Aquapel® Glass Treatment, and Tire Rotation.

- [ ] **Tire Rotation**
  Today's tire manufacturers require tire rotation every 5,000–8,000 miles. Rotating your tires will greatly increase the mileage you will receive from your tire investment.

- [ ] **Prestone® Express Radiator Service**
  Complete antifreeze/coolant replacement service to help maintain the operating temperature of your engine.

- [ ] **T-Tech® Transmission Fluid Service**
  Our T-Tech® Transmission Service will replace 100% of your vehicle's old, dirty fluid with brand new automatic transmission fluid to help your car shift and run smoother.

- [ ] **Fuel Injection Systems Cleaner**
  Our three-part product treatment will penetrate your engine's crankcase, combustion chamber and fuel system for a smoother running engine and better gas mileage.

- [ ] **Gearbox Fluid Change Service**
  Gearbox fluid provides lubrication for the sets of bearings, shafts and gears located in the different types of gearboxes which perform under high stress and extreme heat.

- [ ] **Aquapel® Glass Treatment**
  Your vehicle's windows will become virtually water resistant, repelling rain, sleet and other elements that can blur driver's vision, particularly at night. Best of all, Aquapel® lasts up to 3 months.

EVERY 3000 MILES

jiffy lube® Signature SERVICE

- [x] Service of the Month

0001 PP 1/98

Solving the Compressible Euler Equations in Time Dependent Geometries

Monika Wierse, neé Geiben[1]

ABSTRACT. In this paper we shall present a numerical algorithm to solve the compressible Euler equations in three dimensional geometries with a moving boundary. The numerical algorithm consists of a new cell–centered, upwind, finite volume scheme of higher order on a grid of simplices, and the possibility to refine and coarsen the grid locally appropriate to the approximated solution. We will describe a new criterion to control the refinement and coarsening process. Furthermore we will describe briefly an idea to model the moving boundary without loss of conservation in the flux quantities.

1. FLOW PROBLEM

We want to solve the compressible Euler equations in geometries with inlets and outlets (cylinder model of a two–stroke–engine).

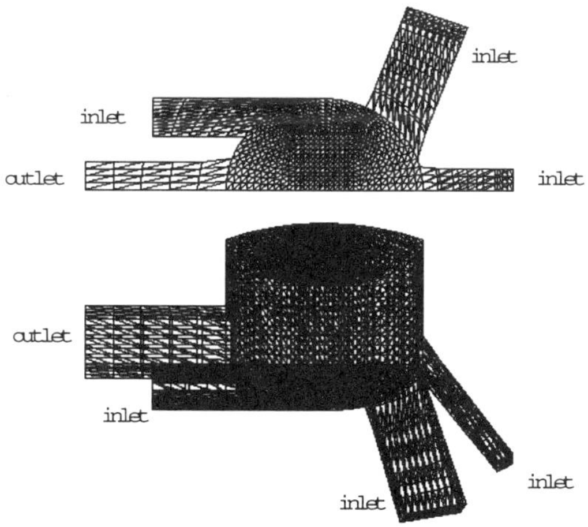

[1]Institut für Angewandte Mathematik, Universität Freiburg, Hermann–Herder–Str. 10, D-79104 Freiburg, Germany. E-mail: mwierse@mathematik.uni-freiburg.de. This work is supported by the Deutsche Forschungsgemeinschaft.

In this geometry consisting of tetrahedra, a piston is moving up and down. The piston motion will cause very high jumps e. g. in the pressure, if an inlet or outlet opens. This motivates the use of an upwind scheme. More of the details for this article can be found in the author's PhD thesis [14].

2. Definition of a New Higher Order Scheme in Space

In this section, we will present an upwind finite volume scheme of higher order with the new limiter function, Ansatz C, on an unstructured grid of simplices. This new limiter function is motivated by the constraints in [10] to prove convergence in the case of scalar conservation laws. We have improved the result in [10] to the effect that the B–triangulation [4] is no longer necessary in order to have an L^∞–bound (maximum principle) for this scheme. The new limiter function, Ansatz C, is constructed such that this bound exists without any additional restriction on the triangulation, and is simpler than the one proposed in [11]. The numerical scheme used is an explicit, upwind, first order in time, cell–centered, finite volume scheme working on simplices T_i:

$$u_i^{n+1} = u_i^n - \frac{\Delta t^n}{|T_i|} \sum_{j=1}^{dim+1} g_{ij}(L_i^n(z_{ij}), L_{ij}^n(z_{ij})), \tag{2.2}$$

where dim denotes the number of spatial dimensions, and g_{ij} denotes numerical flux. The choices $L_i^n(z_{ij}) = u_i^n$ and $L_{ij}^n(z_{ij}) = u_{ij}^n$ for all j, give a first order scheme, while $L_i^n(z_{ij})$, $L_{ij}^n(z_{ij})$ as in the following definition result in a second order scheme.

Definition 2.1. (the limiter function, Ansatz C) Let I be the set of indices labeling the simplices. For each simplex T_i, $i \in I$, construct a linear reconstruction function of the form

$$F_i^n(x) = u_i^n + s_i^n \cdot (x - x_i)$$

(for example with the approach in [1], [5], [7] or [9]). A concrete example will be given in the following note. Let $i \in I$, $n \in \mathbb{N}$ be fixed and $j = 1, \ldots, dim + 1$. Let

$$\begin{aligned} u_i^{max} &= max\{u_i^n, u_{ik}^n, \ldots, u_{idim+1}^n\} \text{ and} \\ u_i^{min} &= min\{u_i^n, u_{ik}^n, \ldots, u_{idim+1}^n\} \end{aligned}$$

denote the maximum and minimum of the values on simplex T_i and its neighbours. We then define the values $L_i^n(z_{ij})$ by

$$L_i^n(z_{ij}) = u_i^n + \beta \cdot term(j),$$

where β and $term(j)$ are defined by the following algorithm.

$$term(j) = \begin{cases} s_i^n \cdot (z_{ij} - x_i), & \text{if } s_i^n \cdot (z_{ij} - x_i) \cdot (u_i^n - u_{ij}^n) \le 0 \\ 0, & \text{otherwise} \end{cases}$$

$$J^+ = \{j, term(j) > 0\} \qquad \text{and} \qquad J^- = \{j, term(j) < 0\}$$

```
if (#J⁺ = 0) or (#J⁻ = 0) then
   β = 0
else {
   if (#J⁺ + #J⁻ ≠ dim + 1) then {
         sum± = Σ_{j∈J±} term(j)
         if (−sum⁻ > sum⁺) then
            γ = −sum⁺/sum⁻
            for all j ∈ J⁻, term(j) = γ · term(j)
         else
            γ = −sum⁻/sum⁺
            for all j ∈ J⁺, term(j) := γ · term(j)
   }
   determine the largest β ∈ [0,1] such that, ∀j and a fixed α ∈ (1/2, 1],
         a) β · |term(j)| ≤ C₁h^α
      and
         b) u_i^min ≤ u_i^n + β · term(j) ≤ u_i^max
}
```

Note 2.1.

i) Since the constant C_1 can not be adjusted automatically by the scheme for every individual problem, the constraint a) in Definition 2.1 is dropped (see [14]).

ii) (example for the implementation of Ansatz C)
Determine values $\tilde{u}_{ij}^n, j = 1, \ldots, dim + 1$ at the vertices P_{ij} of the simplex T_i by averaging the piecewise constant values on the simplices with the vertex P_{ij}. This can be done, for example, with

$$\tilde{u}_{ij}^n = \frac{\sum_{k \in I, P_{ij} \in T_k} \frac{u_k^n}{r_k}}{\sum_{k \in I, P_{ij} \in T_k} \frac{1}{r_k}} \quad \text{where} \quad r_k \equiv |P_{ij} - x_k|.$$

Let $\tilde{u}_{ij+1}^n, \ldots, \tilde{u}_{ij+dim}^n$ be the values at the vertices of the jth edge (2D) or surface (3D) of T_i. Then define

$$F_i^n(z_{ij}) = u_i^n + \frac{\frac{1}{dim}\sum_{m=1}^{dim} \tilde{u}_{ij+m}^n - \tilde{u}_{ij}^n}{dim + 1}. \tag{2.3}$$

This is the reconstruction suggested in [7].

2.1. Numerical Order of Convergence. In order to evaluate the new limiter function Ansatz C, we compared it in [14] with other higher order approaches. Denote by u_h the approximate solution of the first order scheme, and by u_h^C, u_h^S, u_h^F the respective approximate solutions of the second order schemes using Ansatz C, the superbee approach of [5], and the unlimited interpolation approach described in Note 2.1 (ii)). We performed calculations on different kinds of triangulations for problems with smooth solutions and those containing discontinuities. For each, we determined the "experimental order of convergence" (EOC). For the calculations using Burgers equation we use the numerical flux defined by Engquist–Osher [6], and for the system of Euler equations the numerical flux of Steger–Warming [13].

2.1.1. *Approximation of a smooth solution of Burgers equation.* Since there is a dependence of the order of consistency of the scheme on the form of the grid the calculations are performed on different grids with the following triangulations.

Grid 1:

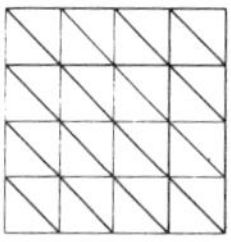

Grid 2:

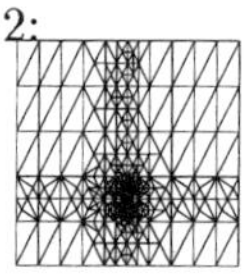

On the structured grid (Grid 1) we obtain an EOC of one for the first order scheme and two for all three second order schemes. The results on the unstructured grid (Grid 2) are presented in the following table. For the second order schemes we also used a second order approximation in time as presented in [12].

h	$\|u-u_h\|_{L^1}$	EOC	$\|u-u_h^C\|_{L^1}$	EOC	$\|u-u_h^S\|_{L^1}$	EOC	$\|u-u_h^F\|_{L^1}$	EOC
0.22	0.0219		0.0065		0.0085		0.0061	
0.11	0.0075	1.54	0.0020	1.70	0.0033	1.36	0.0018	1.77
0.06	0.0031	1.28	0.0006	1.62	0.0013	1.37	0.0005	1.71
0.03	0.0015	1.03	0.0002	1.55	0.0005	1.50	0.0002	1.68

2.1.2. *Solution of the two-dimensional shock tube problem for the system of the Euler equations with one-dimensional structured solution.* In the next table the numerical errors in the density are presented corresponding to the calculations on the following grid.

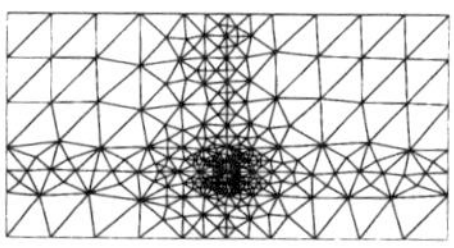

h	$\|\rho-\rho_h\|_{L^1}$	EOC	$\|\rho-\rho_h^C\|_{L^1}$	EOC	$\|\rho-\rho_h^S\|_{L^1}$	EOC
0.15	0.3254		0.1934		0.2704	
0.08	0.2460	0.40	0.0995	0.96	0.1429	0.92
0.04	0.1713	0.52	0.0502	0.99	0.0752	0.93
0.02	0.1138	0.59	0.0267	0.91	0.0394	0.93

3. Local Adaption

We use the local adaption algorithm on a grid of tetrahedra proposed in [2]. With this technique of bisection of tetrahedra it is very easy to project the data between two grids under the constraint of conservation. As criterion to mark regions for refinement and coarsening we used expressions of local residuals.

Definition 3.1. (adaption criterion) Denote by L_i^n the linear reconstruction function for each conservative variable on the tetrahedron T_i determined with the unlimited reconstruction given in (2.3) and with $\mathcal{T}$ the grid of simplices. The approximate solution $\{u_i^n, T_i \in \mathcal{T}\}$ shall be denoted by u_h^n. Set

$$S_i^n = \text{largest face of } T_i, \qquad r_i^n = \frac{u_i^{n+1} - u_i^n}{\Delta t^n} + \sum_{j=1}^{3} f_j'(u_i^n)\partial_j L_i^n(x_i),$$

with f_j the flux vector in the system of the Euler equations. Denote by $(r_i^n)_l$ the lth component of the local residual r_i^n. Define

$$\eta_i(u_h^n) = |T_i|\sqrt{|S_i^n|}\sum_{l=1}^{5}|(r_i^n)_l| \qquad \text{and} \qquad \bar{\eta}^n = \frac{1}{\#\mathcal{T}}\sum_{i=1}^{\#\mathcal{T}}\eta_i(u_h^n).$$

We then mark the tetrahedron T_i for refinement or coarsening as follows:

if $(\eta_i(u_h^n) > ref^n \cdot \bar{\eta}^n)$ mark T_i for refinement
if $(\eta_i(u_h^n) < coa \cdot \bar{\eta}^n)$ mark T_i for coarsening

The additional power of h ($\sqrt{|S_i^n|}$) is motivated in [14] to get the same scale for $||r_i^n||_{L^1(T_i)}$ and $||u(t^n) - u_h^n||_{L^1(T_i)}$. The parameters ref^n ($ref^0 = 0.9$) and $coa = 0.1$ control the adaption process.

4. Modeling Piston Motion in a Conservative Manner

A large amount of effort would be necessary to administrate the cell connectivities at the cell faces between cylinder and ports while the points of an unstructured grid are moving. Furthermore, handling the conservation property of the scheme is complicated because of the interpolation between different meshes and because a grid smoother must be used to avoid the degeneration of the tetrahedra. Therefore we propose a new approach on a grid of tetrahedra; the piston motion is modeled by cutting the piston out of a fixed grid at a position depending on the time. Connected with this approach are two problems. First, the cutting of tetrahedra can cause very small cells. The second problem appears while the piston is moving down; cells of the fixed grid in the previous timestep lying completely inside the piston can be exposed in the next timestep. This means that we have no new values in the new timestep for those exposed cells. In [3] (using the same idea of merging cells on structured grids) these two problems are called the Vanishing–Cell–Problem and the Newborn–Cell–Problem. A solution to these two problems uses **collected cells**, a union of cut and uncut tetrahedra, as shown the following examples.

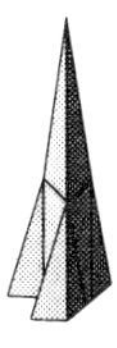

5. Numerical Results

a) Forward facing step problem in 3D

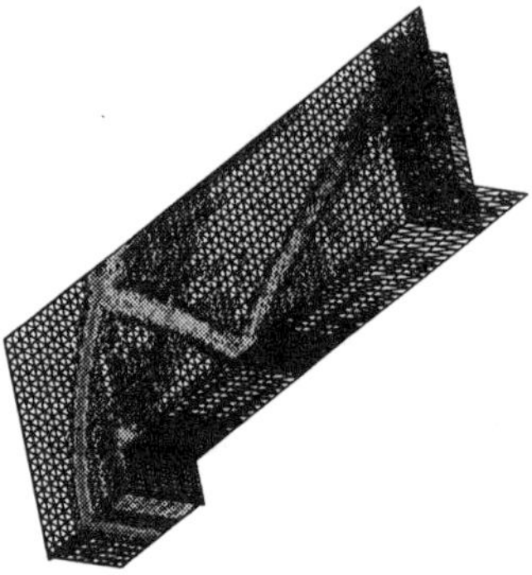

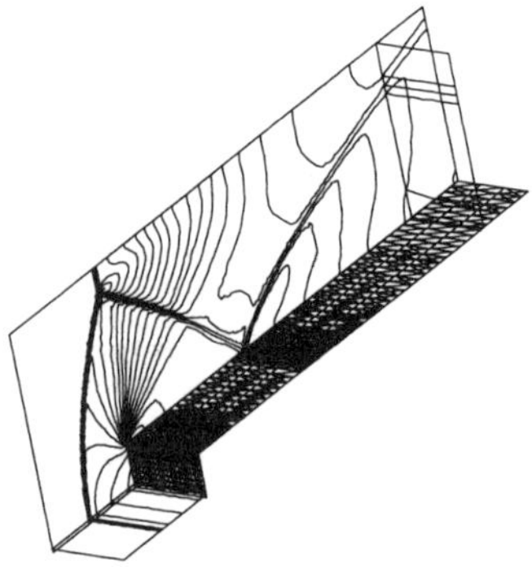

b) Considering the vector field on some cutting planes in the cylinder model of a two–stroke–engine: the cutting plane in the left figure is near the symmetry plane of the geometry

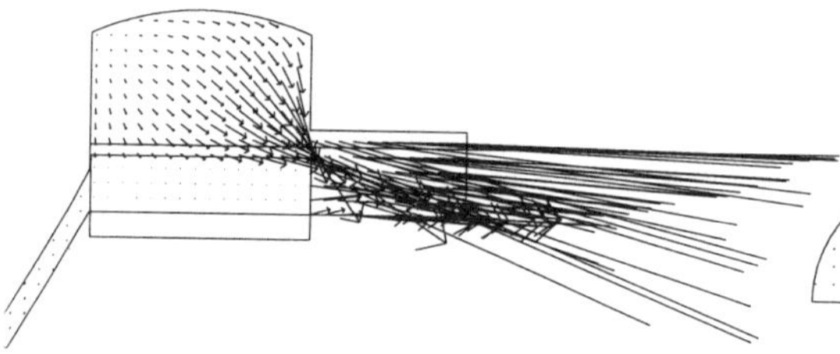

FIGURE 1. Piston motion down.

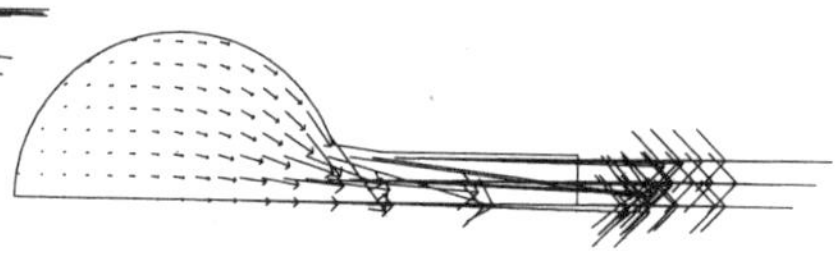

FIGURE 2. Orthogonal plane at z= 0.25.

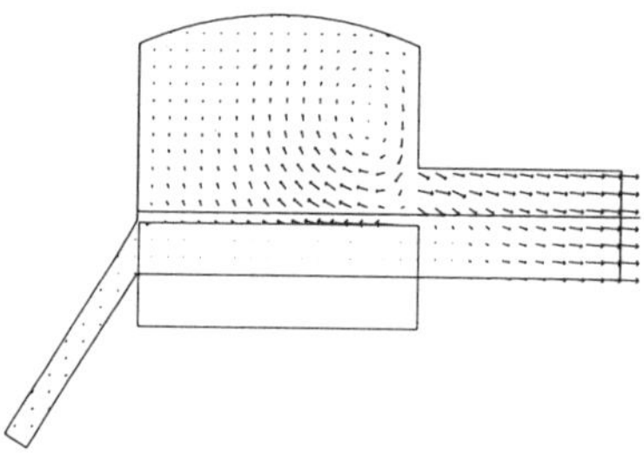

FIGURE 3. Piston motion up.

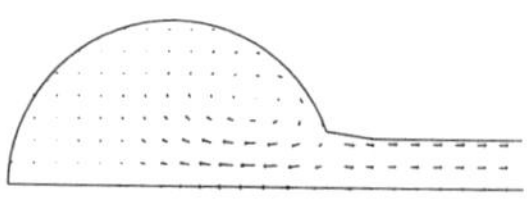

FIGURE 4. Orthogonal plane at z= 0.18.

The little balls lying on the particle traces in the following figure correspond to the position of the particles at the moment where the piston is completely down. The same holds for the depicted velocity field. The particles starting in the upper right corner of the cylinder clarify that this geometry is not optimal with respect to the exchange of "old" and "new" gas.

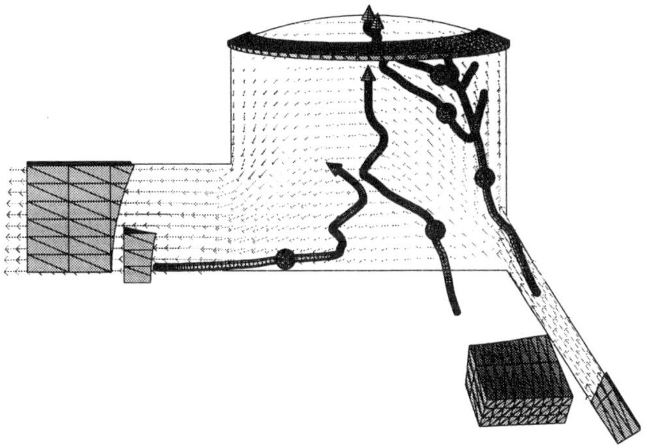

FIGURE 5. Particles during the movement of the piston down and up.

REFERENCES

1. Barth, T. J.; Jesperson, D. C.: *The Design and Application of Upwind Schemes on Unstructured Meshes.* AIAA-98-0366.
2. Bänsch, E.: *An Adaptive Finite-Element Strategy for the Three-dimensional Time-dependent Navier–Stokes equations.* Journal of Computational and Applied Mathematics 36 (1991), 3-28.
3. Bayyuk, S. A.; Powell, K. G.; v. Leer, B.: *A Simulation Technique for 2-D Unsteady Inviscid Flows Around Arbitrarily Moving and Deforming Bodies of Arbitrary Geometry.* AIAA 93-3391
4. Cockburn, B.; Hou, S.; Shu, C.W.: *The Runge-Kutta Local Projection Discontinuous Galerkin Finite Element Method for Conservation Laws IV: The Multidimensional Case.* Math. of Comp. 54 (1990), 545–581.
5. Durlofsky, L. J.; Engquist, B.; Osher, S.: *Triangle Based Adaptive Stencils for the Solution of Hyperbolic Conservation Laws.* J. of Comp. Physics 98 (1992), 64–73.
6. Engquist, B.; Osher, S.: *One Sided Difference Approximations for Nonlinear Conservation Laws.* Math. of Comp. 36 (1981), 321–351.
7. Frink, N. T.; Parikh, P.; Pirzadeh, S.: *A Fast Upwind Solver for the Euler Equations on Three-Dimensional Unstructured Meshes.* AIAA-91-0102.
8. Geiben, M.: *Convergence of MUSCL–Type Upwind Finite Volume Schemes on Unstructured Triangular Grids.* Preprint 318, SFB 256, Bonn 1994.

9. Glinsky, N.; Fézoui, L.; Ciccoli, M. C.; Deśidéri, J.-A.: *Non–Equilibrium Hypersonic Flow Computations by Implicit Second–Order Upwind Finite–Elements.* In: Proceedings of the Eight GAMM– Conference on Numerical Methods in Fluid Mechanics, Notes on Numerical Fluid Mechanics, Vol. 29, Vieweg, Braunschweig, 1990.
10. Kröner, D.; Noelle, S.; Rokyta, M.: *Convergence of Higher Order Finite Volume Schemes on Unstructured Grids for Scalar Conservation Laws in Two Space Dimensions.* Preprint 268, 1993, SFB256, Bonn, Germany.
11. Liu, X.–D.: *A Maximum Principle Satisfying Modification of Triangle Based Adaptive Stencils for the Solution of Scalar Hyperbolic Conservation Laws.* SIAM J. Numer. Anal. 30, (1993), 701–716.
12. Shu, C. W.; Osher, S.: *Efficient Implementation of Essentially Non-Oscillatory Shock-capturing Schemes.* Journal of Computationl Physics, 77 (1988), 439-471.
13. Steger, J. L.; Warming R. F.: *Flux Vector Splitting of the Inviscid Gasdynamic Equations with Application to Finite-Difference Methods.* Journal of Computational Physics 40 (1981), 263-293.
14. Wierse, M.: *Higher Order Upwind Schemes on Unstructured Grids for the Compressible Euler Equations in Timedependent Geometries in 3D.* Dissertation, Freiburg, 1994.

Author Index